Metastable Systems under Pressure

NATO Science for Peace and Security Series

This Series presents the results of scientific meetings supported under the NATO Programme: Science for Peace and Security (SPS).

The NATO SPS Programme supports meetings in the following Key Priority areas: (1) Defence Against Terrorism; (2) Countering other Threats to Security and (3) NATO, Partner and Mediterranean Dialogue Country Priorities. The types of meeting supported are generally "Advanced Study Institutes" and "Advanced Research Workshops". The NATO SPS Series collects together the results of these meetings. The meetings are co-organized by scientists from NATO countries and scientists from NATO's "Partner" or "Mediterranean Dialogue" countries. The observations and recommendations made at the meetings, as well as the contents of the volumes in the Series, reflect those of participants and contributors only; they should not necessarily be regarded as reflecting NATO views or policy.

Advanced Study Institutes (ASI) are high-level tutorial courses intended to convey the latest developments in a subject to an advanced-level audience

Advanced Research Workshops (ARW) are expert meetings where an intense but informal exchange of views at the frontiers of a subject aims at identifying directions for future action

Following a transformation of the programme in 2006 the Series has been re-named and re-organised. Recent volumes on topics not related to security, which result from meetings supported under the programme earlier, may be found in the NATO Science Series.

The Series is published by IOS Press, Amsterdam, and Springer, Dordrecht, in conjunction with the NATO Public Diplomacy Division.

Sub-Series

A.	Chemistry and Biology	Springer
B.	Physics and Biophysics	Springer
C.	Environmental Security	Springer
D.	Information and Communication Security	IOS Press
E.	Human and Societal Dynamics	IOS Press

http://www.nato.int/science
http://www.springer.com
http://www.iospress.nl

Series A: Chemistry and Biology

Metastable Systems under Pressure

edited by

Sylwester Rzoska
Department of Biophysics and Molecular Physics
Institute of Physics, University of Silesia
Katowice, Poland

Aleksandra Drozd-Rzoska
Department of Biophysics and Molecular Physics
Institute of Physics, University of Silesia
Katowice, Poland

and

Victor Mazur
Department of Thermodynamics
Odessa State Academy of Refrigeration (OSAR)
Odessa, Ukraine

Springer

Published in cooperation with NATO Public Diplomacy Division

Proceedings of the NATO Advanced Research Workshop on
Metastable Systems under Pressure: Platform for New Technologies
and Environmental Applications
Odessa, Ukraine
4–8 October 2008

Library of Congress Control Number: 2009934350

ISBN 978-90-481-3407-6 (PB)
ISBN 978-90-481-3406-9 (HB)
ISBN 978-90-481-3408-3 (e-book)

Published by Springer,
P.O. Box 17, 3300 AA Dordrecht, The Netherlands.

www.springer.com

Printed on acid-free paper

TABLE OF CONTENTS

Part II: Liquid crystals

Part III: Near-critical mixtures

Part IV: Water and liquid- liquid transitons

Participants of the ARW NATO "*Metastable Systems under Pressure:Platform for New Technological and Environmental Applications*",
4 – 8 Oct. 2008, Odessa, Ukraine
In the middle: ARW NATO directors (organizers):
Sylwester J. Rzoska (Poland) and Victor Mazur (Ukraine).
Foto in the patio of Hotel Londonskaya, the ARW site.

BELOW- ARW NATO "Odessa 2008 - LIVE":
(i) lecture of Prof. J. Ll. Prof. Tamarit (Spain) on orientational glasses, (ii) rainy night in the front of the ARW site, (iii) Dr El Mekki (France) is waiting for dinner (iv). S. J. Rzoska (Poland) and Prof. K. Shmulovich (Russia) on stairs of Opera (v) lecture of Prof. Nigmatulin (Russia) on negative pressures, cavitation and cold nuclear fusion, (vi) cultural programme: "Chopeniada" in Odessa Opera&Ballet Theatre.

PREFACE:
METASTABLE SYSTEMS UNDER PRESSURE - PLATFORM FOR NOVEL FUNDAMENTAL, TECHNOLOGICAL AND ENVIRONMENTAL APPLICATIONS IN THE 21st CENTURY

[1]SYLWESTER J. RZOSKA, [1]ALEKSANDRA DROZD-RZOSKA, [2]VICTOR MAZUR

[1]*Institute of Physics, University of Silesia, ul. Uniwersytecka 4, 40-007 Katowice, Poland, e-mail:rzoska@us.edu.pl*

[2]*Dept. of Thermodynamics, Academy of Refrigeration, 1/3 Dvoryanskaya Str., 65082 Odessa, Ukraine, e-mail:mazur@paco.net*

Sometimes a matter can be metastable, i.e. heated, compressed or stretched beyond the point at which it normally exhibits a phase change, but without triggering the transition. Recent decades have seen impressive advances in explaining puzzling properties of such metastable states.[1-8] The significance of these studies is supported by the myriad of possible society-relevant applications ranging from the modern material engineering through biochemistry and biotechnology, to the food and pharmaceutical industry and environment-relevant issues within bio-ecologic, atmospheric or deep Earth/planetary sciences.[1-8] Inherently metastable supercooled systems transforming into the glass state are one of the most classical examples here. Surprisingly, despite enormous efforts there seems to be no ultimate models for the glass transition physics, so far.[1,8,9] Hence, novel approaches are of vital importance. The last decade of investigations showed that comprehensive insight linking temperature (*T*) and pressure (*P*) measurements, including their extreme limits, can yield ultimate references for theoretical models in this field. This implies applications of high hydrostatic pressures as well as its negative pressures extension into the isotropically stretched states.[8,9] The same *P-T* studies of complex systems can provoke discoveries of novel stable and metastable phases showing non-conformistic paths of their reaching and indicating how the often unusual properties can be recoverable to ambient conditions. This can yield a surprisingly intermediate intact with commercially relevant quantities and unusual physical properties appropriate for the aforementioned applications.[3-7,10-18] In the case of the glass transition the use of the high hydrostatic pressures enabled the clarifications of fundamental theoretical expectations, for instance related to the secondary, relaxation or yielded a set of "dynamic equations of state", so important in applications.[8,9] Noteworthy are also recently discovered advantages of amorphous forms of medicines/pharmaceutical products which focused a significant part of industry-related efforts on the GFA (Glass Forming Ability) and the glass temperature (T_g) versus pressure dependences.

$$T_g(P)=F(P)D(P)=T_g^0\left(1+\frac{P-P_g^0}{\pi+P_g^0}\right)^{1/b}\exp\left(-\frac{P-P_g^0}{c}\right)$$

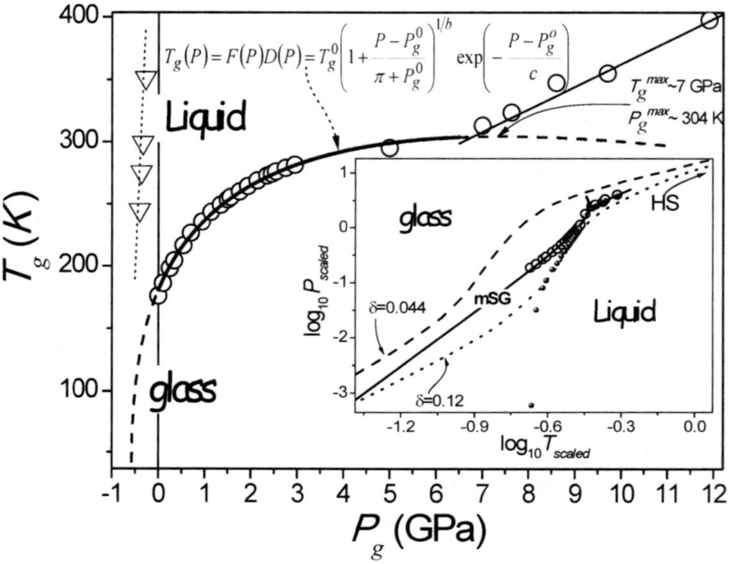

Figure 1. *The pressure evolution of the glass temperature in glycerol.*[19] *The solid curve shows the parameterization of experimental data via the novel, modified Simon-Glatzel type equation, given in the Figure. Contrary to equations applied so far it is governed by pressure invariant coefficient The solid straight line portraying data at extreme pressure can be described by the linear dependence with* $dT_g/dP \approx 18.2\,K/GPa$. *The extrapolations beyond the experimental domain are shown by dashed curve and the dashed line. The dotted line in the negative pressures domain shows the estimated loci of the hypothetical stability limit. The inset recalls the square-well (SW) model and the MCT based analysis of the glass transition evolutions, known for their applicability only for colloidal glasses before, Data presented here in SW model units, namely for glycerol:* $P_{scaled} = P^* = P_g/3.09GPa$ *and* $T_{scaled} = T^* = T_g/826K$.[20] *Note that the same pattern for the molecular liquid, glycerol and for colloid-polymer mixtures*[20] *was obtained due to the pressure data based analysis.*[19]

For instance, studies of $T_g(P)$ evolution up to 12 GPa lead to the possible link between molecular and colloidal glasses, before often considered as separate cases for the vitrification. This issue is discussed in the inset in Fig. 1.[19] The main part of the plot presents one more unusual behaviour – the possible maximum of $T_g(P)$ under extreme pressures. Consequently, the sequence liquid – glass – liquid - (hard sphere) glass on pressurization can be advised in some glass forming systems. The proposal of a common description of systems characterized by $dT_g/dP>0$ and $dT_g/dP<0$, described by pressure invariant coefficients, unavailable before, was also formulated.[19,22,23] All these may illustrate that the application of pressure results in hardly expected phenomena which in turn may create "unifying" factors for properties already known under atmospheric pressure.

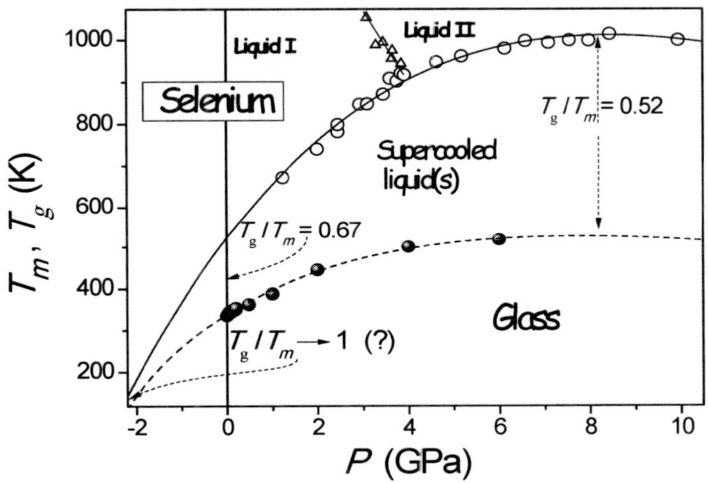

Figure 2. *The pressure evolution of the glass temperature and the melting temperature in selenium. Solid curves show parameterizations via the modified, pressure invariant Simon Glatzel type equation given in Fig. 1. Note the appearance of the maximum at* $P_{g,m}^{max} = 8.3GPa \pm 0.3$ *and the strong changes of the GFA factor on compressing or isotropic stretching of the system.*[22]

The mentioned GFA factor, since it's introducing by Turnbull[21] four decades ago, is a crucial parameter in material engineering applications. Basing on the empirical analysis of hundreds of materials Turnbull proposed the ratio $T_g/T_m \approx 2/3$ as the hallmark of the "good" GFA, i.e. the temperature quench is only near T_m, next a slow cooling is possible down to T_g. The most recent analysis of high pressure data revealed the significant pressure dependence of T_g/T_m, unexpected theoretically, as well as hypothetical significance of negative pressure states.[22,23] For the latter worth mentioning is the statement of Lev D. Landau formulated already in the first edition of his famous monography "*Statistical Physics*"…*There is a basic difference between negative pressures and negative temperatures. The latter are in a natural way unstable hence cannot exist in nature. Negative pressure states can exist in nature, although as metastable ones…*".

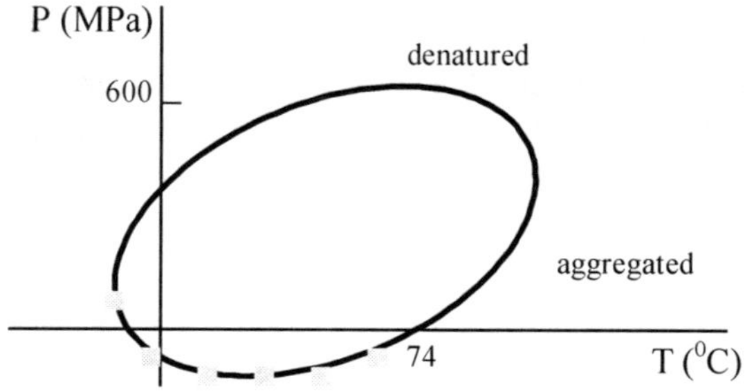

Figure 3. The phase diagramme of an example protein – myoglobine in the "full" pressure space covering both the negative and the positive pressures domain.[12,24]

Very important for applications may appear the case of pressurized and isotropically stretched proteins which may offer a qualitatively new way of the food conservation. It is shown in Fig. 3 that denaturation can be reached both "clasically", by pasteurization (heating up to ca. 80°C), or by strong compressing or by much weaker isotropic stretching (negative pressures). The two latter paths have a fundamental advantage that the coagulation can be almost avoided. Hence, the possibility of food conservation without the taste changes, so uncomfortable for milk, appears. One can also imagine a new type of high pressures/negative pressures related long-term conservation of meat without freezing (!), for instance.[25]

In fact, first commercial applications of such technology have been already launched, although they are still limited, also due to the still poor fundamental insight. It is noteworthy here that the dynamics of proteins is glass-like, what may create new and unexplored tools of monitoring the quality of products conserved in this way.

The next important issue is the quest of water properties, i.e. "the simple but very complex" liquid.[2] It is important for any practical application encountered above, where *P-T* studies revealed not only several forms of ice and water but also the challenging state of amorphous, glassy "dense" water. This issue shows that even in presumably "ordinary" single component liquid the liquid-liquid transition, at first sight beyond the Gibbs phase rule, may exist. This phenomenon seems to explain many anomalous properties of such materials as water, germanium, silicone, phosphorus.[7] However, this can be unambiguously revealed only due to the application of extreme pressures. We did not mention several other significant problems linking pressure and vitrification with critical and near critical phenomena. The latter can occur in multipomponent mixture but also in one component liquids: "ordinary" and mesomorpic (liquid crystalline).[24-27]

The issue which cannot be omitted are deep Earth structures which has a fundamental influence on human life, at least via earthquakes disasters. From recent years investigations one may conclude that the appearance of metastable structures, presumably associated with the pressure induced change of the glass forming ability and the shift of the glass temperature, are important factor which understanding is still at very beginning.[4,28] Also in this case the flow of the recent results obtained within the glass transition physics may be basically important.

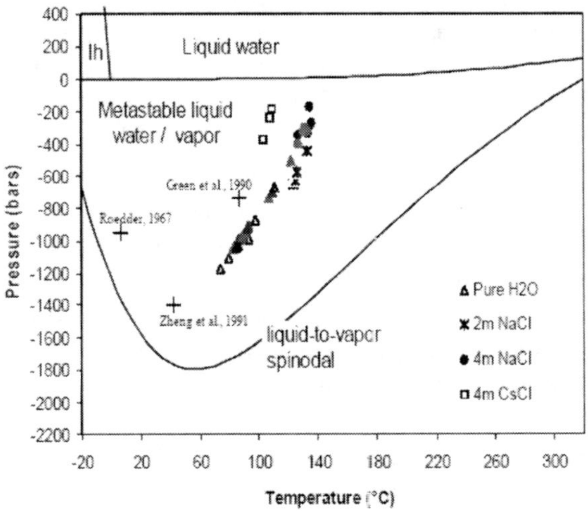

Figure 4. *The P-T phase diagramme of water, including the inherently metastable negative pressure domain.*[28]

One can imagine inherently metastable supercooled vitrifying liquids in the inherently metastable pressure induced states, for instance negative ones, influenced by metastable pretransitional fluctuations. All this can be even more complexed by the complex structure of molecules and addition of nanoparticles, for instance. The smart material processing is often subjected to a variety of thermal and mechanical treatments designed to produce various combinations of stable and metastable phases to reach the desired qualities. For the mentioned multi-metastable systems one can imagine unimaginable implications for future smart, "intelligent materials", with tuned and precisely controlled selection of parameters.

Generally, the metastability is a phenomenon associated with the persistence of the given phase well below the stability domain, bordered by the first order transition, for instance: (*i*) the glass transition phenomenon, (*ii*) metastable systems studies linked to spinodals – absolute stability limits, with particular attention towards the inherently metastable negative pressure domain (*iii*) metastability near a critical point, (*iv*) the quest for the liquid – liquid near-critical transition in one component liquid, (*v*) the issue of liquid crystals where weakly discontinuous phase transitions may coexist with vitrification related phenomena and (*vi*) a myriad of further phenomena for which aforementioned systems can serve as a reference.

The fundamental insight and the technological & environmental relevance of metastable systems recalled above have given a strong impetus from the last decade development of extreme pressures experimental techniques. The ultimate verification of theoretical models and reliable equations for portraying basic properties seems to be possible only when including both temperature and pressure paths into studies. However, the latter should contain also extreme limits, namely very high pressures (GPa) and negative pressures. The emerging possibility of the fast implementation of the fundamental research findings into technological and environment applications stress the importance of the pressure related research of metastable systems. One may speculate that universal patterns discovered in studies on metastable condensed matter/soft matter systems may also serve as a reference for social sciences, economics or communication/informatics analysis. This can be supported by the great success of the physics of critical phenomena in economy, leading to setting up of econophysics.[29]

The ARW NATO "*Metastable Systems under Pressure: Platform for New Techbological and Environmental Applications*", 4-8 Oct. 2008, Odessa, Ukraine created a unique *Forum* at which "fundamental", "technological" and "environmental" researchers could focus on metastability & pressure/negative pressures issues during brainstorming discussions in the inspiring surrounding of XIX century empirial style surrounding of Hotel Londonskaya in Odessa, Ukraine. The poor knowledge-flow between such groups is in our opinion one of the most important artifacts limiting the possible boost associated with metastable systems research & applications.

In the interdisciplinary brainstorming discussion took part 37 researchers from 11 countries, namely: France, Germany, Hungary, Italy, Poland, Russia, Slovenia, Spain, UK, Ukraine and USA, specializing in following areas: **(i)** *solid state and soft matter physics* **(ii)** *earth sci. & geophysics* **(iii)** *biophysics* **(iv)** *environmental protection engineering* **(v)** *polymer physics* **(vi)** *modern material engineering* **(vii)** *telecommunication engineering.* This volume contains both review materials, to facilitate reding, as well as saset of milestone new results.

The ARW NATO directors, Sylwester J. Rzoska (Poland) and Victor Mazur (Ukraine), are very grateful to the NATO Science Programme for the grant which made it possible to arrange this meeting. The editors are also very grateful to Mr. Will Bruins from Springer Verlag for his patience and help.

References

1. Donth, E. (1998) *The glass transition. Relaxation dynamics in liquids and disordered material*, Springer, Series in Material Sci. II, vol. **48** (Springer Verlag, Berlin)
2. Debenedetti, P. G. (1996) *Metastable liquids*, (Springer Verlag, Berlin)
3. Bower, D. I. (2002) *An introduction to polymer physics* (Cambridge Univ. Press, Cambridge,)
4. Poirier, J.-P. (2000) *Introduction to the physics of the earth's interior* (Cambridge Univ. Press., Cambridge)
5. Gruner, S. M. (2004) *Soft materials and biomaterials under pressure. Putting the squeeze on Biology*, in A. Katrusiak and P. McMillan (eds.), *High-Pressure Crystallography*, p. 543 (Kluwer, Dordrecht)
6. Jonas, J. (2000) *High pressure in bioscience*. in. M. H. Manghnani, W. J. Nellis, M. F. Nicol (eds.) *Science and Technology of high pressure*, Universities Press, Hyderabad, India, p. 29
7. McMillan, P. F. (2002) *New Materials from high pressure experiments*, *Nature Materials* **1**, 19
8. Roland, C. M., Hensel-Bielowka, S., Paluch M., and Casalini, R. (2005) Supercooled dynamics of glass-forming liquids and polymers under hydrostatic pressure, *Rep. Prog. Phys.* **68**, 1405
9. Floudas, G. (2004) Effect of pressure on systems with orientational order, *Prog. Polym. Sci.* **29**, 1143
10. Vuataz, G. (2002) *The phase diagram of milk: a new tool for optimizing the drying process*, *Lait* **82**, 485
11. Gorovits, B., and Horovits, P. M. (2002) High hydrostatic pressure can reverse aggregation of protein folding intermediates and facilitate acquisition of native structure, *Biochemistry* **37**, 6132
12. Smeller, L. (1999) Pressure-temperature phase diagrams of biomolecules, *Biochim. Biophys. Acta* **1595**, 4217
13. Arora, A. K. (2000) Pressure-induced amorphization versus decomposition, *Solid State Comm.* **115**, 665
14. Lach, R., Grellmann, W., Schroeter K., and Donth, E. (1999) Temperature dependence of dynamic yield stress in amorphous polymers as indicator for the dynamic glass transition at negative pressures, *Polymer* **40**, 1481
15. Hemley R. J., and Ashcroft, N. W. (1998) The revealing role of pressure in the condensed-matter sciences, *Physics Today* **51**, 26
16. Mishima, O., Calvert, L. D., and Whalley, E. (1984) Melting Ice I at 77 K and 10 kbar: a new method of making amorphous solids, *Nature* **310**, 393
17. Stinecipher M., Campbell, D., Garcia, D., and Idar, D. (2002) Effects of temperature and pressure on the glass transition of plastic bonded

explosives, in *jttp://lib.-www.lanl.gov/la-pubs/00412990.pdf (Los Alamos Nat. Lab.*)

18. Mao, H. K., and Hemley, R. J. (2002) New windows on earth and planetary interiors, *Mineral. Mag.* **66**, 791
19. Drozd-Rzoska, A., Rzoska, S. J., Paluch, M., Imre, A. R., and Roland, C. M. (2007) On the glass temperature under extreme pressures, *J. Chem. Phys.* **126**, 165505
20. Voigtmann, Th., and Poon, W. C. K. (2006) Glass transition under pressure – the link to colloidal science, *J. Phys. Condens.: Matt.* **18**, L465
21. Turnbull, D. (1969) Under what condition can a glass be formed, *Contemp. Phys.* **10**, 437
22. Drozd-Rzoska, A., Rzoska, S. J., and Imre, A. R. (2007) On the pressure evolution of the melting temperature and the glass transition temperature, *J. Non-Cryst. Solids* **353**, 391
23. Drozd-Rzoska, A., Rzoska, S. J., and Roland, C. M. (2008) On the pressure evolution of dynamic properties of supercooled liquids, *J. Phys.: Condens. Matt.* **20**, 244103
24. Imre, A. R., Maris, H. J., and Williams, P. R. (eds.) (2002) Liquids under Negative Pressures, *NATO Sci. Series II*, vol. 84 (Kluwer, Dordrecht)
25. Buldyrev, S. V., Franzese, G., Giovanbattista, N., Malescio, G., Sadr-Lahijany, M. R., Scala, A., Skibinski, A., and Stanley, H. E. (2002) Models for a liquid – liquid transition, *Physica A* **304**, 23
26. Tanaka, H. (2000) General view of a liquid-liquid phase transitions, *Phys. Rev. E* **62**, 6968
27. Mathot, V. B. F., Goderis, B., and Renaers, H. (2003) Metastability in polymers systems studies under extreme conditions: high pressures and scan-iso-T-t ramps, *Fibres & Textiles* **11**, 20
28. Courtessy of Prof. K. Shmulovich
29. Mantegna, R. N., and Stanley, H. E. (2000) *An Introduction to econophysics: correlation and complexity in finance* (Cambridge Univ. Press., Cambridge)

PART I

SUPERCOOLED, GLASSY SYSTEM

THE NATURE OF GLASS: SOMETHINGS ARE CLEAR

[1,2]K.L. NGAI, [2]S. CAPACCIOLI, [2]D. PREVOSTO, [3]M. PALUCH

[1]*Naval Research Laboratory, Washington DC 20375-5320 USA*
[2]*Dipartimento di Fisica, Università di Pisa, Largo Bruno Pontecorvo 3, I-56127, Pisa, Italy and CNR-INFM,polyLab, Largo Bruno Pontecorvo 3, I-56127, Pisa, Italy*
[3]*Institute of Physics, Silesian University, ul. Uniwersytecka 4, 40-007 Katowice, Poland*

Abstract: The long-standing unsolved problem of glass transition has recently been drawn attention to the research community and general public by an article published in the New York Times, entitled "*The Nature of Glass Remains Anything but Clear*". The article mainly samples current and widely different views of some theoreticians, which have led to the conclusion that the situation is anything but clear. We show this pessimistic conclusion is unwarranted because results from recent experimental investigations have not been considered. These results show clearly the importance of the effects of many-body relaxation (or its surrogate, the degree of departure from linear exponential time dependence) and the relation between the structural α-relaxation with a special secondary relaxation. Both aspects have not been taken into account by most theories. If taken into consideration, significance progress can be expected.

Keywords: glass transition, structural relaxation, frequency dispersion, secondary relaxation, many-body relaxation, interacting systems

1. Introduction

The study of vitrification of a liquid to form a glass, commonly referred to as glass transition, has a very long history. Not counting from the ancient times when Babylonians documented the study of glass making, scientific study of glasses and glass transition may be traced back more than 160 years ago to the investigations of electrical and mechanical properties of inorganic glasses and natural polymers by R. Kohlrausch[1,2], a contemporary of C. F. Gauss and W. E. Weber at Göttingen. This is the same Kohlrausch who performed the important Weber-Kohlrausch experiment in 1856 to show somehow a velocity (of light) enters into the electromagnetic picture. The experiment might have some influence on James Clerk Maxwell in the construction of his electromagnetic theory of light in the ensuing years from 1856 to 1868. However, up to the present time, glass transition is still an unsolved problem in condensed matter physics. Another notable historic example of scientific study of glasses is the observation of the effect of aging of silicate glasses thermometer at room temperature over a period of 38.5 years, from April 1844 to December 1882 by James Prescott Joule,[3] which was recently discussed by Nemilov and Johari.[4,5] Aging of the silicate glass caused change of the thermometer scale which in turn was measured by the shift in the 'zero-point temperature'. The T_g of silicate glasses, typically in the range of 680–900 K, is much higher than room temperature, and the structural α-relaxation time would be much longer than centuries. However, structural change of the silicate glass was found by Joule from the change of zero point temperature of about 8 degrees Fahrenheit over 38.5 years. Thus the observed change by Joule cannot be effected by the structural α-relaxation, and his observation already tells us that some relaxation faster than the structural α-relaxation must be considered in the glassy state. The mechanism for this observed spontaneous relaxation of glass at room temperature was attributed to the secondary or β-relaxation which causes changes of local regions in the network glass.[4]

A recent article entitled "*The Nature of Glass Remains Anything but Clear*" by a writer for New York Times[6] has drawn the attention of the scientific community and the general public that this is a challenging but worthwhile problem to be solved. It pays tribute to the importance of the glass transition problem in saying "*Understanding glass would not just solve a longstanding fundamental (and arguably Nobel-worthy) problem and perhaps lead to better glasses. That knowledge might benefit drug makers, for instance. Certain drugs, if they could be made in a stable glass structure instead of a crystalline form, would dissolve more quickly, allowing them to be taken orally instead of being injected. The tools and techniques applied to glass might also provide headway on other problems, in material science, biology and other fields, that look at general properties that arise out of many disordered interactions.*"

Philip W. Anderson, a Nobel Laureate, wrote in 1995:[7] "The deepest and most interesting unsolved problem in solid state theory is probably the theory of the nature of glass and the glass transition, and he added, "This could be the next breakthrough in the coming decade." The NY Times article cited Anderson, but pointed out that the problem remains unsolved with the statement: "*Thirteen years later, scientists still disagree, with some vehemence, about the nature of glass.*" This pessimistic note of the NY Times article, suggesting impasse in the research on glass transition, is the result of the editor sampling the views principally from theoreticians who consider only a few familiar experimental facts and physical quantities all related to the structural α-relaxation. Several important and general experimental facts outside those considered by the consulted theoreticians have been discovered in recent years. These have impact on theoretical understanding and should vastly improve the chance for someone to finally solve the glass transition problem. These facts, not mentioned in the NY Times article and not taken into account by the theoreticians consulted, are discussed in this paper to heighten the awareness for those interested in the glass transition problem. Without incorporating these general experimental facts and providing explanations, any theory of glass transition would be incomplete. Moreover, these additional experimental facts are found in "*other problems, in material science, biology and other fields*".

The NY Times article gave one specific example, which is the problem of improved solubility and bioavailability of drugs in pharmaceutical industry. We supply more examples that include carbohydrates for bio-preservation and for food technology, hydration of biomolecules for functioning in life science, and fuel cell materials for the energy industry. We show the basic features of dynamics of the systems in these other fields are isomorphic to glassformers, and hence solution of the glass transition problem can benefit progress in the other fields.

2. Invariance of the α-dispersion to different combinations of T and P at constant τ_α

The relaxation time at ambient pressure can be maintained constant at elevated pressure P by raising temperature T. Various combinations of P and T can be chosen for which the structural α-relaxation frequency ν_α or time τ_α is the same. One important fact emerging from these pressure studies is that at a constant value of the structural relaxation time τ_α or frequency ν_α, the dispersion of the structural relaxation is constant.[8,9] Very generally it is found that for a given material at a fixed value of τ_α, the frequency dispersion or the time dependent relaxation function is constant, independent of thermodynamic conditions (T, P and their conjugate variables entropy and volume). Alternatively stated, temperature-pressure superpositioning works for the dispersion of the structural α-relaxation at constant τ_α. Lack of superposition may occur at frequencies sufficiently high compared with ν_α. Such deviation can be attributed to the contribution from resolved or unresolved secondary relaxation at higher frequencies or shorter times, whose relaxation strength may not have the same P and T dependences as the α-relaxation. This general experimental fact of glass-formers is supported by experimental data of many different materials, and, for a particular material, by experimental data for several different values of the relaxation time.

In many cases, the dispersion broadens with increasing τ_α on decreasing temperature or increasing pressure, and thus the observed superpositioning cannot be explained trivially by constant dispersion.

The glass-formers include molecular liquids and amorphous polymers of diverse chemical structures,[8,9] room temperature ionic liquids,[10] and binary mixtures of two van der Waals liquids[11,12] or two amorphous polymers.[13] All show the property of temperature-pressure superpositioning of the frequency dispersion of the structural α-relaxation at constant τ_α.

Figure 1 shows an example from electric modulus spectra of the room temperature ionic liquid (an environmental friendly material that have many applications), 1-butyl-1-methylpyrrolidinium bis[oxalato]borate (BMP-BOB) measured over wide temperature (123–300 K) and pressure (0.1–500 MPa) ranges.[10]

Figure 2 shows another example from triphenyl phosphate (TPP), a glass-former with some unusual properties.[11] Notable exceptions are limited to some hydrogen-bonded glass-formers which suffer change in structure at high temperature and pressure, and thus the lack of temperature-pressure superpositioning of the dispersion of the structural α-relaxation is understandable.

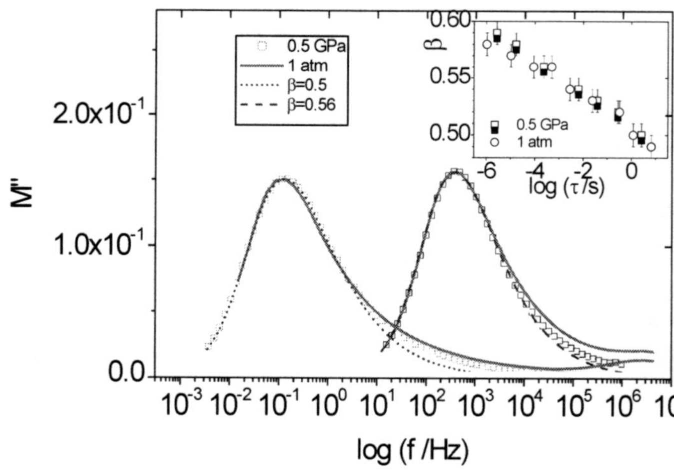

Figure 1. *Electric modulus relaxation spectra (M") of the ionic liquid BMP-BOB at ambient pressure and 231 and 245 K are plotted as solid lines. High pressure M" data (0.5 GPa) at the temperatures that yield relaxation times similar to those of the ambient pressure data, 283 and 308 K, are included in the figure as squares. Data at 0.5 GPa data are slightly shifted in frequency to match perfectly the atmospheric peak frequencies. Long and short dashed lines are fits to a Kohlrausch relaxation function with* $\beta \equiv (1-n) = 0.56$ *and 0.50, respectively. The inset shows the good correspondence between the stretching parameter* β *and the relaxation time at different temperatures and at atmospheric pressure and at 0.5 GPa.*

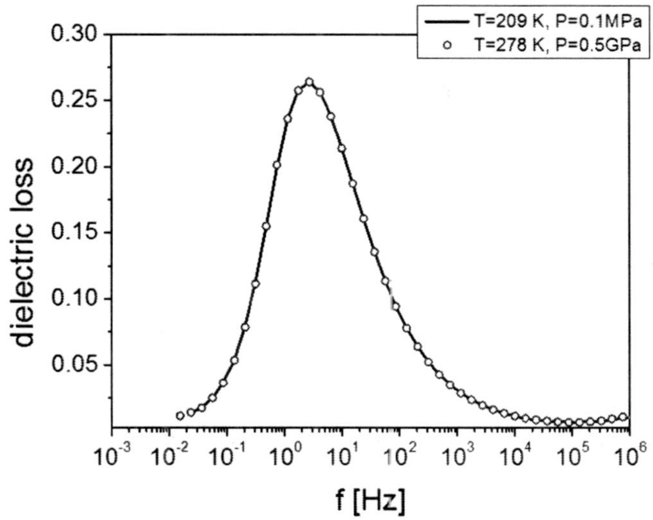

Figure 2. *T,P-superposition of dielectric loss data of liquid triphenyl phosphite at ambient and elevated pressure of 500 MPa.*

The general experimental fact of constant frequency dispersion (or time dependence of the correlation function) of the α-relaxation at constant τ_α for different combinations of T and P has an immense impact on glass transition. Although the data were mostly obtained by dielectric relaxation, the same effect[8,9] was found in some glass-formers by photon correlation spectroscopy.[9] The primary concern of most theories, including those mentioned in the NY Times article, is to explain the temperature and pressure dependences of the structural relaxation time τ_α. In these theories, the dispersion of the structural relaxation is either not addressed, or else considered separately with additional input not involved in arriving at τ_α. Consequently, the frequency dispersion is unrelated to the relaxation time of the structural α-relaxation in these theories, and they are unlikely to be consistent with the T, P-superpositioning property by happenstance.

3. Properties of the α-relaxation are governed by or correlated with the width (or the nonexponentiality) of its dispersion

Not only does the magnitude of τ_α uniquely define the dispersion, as shown herein, but also many properties of τ_α are governed by or correlated with the width of the dispersion of the structural relaxation or the fractional exponent n of the Kohlrausch function, $\phi(t) = \exp[-(t/\tau_\alpha)^{1-n}]$, frequently used to fit the time dependence of the correlation function.[9,15] Here we cite the following examples, the details of which can be found in references.[9,15]

(1) The steepness or 'fragility' index defined by $m \equiv d\log_{10}\tau_\alpha / d(T_g/T)\big|_{T_g/T=1}$, correlates with n for glassformers belonging to the same chemical family. However, we caution that m is a complex quantity. It is determined by the dependence of τ_α not only on volume and entropy but also on the dynamics of many-body relaxation. On the other hand, n reflects only the latter. Hence, the correlation may break down when glassformers having different chemical structures and dependence of volume and entropy on temperature are considered together. Even in the same glassformer, the correlation usually breaks down by elevating pressure. As already mentioned in the previous section, the frequency dispersion of the structural α-relaxation and hence n is constant for various combinations of T and P leading to the same value of τ_α. But it is commonly observed that the fragility of non-associating liquids usually decrease with increasing pressure. For instance, it has been recently found that the value of m of TPP drops from 125 at ambient pressure to 80 at P=0.5 GPa, whereas the shape of α-dielectric loss peak remains unchanged ($n\cong 0.5$). This breakdown of correlation between m and n in the same glassformer is due to altered dependence of τ_α on volume and entropy by elevated pressure.

(2) Quasielastic neutron scattering experiments and molecular dynamics simulations on polymeric and non-polymeric glass-formers have found that the dependence of τ_α on the scattering vector Q is given by $Q^{-2/(1-n)}$. Hence the Q-dependence of τ_α is governed by the breadth of the dispersion or n. Such Q-dependence of relaxation time is also shared by other intermolecularly coupled systems including suspensions of colloidal particles, semidilute polymer solutions, associating polymer solutions, and polymer cluster solutions.

(3) The temperature dependence of τ_α observed over more than twelve decades of time, from sub-nanoseconds to 100 s, cannot be fit by a single Vogel-Fulcher-Tammann-Hesse (VFTH) equation. At short times and temperature higher than a system dependent temperature T_A, τ_α has the Arrhenius dependence. Below T_A, τ_α has a VFTH dependence, $(\text{VFTH})_1$, which is no longer adequate when temperature falls below a second characteristic temperature T_B. A second VFTH equation, $(\text{VFTH})_2$, has to be used to describe τ_α for $T<T_B$. At T_A, $n(T_A)$ is small. There is a significant increase of the rate of change of $n(T)$ with decreasing temperature when crossing T_B.[16a] The difference between $(\text{VFTH})_1$ and $(\text{VFTH})_2$ correlates with the width of the α-relaxation dispersion or $n(T_g)$, if T_g is defined uniformly as the temperature at which τ_α reaches an arbitrarily chosen long time, say 10^2 s.[15]

The crossover from $(\mathrm{VFTH})_1$ to $(\mathrm{VFTH})_2$ was observed isobarically on varying temperature not only at ambient but also at elevated pressures. Also there is a change of τ_α from one pressure dependence to another isothermally on varying pressure on crossing a material specific pressure P_B. The crossover temperature T_B generally increases with applied pressure P, and the crossover pressure P_B increases with temperature, but the value of τ_α or the viscosity at the crossover is the same for a given glass-former.[9,16b,17] An example from phenolphthalein-dimethyl-ether (PDE) is shown in Figure 3, where $\phi_T = (d\log\tau/dT^{-1})^{-0.50}$ and $\phi_P = (d\log\tau/dP)^{-0.50}$ used to linearize the VFTH T-dependence and the analogue P-dependence show the crossover clearly by two straight lines with different slopes.[16] As have already discussed, we have the same dispersion at constant τ_α, independent of T and P. Hence, the dispersion (or n) as well as τ_α is invariant at the crossover from $(\mathrm{VFTH})_1$ to $(\mathrm{VFTH})_2$ or from one P-dependence to another.

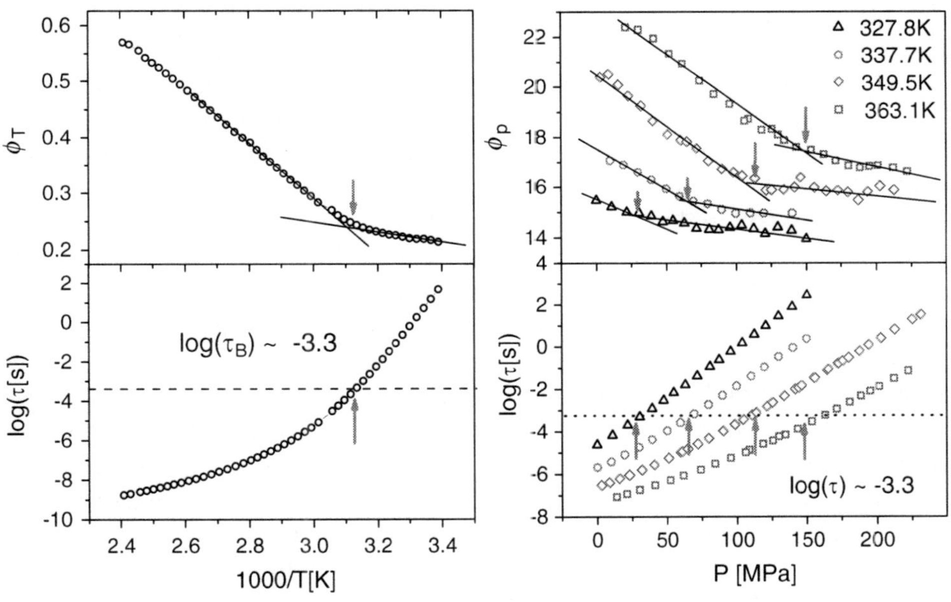

Figure 3. *Phenolphthalein-dimethyl-ether (PDE) data. Left panel, τ_α and derivative function ϕ_T at ambient pressure. Right panel, τ_α and derivative function ϕ_P vs pressure, obtained for isotherms at T=327.8, 337.7, 349.5 and 363.1 K.*

(4) The rotational diffusion coefficient, D_r, of a probe molecule in a glass-former follows the temperature dependence of the Debye-Stokes-Einstein (DSE) equation, $D_r \equiv 1/6\langle\tau_c\rangle = kT/8\pi\eta r_s^{\;3}$. Here η is the shear viscosity, $<\tau_c>$

is the mean rotational correlation time, and r_s the spherical radius of the probe molecule. On the other hand, the translational diffusion coefficient, D_t, of the probe molecule is given by the Stokes-Einstein (SE) relation, $D_t = kT/6\pi\eta r_s$. Thus, the combined SE and the DSE equations predict that the product $D_t\tau_c \equiv (D_t\tau_c)_{SE,DSE}$ should equal $2r_s^2/9$. Measurements of probe translational diffusion and rotational diffusion made in glass-formers have found that the product $D_t\tau_c$ can be much larger than this value, revealing a breakdown of the SE and the DSE relations. There is an enhancement of probe translational diffusion in comparison with rotational diffusion.[18-21] The time dependence of the probe rotational time correlation functions $r(t)$ are well-described by the KWW function, $\exp[-(t/r_c)]^{1-n}$. The ratio $D_t\tau_c/(D_t\tau_c)_{SE,DSE}$ evaluated at $T=T_g$ is a measure of the degree of the breakdown of the SE and DSE relations for various combinations of probes and host glass-formers.[18-21] A strong correlation was observed at $T=T_g$ between the quantity $D_t\tau_c/(D_t\tau_c)_{SE,DSE}$ and n, the width of the dispersion of the probe rotational correlation functions $r(t)$.[20,21] The variation of the dispersion or n of the probe in different hosts was traced to the difference between the probe rotation time and the host structural relaxation time.[21]

Some theories of glass transition mentioned in the NY Times article stress the importance of heterogeneous dynamics. This feature of the dynamics had been used to explain the breakdown of SE and DSE relations. In one version of such explanations,[18-20] presence of spatial regions of differing dynamics is assumed and they give rise to the Kohlrausch relaxation function in ensemble averaging measurements. It was argued that decoupling between self-diffusion and rotation occurs because D_t and τ_c are averages over different moments of the distribution of relaxation times, with $D_t \propto \langle 1/\tau \rangle$ emphasizing fast dynamics, while $\tau_c \propto \langle \tau \rangle$ is determined predominantly by the slowest molecules. This explanation is intuitively appealing and for some times it is believed to be true.

However, in order for this explanation to be consistent with the observed monotonic increases of the products $D_t\eta$ and $D_t\tau_c$ as the temperature is lowered toward T_g, the breadth of the relaxation time distribution has to increase (or the Kohlrausch exponent, $1\text{-}n$, has to decrease) correspondingly. However, for two glassformers, ortho-terphenyl (OTP) and *tris*naphthylbenzene (TNB) which show the breakdown of the SE and DSE relations, Richert and coworkerss[19,22] recently reported that their dielectric spectra are characterized by a temperature independent width (e.g. $1\text{-}n_d$ is constant and is equal to 0.50) from 345–417 K in the case of TNB. The T_g of TNB is 342 K. Photon correlation spectroscopic[23] and NMR [24] measurements all indicate a temperature-independent distribution of relaxation times. Thus, the data of TNB and OTP contradicts the explanation based on spatial heterogeneities. On the other hand, an alternative explanation [21] based on intermolecular coupling (originating from many-molecule relaxation) continues to hold.[25]

(5) The α-relaxation involves cooperative, non-exponential, and heterogeneous dynamics of many molecules (or chain segments for polymers), which at any temperature define a length-scale L_{dh}. The dispersion of the α-relaxation is also a consequence (in parallel with heterogeneous dynamics) of the many-body relaxation. Naturally we expect a larger L_{dh} to be associated with a broader dispersion, because both quantities directly reflect the intermolecularly cooperative dynamics. This correlation is borne out by comparing β_{KWW}≡$(1\text{-}n)$ with L_{dh} for glycerol, ortho-terphenyl and poly(vinylacetate) all obtained by the same technique, the multidimensional ^{13}C solid-state exchange NMR experiment.[25,26]

4. An important class of secondary relaxations bearing strong connection to the α-relaxation

In addition to primary α-relaxation, there are secondary relaxation processes that have transpired at earlier times. Most theories including those cited in the NY Times article have focused their attention on the primary α-relaxation and do not consider any secondary relaxation to be important for glass transition. It turns out secondary relaxation belonging to a special class has various properties indicating that it bears strong connection to the α-relaxation.[9,11,12,27,28,29] Moreover, secondary relaxation of this special class is universal and found in all kinds of glassformers, organic molecular, polymeric, metallic, inorganic, ionic, and plastic crystalline.[9] The most remarkable are the finding of the secondary relaxation in metallic glasses which are atomic particles devoid of rotational degree of freedom, and in plastic crystals which have no translational degree of freedom. These strong connections imply that secondary relaxation in this special class plays a fundamental role in the dynamics leading to glass transition, and theories. Here we briefly mention some of the connections established in the past,[9,27,28] and present some recent data to show the connection

Already shown in 1998 from the data of many glassformers, the relaxation time τ_β at T_g of a special secondary relaxation is strongly correlated with n in the Kohlrausch correlation function, $\exp[-(t/\tau_\alpha)^{1-n}]$, of the α-relaxation,[27] and in approximate agreement with the primitive relaxation of the Coupling Model. Since then, many more experimental investigations have confirmed this and extended it to temperatures above T_g .[30,31] The properties of these secondary relaxations mimic the primary α-relaxation, and are connected to the α-relaxation in various qualitative and quantitative ways.[9,11,12,25,27-30] We call these secondary relaxations the Johari-Goldstein (JG) β-relaxations to honor these authors for the important discovery of secondary relaxation in totally rigid molecules,[32] which belongs to this special glass. It may also be called the primitive relaxation of the Coupling Model[31] since τ_β or τ_{JG} is approximately the same as the primitive relaxation time τ_0.[9-12,17,25,27-30] Here we give a few recent examples to show that the JG β-relaxation is inseparable from the α-relaxation and both have to be taken into account in solving the glass transition problem.

(1) Böhmer and coworkers[33,34] used spin-lattice relaxation weighted stimulated-echo spectroscopy to find evidence for a correlation of the α- and the JG β-relaxation times above the calorimetric glass transition temperature of ortho-terphenyl, D-sorbitol, and cresolphthaleindimethylether (CDE or KDE). They found that the α-relaxation can be modified by suppressing the contributions of some subensembles of the JG β-relaxation in these glassformers. An earlier deuteron NMR experiment also gave indication of a possible correlation of the α- and JG β-relaxations of polystyrene.[35]

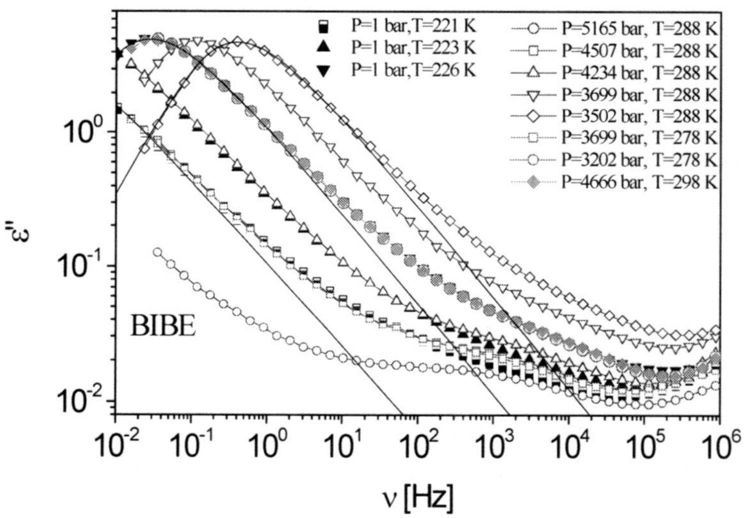

Figure 4. *T-P superposition of both α- and JG β-relaxation of benzoin-isobutylether (BIBE). Note that when the 1 bar data are near coincident with the data at elevated pressures, the former cannot be seen.*

(2) By applying elevated pressure and compensated by raising temperature, a spectacular experimental finding by dielectric relaxation is the invariance of the ratio τ_{JG}/τ_α for different combinations of T and P while keeping τ_α constant. This was found in the neat glassformer, dipropyleneglycol dibenzoate (DPGDB), benzoin-isobutylether (BIBE), polyphenylglycidylether (PPGE), polyvinylacetate (PVAc), and diglycidyl ether of bisphenol A (DGEBA).[36-39]

As an example, this effect is shown by the isothermal dielectric loss spectra of BIBE at ambient pressure and at elevated pressures in Fig. 4.

For DGEBA (Fig. 5), the effect is shown in a different way by the same τ_β at the isobaric glass transition temperatures T_g (for different constant pressures) defined by $\tau_\alpha(T_g)$=10 s, or at the isothermal glass transition pressures P_g (for different constant temperatures) defined by $\tau_\alpha(P_g)$=10 s.

The same effect is found in the dynamics of a component in binary mixtures including *tert*-butylpyridine (TBP),[36] quinaldine (QN),[12] or picoline[8] in mixtures with oligomers of styrene. Shown here in Figs. 6 and 7 are data of 10 wt.% of QN in tristyrene, and in Fig. 8 the data of 25 wt % of 2-picoline in tristyrene.

Previously, it had been shown for many glassformers that the frequency dispersion of the α-relaxation (or n) is invariant to changes of T and P if τ_α is kept constant.[8,9] From this added feature of the JG β-relaxation, we have co-invariance of three quantities, τ_α, n, and τ_{JG}, to widely different T and P combinations involving large variations of specific volume and entropy. This remarkable relation between τ_α and τ_{JG} is another strong evidence that the JG β-relaxation has fundamental significance and its relation to the α-relaxation must be taken into account. However, none of the theories cited in the NY Times article paid any attention to it.

(3) In several cases where the JG β-relaxation is clearly resolved above and below T_g and its relaxation time determined directly without using any assumed fitting procedure, the T-dependence of τ_{JG} is reported to change from Arrhenius dependence below T_g to a stronger T-dependence above T_g .[40-44] The same was observed for the pressure dependence when crossing the glass transition pressure P_g isothermally[12,37-39] (see examples in Figs. 7 and 8). These general experimental facts are indications that the JG β-relaxation is not independent of the α-relaxation, and actually the two are well connected.

(4) The dielectric relaxation strength of the JG β-relaxation, $\Delta\varepsilon_{JG}(T)$, also changes its T-dependence on crossing T_g. It has stronger temperature dependence above T_g than below it. This is found for neat glassformers as well as for a component in binary mixtures[44,45] An example of *tert*-butyl-pyridine (TBP) in tristyrene is shown in Fig. 9.

For DGEBA (Fig. 5), the effect is shown in a different way by the same τ_β at the isobaric glass transition temperatures T_g (for different constant pressures) defined by $\tau_\alpha(T_g)$=10 s, or at the isothermal glass transition pressures P_g (for different constant temperatures) defined by $\tau_\alpha(P_g)$=10 s.

The same effect is found in the dynamics of a component in binary mixtures including *tert*-butylpyridine (TBP),[36] quinaldine (QN),[12] or picoline[8] in mixtures with oligomers of styrene. Shown here in Figs. 6 and 7 are data of 10 wt.% of QN in tristyrene, and in Fig. 8 the data of 25 wt % of 2-picoline in tristyrene.

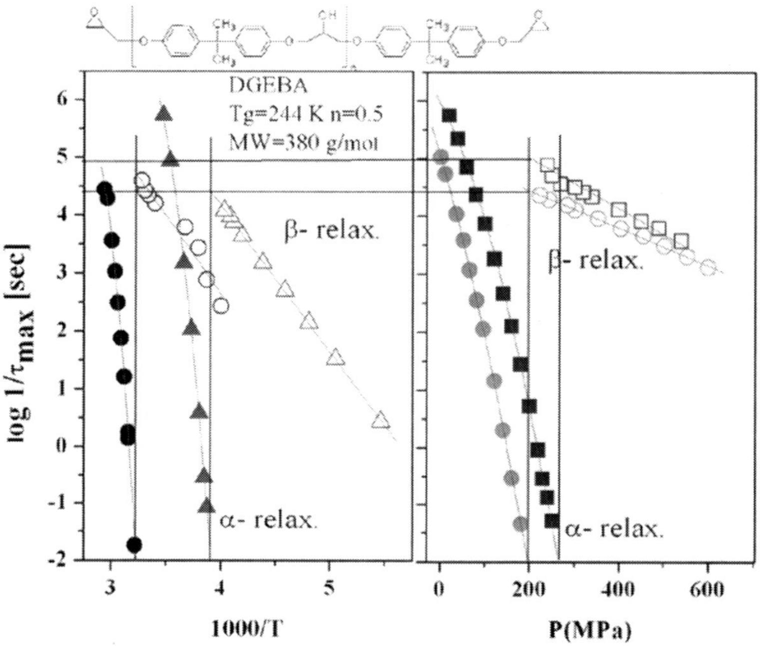

Figure 5. *α- and JG β- relaxation times of DGEBA as a function of temperature at two different pressures, 0.1 and 400 MPa (left), and as a function of pressure at two different temperatures, 293 and 283 K (right). We observed the same* τ_β *at the isobaric glass transition temperature* T_g *or the isothermal glass transition pressure* P_g *defined here by* $\tau_\alpha(T_g)=\tau_\alpha(P_g)=10$ *s.*

(5) In the glassy state at temperatures below T_g, the T-dependence of τ_α is Arrhenius with activation energy E_α. Data of τ_α below T_g are not easy to access because it becomes too long. However, for some glassformers they are either available or can be extracted by analysis[36] based on fictive temperature model. Below T_g, the value of E_α is larger than the activation energy E_{JG} of τ_{JG}. Interestingly the two activation energies are related through n by the equation, $E_{JG}=(1-n)E_\alpha$, as shown in Fig. 10.

Space limitation does not allow us to discuss more properties of the JG β-relaxation that are similar to the α-relaxation. Some of them are simply stated in the following without any further discussion.

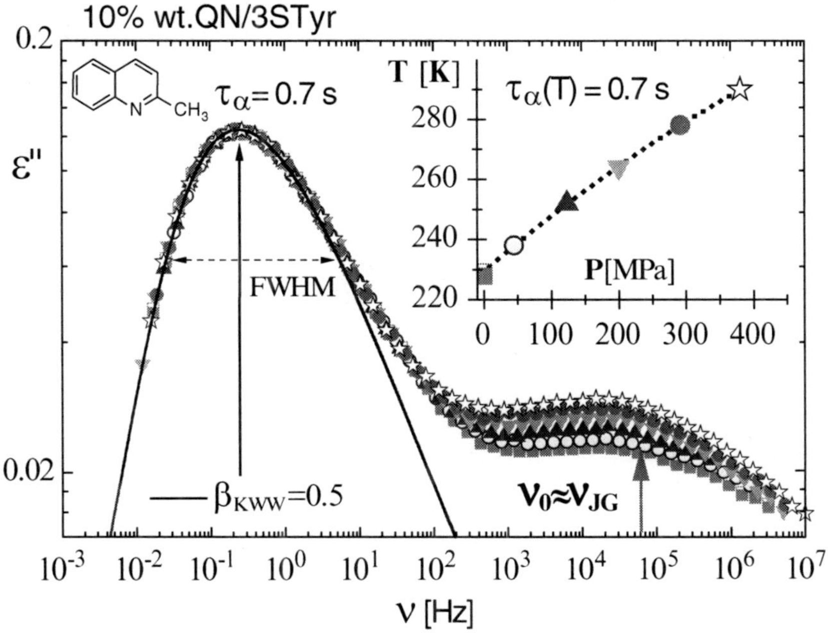

Figure 6. *T-P superposition of loss spectra for 10% QN in tristyrene measured for different T and P combinations but the same* τ_α*= 0.67 s. The line is a Fourier transformed of the Kohlrausch function with* $\beta_{KWW} \equiv (1\text{-}n) = 0.5$*. The results demonstrate the co-invariance of three quantities,* τ_α*, n, and* τ_{JG}*, to widely different combinations of T and P.*

(6) The JG β-relaxation in the glassy state, like the α-relaxation, is sensitive to thermal history, physical aging, and the particular thermodynamic (T,P) path used to arrive at the glassy state.

(7) JG β-relaxation is responsible for structural change deep in the glassy state on aging where the α-relaxation is ineffective because τ_α is far too long. An historic example is the observation of James Prescott Joule[3-5] mentioned in the Introduction.

(8) JG β-relaxation governs the rate of crystal nucleation, the initial process of crystallization.[9]

(9) At times earlier than the onset of JG β-relaxation, all molecules are mutually caged and the caged dynamics are manifested as the nearly constant loss (NCL), which is a more general feature than the so-called β-process predicted by the mode coupling theory.[46] Experimental data have shown that τ_{JG} is not much longer than upper bound the NCL time regime indicating that the JG β-relaxation causes cage decay and terminates the NCL. The intensity of the NCL as a function of temperature also changes slope at T_g .[9,46]

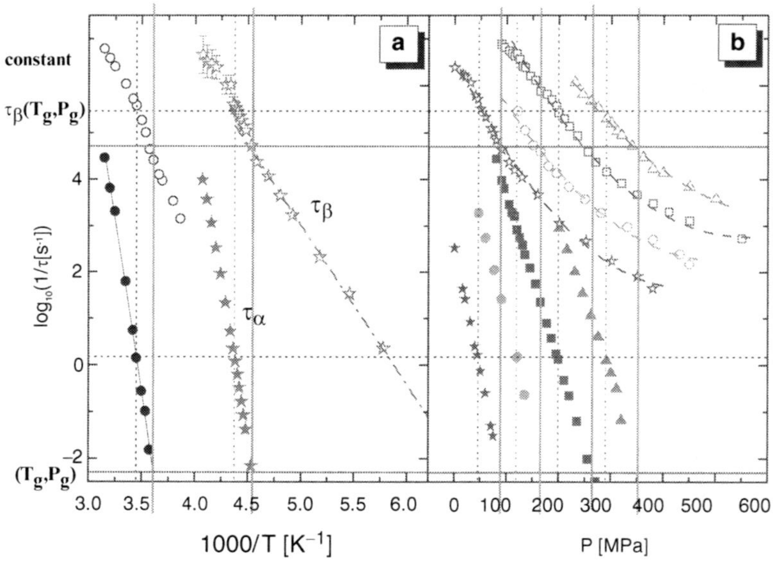

Figure 7. *α- and JG β-relaxation of 10 wt.% of QN in mixture with tri-styrene. Same τ_β at the isobaric glass transition temperature T_g or the isothermal glass transition pressure P_g defined by $\tau_\alpha(T_g)= \tau_\alpha(P_g)=2\times10^2$ s. Note that there is a change of T-dependence and P-dependence of τ_β when crossing T_g and P_g respectively.*

Previously, it had been shown for many glassformers that the frequency dispersion of the α-relaxation (or n) is invariant to changes of T and P if τ_α is kept constant.[8,9] From this added feature of the JG β-relaxation, we have co-invariance of three quantities, τ_α, n, and τ_{JG}, to widely different T and P combinations involving large variations of specific volume and entropy. This remarkable relation between τ_α and τ_{JG} is another strong evidence that the JG β-relaxation has fundamental significance and its relation to the α-relaxation must be taken into account. However, none of the theories cited in the NY Times article paid any attention to it.

5. Knowledge beneficial to other research fields and technology?

The NY Times article[6] pointed out that the knowledge in solving the glass transition problem might benefit other areas of research and technology. The article states: "*The tools and techniques applied to glass might also provide headway on other problems, in material science, biology and other fields, that look at general properties that arise out of many disordered interactions.*". We cannot agree more with this view by showing some examples that knowledge in the glass transition problem has already provided headways in solving the problems in other fields.

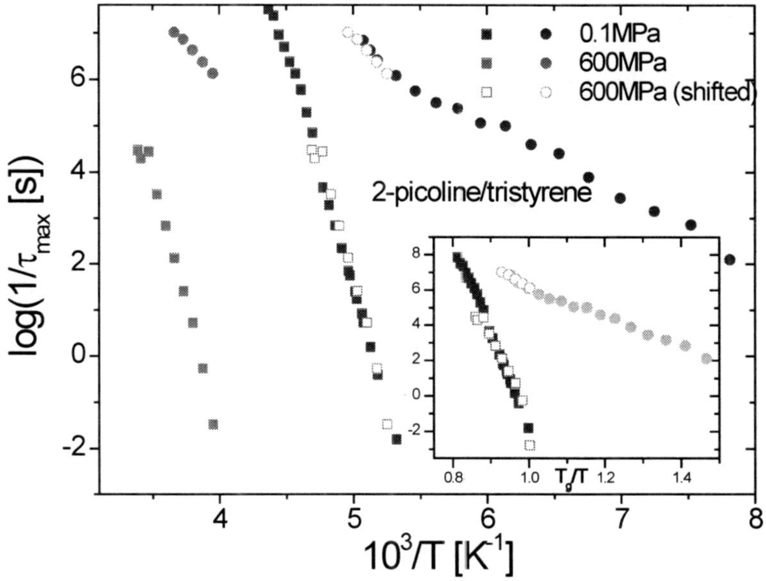

Figure 8. *Relaxation map of τ_α and τ_β of 25 wt % 2-picoline in mixture with tristyrene. Circles are for 600 MPa and squares are for 0.1 MPa. Open circles are the data of 600 MPa after a horizontal shift of 1.3 to the right has been made. In the inset the same data in the $\log\tau$ vs T_g/T representation are shown.*

Pharmaceuticals. Specific mention is made in the NY Times article of the pharmaceutical industry in its effort to make drugs in stable glassy states because the amorphous state of a pharmaceutical has various advantages over the crystalline state such as improved solubility and bioavailability. The JG β-relaxation exists in pharmaceuticals, has the same properties and relations to the α-relaxation as the glass formers at large. Here we show some of the most recent data of the pharmaceutical, (2RS)-2[4-(2-methylpropyl)phenyl]propanoic acid ($C_{13}H_{18}O_2$), commonly known as ibuprofen.[47] Figure 11 show the α-relaxation and two secondary relaxations, the slower one of which is the JG β-relaxation. The figure shows the change of T-dependence of τ_{JG} when crossing T_g in exactly the same manner as other glassformers. Other similarities including the good agreement of the observed τ_{JG} with the calculated primitive relaxation time τ_0, and this is also shown in Fig. 11. More details can be found in the cited reference.

(6) The JG β-relaxation in the glassy state, like the α-relaxation, is sensitive to thermal history, physical aging, and the particular thermodynamic (T,P) path used to arrive at the glassy state.

(7) JG β-relaxation is responsible for structural change deep in the glassy state on aging where the α-relaxation is ineffective because τ_α is far too long. An historic example is the observation of James Prescott Joule[3-5] mentioned in the Introduction.

(8) β-relaxation governs the rate of crystal nucleation, the initial process of crystallization.[9]

(9) At times earlier than the onset of JG β-relaxation, all molecules are mutually caged and the caged dynamics are manifested as the nearly constant loss (NCL), which is a more general feature than the so-called β-process predicted by the mode coupling theory.[46] Experimental data have shown that τ_{JG} is not much longer than upper bound the NCL time regime indicating that the JG β-relaxation causes cage decay and terminates the NCL. The intensity of the NCL as a function of temperature also changes slope at T_g .[9,46]

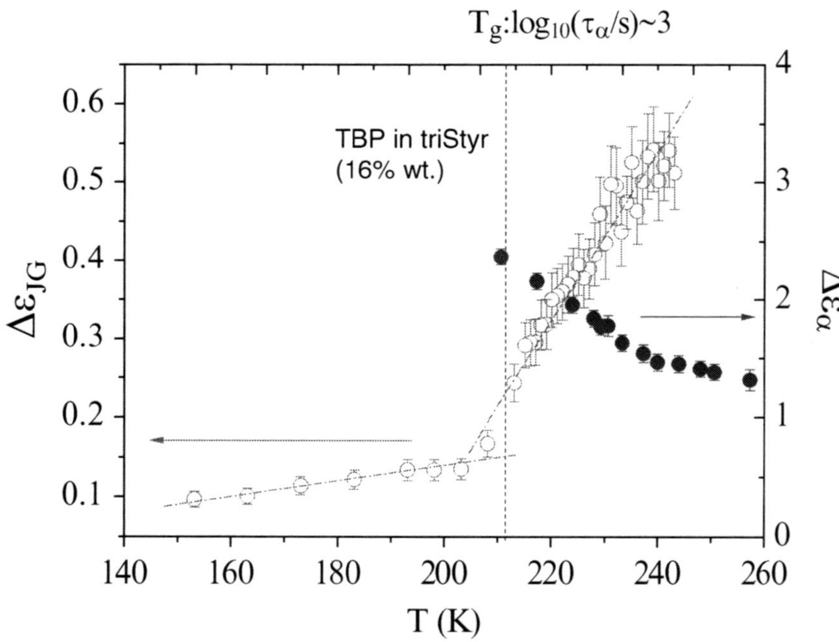

Figure 9. *Dielectric strengths $\Delta\varepsilon_\alpha$ and $\Delta\varepsilon_\beta$ of the mixture of 16 wt.% tert-butyl-pyridine (TBP) in tristyrene. Full symbols are for $\Delta\varepsilon_\alpha$ of the α-relaxation (right axis) and open symbols for $\Delta\varepsilon_\beta$ of the JG β-relaxation (left axis). Dotted lines are linear fits of $\Delta\varepsilon_\beta$ below and above T_g. Dashed vertical line indicates T=211.5 K [where $\log_{10}(\tau_\alpha/s)\sim 3$] and is near the temperature at which occurs the crossover of temperature dependence of $\Delta\varepsilon_\beta$ with an elbow-shape.*

The other pharmaceuticals investigated and showing similar results include: (a) aspirin (acetyl salicylic acid),[48,49] and (b) indomethacin, $C_{19}H_{16}ClNO_4$: (1-(*p*-chlorobenzoyl)-5-methoxy-2-methylindole-3-acetic acid).[50,51,52] As mentioned before, the JG β-relaxation controls the rate of crystal nucleation in the glassy state, and so the knowledge about the properties of this process is very relevant for drug storage and delivery. Some studies of this problem can be found in Refs. [[53,54]].

Carbohydrates. The monosaccharides, glucose, fructose, galactose, sorbose, and ribose, are hydrogen-bonded glass-forming organic substances.[55] They have important applications in food science, medicine, and biology. D-ribose and 2-deoxy-D-ribose are respectively the building blocks of the backbone chains in the nucleic acids DNA (deoxyribonucleic acid) and RNA (ribonucleic acid).[56] The disaccharides such as trehalose, maltose, and leucrose[57,58] are useful in biopreservation[59] and life science, and the polysaccharides are important in other areas. On elevating pressure, fructose, D-ribose[55], 2-deoxy-D-ribose[56].

Knowledge beneficial to other research fields and technology?

The NY Times article[6] pointed out that the knowledge in solving the glass transition problem might benefit other areas of research and technology. The article states: "*The tools and techniques applied to glass might also provide headway on other problems, in material science, biology and other fields, that look at general properties that arise out of many disordered interactions.*" We cannot agree more with this view by showing some examples that knowledge in the glass transition problem has already provided headways in solving the problems in other fields.

Pharmaceuticals. Specific mention is made in the NY Times article of the pharmaceutical industry in its effort to make drugs in stable glassy states because the amorphous state of a pharmaceutical has various advantages over the crystalline state such as improved solubility and bioavailability. The JG β-relaxation exists in pharmaceuticals, has the same properties and relations to the α-relaxation as the glassformers at large. Here we show some of the most recent data of the pharmaceutical, (2RS)-2[4-(2-methylpropyl)phenyl]propanoic acid ($C_{13}H_{18}O_2$), commonly known as ibuprofe.[47] Figure 11 show the α-relaxation and two secondary relaxations, the slower one of which is the JG β-relaxation. The figure shows the change of T-dependence of τ_{JG} when crossing T_g in exactly the same manner as other glassformers. Other similarities including the good agreement of the observed τ_{JG} with the calculated primitive relaxation time τ_0, and this is also shown in Fig. 11. More details can be found in the cited reference. The other pharmaceuticals investigated and showing similar results include: (a) aspirin (acetyl salicylic acid),[48,49] and (b) indomethacin, $C_{19}H_{16}ClNO_4$ (1-(p-chlorobenzoyl)-5-methoxy-2-methylindole-3-acetic acid).[50,51,52] As mentioned before, the JG β-relaxation controls the rate of crystal nucleation in the glassy state, and so the knowledge about the properties of this process is very relevant for drug storage and delivery. Some studies of this problem can be found in Refs. [53,54].

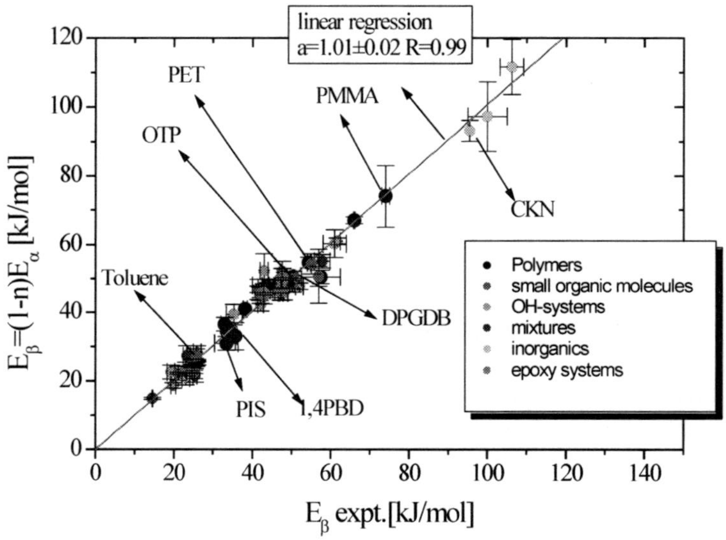

Figure 10. *Linear correlation between the experimental activation energy in the glassy state for the JG β-relaxation process (abscissa) and the activation energy predicted by the Coupling Model for the primitive relaxation (ordinate). Symbols for simple Van der Waals molecules, H-bonded systems, polymers, chlorobenzene/toluene mixture, inorganics and epoxy oligomers are shown in the figure. The solid line is a linear regression of data (linear coefficient 0.99 ± 0.01).*

Carbohydrates. The monosaccharides, glucose, fructose, galactose, sorbose, and ribose, are hydrogen-bonded glass-forming organic substances.[55] They have important applications in food science, medicine, and biology. D-ribose and 2-deoxy-D-ribose are respectively the building blocks of the backbone chains in the nucleic acids DNA (deoxyribonucleic acid) and RNA (ribonucleic acid).[56]

The disaccharides such as trehalose, maltose, and leucrose[57,58] are useful in biopreservation [59] and life science, and the polysaccharides are important in other areas. On elevating pressure, fructose, D-ribose[55], 2-deoxy-D-ribose[56], and leucrose[58] have a secondary relaxation shifting to lower frequencies with applied pressures, mimicking the behavior of the α-relaxation. The one in leucrose is sensitive to the thermodynamic history of measurements. There is also good agreement of the observed relaxation time of the secondary relaxation with the primitive relaxation time calculated from the Coupling Model for D-ribose and 2-deoxy-D-ribose.[56] These results indicate that this secondary relaxation in the mono- and di-saccharides is connected to the α-relaxation in the same way as in ordinary glassformers, and hence it is the JG β-relaxation of these sugars. The knowledge of glass transition can benefit the study of the carbohydrates themselves and in their applications.

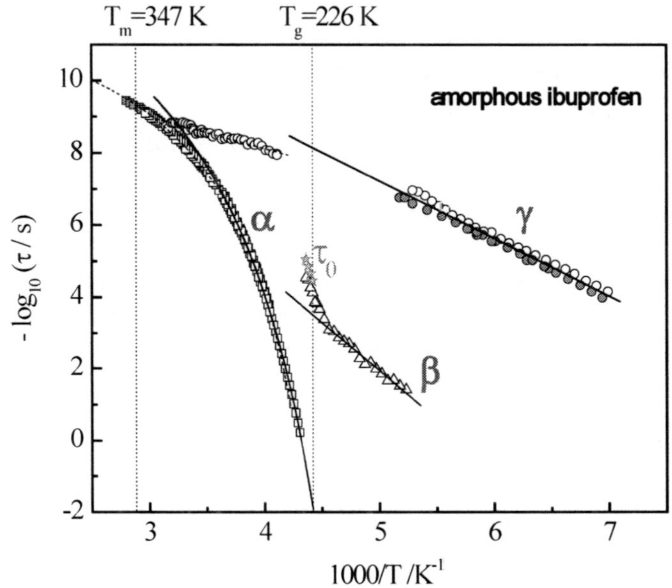

Figure 11. *Logarithm of the relaxation time, $\log_{10}\tau$, versus 1/T for all relaxation processes in ibuprofen: open symbols, τ from obtained isothermal loss data collected during cooling; gray filled symbols, τ from the isochronal plots. Lines are fits of the Arrhenius dependence of β- and γ-relaxation times below T_g, which is indicated by the dotted line. The α-relaxation times require two VFTH formulas to fit: the solid line is the VFTH1 fit and the dashed line is the VFTH2 fit to the data. Light gray stars indicate the JG relaxation time, τ_{JG}, estimated by the primitive relaxation time calculated from Coupling Model. The β-relaxation time (open triangles) change to a stronger T-dependence when temperature is increased above T_g, where there is good agreement with the estimated τ_{JG}. The relaxation of the hydrogen bonded network slower than the α-relaxation is not shown.*

Water and Aqueous Mixtures. Water is the most abundant and important soft matter. Water is important for biological systems to function. The human body contains more than 80% water. It has the simplest molecular structure, but ironically the interpretation of its relaxation dynamics is most controversial in the last many decades of scientific research of soft matter. We have analyzed past and recent experimental data of water in various situations and show the ubiquitous presence of the JG β-relaxation of water that bear exactly the same relation to the α-relaxation as in generic glassformers.[44] Here we show an example from mixtures of 35 wt.% of water with various ethylene glycol oligomers. All mixtures show the presence of the α-relaxation of the solute hydrogen bonded with the water, and a secondary relaxation originating from the water component but also influenced by hydrogen bonding with the solute. The *T*-dependence of the secondary relaxation time, τ_β, is Arrhenius below T_g and has activation enthalpy in the range of 40 to 50 kJ/mol, but it changes to a stronger dependence above T_g (see Fig. 12) Its dielectric strength, $\Delta\varepsilon_\beta$, also exhibits a change to a stronger increase with increasing *T* after crossing T_g (see inset in Fig. 12). Thus, this is the JG β-relaxation of water, and it is found in other aqueous mixtures, nano-confined water,[46] and hydration water of proteins and biomolecules.[46] Because the dynamics are analogous to ordinary glass-formers, the advance in understanding glass transition will promote progress in the aforementioned water related research fields and technology.

Nano Science and Technology. The change of dynamics of glassformers when reducing one or more dimensions to nanometer size is not only of fundamental interest, but also is important for nanotechnology using glassforming materials. The structural α-relaxation involves the cooperative motion of molecules defining a length scale *L* which increases with *n*.[25,26] The size of *L* measured at about ten degrees above T_g is of the order of nanometer for many glassformers. Hence one can expect faster relaxation/diffusion dynamics of glassformers on reducing dimension to nanometers (assuming no chemical bonds at interface) because of diminishing many-body dynamics in the structural α-relaxation. The latter causes the difference between the α-relaxation and the JG β-relaxation to become smaller, the 'fragility' index *m* to decrease, and τ_α to approach τ_{JG}. At sufficiently large reduction of dimension, indeed τ_α has been found to be not much longer than τ_{JG}.[60-63] These changes of the relation of τ_α to τ_{JG} have been experimentally observed. Thus the knowledge of glass transition of bulk materials serves well the interpretation of effects found in nano-science and the development of nanotechnology that use glassformers.

Portable Energy Storage and Fuel Cell Materials. The wide spread use of portable electronics and the future use of hydrogen technology for motor vehicles have generated much research in glassy fast ionic conductors, solid electrolytes, and fuel cell materials. It is by now well known that the ion dynamics in these materials are similar to that of glassformers. These include the Kohlrasuch correlation function, heterogeneous dynamics, primitive relaxation and its relation to caged ion dynamics (NCL), and etc. Perhaps the only difference commonly found is the temperature dependence. The ion relaxation time and conductivity often have Arrhenius dependence, in contrast to the VFTH dependence of glassformers. For examples of glassy fast ionic conductors and fuel cell materials see Refs. [[64-67]]. Local and primitive ion relaxation (counterpart of the JG β-relaxation of glassformers) in glassy ionic conductors has been found by molecular dynamics simulation[65] as well as by experiment.[64,67] The primitive ion motion is strongly connected with the long range diffusion and d.c. conductivity, analogous to the relation between JG β-relaxation and the structural α-relaxation in glassformers.[64-67] As far as computer simulation goes, the glassy ionic conductors show no difference from colloidal particles suspension, and Lennard-Jones liquids in the key quantities including the intermediate scattering function, the van Hove function and the non-Gaussian parameter. The knowledge of glass transition can be passed on to the ion dynamics and vice versa. Thus the transfer of knowledge is not a one way street as the NY Times article may have implied.

6. Discussion

The sentence in the NY Times article that contains the words "*...general properties that arise out of many disordered interactions*" is actually profound and insightful on two counts. First the phrase "*many disordered interactions*" is suggestive that glass transition is a many-body relaxation problem in addition to taking into account the change of thermodynamic variables including volume and entropy on decreasing temperature T or increasing pressure P.[9,39,68] If T and P were fixed, the dynamics are reduced entirely to a many-body relaxation problem. None of the theories mentioned in the NY Times article have considered many-body relaxation either at all or directly. Molecular dynamics simulations starting from some interaction potential as well as Monte Carlo simulations of toy models necessarily have captured the effects of many-body relaxation, but these are computer experiments and not theoretical solution of the problem.

Actually, up to the present time, many-body relaxation is still an unsolved problem in condensed matter physics. In his magical year of 1905,[69] Einstein solved the problem of diffusion of pollen particles in water discovered in 1827 by the botanist, Robert Brown. In this Brownian diffusion problem, the diffusing particles are far apart and do not interact with each other and the correlation function is the linear exponential function, $\exp(-t/\tau)$. It is by far simpler a problem than the interacting many-body relaxation/diffusion problem involved in glass transition. It is a pity that Einstein in 1905 was unaware of the experimental work of R. Kohlrausch and his intriguing stretch exponential relaxation function, $\exp[-(t/\tau)^{1-n}]$, published in 1847 and followed by other publications by his son, F. Kohlrausch.[1,2]

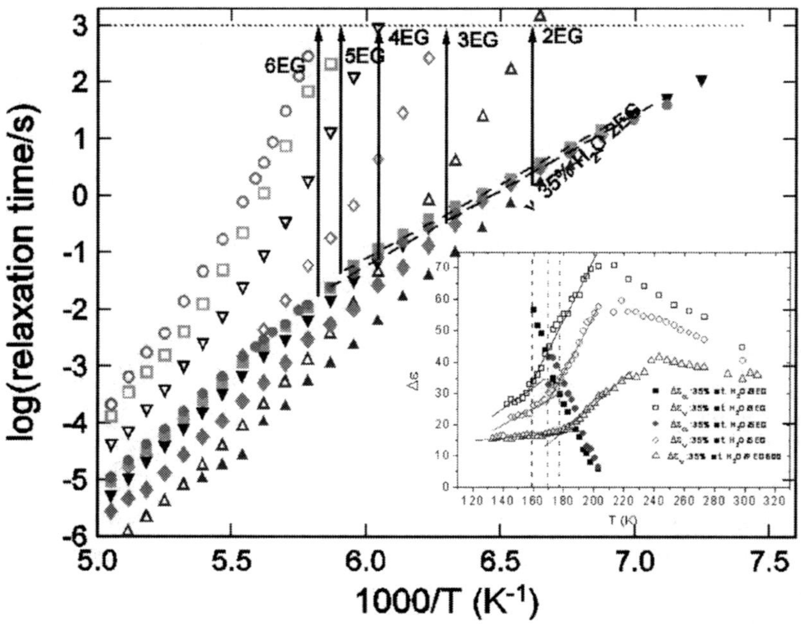

Figure 12. *Temperature dependence of the dielectric relaxation time τ_α (open symbols) and τ_{JG} (corresponding closed symbols) of mixtures of 35 wt.% of water with various ethylene glycol oligomers as indicated. Circles for 6EG. Squares for 5EG. Downward-pointing triangles for 4EG. Diamond for 3EG. Upward-pointing triangles for 2EG. Some of the data of τ_{JG} (closed symbols) overlap and cannot be easily resolved. For this reason, we use the dashed lines to indicate the Arrhenius temperature dependences assumed by τ_{JG} of the mixtures starting approximately at temperatures below T_g of the mixtures defined by $\tau_\alpha(T_g)=10^3$ s located by the vertical arrows drawn. There is change of temperature dependence of τ_{JG} at T_g.*

Had Einstein, the universally recognized genius, tried and attacked the problem emerging from the experiments of Kohlrausch, we may already have a solution for the glass transition problem, and problems of other complex systems having "*many disordered interactions*". It is possible that solution of the many-body relaxation problem may require use of nonlinear Hamiltonian dynamics (i.e. classical chaos) in view of the fact that the "*disordered interactions*" usually originate from anharmonic potentials. This is suggested by the solution of a simple model.[31]Some theories cited in the NY Times article consider heterogeneous dynamic to be the key in the solution of the glass transition problem. Certainly it is an important property but one should not loose sight that it is just one of many parallel consequences of many-body relaxation such as the Kohlrausch function, the invariance of the α-dispersion to different combinations of T and P at constant τ_α, and the various properties of the α-relaxation governed by or correlated with the width of its dispersion discussed before. The breakdown of the SE and DSE relations is another example of consequence of many-body relaxation in parallel with heterogeneous dynamics. It is unsurprising that the explanation of the breakdown of SE and DSE relations by heterogeneous dynamics fails because the two are parallel consequences, they are not inconsistent, but one cannot be derived fully or truthfully from the other.

The second insightful remark of the NY Times article is the mentioning of "*general properties that arise out of many disordered interactions*". This we have demonstrated by showing the similar properties of the dynamics in many different forms of glassforming systems[9,15] as well as in non-glassforming systems such as hydrated proteins and biomolecules,[41] ions in molten, glassy or crystalline matrices,[64-67] semidilute polymer solutions,[70-73] polymer cluster solutions,[74] associating polymer solutions,[75] micelles (water, oil, and surfactant),[76] clay colloidal dispersion such as laponite solutions,[77] polyelectrolytes, and etc. These *general properties*, found in glassformers as well as in many different systems which are not glassforming, tell us that the problem to be solved goes beyond the research field of glass transition. The payoff for the solution of the glass transition problem is immense.

7. Conclusion

Research of glass transition is actually experimentally driven. There is a wide variety of substances and systems exhibiting the glass transition phenomenon in different ways, and yet they exhibit ubiquitous features in the dynamics having general properties. These general properties cannot be understood alone by thermodynamic quantities such as volume and entropy because they originate from many-body relaxation inevitable in disordered interacting systems. Moreover, the manifestations of many-body dynamics cannot be reproduced by mean-field theories, which are the type of conventional theories and models of glass transition popular and used in the literature. Although many-body relaxation is still an unsolved problem in condensed matter physics, theory that neglects it will not be fully successful in explaining all the general properties. Somehow, it has to be incorporated together with volume and entropy into any theory of glass transition in order that the theory becomes viable. Any system that has interactions between the basic relaxation units necessarily involves many-body relaxation in irreversible processes. The glass transition problem is just a special case albeit a prevalent one.

The fact that there are disagreements between theoreticians consulted by the writer of the NY Time article is amply clear, but the substances of their disagreement are limited to the structural α-relaxation, transport coefficients such as diffusion constant and viscosity, and a subset of properties such as dynamic heterogeneity which is just one among many parallel consequences of many-body relaxation. We have seen an example of failure in using one consequence of many-body relaxation to explain another consequence, such as dynamic heterogeneity and the breakdown of SE and DSE relaxation. All the general properties of the α-relaxation now known (some of which are given here) must be explained. Remarkably, none of the theoreticians consulted have considered the frequency dispersion of the α-relaxation to play any fundamental and important role, although this has been repeated shown by experiments, some of which have been discussed in this paper.

The structural α-relaxation that most theories of glass transition focused on is not the only dynamic process having fundamental importance. Experimental evidences have shown beyond any doubt that secondary relaxation belonging to a special class has fundamental importance when considering glass transition because it is inseparable from the α-relaxation in any glassformer. The connection between the two relaxations is mediated by the frequency dispersion width parameter of the α-relaxation. This suggests that it is the evolution of the many-body relaxation with time that leads the secondary relaxation to the α-relaxation and generates their relation. This strong connection between the two relaxations has to be taken into account before the glass transition problem can be truly solved.

Acknowledgements

KLN was supported by the Office of Naval Research. The work at the Università di Pisa was supported by MIUR-FIRB 2003 D.D.2186 grant RBNE03R78E. M. Paluch was supported by the Committee for Scientific Research, Poland KBN, Grant No. N N202007534.

References

1. Kohlrausch, R. (1847) Nachtrag ueber die elastische Nachwirkung beim Cocon und Glasfaden, etc. *Pogg. Ann. Phys.*(III) Vol. **12**, 393-399
2. Kohlrausch, R. (1854) Theorie des electrischen Ruckstandes in der Leidener Flasche, *Pogg. Ann. Phys.* (IV), Vol. **1**, 79-86
3. Joule, J. P. (1867) *Mem. Manchr. Literary Phil. Soc.*, 3rd ser., **3**, 292. (The Scientific Papers of J. P. Joule, 1884) Vol. **1** (London: Physical Society), p. 558
4. Nemilov, S. V., and Johari, G. P. (2003) *Philos. Mag.* **83**, 3117-3132

5. Nemilov, S. V. (1995) *Thermodynamics and Kinetic Aspects of the Vitreous State* (Boca Raton, Florida: CRC Press). (2000) *Glass Phys. Chem.* **26**, 511-530
6. Chang, K., *The Nature of Glass Remains Anything but Clear*, (*The New York Times*, *July* 29, 2008)
7. Anderson, P. W., *Through the Glass Lightly* (1995) *Science* **267**, 1616
8. Ngai, K. L., Casalini, R. Capaccioli S., Paluch, M., and Roland, C. M. (2005) *J. Phys. Chem. B.* **109**, 17356-17360
9. Ngai, K. L., Casalini, R., Capaccioli, S., Paluch, M., and Roland, C. M. (2006) *Adv. Chem. Phys. in Chemical Physics Part B, Fractals, Diffusion and Relaxation in Disordered Complex Systems,* **133B**, 497-582
10. Rivera-Calzada, A., Kaminski, K., Leon, C., and Paluch, M. (2008) *J. Phys. Chem.* B **112**, 3110-3115
11. Mierzwa, M., Pawlus, S., Paluch, M., Kaminska, E., and Ngai, K. L. (2008) *J. Chem. Phys.* **128**, 044512
12. Kessairi, K., Capaccioli, S., Prevosto, D., Lucchesi, M., Sharifi, S., and Rolla, P. A. (2008) *J. Phys. Chem.* B **112**, 4470
13. Alegria, A., Gomez, D., and Colmenero, D. (2002), *Macromolecules* **35**, 2030-2035J. Roland, C. M., McGrath, K. J., and Casalini, R. (2006) *Macromolecules* **39**, 3581
14. Mierzwa, M., Paluch, M., Rzoska, S. J., and Ziolo, J. (2008) *J. Phys. Chem. B* **112**, 10383
15. Ngai, K. L. (2000) *J. Non-Cryst. Solids* **275**, 7
16. (a) For an explanation of the effect, see Casalini, R., Ngai, K. L., and Roland, C. M. (2003) *Phys. Rev. B* **68**, 014201; (b) Casalini, R., Paluch, M., and Roland, C. M. (2003) *J. Chem. Phys.* **118** 5701, (2003) *J. Phys.: Cond. Matt.* **15**, S859
17. Ngai, K. L., and Capaccioli, S. (2008) *J. Phys.: Condens. Matter* **20**, 244101
18. Cicerone, M. T., and Ediger, M. D. (1996) *J. Chem. Phys.* **104**, 7210
19. Mapes, M. K., Swallen S. F., and Ediger, M. D. (2006) *J. Phys. Chem. B* **110**, 507
20. Ediger, M. D. (1998) *J. Non-Cryst. Solids*, **235-237**, 10
21. Ngai, K. L. (1999) *J. Phys. Chem.* **103**, 10684
22. Richert, R., Duvvuri, K., and Duong, L. J. (2003) *J. Chem. Phys.* **118**, 1828; Richert, R. (2005) *J. Chem. Phys.* **123**, 154502
23. Zhu, X. R., and Wang, C. H. (1986) *J. Chem. Phys.* **84**, 6086
24. Zemke, K., Schmidt-Rohr, K., Magill, J. H., Sillescu, H., and Spiess, H. W. (1993) *Mol. Phys.* **80**, 1317
25. Ngai, K. L. (2007) *J. Non-Cryst. Solids* **353**, 709-718

26. Reinsberg, S. A., Heuer, A., Doliwa, B., Zimmermann, H., and Spiess, H. W. (2002) *J. Non-Cryst. Solids* **208**, 307-310
27. Ngai, K. L. (1999) *J. Chem. Phys.* **111**, 3639
28. Ngai, K. L., and Paluch, M. (2004) *J. Chem. Phys.* **120**, 857
29. Ngai, K. L. (2007) *J. Non-Cryst. Solids*, **353**, 4237-4245
30. Ngai, K. L. (2003) *J. Phys.: Condens. Matter*, **15**, S1107
31. Ngai, K. L., and Tsang, K. Y. (1999) *Phys. Rev. E* **60**, 4511
32. Johari, G. P., and Goldstein, M. (1970) *J. Chem. Phys.* **53**, 2372
33. Böhmer, R., Diezemann, G., Geil, B., Hinze, G., Nowaczyk, A., and Winterlich, M. (2006) *Phys. Rev. Lett.* **97**, 135701
34. Nowaczyk, A., Geil, B., Hinze, G., and Böhmer, R. (2006) *Phys. Rev. E* **74**, 041505
35. Leisen, J., Schmidt-Rohr, K., and Spiess, H. W. (1993) *Physica A* **201**, 79
36. Capaccioli, D., Prevosto, M., Lucchesi, P. A., Rolla, R., Casalini, K. L. Ngai, (2005) *J. Non-Cryst. Solids* **351**, 2643
37. Prevosto, D., Capaccioli, S., Sharifi, S., Lucchesi, M., and Rolla, P. A. (2007) *J. Non-Cryst. Solids* **353**, 4278
38. Prevosto, D., Capaccioli, S., Lucchesi, M., Rolla, P. A., and Ngai, K. L. (2008) *J. Non-Cryst. Solids* submitted
39. Ngai, K. L., Prevosto, D., Capaccioli, S., and Roland, C. M. (2008) *J. Phys.: Condens. Matter*, **20**, 244125
40. Paluch, M., Roland, C. M., Pawlus, S., Ziolo, J., and Ngai, K. L. (2003) *Phys. Rev. Lett.* **91**, 115701
41. Blochowicz, T., and Roessler, E.A. (2004) *Phys. Rev. Lett.* **92**, 225701
42. Nogales, A., Sanz, A., and Ezquerra, T. A. (2006) *J. Non-Cryst. Solids* **352**, 4649
43. Capaccioli, S., Kessairi, K., Prevosto, D., Lucchesi, M., and Rolla, P. A. (2007) *J. Phys.:Condens. Matter* **19**, 205133
44. Capaccioli, S., Ngai K. L., and Shinyashiki, N. (2007) *J. Phys. Chem. B* **111**, 8197; Ngai, K. L., Capaccioli, S., and Shinyashiki, N. (2008) *J. Phys. Chem. B* **112**, 3826-3832
45. Johari, G. P., Power, G., and Vij, J. K. (2002) *J. Chem. Phys.* **116**, 5908; *J. Chem. Phys.* **117**, 1714; Power, G., Johari, G. P., and Vij, J. K. (2003) *J. Chem. Phys.* **119**, 435
46. Capaccioli, S., Shahin, T. M., and Ngai, K. L. (2008) *J. Phys. Chem B* Karl Freed Festschrift issue
47. Brás, A. R., Noronha, J. P., Antunes, A. M. M., Cardoso, M. M., Schönhals, A., Affouard, F., Dionísio, M., and Correia, N. T. (2008) *J. Phys. Chem.B.* **112**, 11087-11099
48. Nath, R., El Goresy, T., Geil, B., Zimmermann, H., and Böhmer R. (2006) *Phys. Rev.E* **74**, 021506

49. Nath, R., Nowaczyk, A., Geil, B., and Böhmer, R. (2007) *J. Non-Cryst. Solids*, **353**, 3788
50. Vyazovkin, S., and Dranca, I. (2005) *J. Phys. Chem. B* **109**, 18637
51. Correia, N. T., Ramos, J. J. M., Descamps, M., and Collins, G. (2001) *Pharm. Res.* **18**, 1767
52. Carpentier, L., Decressain, R., Desprez, S., and Descamps, M. (2006) *J. Phys. Chem. B* **110**, 457
53. Johari, G. P., Kim S., and Shanker R. M., (2007) *J. Pharm. Sci.* **96**, 1159-1175
54. Ferrari, C., Tombari, E., Johari G. P., and Shanker R. M. (2008) *J. Phys. Chem. B*, **112**, 10806-10814
55. Kaminski, K., Kaminska, E., Paluch, M., Ziolo, J., and Ngai, K. L. (2006) *J. Phys. Chem. B*, **110**, 25045-25049
56. Kaminski, K., Kaminska, E., Wlodarczyk, P., Paluch, M., Ziolo, J., and Ngai, K. L. (2008) *J. Phys.: Condens. Matter* **20**, 335104
57. Kaminski, K., Kaminska, E., Wlodarczyk, Pawlus, S., Kimla, D., Kasprzycka, A., Paluch, M., Ziolo, J., Szeja, W., and Ngai, K. L. (2008) *J. Phys. Chem. B*, **112**, 12816-12823
58. Kaminski, K., Kaminska, E., Hensel-Bielowka, S., Pawlus, S., Paluch, M., and Ziolo, J. (2008) *J. Chem. Phys.* **129**, 084501
59. Cottone, G., Ciccotti, G., and Cordone, L. (2002) *J. Chem. Phys.* **117**, 9862
60. Ngai, K. L. (1999) *J. Phys: Condens. Matt.* **11**, A119
61. Anastasiadis, S. H., Karatasos, K., Vlachos, G., Manias, E., Giannelis, E. P. (2000) *Phys. Rev. Lett.* **84**, 915
62. Schönhals, A., Goering, H., Schick, Ch., Frick, B., Zorn, R. (2005) *J. Non-Cryst, Solids* **351**, 2668; (2003) *Eur. J. Phys. E Soft Matter* **12**, 173. (2004) *Colloid Polym. Sci.* **282**, 882
63. Ngai, K. L., (2002) *Eur. Phys. J. E.* **8**, 225-235; (2003) *Eur. Phys. J. E* **12**, 93-100; (2002) *Philos. Mag. B*, **82**, 291; (2007) *J. Non-Cryst. Solids* **353**, 4237-4245; (2006) *J. Polym. Sci.: Part B: Polym. Phys.*, **44**, 2980-2995
64. Ngai, K. L., Habasaki, J., León, C., and Rivera, A. (2005) *Z. Phys. Chem.* **219**, 47-70
65. Habasaki, J., and Ngai, K. L. (2006) *J. Non-Cryst. Solids* **352**, 5170-5177
66. Ngai, K. L., Habasaki, J., Hiwatari, Y., and León, C. (2003) *J. Phys.: Condens. Matter* **15**, S1607-S1632; Habasaki, J., Ngai, K. L., and Hiwatari, Y. (2002) *Phys. Rev. E* **66**, 021205
67. (a) Moreno, K. J., Mendoza-Suárez, G., Fuentes, A. F., García-Barriocanal, J., León, C. and Santamaria, J. (2005) *Phys.Rev.B*, **71**, 132301. (b) Díaz-Guillén, M. R., Moreno, K. J., Díaz-Guillén, J. A.,

Fuentes, A. F., Ngai, K. L., García-Barriocanal, León, C., and Santamaria, J. (2008) *Phys.Rev. B to appear*

68. Ngai, K. L. (2005) *J. Non-Cryst. Solids*, **351**, 2635-2642
69. Einstein, A. (1956) *Investigations on the Theory of Brownian Movement*, Dover, New York
70. Ngai, K. L. (1996) *Advan. Colloid Interf. Sci.*, **64**, 1-43
71. Adam, M., Delasanti, M., Munch, J. P., and Durand, D. (1988) *Phys. Rev. Lett.* **61**, 706
72. Ren, S. Z., Shi, W. F., Zhang, W. B., and Sorensen, B. (1992) *Phys. Rev. A* **45**, 2416
73. Phillies, G. D. J., Richardson, C., Quinlan, C. A., and Ren, S. Z. (1993) *Macromolecules* **26**, 6849
74. Ngai, K. L., and Phillies, G. D. J. (1996) *J. Chem. Phys.* **105**, 8385
75. Tsianou, M., Kjøniksen, A.-L., Thuresson, K., and Nystrom, B. (1999) *Macromolecules* **32**, 2974-2982
76. Ngai, K. L. (1992) in *NATO ASI Series C* **369**, 221-232
77. Zulian, L., Ruzicka, B., Ruocco, G., and Capaccioli, S. (2007) *J. Non-Cryst. Solids,* **353**, 388

THE LINK BETWEEN THE PRESSURE EVOLUTION OF THE GLASS TEMPERATURE IN COLLOIDAL AND MOLECULAR GLASS FORMERS

[1]SYLWESTER J. RZOSKA, [1]ALEKSANDRA DROZD-RZOSKA, [2]ATTILA R. IMRE

[1]Institute of Physics, University of Silesia, ul. Uniwersytecka 4, 40-007 Katowice, Poland
[2]KFKI Atomic Energy Research Institute, H-1525 Budapest, POB 49, Hungary

Abstract: Recently, Voigtmann [*Phys. Rev. Lett.* **101**, 095701 (2008)] suggested the existence of the universal "generic steep" behavior for the glass pressure vs. temperature dependence in molecular glass formers. We indicate that such behavior disappear when the absolute stability limit in the negative pressures domain is taken as the reference. It is a parasitic artifact of omitting the stability limit in the negative pressures domain for the log-log scale plot. Results presented suggest a totally common pattern for the evolution of the glass temperature in colloidal fluids and molecular liquids. However, for molecular liquids both positive and negatives pressures domains have to be taken into account. Consequently the pattern for colloidal glass formers introduced by Sciortino "*One Liquid, Two Glasses*" , [*Nature Materials* **1**, 1-3 (2002)] may appear to be valid also for molecular glass formers.

The pressure evolution of the glass temperature, T_g vs. P_g, is one of basic problems for the glass transition physics.[1,2] It is essential for fundamental, technological and geophysical application.[1-4] Regarding the latter it is one of key tools for predicting deep Earth properties.[4] Unfortunately, experimental data for $T_g(P)$ behavior are still hardly available, due to enormous experimental difficulties.[1,3.5-7] They are often were portrayed via a polynomial:[1,3,5]

$$T_g(P) = T_g^0 + aP + bP^2 \tag{1}$$

However, this is a "formal" way of parameterization, which without any physics behind. A decade ago Andersson and Andersson[8] proposed for $T_g(P)$ a description parallel to the Simon – Glatzel (SG)[3] equation, since decades a key tool for describing the evolution of the melting temperature $T_m(P)$. Basing on this can wrote a common Simon-Glatzel-Andersson relation:[3,8]

$$T_{g,m}(P) = T_0^{g,m}\left(1 + \frac{P}{\Pi}\right)^{1/b} \tag{2}$$

where b and Π are material dependent empirical coefficients for $T_g(P)$ and $T_m(P)$ data, respectively.

Since decades the SG equation is a fundamental tool for describing $T_m(P)$ behavior.[3] Then, the boost in using the Andersson equation for $T_g(P)$ portrayal cannot be surprised.[8] The common form of SG and Andersson equations may be supported by the empirical relationship, first noted by David Turnbull: $0.5 < T_g/T_m < 1$. Values above $T_g/T_m > 2/3$ are considered as a hallmark that a given material can be easily supercooled below T_m and reach T_g even for a "slow" cooling.[2] In fact there materials, for instance glycerol or epoxy resins, for which supercooling is easier than crystallization at T_m. Nevertheless, there are fundamental, although hardly recalled, problems of eq. (2). Eq. (2) can suggest that T_m and T_g always increase on pressuring. Nevertheless, there is a group of materials where the reverse behavior occurs, i.e. in which domains $dT_m/dP < 0$ and $dT_g/dP < 0$ can exists. For vitrification the evidence of such behavior is clear although still restricted only to few cases, such as some magmatic silicate melts or strongly ionic glasses[5-7]

Recently, Voigtmann reported a fundamental universality for the pressure induced vitrification in molecular glasses.[9,10] First, in ref. [[9]] he tested such dependence for a hard sphere model fluid with a square-well (SW) potential. Such system is known as a theoretical parallel for colloidal fluids. Voigtmann successfully scaled theoretical and experimental data using $log_{10}P_g$ vs. $log_{10}T_g$ plot.[9] The latter was obtained via rescaling of concentration – temperature experimental data for vitrifying colloidal fluids, since there are no "real" pressure results for such systems so far. Then Voigtmann indicated the existence of three basic domains/regions. For extremely compressed liquids the obtained behavior approached to the hard sphere limit, manifesting by a linear behavior in the mentioned scale (region I). For intermediate pressures the "S" shape behavior, denoted as "region II", appeared. Further lowering of pressure once more lead to a linear behavior, in the mentioned scale (region III). Subsequently, Voigtmann added to the plot experimental data for 6 low molecular weight glass formers using experimental data $P^* = P_g/P_{model}$ and $T^* = T_g/T_{model}$, where P_{model} and T_{model} are constant parameters introduced to make the common plot of theoretical and experimental" systems possible.[9] For all molecular glass formers the domain III was inherently inaccessible. In domain II, for $log_{10}P^*$ vs. $log_{10}T^*$, a "generic steep behavior" with exactly the same loci of the anomaly for any molecular liquid glass former was obtained (!).[9] This unusual and "universal" behavior, was next explored in ref. [[10]] for Lennard-Jones model-fluids with a purely repulsive interaction (LJR) and for the general case with an attraction term (LJ). Voigtmann reported a qualitatively different behavior for LJ and LJR fluids at the $\log_{10} P^*$ vs. $\log_{10} T^*$ plot. For LJR fluid an almost linear dependence occurred which coincided with LJ fluid behavior at very high pressures (regime I). For the medium pressures domain (regime II) he found a "generic steep" behavior for the LJ system for $T \to T^*_{crit.} \approx 1.4$, where T^* is the model-scaled glass temperature.[9] Such behavior was suggested as universal for molecular glass formers. This singularity was absent for vitrified colloids, well portrayed by the square-well (SW) model, for which the low-density domain (regime III) was available.[9,10]

In the opinion of the authors the part of analysis related to molecular liquids in refs. [[9,10]] is inherently inconsistent. For real liquids the stability limit is hidden in the negative pressures domain for $P << 0$ should be taken into account. It is existence is well proved experimentally. There is also a clear evidence that passing $P = 0$ does not introduce any artifacts in liquids.[11] In fact, in refs. [[9,10]] the same value of $P = 0$ for the absolute stability limit was taken for all discussed experimental data, leading to a "universal" generic steep anomaly, with exactly the same loci for any discussed molecular liquid when using $\log_{10} P^*$ vs. $\log_{10} T^*$ plot. (!).[9,10] In fact one can add here the variety of available $T_m(P)$, although such action has a limited sense, and one obtain the same "generic steep" universal behavior (!). Unfortunately, this will be only the artifact of the log-log scale and erroneous reference.

The experimental, common, measure of pressure with $P = 0$ as a reference is valid only for gases. For liquids the stretching down to the negative pressure limit has to be taking into account.[11]

Figure 1 recalls these results [[9,10]], including the analysis of experimental data for glycerol [[6]]. In our opinion the "generic steep behavior" for molecular glass formers does not exist. It is an artifact associated with taking $P = 0$ as the reference, what leads to a "quasi-anomaly" in the log-log plot. However, the liquid state is not bordered by $P = 0$ but by the liquid-gas spinodal, partly hidden in the negative pressure domain.[5-7,11] This is taken into account by the relation recently proposed by A. Drozd-Rzoska (ADR) et al.[5-7]

$$T_{g,m}(P)=T_g^o\left(1+\frac{\Delta P}{\Pi}\right)^{1/b}\exp\left(\frac{\Delta P}{c}\right)=T_0^g\left(1+\frac{P-P_g^0}{\pi+P_0^g}\right)^{1/b}\exp\left(\frac{P-P_g^0}{c}\right) \quad (3)$$

where $T_g^0\left(P_g^0\right)$ is the starting pressure, $-\pi$ is the negative pressure asymptote and c is the damping pressure coefficient.

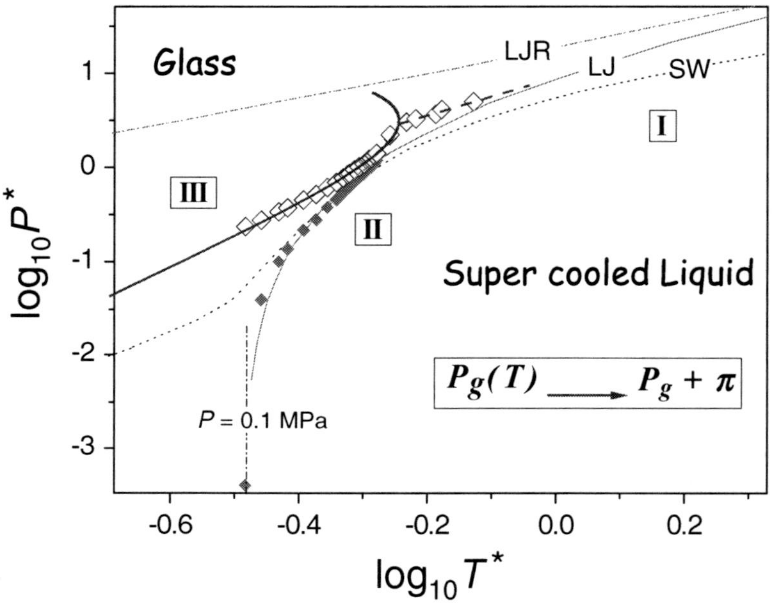

Figure 1. *The pressure dependence of the glass temperature in model and experimental fluids. The dashed-dotted curve (red) is for the Lennard-Jones with a purely repulsive interaction (LJR) fluid, thin solid curve (red) is for the Lennard-Jones fluid (LJ) and the black dotted curve is for the hard spheres fluid with a square-well (SW) potential. These curve recall results of refs. [9,10]. The full diamonds are for glycerol, assuming* $T^*=T_g/(\varepsilon/k_B)=T/500K$ *and* $P^*=P_g/\left(\varepsilon/\sigma^3\right)=P/2.5GPa$*, as in refs. [9,10]. The transformation was introduced for the common presentation of model fluids and experimental liquids. Note that direct experimental values of the glass transition pressure were taken here. The open diamond are for the same glycerol data but taking into account the existence of the absolute stability limit hidden in the negative pressure domain, as estimated in ref. [6], namely:* $-\pi=-0.8GPa$, $b=3.5$ *and* $c=3GPa$*. The thick solid curve (in blue) is for the parameterization based on ADR eq. (1), with mentioned parameters. Experimental data for glycerol were taken from ref. [6]. See also the Preface in this volume and ref. [6] for further discussion.*

The "ADR" equation shows the asymptotic behavior in the negative pressures domain and can be used for materials where both $T^g / dP > 0$ and/or $T^g / dP < 0$. Coefficients π, b and c are pressure-invariant,[4,6] what can validate the extension of $T_g(P)$ beyond the experimental range. Optimal values of coefficients π, b and c can be estimated from the preliminary linearized derivative based analysis, and consequently eq. (3) can be used without any further fitting.[5-7] For a "real" liquid the transformation $P \to P - (-\pi)$ is needed.[5-7] In refs. [9,10] by Voigtmann, the unphysical assumption $\pi = 0$, the same for any liquid, was introduced. Such a problem cannot appear for vitrifying colloidal fluids where the transformation of variables from concentration to pressure variables was carried out in ref. [9]. In fact, there is no universal "generic steep" anomaly for colloidal fluids as can be seen from data presentation in ref. [9]. We do not discuss here reasons of appearance of the "generic steep" universal behavior for the Lennard – Jones model liquid in ref. [10]. Nevertheless, assumptions leading the artificial "universal, generic steep anomaly" should be seriously re-examined.

In conclusion, basing on results shown in Fig. 1 we are claiming that there are no generic steep anomaly for molecular glass formers and all "regimes" indicated in refs. [9,10] are accessible both for colloidal and molecular glass formers. Fig. 1 lead also to a conclusion that of existing of two kinds of supercooled liquids related to $dT_g / dP > 0$ and $dT_g / dP < 0$, and two glasses: "soft" and "hard". The location and availability of these domains depends on the interplay between interactions. It is noteworthy that a similar picture was recently proposed for colloidal glass formers where attraction can be facilitated by a polymer addition.[12] Consequently, in such colloids two distinct kinds of glasses were formed: an attractive and a repulsive one. The devitrification on increasing the concentration is associated with breaking of strong inter-colloid interactions and creation of a percolation path.[12] One may expect that a similar mechanism occurs in strongly bonded covalent glasses, such a silicate albite or some ionic glasses.[5-7] For "simple" molecular glass forming liquids possibilities of tuning molecular interactions are much more restricted than in colloids with the depletion (polymer related) attractive interactions. In fact, in molecular liquids key parameters of interactions are constant and can be only slightly moderated by compressing. This shows that ionic glasses, where tuning such possibilities are enormous, are probably the most promising type of systems for obtaining systems with domains $dT_g / dP > 0$ and $dT_g / dP > 0$, suitable for further pressure tests.

Voigtmann suggested different patterns of the pressure evolution of the glass temperature and colloidal and molecular glasses.[9,10]

We claim that evolutions of the glass temperature in vitrifying colloidal fluids and molecular liquids are analogous , even including the possibility of the reverse vitrification or devitrification on pressuring. However, for molecular liquids the negative pressures domain of isotropically stretched liquid has to be taken into account. Then, a link between molecular and colloidal glasses, often considered as separate cases so far, seems to be possible

Few years ago Sciortino[12] entitled his paper in Nature "*One Liquid, Two Glasses*" to stress the possible general pattern for colloidal glasses. We suggest that such picture may also emerge from pressure studies on molecular glass forming liquids.

References

1. Donth, E. (1998) *The Glass Transition: Relaxation Dynamics in Liquids and Disordered Materials*, Springer Ser. in Mat. Sci. II, vol. **48**, (Springer, Berlin)
2. Angell, C. A. (2008) Glass-formers and viscous liquid slowdown since David Turnbull: enduring puzzles and new twists, *MRS Bulletin* **33**, 111
3. Skripov, V. P., and Faizulin, M. Z. (2006) *Crystal-Liquid-Gas Phase Transitions and Thermodynamic Stability* (Wiley-VCH, Weinheim)
4. Poirier, J. -P. (2000) *Introduction to the Physics of the Earth's Interior*, (Cambridge Univ. Press., Cambridge UK)
5. Drozd-Rzoska, A., Rzoska, S. J., and Roland, C. M. (2008) On the pressure evolution of dynamic properties in supercooled liquids, *J. Phys.: Condens. Matter* **20**, 244103
6. Drozd-Rzoska, A., Rzoska, S. J., Paluch, M., Imre, A. R., and Roland, C. M. (2007) On the glass temperature under extreme pressures, *J. Chem. Phys.* **126**, 165505
7. Drozd-Rzoska, A., Rzoska, S. J., and Imre, A. R. (2007) On the pressure evolution of the melting temperature and the glass transition temperature, *J. Non-Cryst. Solids* **353**, 3915-3923
8. Andersson, S. P., and Andersson, O. (1998) Relaxation studies of poly(propylene glycol) under high pressure, *Macromolecules* **31**, 2999
9. Voigtmann, Th., and Poon, W. C. K. (2006) Glasses under high pressure: a link to colloidal science? *J. Phys.: Condens. Matter* **18**, L465-469
10. Voigtmann, Th. (2008) Idealized Glass Transitions under Pressure: Dynamics versus Thermodynamics, *Phys. Rev. Lett.* **101**, 095701

11. Imre, A. R., Maris, H. J., and Williams, P. R. (eds.) (2002) *Liquids under Negative Pressures,* NATO Sci. Ser. II, vol. **84** (Kluwer, Dordrecht)
12. Sciortino, F. (2002) One Liquid, Two Glasses, *Nature Materials* **1**, 1-3

EVIDENCES OF A COMMON SCALING UNDER COOLING AND COMPRESSION FOR SLOW AND FAST RELAXATIONS: RELEVANCE OF LOCAL MODES FOR THE GLASS TRANSITION

S. CAPACCIOLI, K. KESSAIRI, D. PREVOSTO, Md. SHAHIN THAYYIL, M. LUCCHESI, P.A. ROLLA

Dipartimento di Fisica, Università di Pisa, Largo B. Pontecorvo 3, I-56127, Pisa, Italy and CNR-INFM,polyLab, Largo B. Pontecorvo 3, I-56127, Pisa, Italy

Abstract: The present study demonstrates, by means of broadband dielectric measurements, that the primary α- and the secondary Johari-Goldstein (JG) β-processes are strongly correlated, in contrast with the widespread opinion of statistical independence of these processes. This occurs for different glass-forming systems, over a wide temperature and pressure range. In fact, we found that the ratio of the α- and β- relaxation times is invariant when calculated at different combinations of P and T that maintain either the primary or the JG relaxation times constant. The α-β interdependence is quantitatively confirmed by the clear dynamic scenario of two master curves (one for α-, one for β-relaxation) obtained when different isothermal and isobaric data are plotted together versus the reduced variable $T_g(P)/T$, where T_g is the glass transition temperature. Additionally, the α-β mutual dependence is confirmed by the overall superposition of spectra measured at different T-P combinations but with an invariant α-relaxation time.

Keywords: glass transition, pressure, structural relaxation, secondary relaxation, inter-molecular relaxation, binary mixtures

1. Introduction

Various length and time scales are involved by the motions characteristic of the dynamics of glass-forming systems: *(a)* cooperative motions originating the structural α-relaxation involve an increasing number of molecules, slowing down dramatically on approaching the glass transition by decreasing temperature T or increasing pressure P; *(b)* non-cooperative local dynamic processes involving single or few molecules, or even parts of them, like secondary relaxations, are faster and less dependent on T and P than the α-relaxation. Recently there is an increasing interest on the possible fundamental role played by the secondary relaxation in glass transition, whose origin is usually ascribed only to the much slower primary structural relaxation. Such possible relevant role was suggested by relations and correlations found between various properties of secondary β- and primary α-relaxations.[1,2,3,4,5,6] Usually intramolecular secondary relaxations do not have any relation to the primary relaxation, and hence they are not relevant for glass transition. The β-relaxation showing correlation to the α-one[4] are now referred to as the Johari-Goldstein (JG) secondary or JG β-relaxations to honor the important discovery done by these two scientists that, almost 40 years ago, who found a secondary relaxation even in glassy dynamics associated to totally rigid molecules that have no internal degree of freedom.[7] The origin of this kind of β-relaxation is truly intermolecular: it cannot be explain from time to time on the base of molecular details characteristic of the selected systems. On the contrary, JG β-relaxation has a rather universal character, as demonstrated by its ubiquitous presence in wide classes of glass-formers including simple organic molecular liquids, polymers, molten inorganic salts, plastic crystals and metallic glasses.[8,9]

For these reasons Goldstein defined β-relaxation as "*a universal feature of amorphous packing*".[10] The mechanism of JG relaxation would be related to the overcoming, by a single unit (whole molecule or segment of macromolecular chain), of the intermolecular barrier imposed by the caging neighbours.[7] For these reasons since the seventies, some interest was expressed about the pressure (or density) dependence of the JG β-relaxation.[11,12] As pointed out, for instance, by G. D. Patterson: *"if it (JG) were purely intramolecular there would probably be a much smaller pressure dependence to that transition temperature than it is actually observed. It is much less than the glass transition but it's not zero, and it's a good deal larger than you would expect for simple molecules….There must be an intermolecular contribution because it shows up in a fairly strong pressure dependence of the transition temperature"*.[10] Despite that, we had to wait until few years ago before that a systematic investigation of the pressure dependence of the secondary relaxation started.[4]

On the theoretical point of view, only few models assume a correlation between α- and JG β- relaxation to account for glass-forming dynamics.[13, 14] In particular, the Coupling Model (CM)[13,15] provided a quantitative relation linking the α-process dispersion and the time scale of α- and β-relaxation. In fact, according to CM, the JG β-process time scale can be identified with that of the primitive relaxation, a local motion acting as the precursor of the α-relaxation:[1,8,9,13] $\tau_{JG}\approx\tau_0$. CM gives a quantitative relation, due to many-molecules dynamics, between the primitive τ_0 and the α-relaxation time τ_α.[1,15,16] So, the following relation is predicted:

$$\tau_{JG} \approx \tau_0 = \tau_\alpha^{(1-n)} \frac{t_c^n}{(1-n)} \tag{1}$$

where t_c=2 ps and n is the coupling parameter, which is related to the stretching parameter of KWW function reproducing the structural peak, n=1-β_{KWW}:

$$\phi(t) = \exp\left[-\left(t/\tau_\alpha\right)^{1-n}\right],\ 0<n<1 \tag{2}$$

n is a measure of how non-exponential is the relaxation and it is related to the intermolecular coupling. According the predictions of CM, the time scales of slow (τ_α) and fast (τ_{JG}) relaxation and the dispersion of structural dynamics (β_{KWW}) are all strongly related. Higher is the intermolecular coupling (n) larger is the time scale separation of the JG peak from the structural one, i.e. the ratio τ_α/τ_{JG}. It is tempting to check these correlations over a wide range of P and T, since eq.(1) is expected to hold for any thermodynamic condition in the supercooled liquid state. Moreover, it has been shown for several glass-former that, once τ_α(T,P) is chosen constant for different combinations of T and P, then the stretching parameter β_{KWW} (and so n) is constant.[17] So, once fixed τ_α(T,P) for different combinations of T and P, then the coupling parameter n is constant, and consequently from eq. (1) τ_{JG} is also expected to be constant. A preliminary successful test of this relation has been recently published.[18] Additionally, the strong correlation of eq. (1) between τ_α and τ_{JG} can explain the anomalous change of dynamics at T_g recently reported in the T-P behavior of JG relaxation. For instance, the well known crossover of T-dependence of τ_α from Vogel-Fulcher above T_g to Arrhenius below T_g is transferred from τ_α to τ_{JG} (or vice versa) in a qualitative manner: in fact the T-P dependence of τ_{JG} mimics that of τ_α, although weaker due to the exponent (1-n). This issue has been extensively discussed in some recent papers.[19,20] Ref. [[21]] provides a comprehensive review of the fundamental role played by JG relaxation in glass transition, as well as a rationale for the experimental findings in the framework of CM. The present paper reports new experimental results on different glass-formers (mainly binary mixtures) over wide range of T and P, providing clear evidences that a strong correlation between slow (α-) and fast (JG β-) dynamics exists. In fact, the ratio of their relaxation times was found invariant to different combinations of P and T keeping the primary relaxation time constant. In addition, there is a fairly good superposition of isochronal spectra, obtained with different T-P combinations.

2. Experimental details

Binary mixtures of styrene oligomers with polar aromatic molecules have very good glass-forming and mixing properties. They are ideal systems for investigating selective dynamics of the polar component by dielectric spectroscopy because the dipole moment of the styrene repeat unit is very small compared to the strong dipole moment of the soulte. We investigated the dynamics of mixtures containing the polar rigid molecules cyanobenzene (CNBz) and quinaldine (QN) (T_g=180 K). All these compounds were purchased from Aldrich and used as received. Quinaldine was mixed, at concentration ranging from 5 to 100 wt.%, with tristyrene (3St) (T_g =234 K) (obtained from PSS) and oligostyrene Mw=800g/mol (PS800) (T_g=286 K, obtained from Scientific Polymer Product). In this paper we will present only the low concentration cases (5-10% wt.) in order to have negligible fluctuation concentration effects, that, for higher concentrations, gives an additional broadening to the α-relaxation peak, preventing a correct calculation of n.[22] Cyanobenzene was mixed with tristyrene at the concentration of 5% wt. Additionally, data for the neat system poly-phenyl-glycidyl-ether (PPGE) (T_g =262 K) were also compared.

A Novocontrol Alpha-Analyzer ($\nu=10^{-5}$-10^{7} Hz) was used for dielectric measurements, both at atmospheric and at high pressure. For atmospheric pressure measurements, a parallel plate capacitor separated by a quartz spacer (empty capacitance ~ 90 pF) and filled by the sample was placed in the nitrogen flow Quattro cryostat (T=100-360 K). For high pressure measurements, a sample-holder multi-layer capacitor (empty capacitance ~ 30pF) was separated from the pressurizing fluid (silicon oil) by a Teflon membrane. The high pressure chamber (Cu-Be alloy), provided by UNIPRESS, was connected to a hydraulic pump able to reach 700 MPa, and controlled in the interval 195–360 K within 0.1 K by means of a thermally conditioned liquid flow.

3. Results and Discussion

Dielectric spectroscopy measurements of binary mixtures of rigid polar molecule in apolar matrices is a selective technique that provides the relaxation dynamics of the component having the larger dipole moment. Even at the lowest concentration used here, only the polar molecules are observed in the dielectric spectra because the dipole moment of the host molecules (oligomers or styrene) is so small that their motions make no contribution. The dipolar probe molecule of the solute provides the overall contribution to both α- and β-relaxation. Measurements of the dielectric loss spectra of the mixture CNBz/tristyrene (5% wt.) at different temperatures above and below T_g are shown in Fig. 1. A bimodal relaxation scenario is visible above T_g, where the α-relaxation loss peak, located at lower frequencies and more intense, is more strongly affected by cooling than the β-relaxation. Below T_g the β-process appears symmetric in shape and decreasing on cooling in intensity. Near T_g, the two relaxations are both visible and well separated so that the α-loss peak can be separately well fitted by the Fourier transform of the KWW function (Eq. 2) with n=0.55, as shown by the dotted line in Fig. 1. Alternatively, dielectric spectra can be well fitted in the whole temperature range by a superposition of Havriliak-Negami (for α-process) and Cole-Cole (β-process) functions. In this way the contribution of each process has been singled out by a simple superposition fitting procedure,[23] and the relaxation times of the single processes, τ_α, τ_β, were obtained by calculating $(2\pi \nu_m)^{-1}$, the reciprocal of the loss peak maximum angular frequencies related to the α- and β- relaxation, respectively.

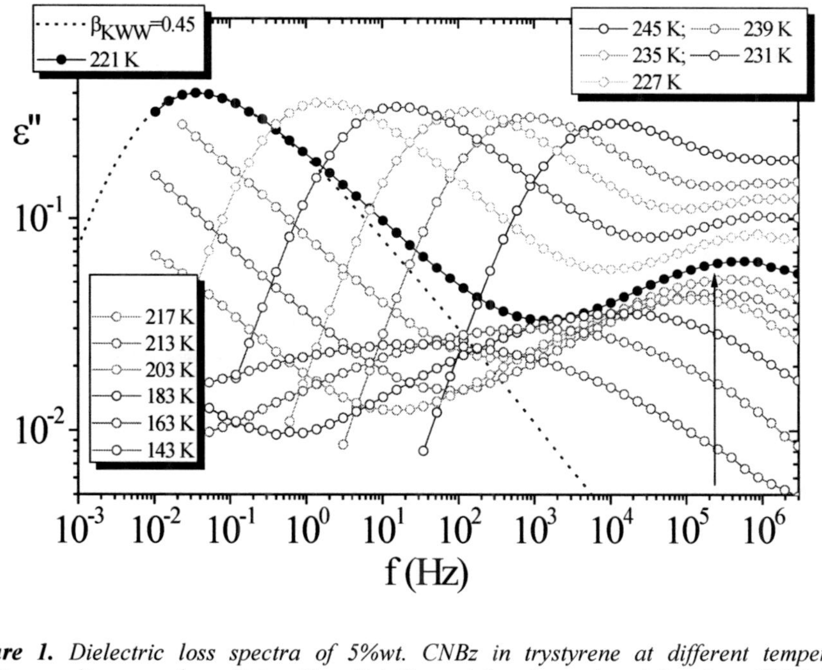

Figure 1. *Dielectric loss spectra of 5%wt. CNBz in trystyrene at different temperatures. Continuous lines are from fitting Havriliak-Negami functions, dotted line is a KWW function (n=0.55) fitted to the spectrum at T=221 K. Vertical arrow shows the location for the characteristic frequency of the JG process,* $\nu_{JG}=1/(2\pi\tau_{JG})$ *as predicted by Eq.1.*

In order to check the validity of CM predictions, the frequency of primitive relaxation $\nu_0=(2\pi \tau_0)^{-1}$ was calculated according to Eq. (1), where α-loss peak frequency ν_α was directly determined from the maximum, and n from the KWW function fit. The arrow in Fig. 1 indicates the location of the calculated frequency ν_0 of the primitive process, which is in good correspondence with the loss maximum frequency of the β-process, ν_β. The peak of the JG β-relaxation is well separated from the α-processes at the selected temperature, so allowing a reliable evaluation without any deconvolution procedure. The same procedure was repeated for different spectra. The fact that the CM predictions are in agreement with the experimental results is not surprising, as CNBz is a rigid dipolar probe and the secondary relaxation occurring in the glassy state cannot be ascribed to nothing but to an intermolecular process. If this is the case, a strong pressure dependence of the β-relaxation time is expected, as well as a tight correlation to the α-one. For this reason, measurements at different temperature above and below T_g were repeated under other 3 isobaric conditions also for higher pressure, up to 500 MPa (not shown). A scenario similar to that in Fig. 1 was found, with both processes shifting to lower frequency with cooling. The glass transition temperature T_g was found to increase with increasing pressure. The relaxation map of Fig. 2 can provide an overall representation of α- and β- dynamics under a wide range of P and T.

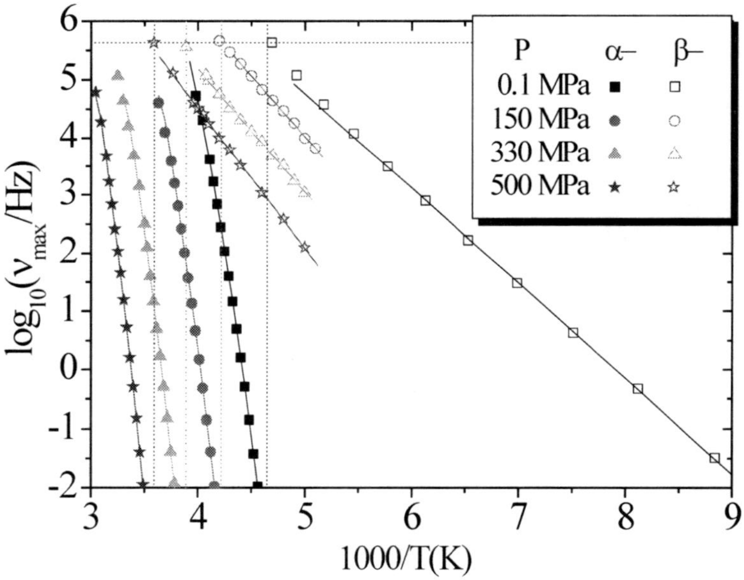

Figure 2. *Relaxation map of α- and JG β-relaxation of 5 wt.% of CNBz in mixture with tri-styrene. Continuous lines are VFT for ν_α above T_g and linear regressions for ν_β in the glassy state. Vertical dotted lines mark the $T_g(P)$ crossing for each isobar. Horizontal dotted line marks the position of JG β-relaxation at $T_g(P)$.*

All the isobaric sets of data have some common features: the T behaviour of ν_α can be represented by $\nu_\alpha=\nu_\infty\exp[-B/(T-T_0)]$, the Vogel-Fulcher-Tamman (VFT) function, whereas that of ν_{JG} is of the Arrhenius type $\nu_\beta=\nu_\infty\exp(-E_{JG}/RT)$ in the glassy state but it crosses over to stronger temperature dependence above T_g, as already reported in literature for other systems.[24,25,26] We found that ν_∞ is almost pressure independent, where E_{JG} was strongly changing with P. Moreover, it is noteworthy that the frequency location of the JG β-relaxation ν_β at the glass transition (i.e. where $\tau_\alpha(T_g,P_g)=10^2$ s) was found constant, implying a constant ratio τ_α/τ_{JG} at the glass transition. This result has been found in several other glass-forming systems, both neat and mixtures.[25,27]

Another interesting feature is emerging when spectra obtained at different T and P but with the same structural α-relaxation frequency ν_α or time τ_α (isochronal spectra) are compared. In our experiment, both temperature T and applied pressure P were varied over wide ranges. Elevated pressure slows down the α-relaxation and increases its relaxation time τ_α, but the increase can be compensated by raising temperature. Naturally, widely different combinations of P and T can be found to have the same α-relaxation time τ_α, although there are significant variations in the density. A selection of loss spectra of 5%wt. CNBz in tristyrene obtained for different combinations of T_X and P_X is shown in Fig. 3, as well as a diagram of T_X vs. P_X, keeping constant the frequency of the structural process f_α=1.83 Hz. It is noteworthy that, for the same τ_α, also the shape (or the dispersion) of the α-peak is invariant in that condition, a fact already reported in ref. [[20,21]] for several glass-forming systems. The KWW function fits very well the loss peak with β_{KWW}=0.45. Unexpectedly, also the JG β-relaxation (that shifts to lower frequency on increasing pressure) has the same frequency location of the maximum and almost the same shape, although its dielectric strength is slightly different. This results is not at all trivial: first, the position of ν_β is kept constant despite the temperature is increased by 1.3 times and that means that pressure (and so density) strongly affects the β-relaxation dynamics; moreover, the fact that the β-relaxation has Arrhenius behaviour with ν_∞ pressure independent and the ratio τ_α/τ_{JG} is constant implies that on increasing pressure the activation energy E_{JG} should linearly scale with T_X.

Summarizing, pressure and temperature seem to have similar effect on α- and β- dynamics, although the *T-P* dependence for both processes are different. Therefore, these results could imply a strong correlation between the two processes and lend strong support to the interdependency of the JG β-relaxation and the α-relaxation, suggested by a recent NMR study.[5] The same scenario occurring for CNBz/tristyrene mixture shown in Fig. 2 and 3 was found in the dynamics of a component in binary mixtures including *tert*-butylpyridine (TBP),[28] quinaldine (QN)[29], or picoline[30] in mixtures with oligomers of styrene. In another paper of this volume,[27] there are some examples related to QN dynamics in styrene oligomers. Fig. 6 of ref. [[27]] shows selected isochronal spectra of 10 wt.% of QN in tristyrene, obtained at different T and P, showing a good superposition of both α- and β- process. Fig. 7 shows, for different isobaric and isothermal conditions, the relaxation map for data of 10 wt.% of QN in tristyrene, and in Fig. 8 that for the data of 25 wt % of 2-picoline in tristyrene. Also in these cases a change in dynamics of β-relaxation occurs on crossing the glass transition.

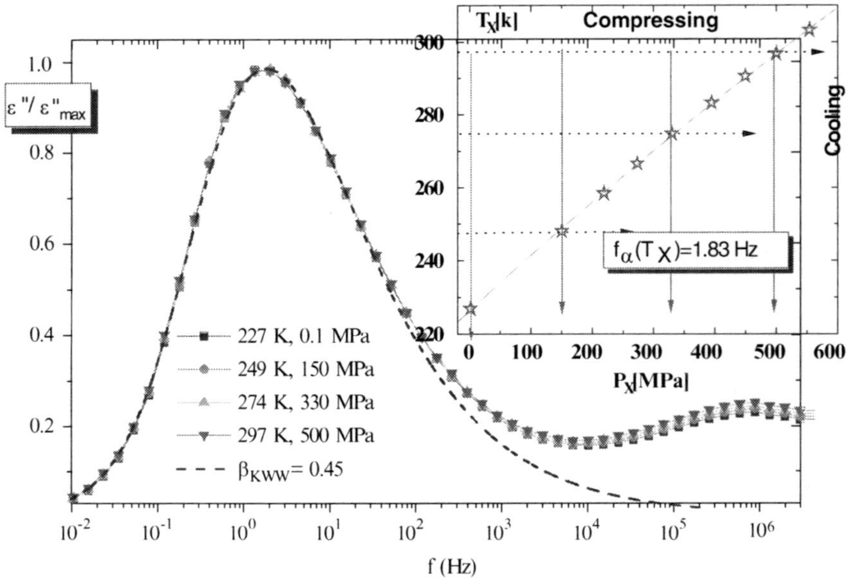

Figure 3. *T-P superposition of loss spectra for 5% CNBz in tristyrene measured for different T and P combinations but the same frequency f_α=1.83 Hz. The line is a Fourier transformed of the Kohlrausch function with $\beta_{KWW} \equiv (1\text{-}n) = 0.45$. The results demonstrate the co-invariance of three quantities, τ_α, n, and τ_{JG}, to widely different combinations of T and P. Inset shows the Tx vs. Px combinations that keep the frequency f_α=1.83 Hz.*

Moreover, a constant ratio τ_α/τ_{JG} is found for any combination taking τ_α constant. Again, this property is related to the invariant shape of the α-relaxation for isochronal spectra. As another example, a selection of loss spectra of 5%wt. QN in trystirene obtained along two different thermodynamic paths is shown in Fig. 4. The α- and β- relaxation peaks are well resolved in the dielectric spectra for any T-P conditions. For some of them, the T-P condition is able to yield the same frequency of the loss maximum (and so τ_α) for structural process. As in the case of CNBz/3St shown in Fig. 3, once τ_α is fixed, also the shape of the α-peak is invariant and as well as the frequency location of the β-relaxation. The two processes appear again strongly related. From all the data acquired up to now, we can claim that the ratio τ_α/τ_{JG} is invariant to changes in the combinations of P and T that keep τ_α constant.

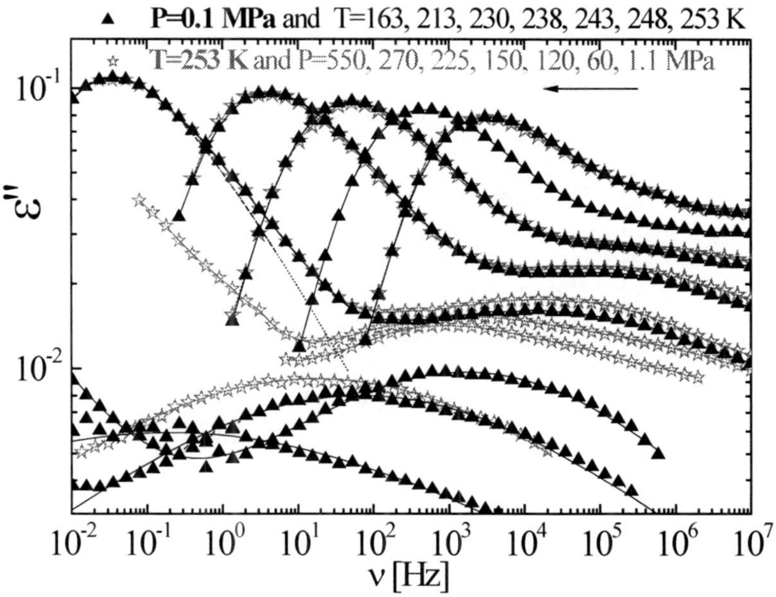

Figure 4. *Selected loss spectra for mixtures of the polar molecule (5 wt.%) QN in tristyrene. Open circles and full triangles represent the isobaric (P=0.1 MPa) and the isothermal (253 K) measurement respectively. The arrow indicates the direction of increasing pressure or decreasing temperature. Continuous lines are HN equation fits. Dotted line is a KWW fit.*

The invariance is valid not only at T_g (i.e. for τ_α=100 s), but for any chosen value of τ_α or τ_{JG}. Moreover, also the frequency dispersion of the α-relaxation (or n) is invariant to changes of T and P if τ_α is kept constant: in other terms, the coupling parameter n is a function of τ_α only, i.e. $n(\tau_\alpha)$. Therefore we have co-invariance of three quantities, τ_α, n, and τ_{JG}, to widely different T and P combinations involving large variations of specific volume, entropy and enthalpy. There is no way that this could happen only by chance for so many different systems. This is an indication that the JG β-relaxation not only is not independent of the α-relaxation, but actually the two processes are well connected. These considerations suggest a way to rescale the overall α-β dynamics by plotting the logarithm of the maximum frequency versus a temperature scale peculiar to the system under study. For instance, one could use the value of the glass transition temperature attained at the value P at which the measurement is performed (for both isothermal and isobaric measurements) or, more generally, the value of a reference temperature $T_{ref}(P)$ related to a fixed value of ν_α.

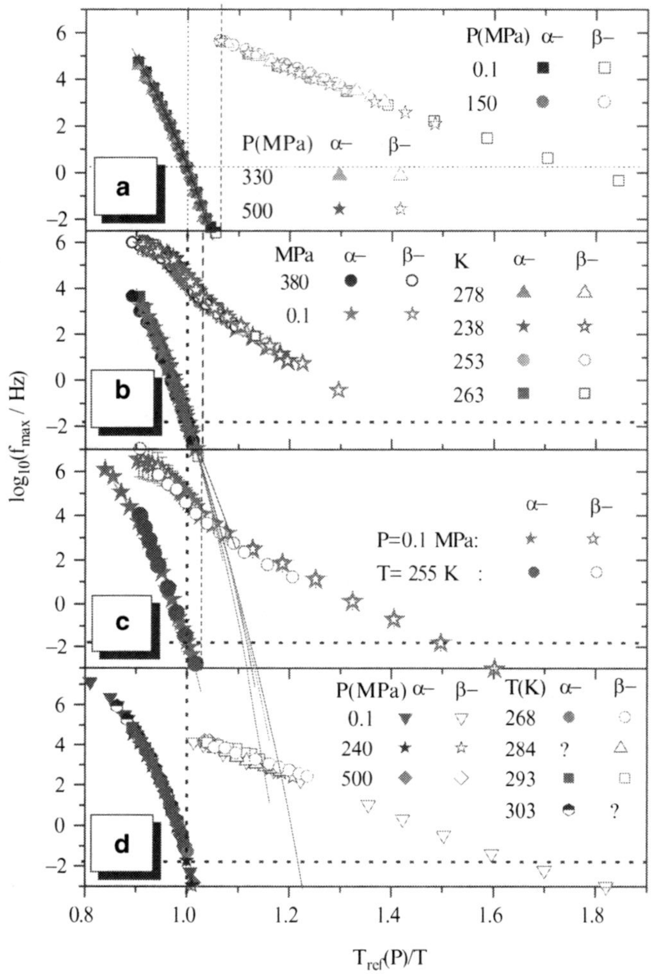

Figure 5. *Logarithm of maximum frequency of α- and JG β-relaxation versus $T_{ref}(P)/T$ for (a) 5% wt. CNBz in tristyrene, (b) 10% wt. QN in tristyrene, (c) 5% wt. QN in tristyrene, (d) PPGE. Vertical dotted lines mark T_{ref}, $\nu_\alpha(T_{ref})$=1.8 Hz (a) and 0.014 Hz (b-c-d). Vertical dashed lines indicate the value of $T_g(P)/T$.*

Figure 5 shows this plot for 3 different binary mixtures, as well as for a neat system (the epoxy resin PPGE). A good superposition for the α-relaxation frequency is obtained, as it is expected, since the steepness index of $\tau_\alpha(T)$ m (fragility according to Angell definition[31]) is almost constant under pressure variation (it actually decreases by less than 10%). Also n does not change appreciably close to T_g. Therefore, Eq. (1) predicts the master curve for τ_{JG} shown by the results in Fig. 5. Moreover, from the plots in Fig. 5 it is also clear that the change of T-P behavior reported on crossing T_g or P_g line for several glass-formers is reflecting the freezing of the structural process. The master curve obtained for the overall dynamics in Fig. 5 is noteworthy because τ_α or τ_{JG} look related not only above T_g, where a clear prediction comes from Eq. (1), but also in the glassy state. A quantitative check of CM is actually not possible below T_g, since τ_α is not directly accessible and it can only be extrapolated from data obtained in the liquid state.

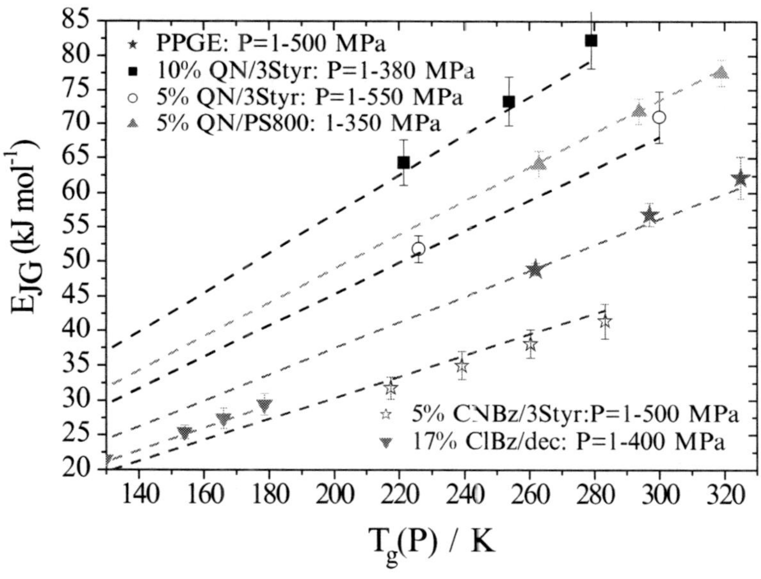

Figure 6. *Activation energy of the JG relaxation in the glassy state for different isobaric measurements plotted versus the related $T_g(P)$. QN in tristyrene (5% and 10% wt. are indicated by open circles and solid squares, respectively), 5% wt. QN in PS800 (solid triangles), PPGE (solid stars), 5% wt CNBz in tristyrene (open stars), 17% ClBz in decalin from ref.[34] (solid down triangles). Dashed lines are the predictions according to Eq. (3).*

A rationale could come from a relation previously derived for glass-forming systems at ambient pressure. The ratio between the activation energy E_{JG} of the JG β-relaxation time τ_{JG} in the glassy state and the glass transition temperature T_g (that is a quantity related to the α-process) is slightly constant for several systems.[32] This result is very general and was extended to several other systems and rationalized by using CM prediction.[33] The ratio was found to depend on the coupling parameter n as:

$$E_{JG}/RT_g = 2.303(2 - 13.7n - \log_{10}\tau_\infty) \tag{3}$$

where τ_∞ is the prefactor of the Arrhenius fit to JG times. Eq. (3) can be extended also to data obtained at high pressure. As $log_{10}(\tau_\infty)$ and $n(T_g)$ are almost pressure independent, a constant value for the ratio E_{JG}/RT_g is expected for isobaric measurements also at high pressure. This is evident from Fig. 5-a-b-d, where the data in the glassy state are quite superposed. The ratio has been found constant under pressure variation also for the other systems investigated (having different values of $n(T_g)$ and τ_∞), as well as for the prototypical binary mixture 17% chlorobenzene in decalin from ref. [[34]] (see Fig. 6).

The master plots of Fig. 5 show a remarkable relation between structural and JG relaxation both above and below T_g, demonstrating the interdependency of the two processes. Eq. (1) provides a rationale for the quantitative relation above T_g. As far as concerns below T_g, the master curve is possible since the activation energy E_{JG} of JG relaxation scales with P linearly with $T_g(P)$. The activation energy of the intermolecular secondary process *in the glassy state* is scaling with the glass transition temperature, that is characteristic of the primary relaxation of the *supercooled liquid.*

4. Conclusion

Dynamics of polar rigid molecules dissolved in apolar solvents and of a neat epoxy system were studied by means of broadband dielectric spectroscopy. It was possible to resolve α- and JG β-relaxation peaks in the loss spectra and to acquire their evolution under temperature and pressure variation. As expected, for each investigated system the JG β-relaxation was strongly affected by pressure both in the supercooled liquid and glassy state. Moreover, analyzing the T and P behavior of α- and β- processes, a clear correlation was found between the maximum frequency of structural and JG relaxation and the dispersion of the structural relaxation *n*. In particular, over a broad *T-P* range, the dispersion (i.e. the stretching parameter) of α-relaxation was found constant for a fixed value of the α-relaxation time, independent of thermodynamic (*T-P*) conditions. If this result agrees with what recently found for many glass-formers,[21] in our experiment also the ratio of α- and β- relaxation times was found invariant to different combinations of *P* and *T* keeping the primary relaxation time constant. The interdependence between α- and β- relaxation is well demonstrated by two main results: *i)* spectra obtained at very different T-P conditions but with the same frequency of loss maxima almost superpose in both α- and β- time scale range; *ii)* data of α- and β- relaxation, plotted versus a single reduced parameter (i.e. $T_g(P)/T$)), collapse in two master curves. These results indicate the fundamental role played by the JG relaxation as a precursor of the structural relaxation in glass transition, a role often overlooked by common theories. On the other hand, all these evidences can be rationalized in the framework of the Coupling Model (CM).[13,15]

Acknowledgements

The authors warmly thank Sergy Anchebrak for the help in pressure experiments, Dr. Monica Bertoldo for assistance in the preparation of materials, Marco Bianucci for the development of the pressure cell. This work was supported by MIUR-FIRB 2003 D.D.2186 grant RBNE03R78E.

References

1. Ngai, K. L. (1998) *J. Chem. Phys.* **109**, 6982
2. Johari, G. P., Power, G., and Vij, J. K. (2002) *J. Chem. Phys.* **116**, 5908; *J. Chem. Phys.* **117**, 1714. Power, G., Johari, G. P., and Vij, J. K. (2003) *J. Chem. Phys.* **119**, 435
3. Brand, R., Lunkenheimer, P., and Loidl, A. (2002) *J. Chem. Phys.* **116**, 10386
4. Ngai, K. L., and Paluch, M. (2004) *J. Chem. Phys.* **120**, 857
5. Böhmer, R., Diezemann, G., Geil, B., Hinze, G., Nowaczyk, A., and Winterlich, M. (2006) *Phys. Rev. Lett.* **97**, 135701
6. Casalini, R., and Roland, C. M. (2003) *Phys. Rev. Lett.* **91**, 015702
7. Johari, G. P., and Goldstein, M. (1970) *J. Chem. Phys.* **53**, 2372
8. Ngai, K. L. (1998) *Physica A* **261**, 36.
9. Ngai, K. L. (2005) *J. Non-Cryst. Solids* **351**, 2635 and references therein
10. Eisenberg, A. (1976) Discussion in *Annals of the New York Academy of Science*, **279**, 141-149
11. Johari, G. P., and Whalley, E. (1972) *Faraday Symp. Chem. Soc.* **6**, 23
12. Williams, G. (1964) *Trans. Faraday Soc.* **60**, 1548
13. Ngai, K. L. (2003) *J. Phys.: Condens. Matter*, **15**, S1107
14. Cavaille, J. Y., Perez, J., and Johari, G. P. (1989) *Phys. Rev. B* **39**, 2411
15. Ngai, K. L., and Capaccioli, S. (2008) *J. Phys.: Condens. Matter* **20**, 244101
16. Ngai, K. L. (1979) *Comm. Solid State Phys.* **9**, 141
17. Ngai, K. L., Casalini, R., Capaccioli, S., Paluch, M, and Roland, C. M. (2005) *J. Phys. Chem. B.* **109**, 17356-17360
18. Kessairi, K., Capaccioli, S., Prevosto, D., Lucchesi, M., Sharifi S., and Rolla, P. A. (2008) *J. Phys. Chem. B* **112**, 4470
19. Capaccioli, S., Ngai, K. L., and Shinyashiki, N. (2007) *J. Phys. Chem. B* **111**, 8197

20. Ngai, K. L., Capaccioli, S., and Shinyashiki, N. (2008) *J. Phys. Chem. B* **112**, 3826-3832
21. Ngai, K. L., Casalini, R., Capaccioli, S., Paluch, M., and Roland, C. M. (2006) *Adv. Chem. Phys. in Chemical Physics Part B, Fractals, Diffusion and Relaxation in Disordered Complex Systems,* **133B**, 497-582
22. Capaccioli, S., and Ngai, K. L. (2005) *J. Phys. Chem. B* **109**, 9727
23. Corezzi, S., Beiner, M., Hu, H., Schröter, K., Capaccioli, S., Casalini, R., Fioretto, D., and Donth, E. (2002) *J. Chem. Phys.* **117**, 2435
24. Blochowicz, T., Roessler, E. A. (2004) *Phys. Rev. Lett.* **92**, 225701
25. Capaccioli, S., Kessairi, K., Prevosto, D., Lucchesi, M., and Rolla, P. A. (2007) *J. Phys.: Condens. Matter* **19**, 205133
26. Paluch, M., Roland, C. M., Pawlus, S., Ziolo, J., and Ngai, K. L. (2003) *Phys. Rev. Lett.* **91**, 115701
27. Ngai, K. L., Capaccioli, S., Prevosto, D., and Paluch, M. (2009) *"The Nature of Glass: somethings are clear"* in NATO ARW Series, this volume
28. Capaccioli, S., Prevosto, D., Lucchesi, M., Rolla, P. A., Casalini, R., and Ngai, K. L. (2005) *J. Non-Cryst. Solids* **351**, 2643
29. Kessairi, K., Capaccioli, S., Prevosto, D., Lucchesi, M., Sharifi, S., and Rolla, P. A. (2008) *J. Phys. Chem.* B **112**, 4470
30. Mierzwa, M., Pawlus. S., Paluch, M., Kaminska, E., and Ngai, K. L. (2008) *J. Chem. Phys.* **128**, 044512
31. Angell, C. A., (1997) *Polymer* **38**, 6261
32. Kudlik, A., Tschirwitz, C., Benkhof, S., Blochowicz, T., and Rössler, E. (1997) *Europhys. Lett.* **40**, 649
33. Ngai, K. L. and Capaccioli, S. (2004), *Phys. Rev. E*, **69**, 031501
34. Köplinger, J., Kasper, G., and Hunklinger, S. (2000) *J. Chem. Phys.* **113**, 4701

REORIENTATIONAL RELAXATION TIME AT THE ONSET OF INTERMOLECULAR COOPERATIVITY

[1*]C. M. ROLAND* AND [2]R. CASALINI

[1]*Naval Research Lab, Code 6120, Washington DC 20375-534, USA;* [2]*Naval Research Lab, Code 6120, Washington DC 20375-5342 and Chemistry Department, George Mason University, Fairfax VA 22030, USA*

Abstract: For three liquids, salol, propylene carbonate, and *o*-terphenyl, we show that the relaxation time or the viscosity at the onset of Arrhenius behavior is a material constant. Thus, while the temperature of this transition can be altered by the application of pressure, the time scale of the dynamics retains a characteristic, pressure-independent value. Since the onset of an Arrhenius temperature-dependence and the related Debye relaxation behavior signify the loss of intermolecular constraints on the dynamics, our result indicates that intermolecular cooperativity effects are governed by the time scale for structural relaxation.

Keywords: arrhenius behavior, relaxation time, viscosity, thermodynamic scaling

1. Introduction

The physical and mechanical properties of materials are the primary concern for their use in applications. However, these properties reflect the motion of the constituent molecules, which makes study of the latter essential to the fundamental understanding necessary for developing new technologies. The structural dynamics is quantified by a time constant, τ, which is a measure of the time scale for reorientation of a small molecule or the correlated conformational transitions of a few backbone bonds in a polymer. For both liquids and polymer melts the structural relaxation time (and viscosity, η, which is roughly proportional to τ) varies with temperature, with Arrhenius behavior

$$\tau = \tau_0 \exp(E_a / RT) \tag{1}$$

* To whom correspondence should be addressed.

obtained only at high temperature. In eq.(1) R is the gas constant and the prefactor τ_0 and activation energy E_a are material constants. The onset of an Arrhenius temperature dependence is usually associated with the relaxation function assuming the simple exponential form ("Debye behavior"). These two properties signify that thermal energy and the unoccupied volume have become sufficiently large that molecular motions proceed unaffected by motion of neighboring molecules or segments; that is, the dynamics are noncooperative.

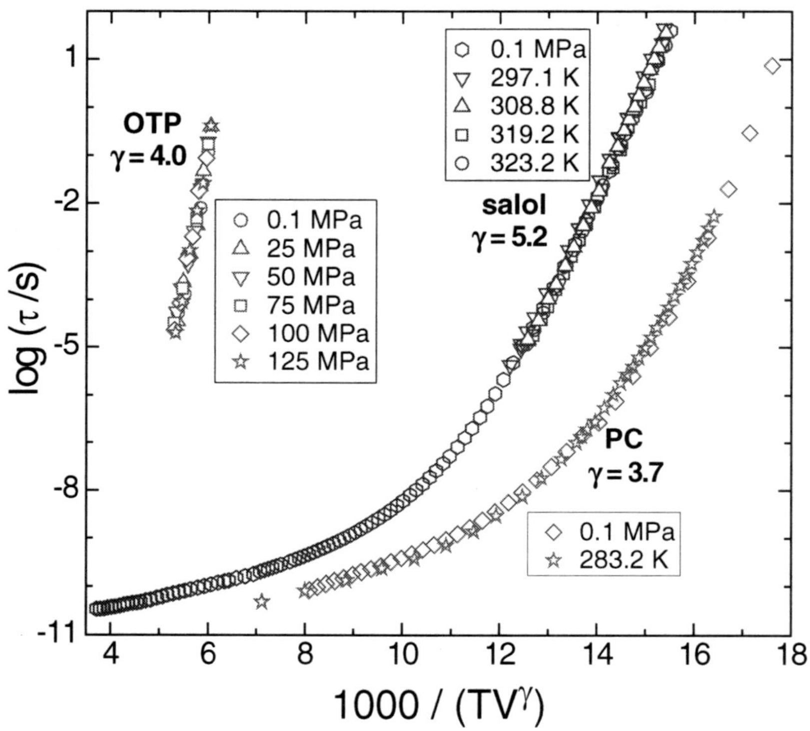

Figure 1. *Scaled plots of the relaxation times of OTP,*[5] *salol*[3] *and propylene carbonate*[12].

Measurements of the temperature, T_A, demarcating the departure from Debye and Arrhenius behaviors are sparse, but the available data indicate that neither T_A nor the value of the relaxation time at T_A, τ_A, is a universal constant.[1] The question examined herein is whether τ_A is a material constant; that is, while T_A increases with increasing pressure, does the relaxation time at this characteristic temperature remain invariant? High pressure measurements are required to address this question; however, despite the enormous number of results published in recent years describing the effect of pressure on relaxation times[2], for experimental reasons these $\tau(P)$ rarely extend to temperatures above T_A.

To circumvent this limitation in the available data, we make use of a recently discovered scaling relationship[3,4,5]

$$\tau = f(TV^{\gamma}) \tag{2}$$

in which f is a function, V is the specific volume, and the scaling exponent γ a material constant. Relaxation times obtained by various methods including dielectric spectroscopy,[2,3,6,7] light scattering,[5] and simulations[8,9,10] as well as viscosity data[11] have been shown to conform to eq. (2). We illustrate this herein in Figure 1 with results for three liquids, salol (phenol salicylate),[3] propylene carbonate (PC),12 and *o*-terphenyl (OTP),[5] for which γ = 5.2, 3.7, and 4.0 respectively.

The broadest range of dynamic data generally is for atmospheric pressure, since such measurements are easiest. However, whenever γ is known, eq. (2) can be used to calculate τ or η for any thermodynamic condition. The calculation requires the equation of state (EOS), but that must already be available for the determination of γ. The procedure is to find the T for any arbitrary P that yields a value of TV^{γ} lying within the range of this quantity for the experimentally measured P = 0.1 MPa data. τ_A is then known for this higher P condition, since it is a unique function of TV^{γ}. In this paper we use this method to obtain the relaxation time or viscosity at T_A for different P, in order to determine if τ_A is a material constant, independent of P and V. The calculation requires two pieces of information, the scaling exponent and T_A for one pressure (i.e., atmospheric), which limits the analysis to three materials, salol, PC, and OTP.

2. Results

In Figure 2 (lower panel) the relaxation times of salol for P = 0.1 MPa are displayed versus reciprocal temperature[13]. Although there is the suggestion of linear behavior at the higher temperatures, a reliable value of T_A cannot be extracted from the plot. Accordingly we employ the derivative method of Stickel et al.[14] plotting in the upper panel the quantity $\left(d\log\tau / dT^{-1}\right)^{-1/2}$. The onset of an Arrhenius temperature dependence is clearly seen, with T_A determined from the intersection of the two linear regions. We now choose two arbitrary pressures, 100 and 500 MPa. The EOS for salol[15] can be expressed using the Tait form

$$V(T,P) = V(T,0)\left[1 - C\ln\left(1 + P / b_0 \exp\left[-b_1 T\right]\right)\right] \tag{3}$$

with the atmospheric pressure specific volume given by

$$V(T,0) = v_0 + v_1 T + v_2 T^2 \tag{4}$$

The EOS parameters C, b_0, b_1, v_0, v_1, and v_2 for salol are listed in Table 1.

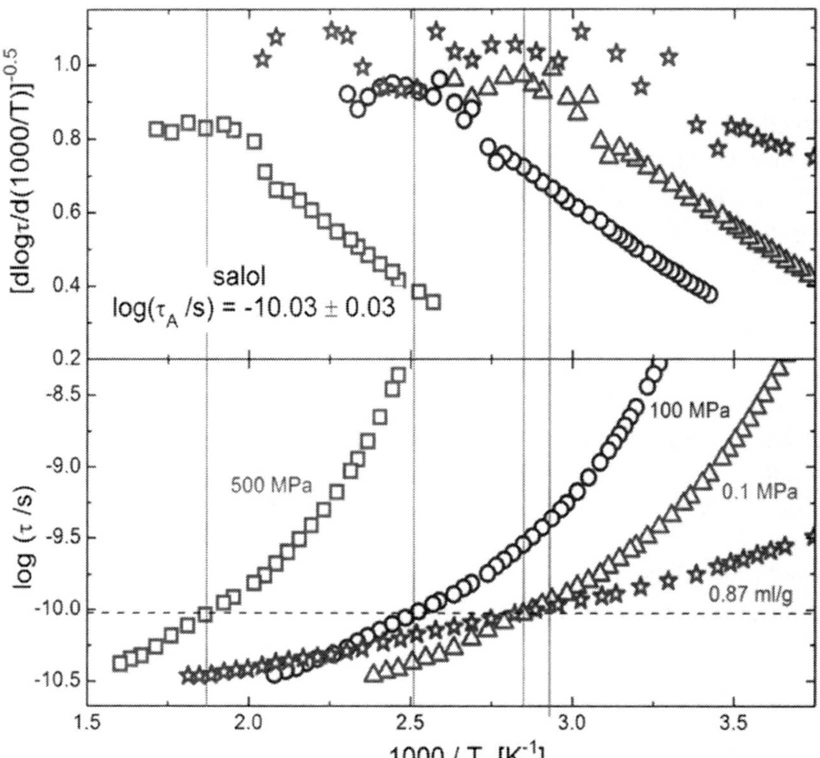

Figure 2. *(lower panel) Relaxation times of salol measured at atmospheric pressure*[13] *and calculated for higher pressure or constant volume using eq.(2). (upper panel) Corresponding Stickel plots of τ, with the intersection of the* linear segments yielding the value of T_A indicated by the vertical lines. The obtained τ_A is denoted by the horizontal dashed line.

Using eq. (3) the temperatures are determined that will yield a value of the quantity TV^{γ} at the higher P equal to values for the atmospheric pressure data. The resulting curves are shown in Figure 2, along with the corresponding derivative plots. From the intersection of the linear segments, we obtain the T_A for each pressure and hence $\tau_A(P)$. The latter, listed in Table 1, is invariant to pressure. We also include in Fig. 2 an isochoric curve at an arbitrary V. Likewise, the transition to Arrhenius behavior under constant volume is associated with the same value of τ_A. A similar analysis was applied to atmospheric pressure data for PC (Figure 3), with the EOS[12] parameters listed in the table. Interestingly the high temperature linear segments of the derivative curves for this liquid do not have zero slope, as would be expected for strictly Arrhenius behavior.[16]

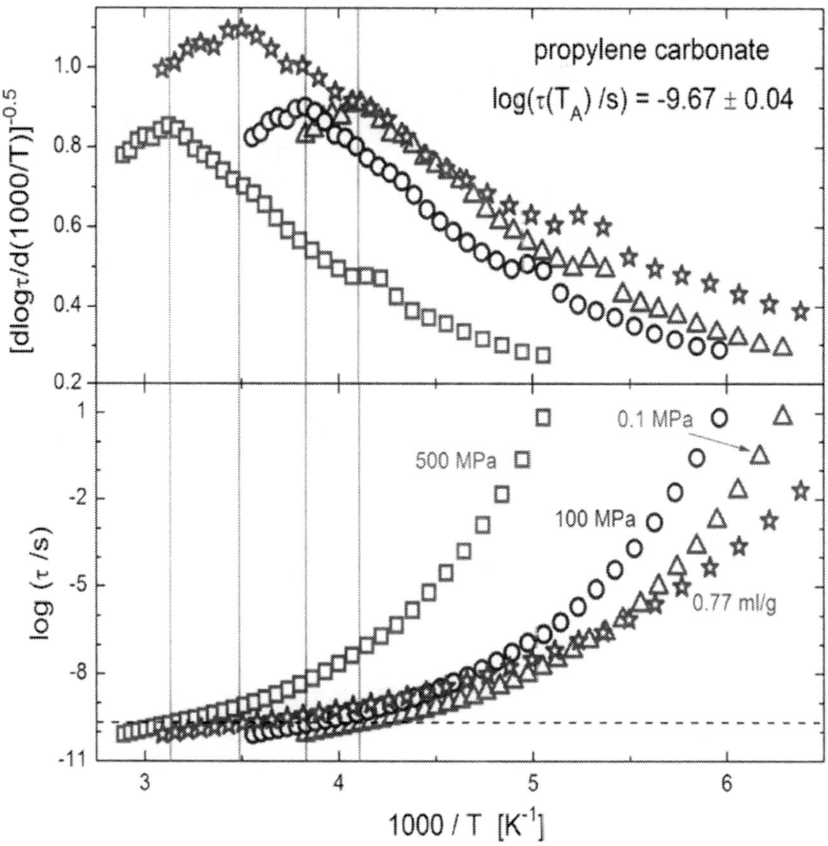

Figure 3. (lower panel) Relaxation times of propylene carbonate measured at atmospheric pressure[12] and calculated for higher pressure or constant volume using eq.(2). (upper panel) Corresponding Stickel plots of τ, with the intersection of the linear segments yielding the value of T_A indicated by the vertical lines. The obtained τ_A is denoted by the horizontal dashed line.

Nevertheless we determine T_A in the same manner as for salol. Although τ_A is more than four times larger than for salol, it is again constant for isobaric conditions at any P or isochoric conditions at any V (Table 1).

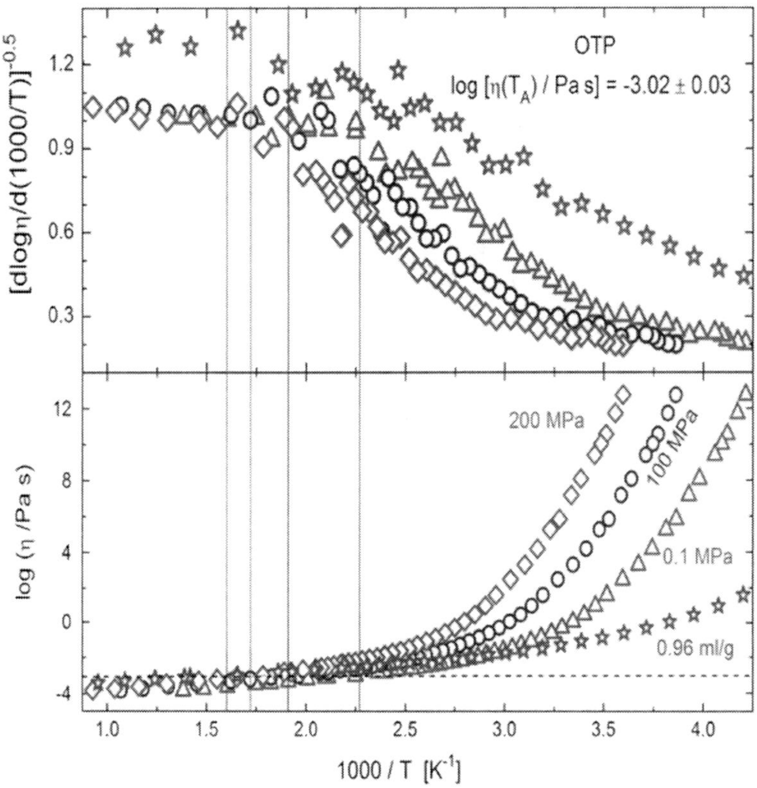

Figure 4. *(lower panel) Viscosities of OTP measured at atmospheric pressure and calculated for higher pressure or constant volume using eq.(2). (upper panel) Corresponding Stickel plots of η, with the intersection of the linear segments yielding the value of T_A indicated by the vertical lines. The obtained η_A is denoted by the horizontal dashed line.*[19]

Another prototypical glass-forming liquid is OTP. In addition to the dynamic light scattering results[5] in Fig. 1, the thermodynamic scaling of eq. (2) has been applied to data from quasi-elastic neutron scattering and viscosity measurements;[17,5] all yield $\gamma = 4$. From dielectric data,[18] a slightly higher value of $\gamma = 4.25$ was reported.[5] The broadest range of relaxation data for OTP is from viscosity measurements;[19] these are displayed in Figure 4. Arrhenius behavior is evident at the higher temperatures. To determine accurately η at the onset of Arrhenius behavior we again employ the derivative analysis, which yields $\eta(T_A)$. Using published EOS parameters (Table 1),[5] we calculate viscosities for one isochore and for two higher pressures, 100 and 200 MPa. The latter is lower than the (arbitrary) value used for salol and propylene carbonate, in order to restrict the temperatures to the range where the EOS is valid. The obtained $\eta(T_A)$ = 9.6 (±0.6)×10^{-4} Pa s, is independent of pressure. Using the Maxwell relation we estimate τ_A to be about 1 ps, assuming the limiting value of the shear modulus at high frequency equals 10^8 Pa.

Table 1 Values of fitted parameters for EOS (eq. (3)) and time-scales for the dynamic crossover from the Arrhenius domain in tested glass forming liquids.

	salol	propylene carbonate	o-terphenyl
v_o	0.654	0.824	0.911
v_1	6.21×10^{-4}	6.82×10^{-4}	6.40×10^{-4}
v_{o2}	0	7.5×10^{-7}	5.5×10^{-7}
C	0.0870	0.0894	0.0894
b_0	790	219	189
b_1	4.69×10^{-3}	5.98×10^{-3}	4.56×10^{-3}
γ	5.2	3.7	4.0
$\log(\tau(T_A)\ /s)$	-10.03 ± 0.03	-9.67 ± 0.03	—
$\log(\eta(T_A)\ /Pa\ s)$	—	—	— -3.02 ± 0.03

3. Conclusions

Using the scaling relation to deduce the relaxation times or viscosities at elevated pressure, we find that the characteristic value of τ or η at the onset of Arrhenius behavior is invariant to pressure.

Since this Arrhenius temperature dependence implies that the molecular motions are unimpeded by intermolecular constraints, our finding reveals that the loss of intermolecular cooperativity is governed by the dynamics, or at least they have the same control parameter. Since the relaxation time determines the shape of the relaxation function[20,21], we can conclude from the results herein that at the onset of intermolecular cooperativity the breadth of the relaxation function is constant, independent of pressure. Of course, since in the absence of intermolecular constraints on the molecular motions we expect Debye behavior, the shape of the relaxation function should also be invariant for all T greater than T_A.

Acknowledgement

We thank D.J. Plazek and C. Dreyfus for providing viscosity and light scattering data respectively for OTP. This work was supported by the Office of Naval Research.

References

1. Roland, C.M. (2008) Characteristic relaxation times and their invariance to thermodynamic conditions, *Soft Matter* **4**, 2316
2. Roland, C.M., Hensel-Bielowka, S., Paluch, M., and Casalini, R. (2005) Supercooled dynamics of glass-forming liquids and polymers under hydrostatic pressure, *Rep. Prog. Phys.* **68**, 1405
3. Casalini, R., and Roland, C.M. (2004) Thermodynamical scaling of the glass transition dynamics, *Phys. Rev. E* **69**, 062501
4. Alba-Simionesco, C., Cailliaux, A., Alegria A., and Tarjus, G. (2004) Scaling out the density dependence of the α relaxation in glass-forming polymers, *Europhys. Lett.* **68**, 58
5. Dreyfus, C., Le Grand, A., Gapinski, J., Steffen, W., and Patkowski, A. (2004) Scaling the α-relaxation time of supercooled fragile organic liquids, *Eur. J. Phys.* **42**, 309
6. Win, K.Z., and Menon, N. (2006) Glass transition of glycerol in the volume-temperature plane, *Phys. Rev. E* **73**, 040501
7. Urban, S., and Würflinger, A. (2005) Thermodynamical scaling of the low frequency relaxation time in liquid crystalline phases, *Phys. Rev. E* **72**, 021707
8. Tsolou, G., Harmandaris, V.A., and Mavrantzas, V.G. (2006) Atomistic molecular dynamics simulation of the temperature and pressure dependences of local and terminal relaxations in *cis*-1,4-polybutadiene, *J. Chem. Phys.* **124**, 084906
9. Budzien, J., McCoy, J.D., and Adolf, D.B. (2004) General relationships between the mobility of a chain fluid and various computed scalar metrics, *J. Chem. Phys.* **121**, 10291
10. Coslovich, D., and Roland, C.M. (2008) Thermodynamic scaling of the diffusion coefficient in supercooled Lennard-Jones liquids *J. Phys. Chem. B* **112**, 1329
11. Roland, C.M., Bair, S., and Casalini, R. (2006) Thermodynamic scaling of the viscosity of van der Waals, H-bonded, and ionic liquids, *J. Chem. Phys.* **125**, 124508
12. Pawlus, S., Casalini, R., Roland, C.M., Paluch, M., Rzoska, S.J., and Ziolo, J. (2004) Temperature and volume effects on the change of dynamics in propylene carbonate, *Phys. Rev. E* **70**, 061501

13. Ngai, K.L., and Tsang, K.Y. (1999) Similarity of relaxation in supercooled liquids and interacting arrays of oscillators, *Phys. Rev. E* **60**, 4511
14. Stickel, F., Fischer, E.W., and Richert, R. (1996) Dynamics of glass-forming liquids. II. Detailed comparison of dielectric relaxation, dc-conductivity, and viscosity data, *J. Chem. Phys.* **104**, 2043
15. Comez, L., Fioretto, E., Kriegs, H., and Steffen, W. (2004) Slow dynamics of salol: A pressure- and temperature-dependent light scattering study, *Phys. Rev. E* **70**, 011504
16. Hansen, A., Stickel, F., Richert, R., and Fischer, E. W. (1998) Dynamics of glass-forming liquids. IV. True activated behavior above 2GHz in the dielectric a-relaxation of organic liquids, *J. Chem. Phys.* **108**, 6408
17. Tolle, A. (2001) Neutron scattering studies of the model glass former ortho-terphenyl, *Rep. Prog. Phys.* **64**, 1473
18. Naoki, M., Endou, H., and Matsumoto, K. (1987) Pressure effects on dielectric-relaxation of supercooled ortho-terphenyl, *J. Phys. Chem.* **91**, 4169
19. Hansen, C., Stickel, F., Berger, T., Richert, R., and Fischer, E.W. (1997) Dynamics of glass-forming liquids. III. Comparing the dielectric α- and β- relaxation of 1-propanol and *o*-terphenyl, *J. Chem. Phys.* **107** (4); Greet, R.J., and Magill, J.H. (1967). An empirical corresponding-states relationship for liquid viscosity, *J. Phys. Chem.* **71**, 1746
20. Roland, C.M., Casalini, R., and Paluch, M. (2003) Isochronal temperature-pressure superpositioning of the α-relaxation in type-A glass formers, *Chem. Phys. Lett.* **367**, 259
21. Ngai, K.L., Casalini, R., Capaccioli, S., Paluch, M., and Roland, C.M. (2005) Do theories of the glass transition, in which the structural relaxation time does not define the dispersion of the structural relaxation, need revision?, *J. Phys. Chem. B* **109**, 17356

NEUTRON DIFFRACTION AS A TOOL TO EXPLORE THE FREE ENERGY LANDSCAPE IN ORIENTATIONALLY DISORDERED PHASES

MURIEL ROVIRA-ESTEVA, LUIS C. PARDO, JOSEP LL. TAMARIT*

Group of Characterization of Materials, Department of Physics and Nuclear Engineering, ETSEIB, Diagonal 647, 08028 Barcelona. Universitat Politècnica de Catalunya, Catalonia, Spain

F. JAVIER BERMEJO

CSIC, Instituto de Estructura de la Materia and Departamento de Electricidad y Electrónica, Facultad de Ciencia y Tecnología, Universidad del País Vasco, P.O. Box 48080 Bilbao, Spain

Abstract: The temperature dependence of structural parameters of orientational glasses of the halogenomethane family, Freon 112 (FCl_2C)-(CCl_2F)) and Freon 112a (F_2ClC)-(CCl_3)) are studied at short- (molecular) intermediate- (orientational correlations) and long-range (lattice parameters) scales by means of neutron diffraction. The two materials which are chemical isomers display strikingly different properties in their ordering patterns resulting from a shift in balance between electrostatic and excluded-volume interaction. The relevance of these findings to our understanding of glassy phenomena is discussed.

Keywords: glass transition, orientational disorder, conformational disorder, short range order, neutron scattering, ODIC, 1,1,2,2-tetrachloro-1,2-difluoroethane, 1,1,1,2-tetrachloro-2,2-difluoroethane

1. Introduction

Liquids are systems devoid of both long-range translational and orientational order whereas short-range order still remains at molecular scales resulting from the subtleties of forces acting on their constituent molecules. In turn, rotator-phase (plastic) crystals are liquid-like in the sense that molecules

* Author to whom correspondence should be addressed. E-mail: jose.luis.tamarit@upc.edu

may rotate rapidly about the nodes of a crystal lattice defined by the molecular centers of mass. Rapid cooling of a liquid leads to a system where disorder is now of static nature, that is an amorphous material or a glass. Rather similar phenomena are found when cooling many plastic crystals, where the transition into a glassy state, now termed an orientational glass, involves freezing the rotational degrees of freedom leading to a system with static orientational disorder (Fig. 1).[1-3]

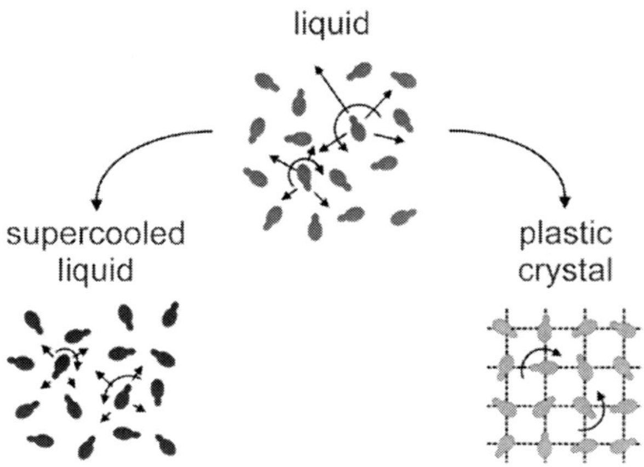

Figure 1. *In a plastic crystal the molecular centers of mass are placed in a lattice but the molecules can rotate, more or less freely. [From Brand et al.[3]]*

A glass transition may thus take place resulting from the dynamical arrest of one or more degrees of freedom due to the action of an external field such as a rapidly decreasing temperature or an increase in pressure. By convention it is considered that a glassy state has been achieved when the relevant relaxation time reaches value of 10^2 s.

In addition to positional and orientational degrees of freedom characteristic of rigid-bodies, further degrees of freedom need to be considered for materials composed of molecules having internal degrees of freedom with characteristic energy levels not too far above k_BT. These usually concern motions involving molecular internal rotation which lead to different molecular conformations and in this sense one terms “conformational glasses” to systems with con-conformational disorder.

Here we focus on the compound 1,1,2,2-tetrachloro-1,2-difluoroethane (Freon 112) which has a transition from a liquid to a bcc plastic phase at 299 K and an orientational glass transition at ca. 90 K, the transition to the completely ordered phase being extremely slow.[4] Freon 112 has two conformations energetically nonequivalent named *trans* and *gauche* (Fig. 2). The *trans* conformation (with C_{2h} symmetry) is somewhat more stable and has vanishing dipole moment while the *gauche* conformation (with C_2 symmetry) does (0.26 D), the proportion between them being a function of temperature.[5-6] Because a glass is defined by its dynamical properties, special attention must be drawn to them when characterizing it. The fragility, for instance, provides a measure of the temperature dependence of dynamical properties such as the relaxation time associated to the macroscopic viscosity.[7] In this respect, it turns out that most of the orientational glasses are rather strong, showing an exponential temperature dependence of their relaxation time, but Freon 112 is quite fragile. In fact, is the most fragile plastic crystal known so far (Fig. 3).[3,8-12]

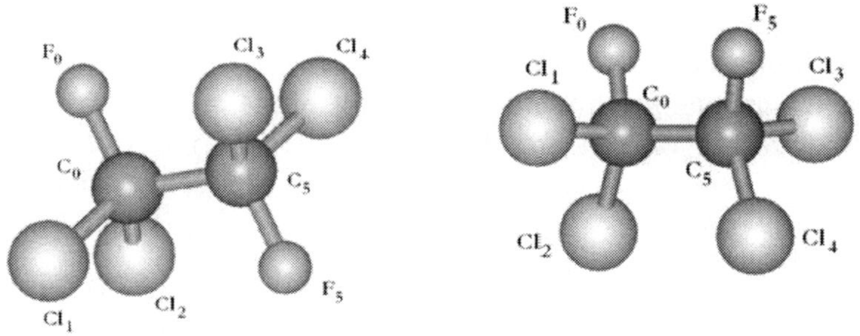

Figure 2. Freon 112, *trans* (left) and *gauche* (right) conformations.

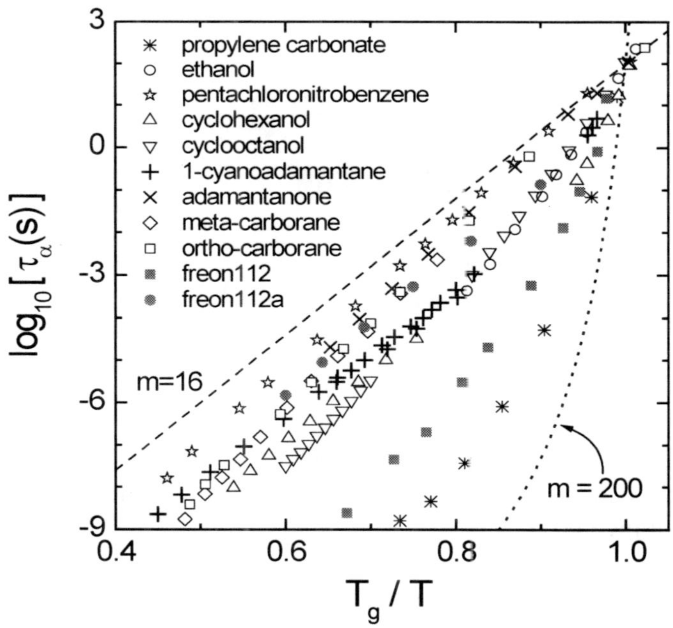

Figure 3. *Angell plot of several orientational glasses, including Freon 112 and 112a, and a structural fragile glass (propylene carbonate) for comparison. Strong glasses are closer to an Arrhenius behavior (m=16) while fragile glasses depart from it. [From Pardo et al.[12]]*

To ascertain the reasons leading to this extreme fragility, the study of the static, that is time-averaged, properties of Freon 112 is a must. To such an end, we have taken the endeavour of a full characterization of its structural properties, which apart from the intramolecular structure comprise the conformational disorder, as well as the short-, intermediate and long-range order.

Neutron scattering is a powerful technique to get information from the structure as well as from the molecular dynamics. In particular, neutron diffraction has several advantages over other diffraction methods for studying liquid and amorphous structures which stem from the fact that neutrons interact mainly with atomic nuclei via the strong nuclear force (every isotope having a different scattering length that quantifies the strength of the interaction). As a result, the information contained in a diffraction pattern can be directly related to the internuclear (or interatomic) structure of the system.[13]

In this work we will also study the compound 1,1,1,2-tetrachloro-2,2-difluoroethane (Freon 112a), chemically very close to Freon 112 but with only one relevant conformation for temperatures of interest (Fig. 4) and displaying a permanent dipole moment.

This substance also shows a transition from a liquid to a bcc plastic phase at 309 K but, unlike Freon 112, transits into a more stable ordered phase at 158 K and then a glass transition at 90 K[12], the same temperature as the transition of Freon 112. But, like most orientational glasses, it is rather strong.[3,12,14-15]

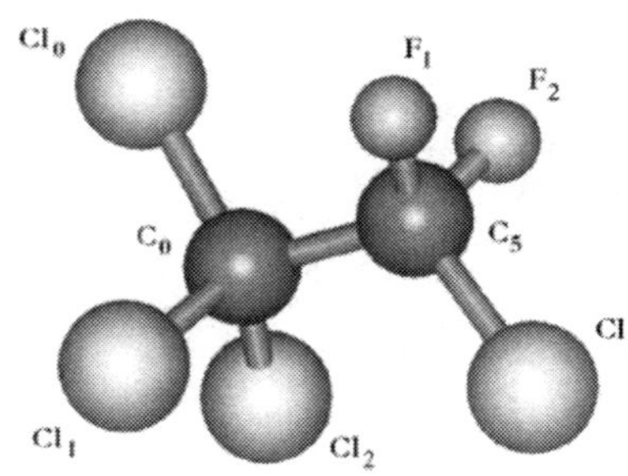

Figure 4. *Freon 112a.*

In what follows we will thus compare the properties and behavior of Freon 112a to Freon 112, especially in liquid and bcc plastic phases, because its study may shed some light on the causes that bring about some of the peculiar characteristics of the latter.

2. Experiments and data reduction

The main objective of neutron diffraction is the determination of structure in terms of the radial distribution function, $g_{\alpha\beta}(\boldsymbol{r})$. This function is related to the probability of finding an atom β at position $\boldsymbol{r}$, relative to a reference atom α taken to be at the origin.

But what is measured in fact in neutron scattering is the differential scattering cross-section, $d\sigma/d\Omega\,(q)$, which is defined as the number of neutrons scattered per second towards a detector in a certain direction per incident beam flux and solid angle. In the case of a liquid or a glass sample for which the average structure is isotropic, only the vector norms ($r = |\boldsymbol{r}|$ and $q = |\boldsymbol{q}|$) are relevant.

The single differential scattering cross-section (1) can be split (in the static approximation) into its incoherent and coherent contributions.[13]

$$\left[\frac{d\sigma}{d\Omega}(q)\right] = \left[\frac{d\sigma}{d\Omega}(q)\right]^{incoh} + \left[\frac{d\sigma}{d\Omega}(q)\right]^{coh} \quad (1)$$

The first term is independent of spatial correlation of the atomic sites and depends only on the distribution of scattering lengths present in the sample (2), leading to an isotropic (angle-independent) diffraction.

$$\left[\frac{d\sigma}{d\Omega}(q)\right]^{incoh} = N_{mol}\sum_{i}^{m} b_{incoh,i}^{2} \quad (2)$$

where N_{mol} is the number of molecules in the sample, i are the atomic positions within a single molecule and $b_{incoh,i}$ is the incoherent scattering length of the chemical species.

The second term concerns diffraction from all atomic sites (including *self-scattering* from a single atom), but is independent from the distribution of scattering lengths, including only an average of them.

The coherent contribution can be further split into its self part (4), which does not give rise to any interference, and its distinct part (6), giving rise to interference due to the atoms within the same molecule and also from different molecules (5).

$$\left[\frac{d\sigma}{d\Omega}(q)\right]^{coh} = \left[\frac{d\sigma}{d\Omega}(q)\right]^{coh}_{self} + \left[\frac{d\sigma}{d\Omega}(q)\right]^{coh}_{distinct} \quad (3)$$

$$\left[\frac{d\sigma}{d\Omega}(q)\right]^{coh}_{self} = N_{mol}\sum_{i}^{m} b_{coh,i}^{2} \quad (4)$$

where $b_{coh,i}$ is the coherent scattering length of the chemical species at site i.

$$\left[\frac{d\sigma}{d\Omega}(q)\right]^{coh}_{distinct} = \left[\frac{d\sigma}{d\Omega}(q)\right]^{coh}_{intra} + \left[\frac{d\sigma}{d\Omega}(q)\right]^{coh}_{inter} \quad (5)$$

The distinct intramolecular and intermolecular contributions of the differential cross-section (5) are related to the sum of all partial structure factors, which is *essentially* proportional to the differential scattering cross-section, weighted by the respective coherent neutron scattering lengths (6).

$$\left[\frac{d\sigma}{d\Omega}(q)\right]^{coh}_{distinct} = \sum_{\alpha,\beta}^{n} b_{coh,\alpha} b_{coh,\beta}\left[S_{\alpha\beta}(q)-1\right] \quad (6)$$

$$= NF(q) \quad (7)$$

where n is the number of atoms within the molecules of the sample, N de total number of atoms and $F(q)$ is the total interference function.

Once the total interference function, corresponding to the weighted average on the summation (6) has been obtained, a simple Fourier transform of this reciprocal-space function will lead to the total pair-correlation function (8).

$$G(r) \equiv \frac{1}{2\pi^2 r\rho_0}\int_0^\infty qF(q)\sin(qr)dq \quad (8)$$

3. Experiments

To obtain precise information on the molecular correlations at atomic scales together with information about the crystal lattice parameters use is made of two neutron diffractometers, D1b and D4 at the Institut Laue-Langevin (Grenoble, France). The first instrument (D1b) employed thermal neutrons with λ= 1.28 Å chosen to determine with maximal precision the variation of the crystal lattice parameters by means of the positions of the main Bragg peaks. The second set (D4) was carried out using a diffractometer on a hot source with neutrons having a wavelength λ = 0.502 Å, shorter than the first, and thus with a broader momentum transfer range (q_{max}= 23 Å^{-1}), then enabling the determination of the smaller distances corresponding to intramolecular structure and short range order of the compounds.

4. Corrections and data reduction

The normalization of the diffraction intensity pattern of a sample to an absolute cross section can be done through the comparison with another sample of known cross-section and volume with respect to the first. In our case, a vanadium solid cylinder was also measured in the experiments for this purpose.

The computer code Correct[16] has been used to perform this normalization and also the background, multiple scattering and container attenuation corrections to the neutron diffraction data.

When the energy exchange between the neutron and the sample becomes comparable to the incident energy of the neutron, an inelastic correction becomes necessary due to the breakdown of the static approximation[13]. This is of greater magnitude with increasing q values (in a reactor source) and for lighter atoms, since a neutron striking an atom of small mass will transfer more energy.

A Placzek correction[17] to remove the inelastic effects has been performed by fitting a polynomial (9) to the higher range of q linearly weighting the data to account for the higher effect at higher values of q.

$$F(q) = p_0 + p_2 q^2 + p_4 q^4 + p_6 q^6 \tag{9}$$

Once the corrected total interference function was obtained, a Fourier transform had to be done to the experimental total interference function to obtain the total pair-correlation function.

Another difficulty with which we must deal with in diffraction experiments is that any instrumental setup has a maximum accessible momentum transfer, q_{max}, and the Fourier transformation of that finite pattern leads to peak broadening in real space as well as to non-physical oscillations in $G(r)$ and its related functions. Since those ripples can be confused with the physical diffraction peaks, especially in the range of smaller distances, they must be avoided to obtain a reliable analysis.

A method commonly used to deal with that problem is to modulate the experimental total interference function by a damping window function before applying the Fourier transform[18] instead of just using the step function.

In this work the normalized *sinc* function (10) has been used for that purpose.

$$\mathrm{sinc}(x) \equiv \frac{\sin \pi x}{\pi x} \tag{10}$$

where $x = q/q_{\max}$.

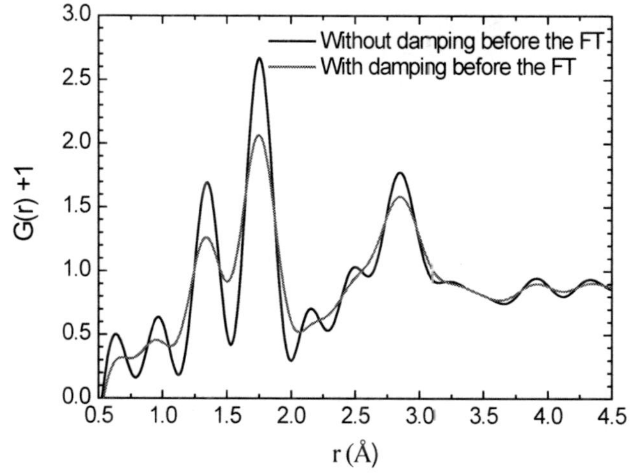

Figure 5. *Comparison of the obtained total pair-correlation function of Freon 112a at 320 K when doing the FT with or without a sinc window function.*

Although the use of this function preserves the area of the peaks in the profile of the total pair-correlation function, one of the drawbacks of this method is that not only the unphysical ripples are damped but all the oscillations are, with the consequent loss of information (Fig. 5).

An alternative method consists on Fourier-transforming the step function that describes the experimental q range and convolute it with the theoretical r-space functions before comparing or fitting them to the Fourier-transformed data.[19]

Although this is a more accurate method is somewhat more sophisticated and time consuming and it was not considered necessary in the present work.

5. Results and Discussion

The first result coming out from the powder diffraction measurement carried out at D1B yields the temperature dependence of the molecular density of the compounds and is shown in Figs. 6 and 7. The density is estimated within the bcc phase from the position of the main Bragg peaks in the diffraction pattern. The molecular densities for the liquid phase were previously measured by means of densitometry. In Freon 112 a clear change at 90 K in the slope of the molecular density yields a first signature of the main glass transition. A more subtle change in the tendency of the density is also found at 130 K. This change can be clearly seen in the inset of Fig. 6 and it corresponds to the freezing of the *gauche-trans* conformational disorder as proposed by Kishimoto et al.[4] according to specific heat measurements.

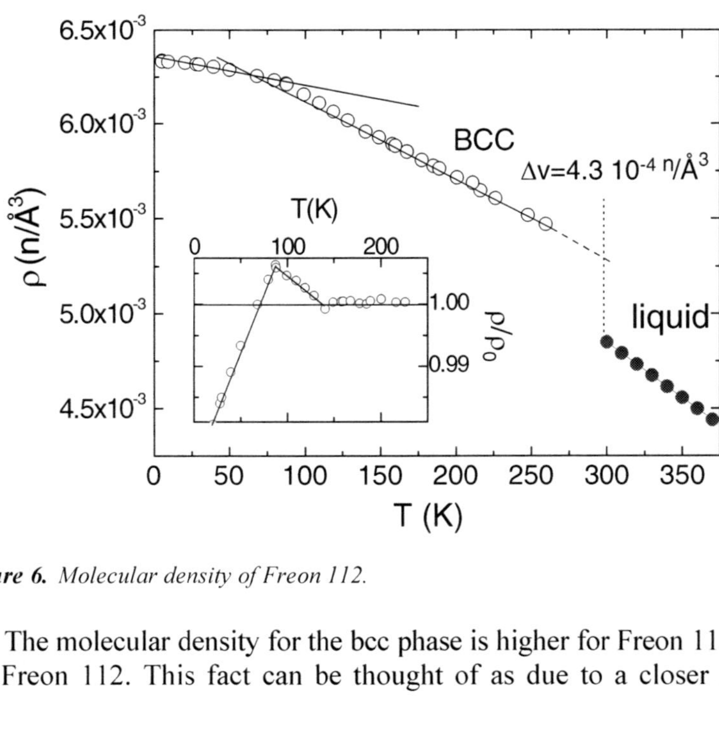

Figure 6. *Molecular density of Freon 112.*

The molecular density for the bcc phase is higher for Freon 112a (Fig. 7) than for Freon 112. This fact can be thought of as due to a closer packing in the former thanks to an easier arrangement of molecules when there is only one conformation in respect to the case when there are several.

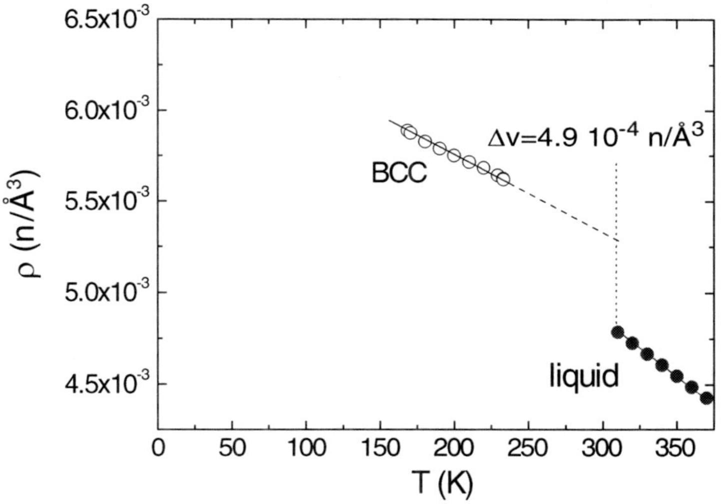

Figure 7. *Molecular density of Freon 112a.*

A series of molecular mechanics and ab initio calculations of an isolated single molecule have been performed with the Gaussian software package (using MM+ and STO-3G, 3-21G, 6-31G* and 6-31G** basis sets, respectively) to explore the expected positions of the peaks corresponding to intramolecular distances and also to determine the optimal procedure to find them.

Every calculation has been made using the results of the previous one as the initial conditions to reduce the computing time, except for the last one (6-31G**) were the results from the MM+ calculation were also used to check the robustness of the method.

Since we have considered 0.01 Å as a lower bond for the minimum distance between peaks that we can resolve, we have concluded that in future works it's not worth it to increase the computing time using the larger basis set. Within the needed precision, the use of 6-31G* will yield the best result with an optimum computing time.

The experimental intramolecular part of the total pair-correlation function at different temperatures has been plotted together with the ab initio calculation to provide a visual aid for the intramolecular distances assignment to the *G(r)* peaks for both Freon 112 and Freon 112a and is shown in Fig. 8 and Fig. 9 respectively. The total pair-correlation function of Freon 112 contains contributions from both the *trans* and *gauche* conformations, and so we expect major changes in the shape of the peaks as the concentration of each conformation changes with temperature, as indeed is observed. A closer look into the change of the position of certain peaks reveals a prominent variation of some relevant angles and distances within the molecule, occurring at the temperature of conformational freezing (Fig. 10).

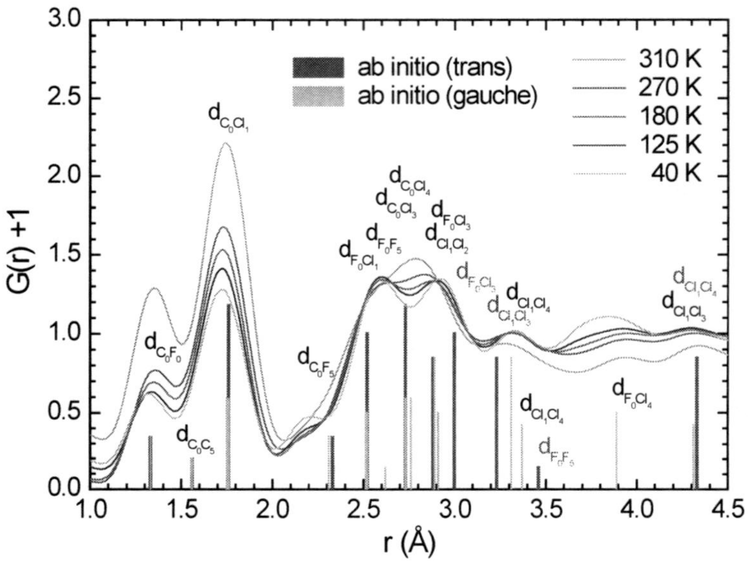

Figure 8. *Total pair-correlation function for Freon 112 together with the ab initio calculation of intramolecular distances.*

The distances C-Cl and C-F, corresponding to directly bonded pairs, and the angle Cl-C-F, show this kind of behavior, while the angle Cl-C-Cl does not seem to be sensitive to that particular temperature. The use of the relative distances within the crystalline bcc phase, $(d-d_0)/(d_T-d_0)$, instead of the plain distances, enables to scale the y-axis for an easier comparison. The spatial extent of the total pair-correlation functions corresponding to purely intermolecular distances has also been plotted for Freon 112 and 112a to show the different short range order features and is shown in Figs. 11 and 12, respectively. Rather significant changes on the total pair-correlation function profile of Freon 112 can be observed for all the temperature range. The prominent shoulder of the main peak merges with the latter at 130 K where the *gauche-trans* freezing occurs; and there is a strong variation in the position of the peak around 9 Å, that has a change in the tendency precisely at 130 K, probably indicating substantial modifications in the short range order of the compound due to the conformational disorder (see the inset on Fig. 11). In stark contrast, the changes with temperature of the short-range order pattern for Freon 112a are far milder.

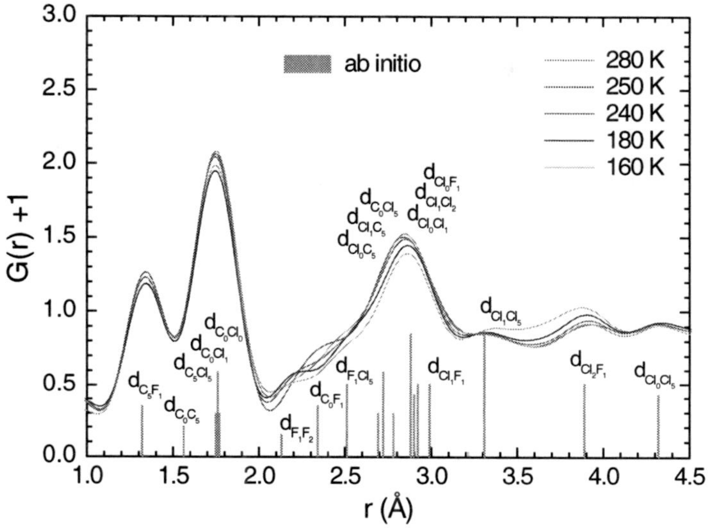

Figure 9. *Total pair-correlation function for Freon 112a together with the ab initio calculation of intramolecular distances.*

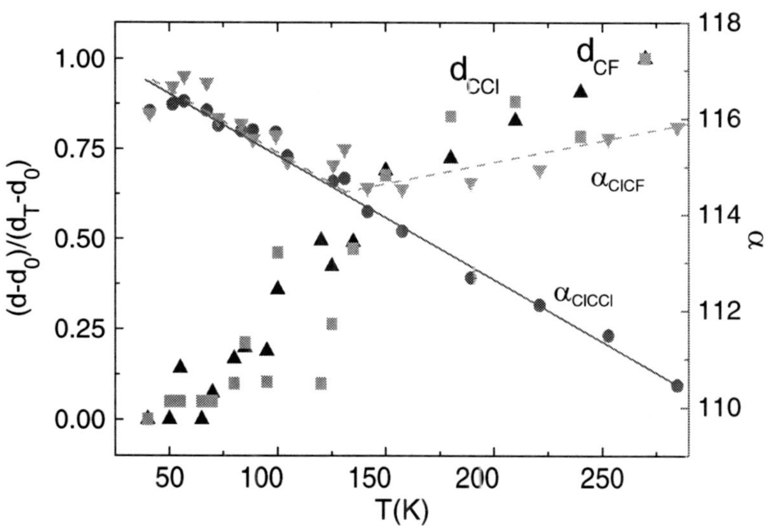

Figure 10. *Reduced distances C-Cl and C-F and angles Cl-C-Cl and Cl-C-F (d_0 and d_T are the distances at 40 K and 275 K, respectively).*

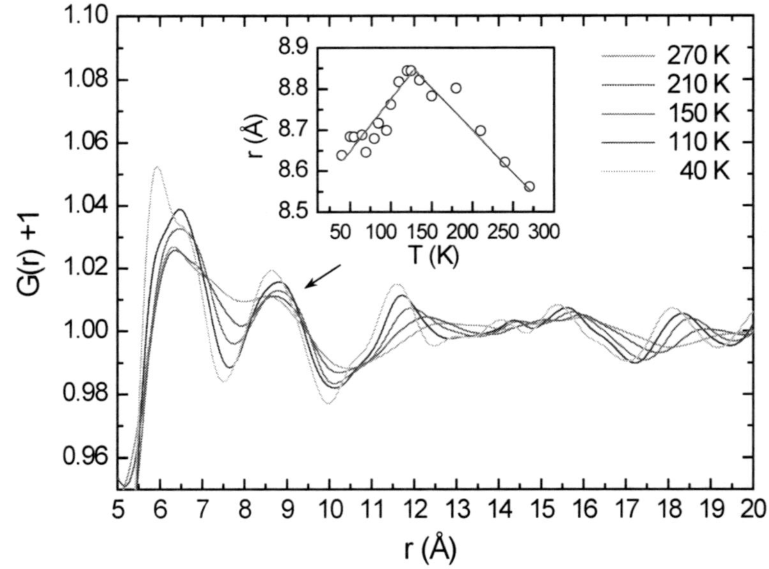

Figure 11. *Total pair-correlation function for Freon 112 in the range of intermolecular distances. Temperature dependent position of the peak around 9 Å on the inset.*

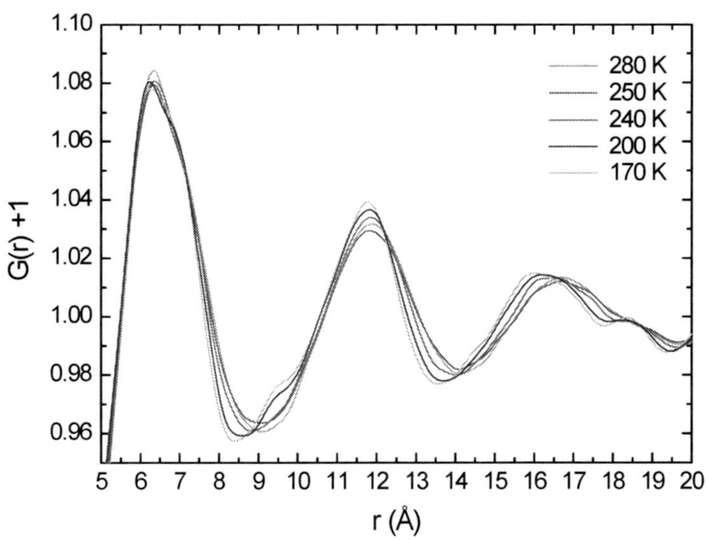

Figure 12. *Total pair-correlation function for Freon 112a in the range of intermolecular distances.*

Its total pair-correlation function appears with well-defined peaks that change only slightly its position with temperature. Besides, there are bigger and less numerous peaks in the total pair-correlation function of Freon 112a which, altogether, denotes a simpler and rather unvarying features on its short-range order.

The high fragility of glasses is linked to the idea of a more complex energy landscape.[20-21] The fact that orientational glasses are in a regular lattice (disorder being only of orientational character) usually accounts for the observation that most of them are rather strong. The case of Freon 112 is quite different because the conformational disorder provides an additional source of intramolecular and intermolecular competing interactions, giving rise to a higher complexity that could explain its outstanding fragility among the orientational glasses.

6. Conclusions

The structural changes observed for Freon 112 at all length scales provide some clues for a better understanding its outstanding properties. In particular, the main glass transition at 90 K of Freon 112 shows a clear signature in the temperature dependence of the molecular density arising from the reduced volume expansivity within the glassy state. On the other hand, the molecular density, short range order and molecular structure for Freon 112 show a sharp change at 130 K, the temperature at which the *gauche-trans* conformational freezing is taking place. In contrast, Freon 112a, with no conformational disorder, shows a mild variation of its structural properties with much more subtle changes. These results suggest that the conformational disorder of this compound plays a major role on explaining its complexity.

The differences in behaviour between the two compounds have to be ascribed to the presence in Freon 112a of relatively strong directional interactions due to its dipole moment as well as to the different molecular shapes of the two chemical isomers. These distinct behaviors unveil large differences between the potential energy surfaces of both isomers and thus exemplify how a shift in the balance between highly directional (electrostatic) and excluded-volume (vdW) interactions induced by a change in molecular topology gives rise to a whole set of differences in structure and thermodynamics.

The present results constitute a step forward in our understanding of the microscopic details that may lead to disparate behaviour in macroscopic glassy properties such as the fragility and also come into line with results given in Ref. 22 on a study of the isomeric effect concerning another glassy material.

In order to determine exactly which changes on short-range order are taking place and to determine structural molecular scenarios compatible with the experimental data, Monte Carlo and molecular dynamics simulations are being performed to try to reproduce some characteristics of its behaviour and to further understand the glass transition on this kind of materials.

Acknowledgements

This work was supported by grants from the Spanish Ministry of Science and Innovation (MICINN) (FIS2008-00837) and from the Generalitat de Catalunya (2005SGR-00535). One of us (MRE) acknowledges the PhD fellowship from MICINN. NATO Collaborative Linkage Grant CBP NUKR. CLG 982312 is also acknowledged.

References

1. Pardo, L. C., Veglio, N., Bermejo, F. J., Tamarit, J. Ll., and Cuello, G. J. (2005) *Phys. Rev. B* **72(1)**, 014206
2. Veglio, N., Bermejo, F. J., Pardo, L. C., Tamarit, J. Ll., and Cuello, G. J. (2005) *Phys. Rev. E* **72(3)**, 031502
3. Brand, R., Lunkenheimer, P., and Loidl, A. (2002) *J. Chem. Phys.* **116(23)**, 10386
4. Kishimoto, K., Suga, H., and Seki, S. (1978) *Bull. Chem. Soc. Jpn.* **51(6)**, 1691
5. Iwasaki, M., Nagase, S., and Kojima, R. (1957) *Bull. Chem. Soc. Jpn.* **30(3)**, 230
6. Kagarise, R. E., and Daasch, L. W. (1955) *J. Chem. Phys.* **23**, 113
7. Angell, C. A. (1988) *J. Phys. Chem. Solids* **49**(8), 863
8. Angell, C. A., Dworkin, A., Figuiere, P., Fuchs, A., and Szwarc, H. (1985) *J. Chim. Phys.* **82**(7-8), 773
9. Drozd-Rzoska, A., Rzoska, S. J., Pawlus, S., and Tamarit, J. Ll. (2006) *Phys. Rev. B* **73**(22), 224205
10. Drozd-Rzoska, A., Rzoska, S. J., Pawlus, S., and Tamarit, J. Ll. (2006) *Phys. Rev. B* **74**(6), 064201
11. Mondal, P., Lunkenheimer, P., Böhmer, R., Loidl, A., Gugenberger, F., Adelmann, P., and Meingast, C. (1994) *J. Non-Cryst. Solids* **172**, 468
12. Pardo, L. C., Lunkenheimer, P., and Loidl, A. (2006) *J. Chem. Phys.* **124**(12), 4911
13. Fischer, H. E., Barnes, A. C., and Salmon, P. S. (2006) *Rep. Prog. Phys.* **69**(1), 233
14. Puertas, R., Rute, M. A., Salud, J., López, D. O., Diez, S., Kees van Miltenburg, J., Pardo, L. C., Tamarit, J. Ll., and Barrio, M. (2004) *Phys. Rev. B* **69**(22), 224202

15. Puertas, R., Salud, J., López, D. O., Rute, M. A., Diez, S., Tamarit, J. Ll., Barrio, M., Pérez-Jubindo, M. A., de la Fuente, M. R., and Pardo, L. C. (2005) *Chem. Phys. Lett.* **401**(4-6), 368
16. Howe, M., McGreevy, R., and Zetterström, P. (1966) Computer code CORRECT, Correction program for neutron diffraction data, in *NFL Studsvik internal report*
17. Placzek, G. (1952) *Phys. Rev.* **86**(3), 377
18. Lorch, E. (1969) *J. Phys. C Solid State Phys.* **2**, 229
19. Petri, I., Salmon, P. S., and Fischer, H. E. (2000) *Phys. Rev. Lett.* **84**(11), 2413
20. Debenedetti, P. G., and Stillinger, F. H. (2001) *Nature* **410**, 259
21. Shintani, H., and Tanaka, H. (2006) *Nature Phys.* **2**(3), 200
22. Talón, C., F. J., Bermejo, Cabrillo, C., Cuello, G .J., González, M., Richardson Jr. J. W., Criado, A., Ramos M. A., Vieira, S., Cumbrera, F., and González, L. M. (2002) *Phys. Rev. Lett.* **88**, 0115506

A PROCEDURE TO QUANTIFY THE SHORT RANGE ORDER OF DISORDERED PHASES

LUIS CARLOS PARDO, MURIEL ROVIRA-ESTEVA, JOSEP LLUIS TAMARIT, NESTOR VEGLIO
Group of Characterization of Materials, Department of Physics and Nuclear Engineering, ETSEIB, Diagonal 647, 08028 Barcelona. Universitat Politècnica de Catalunya, Catalonia, Spain

FRANCISCO JAVIER BERMEJO
CSIC–Department Electricity and Electronics, UPV/EHU, Box 644, 4880 Bilbao, Spain

GABRIEL JULIO CUELLO
Institut Laue Langevin, 6 Rue Jules Horowitz, Boîte Postal 156x, F-38042 Grenoble Cedex 9, France

Abstract: Determination of the short- and intermediate-range structure of disordered materials is a necessary step to fully understand their properties. Despite of this, no generally accepted procedure exists to date to extract structural information from diffraction data. In this paper we describe a method which enables determination of the short-range structure of disordered molecular phases. This general method is applied to one of the first studied molecular liquids, carbon tetrachloride, and to its plastic phase being able to unravel the so called local density paradox: although molecules are closer in the liquid than in the plastic phase, the density of the former is lower than that of the later. The analysis of the short range order in both phases shows that although the minimal energy configuration allows a closer approach of molecules, it hinders the formation of the face centered cubic long range ordered lattice due to the difficulty of molecules to form stacked structures.

Keywords: short range order, liquid structure, plastic phase, disordered phases, neutron diffraction

1. Introduction

Crystallography is a well established science that allows, among other things, to extract structural information of an ordered arrangement of atoms from a diffraction experiment. In fact, the knowledge of its structure is fundamental in order to characterize a crystalline phase and this helps in the comprehension of a large amount of physical data, from dynamics to thermodynamics. This is done by means of well established procedures (as Rietveld refinement) supported by a huge amount of software. Astonishingly enough this is not the case for disordered phases such as liquids or plastic phases (a phase where, although the centers of mass of the molecules are in long range ordered positions, the molecules rotate more or less freely) and their non-ergodic associated states, or in other words, their glasses. In these cases physical data is usually interpreted not taking into account the Short Range Order (SRO) structure, probably simply due to the lack of a well established procedure to determine their structures.

In this chapter we offer a method to fully determine the structure of disordered phases, from the molecular structure to the SRO, and we apply it to the determination of the structure of one of the first molecular liquids studied ever, carbon tetrachloride, and its not so well studied plastic phase. Concerning the SRO of the liquid phase, regarded as the most probable dimmer configuration between two "close" molecules, no consensus exists so far. Considering the CCl_4 molecule as a tetrahedron, where the chlorine atoms are sitting in its corners and the central carbon atom is equidistant to those corners, we find in the literature the configurations face to face,[1] corner to face (also called Apollo),[2] corner to corner,[3] and edge to edge or interlocked.[4] However a recent work of Rey[5], where for the first time a clear quantitative definition of the aforementioned configurations is presented, demonstrates that although the edge to edge (or interlocked) configuration (as defined there) clearly dominates, other configurations such as face to face are also possible at very short distances between molecules. Concerning the plastic phase, all the members of the methyl-halogenomethane family ($CCl_n(CH_3)_{(4-n)}$ n = 0,1,2,3,4 including CCl_4 (n=4) has a phase transition from the liquid phase to a plastic phase with a high-symmetry lattice (cubic or rhombohedral),[6] which in the case of CCl_4 is Face Centered Cubic (FCC).[7]

2. Experiments and data treatment

A series of neutron diffraction experiments were carried out using the D1b diffractometer at the Institute Laue Langevin, Grenoble, France. The instrument is a general purpose powder diffractometer which uses a banana-shaped detector covering a wide angular range. The measurements were performed using a wavelength λ=1.2805 Å which, combined with data acquisition at two different detector positions, allowed us to cover a q-range up to 8 $Å^{-1}$ large enough to study the structure factor, account made of the small intermolecular distances. Details concerning the instrument settings and data correction procedures (inelastic contributions, multiple scattering, detector efficiency, self absorption, and normalization to a known vanadium sample) are given elsewhere.[8] Concerning the plastic phase, the growth of a polycrystalline

FCC phase was ascertained by the emergence of a set of crystalline Bragg peaks as described previously.[8] Experimental results for the liquid phase are shown in Fig. 1a, as well as the total radial distribution function for the FCC phase (calculated as $G(r) = 1 + \rho_o^{-1}(2\pi)^{-3}$ FT [S(q) − 1], where FT means Fourier Transform). For a description of data treatment see in this series the paper entitled "Neutron diffraction as a tool to explore the free energy landscape in orientationally disordered phases" or ref. ([9]).

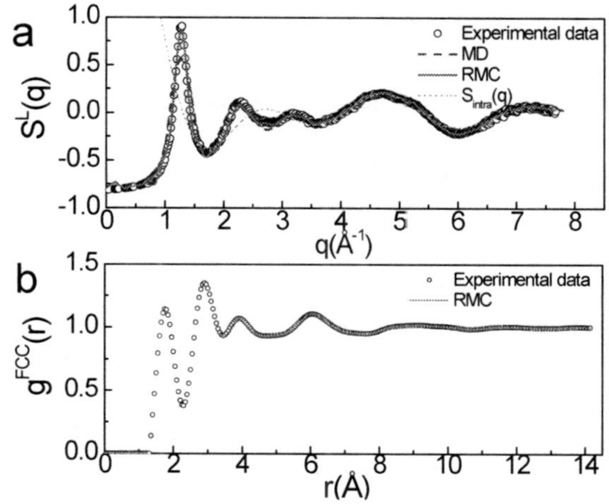

Figure 1. *a) Scattering function for liquid Carbon Tetrachloride together with its determination by means of molecular dynamics and Reverse Monte Carlo (RMC). With dotted lines we show the determination of the intramolecular structural parameters determined by the Bayesian method described in the text. b) Total radial distribution function for the FCC phase, together with the results of RMC.*

3. Extracting the molecular structure from data

Total scattering function S(q) has two terms corresponding to distances between atoms at two different length scales: a long scale contribution from intermolecular distances that is the most important contribution in the low-q region of S(q), and a short scale contribution which will mainly contribute in the high-q region of the scattering function. A common method to determine the molecular structure is to fit the scattering function using the expression:

$$S(q) = \sum_{i,j}^{m} b_i b_j \cdot \frac{\sin(qr_{ij})}{qr_{ij}} \cdot e^{-u_{ij}^2 q^2 / 2} \qquad (1)$$

where r_{ij} are the intramolecular distances and u^2_{ij} are the mean square displacements between i and j atoms (the Debye-Waller term, see ref. (([9])).

In order to fit experimental data, a standard Levenberg-Marquardt (LM) method for minimizing χ^2 is usually performed. This method has two main drawbacks:

it can get stuck in local minima of the parameter-χ^2 space and therefore must be initialized using values close enough to the final parameters somehow inferred beforehand. In addition the method supposes that there is only one minimum in the parameter-χ^2 space (i.e. the solution is not multimodal) and that this minimum must have a quadratic form on all parameters involved. The first problem can be avoided using a "shake" algorithm, i.e., once the LM procedure is stopped at some point, the fitting is repeated from a close set of parameters in order to assure the robustness of the result. The second problem is unavoidable using a minimum χ^2 approach. It implies that LM can only deal with symmetric errors around the highest probable value of parameters, and that the correlation between parameters can only be lineal. Moreover, even if the χ^2 minimum is in fact quadratic, the procedure makes very difficult to take into consideration the correlation between parameters. To obtain them the covariant matrix should be diagonalized, being the eigenvalues the real error of the parameters along the eigenvectors defined by linear combinations of parameters. This implies that in many works errors are calculated under the hypothesis that parameters are independent, and for this reason are underestimated.

An alternative way to reach the minimum of χ^2 is the use of the Bayesian approach, which deals with the direct determination of the Probability Distribution Function (PDF) for the final parameters. This method has the advantage that all the parameter space compatible with the experimental error is explored and therefore correlation between parameters and multimodal minima in the parameter-χ^2 space are naturally taken into account. For a review on Bayesian methods the reader is refereed to the excellent monograph of Sivia et al. (Ref.[10]), while in this work we will only briefly explain the method used for our specific problem.

In order to obtain the PDF for the parameters, we have used a Markov Chain Monte Carlo method to explore the parameter space. This method is similar to a classical Monte Carlo simulation where the distance between calculated and experimental data ($(y_{calc}-y_{exp})^2$) plays the role of energy and the experimental error (σ) plays the role of temperature. The method (as employed in this work) is based on the hypothesis that experimental data have a Gaussian distribution around the real value. This is true for a counting experiment with large enough amount of counts, because in this case a Poisson distribution can be approximated by a Gaussian one. Therefore the probability that a function with given fitting parameters (D_k) (what is called the "Hypothesis" in Bayes theorem) is describing your n experimental data points (x_k) with an error σ_k (i.e. data $\mathbf{x_k}$ supposed to be normally distributed around D_k with a standard deviation of σ_k) can be expressed as:

$$L \propto prob(x_k / D_k) = \prod_k^n e^{-\frac{1}{2}\left(\frac{x_k - D_k}{\sigma_k}\right)^2} = e^{-\frac{1}{2}\sum_k^n\left(\frac{x_k - D_k}{\sigma_k}\right)^2} = e^{-\frac{\chi^2}{2}} \quad (2)$$

This is usually called the likelihood of the hypothesis, or in other words the probability that the fitted function with a given set of parameters is describing the data within the experimental error. In regard to the previous expression, it is easy to see that in fact minimizing χ^2 is just maximizing the likelihood ($\log L \propto -\chi^2/2$) when a Gaussian distribution is supposed for data. However, the previous method for fitting functions is more general because other distributions than Gaussian can also be used, as in the case of an experiment with a low count rate where the Gaussian approximation to the Poisson distribution is no more applicable.

In order to obtain the PDFs, starting from a set of parameters that minimizes χ^2, we generate randomly a new set of them, being the change between the new and old accepted having in to account the likelihoods of the two parameter sets

$$\frac{prob\left(x_k / D_k^{new}\right)}{prob\left(x_k / D_k^{old}\right)} = e^{-\frac{(\chi^2_{new}-\chi^2_{old})}{2}} \tag{3}$$

where $D_k^{new(old)}$ are the points generated using the new (old) set of parameters.

In Fig. 2 we show the PDFs obtained for the parameters fitted to the intra-molecular structure of CCl_4 using equation (1): the distance between the carbon and chlorine atom, and the mean square displacement between the C-Cl and Cl-Cl atom pairs (Cl-Cl distance can be calculated form the tetrahedral geometry of the molecule). In Fig. 1a we show the fitted function together with experimental data, taking the highest probable parameters obtained by the proposed Bayesian method: d_{CCl}=1.768±0.004 Å; u_{ClCl}=0.18±0.01 Å^2 and u_{CCl}=0.06±0.02 Å^2. Having a careful look at the last parameter u_{CCl} depicted in Fig. 2 we can see that the probability distribution is not symmetric, being more probable for this parameter to be smaller than larger with respect to the highest probable one. In the case of the simple molecule studied in this work, the proposed method gives the real error of parameters and improves the robustness of their determination in comparison with the standard minimum χ^2 method, but will give the same final parameters as the LM algorithm. We have seen however that for more complicated molecules only the Bayesian method exposed in this work is able to give reasonable results, avoiding the aforementioned danger of LM algorithm being stuck at a local minimum. (as the ones studied in the article "neutron diffraction as a tool to explore the free energy landscape in orientationally disordered phases" in this series, and ref. ([11])

4. Extracting the short range order from data

In order to extract the configurations to be analyzed in the next section, we have both taken the results obtained in a previous Molecular Dynamics (MD) simulation (details on the simulation are given elsewhere[4], and also performed a Reverse Monte Carlo (RMC) analysis of the obtained data on both phases, liquid and ODIC.[8] It should be kept in mind that the two methods are from first principles completely different, needing the first one (MD) a priori information about the intermolecular interaction potentials.

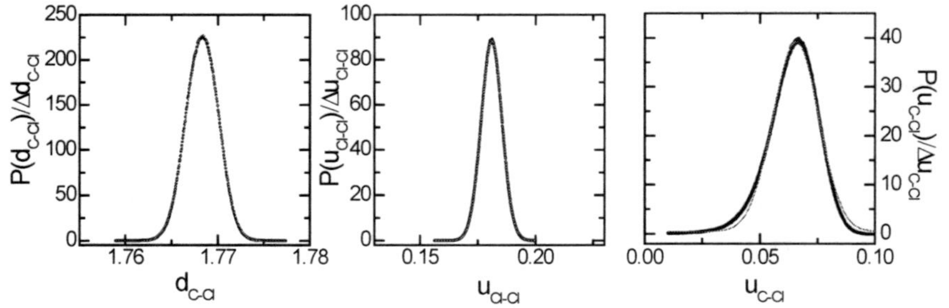

Figure 2. *Normalized PDFs for the intramolecular structural parameters, together with the fitting of a Gaussian function. The maximum of each PDF, i.e. the most probable parameter value, has been used to calculate the intramolecular structure function of Fig. 1.*

In the case of RMC no initial information is needed (except from macroscopic density), and only the experimental result is driving the algorithm to find a final configuration compatible with the experimental data. It should also be pointed out that a third way combining the advantages of the two methods is possible, the so called Empirical Potential Structure Refinement (EPSR)[13], but this method has not been used in the present work. Flexible molecules have been used in both cases (MD and RMC), being their initial geometry determined by ab initio calculations.[8] The total structure factor for the liquid at 298 K was analyzed by RMC method using a simulation box composed by 1000 molecules, with dimensions settled to reproduce the experimental density of the liquid (L=54.34 Å). As it has been shown in Fig. 1, the agreement between the spectra simulated from the RMC and MD configurations and the experimental S(q) is excellent. In what concerns the ODIC phase, a RMC simulation at 240 K has been performed using a box containing 6x6x6 cells, with a length extracted from the Bragg peaks appearing in the spectra (L=50 Å). In this case, the fitting was performed using the total radial distribution function as in ref. 14. The RMC fitting has been performed allowing only small-amplitude motions of the molecular centers about the lattice sites defined by the Bragg peaks, and changing the orientations of the molecules. The agreement between fitted and experimental data can be seen in Fig. 1(b) (for further details, see Ref. 8).

5. Analysis of final configurations

Positional ordering

In Fig. 3 we have plotted the Partial Radial Distribution Function (PRDF) $g_{CC}(r)$ for the molecular centers ($\rho g_{\alpha\beta}(r)dr$ is defined as the probability of finding a molecule β within the shell rdr surrounding an atom α, being ρ the macroscopic density). Astonishingly enough, although the macroscopic density is larger in the FCC plastic phase than in the liquid phase, molecules are closer in the second phase. Moreover, the maximum of the PRDF is located at higher distances independently of the used method to obtain the molecular centers PRDF. This "local density paradox" adds new interest in studying the SRO of the two disordered phases, and only a careful method taking into account the 3D SRO would lead to a correct answer.

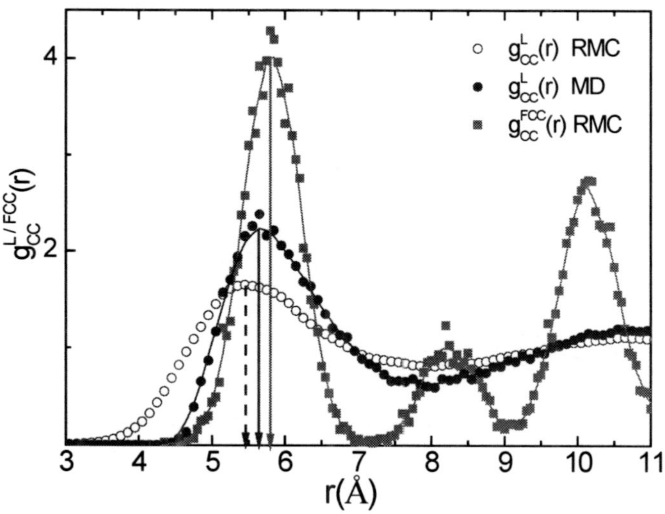

Figure 3. *Molecular Coordination Number (MCN) as a function of the distance, for the liquid phase (filled circles MD, empty circles RMC) and the FCC phase (squares). Arrows show the maxima of MCN for the aforementioned phases, i.e., the maximum of local density.*

To obtain the SRO, we have taken the idea of the bivariate analysis, used to study the molecular ordering at interfaces between liquids and vapors [15], and we have applied it to the study of liquid and plastic phases of CCl_4. In order to locate the position of a second molecule from a central one, three orthogonal axis must be defined in relation with the molecular structure. In our case the z-axis is set along the direction of a C-Cl bond, being another C-Cl bond in the zy-plane (see Fig. 4). Using this convention we can calculate the azimuthal angle θ as the scalar product between the unit vector along the intramolecular C-Cl bond defined as the z-axis and the intermolecular C-C distance joining the molecular centers of two molecules.

$$\theta = \vec{r}_{C_1-Cl_{1i}} \cdot \vec{r}_{C_1-C_2} \tag{4}$$

being C_i (i=1,2) the carbon atom from the reference molecule or from the molecule the position of which is to be calculated respectively, and Cl_{ij} (i=1,2, j=1,4) one of the four *i(j)* chlorine atoms of molecule *i(j)*.

In the same way we can define the equatorial angle φ as the scalar product between the unit vectors perpendicular to two planes: the zy-plane defined by two different intramolecular C-Cl vectors and the plane defined by the z-axis and the intermolecular C-C vector:

$$\phi = \vec{r}_{C_1-Cl_{1i}} \times \vec{r}_{C_1-Cl_{1j}} \cdot \vec{r}_{C_1-Cl_{1i}} \times \vec{r}_{C_1-C_2} \tag{5}$$

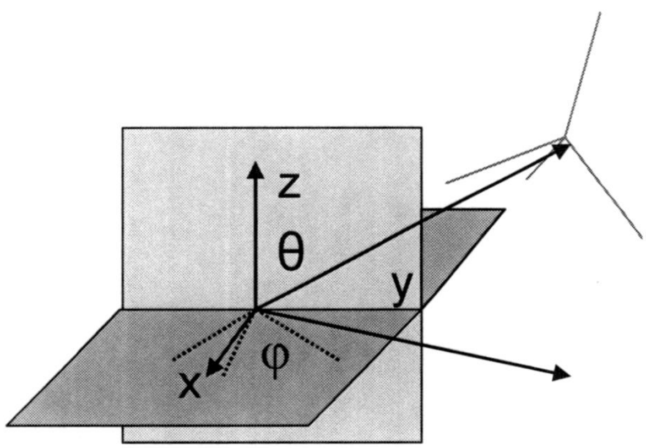

Figure 4. *Chosen axis in order to calculate the positional ordering of two CCl_4 molecules.*

In Fig. 5 we show the probability of finding a molecule at a position determined by the equatorial and azimuthal angles (ϕ,$cos(\theta)$) for the first four neighbors (a,c) and therefore molecular distances r_{CC} less than 5.73 Å and 5.63 Å for MD and RMC configurations respectively. The same is depicted in Figs. 5b and 5d also for MD and RMC but in this case for the next four-molecule shell ($5.63<r_{CC}<6.5$ Å and $5.63<r_{CC}<6.6$ Å). As it can be immediately appreciated, although RMC gives a much more disordered distributions of molecules, for the first molecules the SRO obtained using the two methods is the same, being both completely independent. Unfortunately, because RMC method is a Maximum Entropy method, i.e. it gives the most disordered solution compatible with experiment and therefore the most probable, no clear SRO is appreciated for long distances, and therefore for the liquid phase the discussion will be focused on the MD configurations (see Fig. 6). On the contrary, in the case of the FCC phase, because a positional ordering already exists, the SRO is clearly determined by the RMC method (see Fig. 6c).

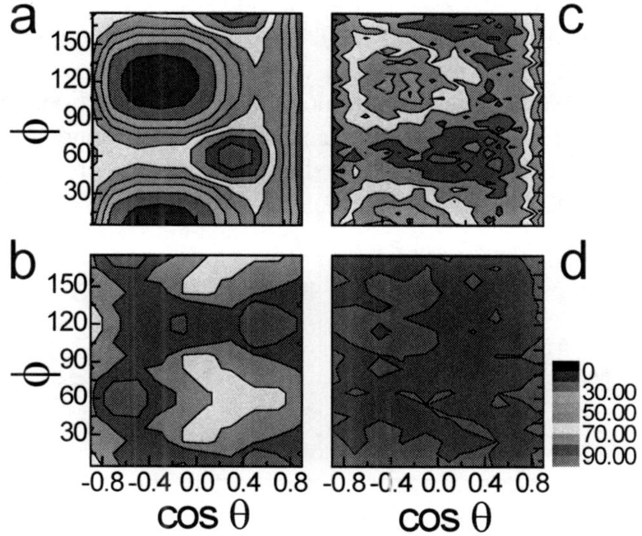

Figure 5. *(Color online) Comparison of the ordering obtained for the liquid phase by means of MD (a,b) and RMC (c,d) for the first four molecules surrounding a central one (a MD,c RMC), and for the next four molecules (b MD, d RMC). The color scale represents the normalized probability of finding the molecule at a given position ($P(\phi,cos(\theta))/P_{max}$).*

We define the First Molecular Coordination Shell (FMCS) as the molecules within the first peak of the C-C PRDF, i.e. for distances r_{cc}<7.5 Å (see Fig. 2). In Fig. 6 we have depicted the probability of finding a molecule at a position ($\phi,cos(\theta)$) for successive shells surrounding a central molecule within the FMCS containing only four molecules each: in Fig. 6a we have therefore the positional ordering for the first four molecules, in Fig. 6b for the next four molecules and in Fig. 6c for the last four molecules inside the FMCS. A glance to Fig. 6 clearly reveals that the distribution of molecules changes as a function of their distance to the central molecule even inside the FMCS. Moreover, molecules in a shell tend to fill the gaps left by the molecules in a shell closer to the central molecule. Taking into account the tetrahedral symmetry of the molecule, the first four neighbors are sitting in the faces of the central molecule (Fig. 6a), the next four neighbors in the edges and a small fraction in the corners (Figs. 6a,b), and the last four neighbors in the corners of the central molecule (Fig. 6c). On the other hand, for the FCC phase, due to its positional ordered nature, we must consider the 12 closer molecules at a distance $r\approx a/\sqrt{2}$. In fact what is represented in Fig. 6c is the relative orientation of the molecule with respect to the lattice axes, since the positional ordering of molecular centers is fixed in the FCC lattice. Our results are in complete agreement with previous MD simulations[2,16] where it is asserted that C-Cl vectors lie along [110] and [100] directions. To make this point clear, we can see in Fig. 6c spots at $(cos\theta,\phi) = (-0.33,0°)$ and $(0.33,60°)$, which means that molecules are placed in the corners and faces with respect of the first neighbors, which lie in the [110] directions. Molecules oriented along the [100] directions are represented at the spots (0.7,0°) and (−0.7,60°) and would correspond to the molecules placed in the edge of the molecules in the liquid phase represented by the large spots at about (0.58,0°) and (−0.58,60°).

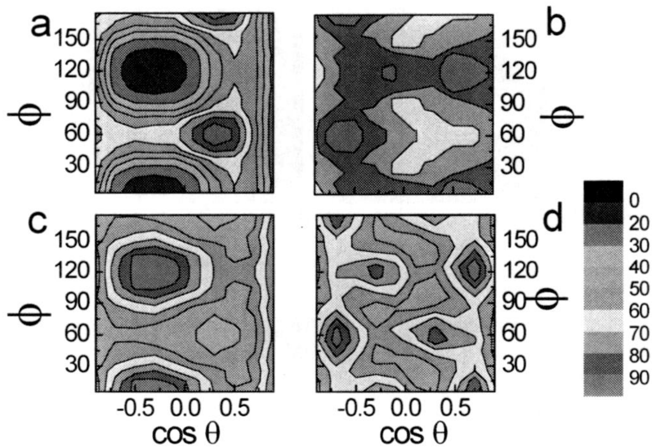

Figure 6. (Color online) Comparison of the positional ordering obtained for the liquid phase (a,b,c) and for the FCC phase (d) by means of MD simulation. Color scale is defined as in Fig. 5.

Orientational ordering

In order to extract the maximum quantitative information about the relative orientation of molecules we have calculated an histogram of the angle between all possible combinations of C-Cl vectors of two different molecules

$$\alpha = \vec{r}_{C_1-Cl_{1i}} \cdot \vec{r}_{C_2-Cl_{2j}} \qquad (5)$$

Nevertheless if only the probability $P(\alpha)$ of finding an angle α between two C-Cl vectors is studied, the information about the position of molecules is lost, and therefore we would add orientational information for molecules that are placed at different points. For this reason in Fig. 7 we have plotted the probability $P(\alpha)$ as a function of the cosine of their azimuthal angle $P(cos\theta,cos\alpha)$, and therefore we would be eventually able to distinguish between different orientations of molecules sitting in different places. Nevertheless that is not the case as it can be seen in Fig. 7, where the spots found in the $cos\alpha$ are independent of that found in the $cos\theta$, that is, the position. Lets now define molecules that are "parallel" oriented as those with C-Cl vectors of different molecules pointing in the same direction (therefore $cos\alpha$ =1), and "antiparallel" as those with C-Cl vectors pointing in opposite directions (therefore $cos\alpha$ =-1). Then the first four molecules surrounding the central one are antiparallel oriented, the next four parallel, and the last four molecules within the FMCS are again antiparallel oriented, standing for the spots at $cos\alpha$ = -1, $cos\alpha$ = 1 and $cos\alpha$ = -1, respectively. For the case of the FCC phase, molecules are parallel oriented irrespective of their position, as can be seen in Fig. 7d.

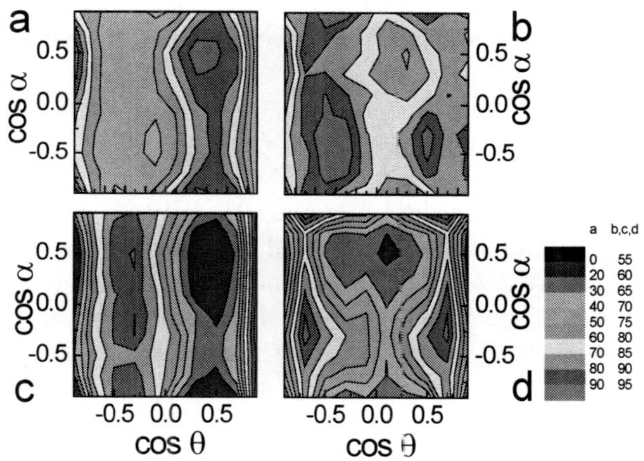

Figure 7. *Comparison of the orientational ordering obtained for the liquid phase (a,b,c) and for the FCC phase (d) by means of MD simulation. Color scale is defined as in Fig. 5.*

Regarding the previous works carried out on CCl_4 in which different or even contradictory molecular arrangements are proposed, and taking into account the present analysis, we can understand the origins of the controversy on the molecular arrangement of this simple molecular liquid. First, the structure of the liquid is distance-dependent and an analysis of the whole first FMCS will lead to wrong results, and second, a bivariate analysis, or any other analysis taking into account the three spatial degrees of freedom of the SRO, must be performed to obtain SRO and must avoid collapsing information in 1-D representations. Joining the aforementioned results obtained for positional and orientational ordering of CCl_4 molecules in the liquid phase we can tentatively assign some configurations found in the literature to molecular arrangements although, as pointed out by Rey[4] only a quantitative definition of those molecular arrangements makes fully sense. Then we can characterize the molecular arrangement of the first four molecules in the liquid as the face to face configuration [1], the next four molecules would be in an Apollo [2] or edge to edge configuration (interlocked of ref.[4]), and the last four molecules of the FMCS in a corner to corner configuration [3]. On the contrary, for the case of the plastic phase, no matter how molecules are positionally arranged, C-Cl vectors are parallel giving rise to Apollo or edge to edge like configurations. This fact provides us a hint to solve the aforementioned "local density paradox". The minimum energy dimmer arrangement is so that faces of the tetrahedra are touching each other (face to face arrangement), but such an arrangement avoids the possibility of a long range ordered lattice. On the contrary for the FCC phase, although closest molecules are not arranged in a minimum energy configuration, they are in all cases arranged in a parallel way, allowing therefore the formation of molecular stacking. It must be borne in mind that this ordering is dynamic in nature, but favoring the aforementioned parallel orientation of molecules.

6. Conclusions

In this paper we have described a method to fully characterize the SRO in disordered phases. The Bayesian method exposed in the first part of this work allows a robust determination of molecular parameters, and a clearly defined calculation of the errors on the basis of probability theory. In addition we offer an easy method to extract from molecular configurations the SRO (positional and orientational). Using these two methods we have been able to unravel the structure of the liquid phase for carbon tetrachloride, which has been revealed to be richer than previously thought within the first molecular coordination shell. In this phase positional and orientational short range ordering of molecules simply tries to minimize the energy, the molecules filling the gaps left by successive shells of molecules, and therefore changing the molecular arrangement along shell distance. The situation is different for the plastic phase, for which not only minimizing the energy plays a role in the SRO, but also the possibility of forming molecular stacking, and therefore allowing the formation of a positional long range ordered structure.

Acknowledgements

We would like to acknowledge R. Rey for instructive discussions, and also to allowing us to use the MD configurations obtained in ref. 4. This work was supported by grants from the Spanish Ministry of Science and Innovation (MICINN) (FIS2008-00837) and from the Generalitat de Catalunya (2005SGR-00535). One of us (MRE) acknowledges the PhD fellowship from MICINN. NATO Collaborative Linkage Grant CBP NUKR. CLG 982312 is also acknowledged. We thank ILL and Spanish CRG-D1B for allocating neutron beam time.

References

1. Lowden L. J., and Chandler D. (1974) *J. Chem. Phys.* **61**(12), 5228; Jedlovszky P. (1997) *J. Chem. Phys.* **107**(18), 7433
2. McDonald I., Bounds D. G., and Klein M. L. (1982) *Mol. Phys.* **45**(3), 521
3. Jovari P., Meszaros G., Pusztai L., and Svab E. (2001), *J. Chem. Phys.* **114**, 8082
4. Rey R., Pardo L.C., Llanta E., Ando K., López D. O., Tamarit J. Ll., and Barrio M. (2001) *J. Chem. Phys.* **112**(17), 7505
5. Rey R. J. (2007) *Chem. Phys.* **126**(16), 164506
6. Pardo L. C., Barrio M., Tamarit J. Ll., López D. O., Salud J., and Oonk H. A. J. (2005) *Chem. Mater.* **17**, 6146; Pardo L. C., Barrio M., Tamarit J. Ll., López D. O., Salud J., Negrier P., and Mondieig D. (2001) *J. Phys. Chem.* **105**, 10326
7. McDonald I., Bounds D. G., and Klein M. L. (1982) *Mol. Phys.* 45, 521; Breymann W. and Pick R. M. (1989) *J. Chem. Phys.* **91**, 3119; More M., Lefebvre J., Hennion B., Powell B. M., and Zeyen C. M. E. (1980) *J. Phys. C* **13**, 2833
8. Veglio N., Bermejo F. J., Pardo L. C., Tamarit J. L., and Cuello G. J. (2005) *Phys. Rev. E* **72**, 031502; Pardo L. C., Veglio N., Bermejo F. J., Tamarit J. L., and Cuello G. J. (2005) *Phys. Rev. B* **72**, 014206
9. Cuello G. J. (2008), *J. Phys.: Cond. Matt.* **20**, 244109; Fischer E, Barnes A. C., and Salmon P. S. (2006) *Rep. Prog. Phys.* **69**, 233-299; Talón C., Bermejo F. J., Cabrillo C., Cuello G. J., González M. A., Richardson Jr J. W., Criado A., Ramos M. A., Vieira S., Cumbrera F. L., and González L. M. (2002), *Phys. Rev. Lett.* **88**, 115506-1-4
10. Sivia D., and Skilling J., Data Analysis: A Bayesian Tutorial, Oxford University Press isbn: 978-0-19-856832-2 (2006)
11. Pardo L. C., Bermejo F. J., Tamarit J. Ll., Cuello G. J., Lunkenheimer P., and Loidl A. (2007) *J. Non-Crys. Sol.* **353** (8-10), 999-1001
12. For reviews on applications of the method see McGreevy R. L. (2001) *J. Phys.: Condens. Matter* **13**, R877; Evrard G., and Pusztai L. (2005) *J. Phys.: Condens. Matter (Special Issue)* **17**, S1
13. see http://www.isis.rl.ac.uk/Disordered/DMGroup/DM_epsr.htm and references therein
14. Karlsson L., and McGreevy R. L. (1997) *Physica B*, **100**, 234-236,
15. Jedlovszky P., Vincze A., and Horvai G. (2004) *Phys. Chem. Chem. Phys.* **6**, 1874
16. Breymann W., and Pick R. M. (1989) *J. Chem. Phys.* **91**, 3119; More M., Lefebvre J., Hennion B., Powell B. M., and Zeyen, C.M. E. (1980) *J. Phys. C* **13**, 2833; Rey R. (2008) *J. Phys. Chem. B*, **112**(2), 344-357

CONSISTENCY OF THE VOGEL – FULCHER – TAMMANN (VFT) EQUATIONS FOR THE TEMPERATURE-, PRESSURE-, VOLUME- AND DENSITY- RELATED EVOLUTIONS OF DYNAMIC PROPERTIES IN SUPERCOOLED AND SUPERPRESSED GLASS FORMING LIQUIDS/SYSTEMS

ALEKSANDRA DROZD-RZOSKA AND SYLWESTER J. RZOSKA
Institute of Physics, University of Silesia, ul. Uniwersytecka 4, 40-007 Katowice, Poland; e-mail: rzoska@us.edu.pl

Abstract: A consistent set of temperature- (T), pressure- (P), volume- (V) and density- (ρ) related VFT-type equations for portraying the evolution of the structural relaxation time or viscosity is presented, namely:

$$\tau(P) = \tau_0 \exp\left[D_P\left(P - P_{SL}\right)/\left(P_0 - P\right)\right]$$

$$\tau(T) = \tau_0 \exp\left[D_T\left(T_{SL} - T\right)\left(T_0/T_{SL}\right)/\left(T - T_0\right)\right],$$

$$\tau(\rho) = \tau_0 \exp\left[D_\rho\left(\rho - \rho_{SL}\right)/\left(\rho_0 - \rho\right)\right]$$

and $\tau(V) = \tau_0 \exp\left[D_T\left(V_{SL} - V\right)\left(V_0/V_{SL}\right)/\left(V - V_0\right)\right]$, where T_0, P_0, V_0 and ρ_0 are VFT estimates of the ideal glass loci and T_{SL}, P_{SL}, V_{SL} and ρ_{SL} are estimates of the location of the absolute stability limit, partially hidden in the negative pressures domain ($P < 0$). For these equations prefactors are well defined via $\tau_0 = \tau\left(T_{SL}, P_{SL}, V_{SL}, \rho_{SL}\right)$, i.e. they are linked to the absolute stability limit loci (gas-liquid spinodal). Noteworthy is their smooth transformation into VFT-type equations, used so far, on approaching the glass transition, and into Arrhenius-type equations remote from the glass transition, on approaching the absolute stability limit. The latter may suggest the re-examination of experimental data suggesting the VFT-to-Arrhenius crossover far away from the glass transition. Novel VFT counterparts also lead to the consistent set of fragility strength coefficients (D_T, D_P, D_V, D_ρ) and fragilities associated with the slope (steepness index) at appropriate "Angell plot" counterparts.

Keywords: glass transition, dynamics, Vogel-Fulcher-Tammann counterparts, negative pressures, fragility

1. Introduction

Glass transition physics is one greatest challenges of the soft condensed matter physics and the modern material engineering. The basic artifact of this phenomenon are dynamics-related precursors appearing already well above the glass temperature (T_g), in the metastable, supercooled liquid state. This "pretransitional" behavior shows a variety of "universal" features, shared amongst microscopically different systems.[1-7] The possibility of portraying the temperature evolution of dynamic properties via the Williams-Landolt-Ferry or equivalently the Vogel-Fulcher-Tammann (VFT), known also as Vogel-Fulcher-Tammann-Hesse (VFTH), dependences is the often suggested as an example of such behavior.[1-6] At present, the VFT equation is applied in the form:[1-10]

$$\tau(T)=\tau_0 \exp\left[\frac{B_{VFT}}{T-T_0}\right]=\tau_0 \exp\left[\frac{D_T T_0}{T-T_0}\right] , \; T>T_g \;\; \text{or} \;\; \tau<\tau(T_g)\approx 100s \qquad (1)$$

where T_0 is the VFT approximation of the ideal glass temperature, D_T is the fragility strength coefficient and T_g denotes the glass temperature.

This relation may be considered as the Arrhenius-like equation with the apparent, temperature dependent, activation energy, namely: $\tau(T)=\tau_0 \exp[E_a(T)/k_B T]$ with $E_a(T)=k_B T/(T-T_0)$.[6] Since the derivation of the VFT equation as a simple consequence of the basic free volume model,[1,11,12] it is considered as a checkpoint for the glass transition models.[1-4,11-18] Nevertheless, the question arises whether the VFT equation is indeed the optimal output of theoretical models or only a result of efforts to approach the most popular experimental dependence. Worth recalling is the link between the VFT equation and the fragility concept. The latter is one of key ideas of the "glassy physics".[1-6] It is associated with the plot introduced by Austin Angell, $\log_{10}\tau$ or $\log_{10}\eta$ vs. T/T_g,[1] enabling a system independent presentation of dynamics on approaching the glass temperature.[1] For describing the distortion from the Arrhenius behavior, manifested by a linear dependence at the Angell plot, the concept of fragility was introduced.[1,4,19 and refs. therein] For its metric the steepness index m, i.e. the slope at the Angell plot for $T\to T_g$ was proposed. Basing on the VFT eq. (1) and the definition of fragility via the steepness index it was shown that:[20]

$$m=m_P(T\to T_g)=\left[\frac{\partial \log\tau}{\partial(T_g/T)}\right]_{T\to T_g}^{P=const}=\frac{1}{\log_{10}(e)}\frac{D_T(T_0/T_g)}{(1-T_0/T_g)^2}=A+\frac{A'^2}{D_T} \qquad (2)$$

In fact the equivalence of m and D_T as fragility metrics was first indicated by Böhmer et al. ([19]), who introduced a relation often recalled in this respect:

$$m=m_{\min}+m_{\min}^2/D_T=16+590/D_T \qquad (3)$$

where $m_{\min} = \log_{10} \tau(T_g)/\tau_0^T = 16$, obtained assuming $\tau(T_g) = 100s$ and the same value of the prefactor for any liquid: $\tau_0 = 10^{-14} s$.[19]

For $m > 30$ and $D_T < 10$ liquids are considered fragile whereas $m < 30$ and $D_T > 10$ there are encountered to strong glass formers, with dynamics closer to the Arrhenius pattern.[1-4,19] In supercooled liquids at least two "dynamical domains" with the crossover at $\tau(T_B) \approx 10^{-7\pm1} s$ are considered.[21,22 and refs. therein] This crossover is associated with the change of parameters in the VFT eq. (1). In the given dynamical domain the value of D_T is constant but the steepness index $m_P(T)$ permanently increases on cooling.[22] Hence, D_T may be considered as the rate of change of the $m_P(T)$ on cooling in the given domain.[22] Alternatively, D_T may be considered as a tool for estimated fragility well before reaching the glass temperature, what is necessary when using solely the steepnes index *m*. At high temperatures the existence of a second crossover to the Arrhenius behavior is often claimed.[1-4]

The structural relaxation time is most often determined from the peak frequency of dielectric loss curves, $\tau = 1/2\pi f_{peak}$, obtained from the broad band dielectric spectroscopy (*BDS*) measurements. *BDS* enables an insight into ca. 15 decades in frequency or time in a single experiment. This feature is basic for vitrification studies where the anomalous change in dynamic on cooling towards T_g is the key artifact. It should be stressed that the structural (α-, *main-*) relaxation time (τ), is associated mainly with dynamics of permanent dipole moments linked to molecules. This dynamics can be influenced by molecular interactions, steric hindrances or appearance/disappearance of heterogeneities in a *fluidlike* surrounding.[2] The latter can be associated with local density and/or structure fluctuations.

Searching for the pressure equivalent of the VFT equation is a long standing problem, important also in industrial and geophysical applications. Only in 1998 the relation claimed to be a pressure counterpart of the VFT equation (PVFT) was proposed, namely:[23]

$$\tau(P) = \tau_0 \exp\left(\frac{B_P}{P_o - P}\right) \tau_0 \exp\left[\frac{D_P P}{P_0 - P}\right], \quad \text{for } P < P_g \text{ and } T = const \tag{4}$$

where P_0 is the estimation of the ideal glass pressure, P_g denotes the glass transition pressure, $\tau(P_g) = 100s$, and D_P is the pressure-related fragility strength coefficient.

In the early seventies a similar equation was used for portraying the pressure evolution of viscosity in glycerol by Johari, although with the pressure independent coefficient $B_P = const$.[24] Consequently the equation by Johari could portray dynamical properties only for "strong" glass formers.

Linking temperature and pressure related *BDS* or viscosity studies with *PVT* or *PρT* thermodynamic data one can obtain the volume- (V)[25-28] or density-related (ρ)[28-30] dependences. For portraying such behavior appropriate counterparts of the VFT equation were postulated, namely:[25-30]

$$\tau(V) = \tau_0 \exp\left(\frac{C_V}{V_0 - V}\right) = \tau_0 \exp\left(\frac{D_V V_0}{V_0 - V}\right) \qquad \text{for} \qquad V > V_g \tag{5}$$

$$\tau(\rho) = \tau_0 \exp\left(\frac{C_\rho}{\rho - \rho_0}\right) = \tau_0 \exp\left(\frac{D_\rho \rho}{\rho - \rho_0}\right) \qquad \text{for} \qquad \rho < \rho_g \tag{6}$$

where V_0 and ρ_0 are VFT estimations of the ideal glass volume and density estimates, V_g and ρ_g are volume and density associated with the glass transition: $\tau(V_g, \rho_g) = 100s$. Parameters D_V and D_ρ denote fragility strengths related to the volume and density changes.

Equation (5) was introduced as a simple consequence of the Doolittle equation, which is the basic artifact of free volume models for the glass transition.[11, 12] However, its final form with the volume related fragility strength coefficient D_V was only recently proposed.[25-28] Eq. (6) was postulated ad hoc, by analogy to eq. (4), since the pressurization was coincided with densification.[28-30]

The mentioned above counterparts of the VFT eq. (1) for the pressure (eq. (4)), volume (eq. (5)) and density (eq. (6)) paths of approaching the glass transition have became basic tools in data analysis in the last decade.[22, 25-30 and refs therein] Worth recalling is the discussion linking fragility expressed via D_V, D_ρ, D_V and D_ρ coefficients and steepness indexes defined via eq. (2), namely:[25-30]

$$m_T = m_T(P \to P_g) = \frac{d \log_{10} \tau}{d(P/P_g)}, \qquad m_V = m_V(V \to V_g) = \frac{d \log_{10} \tau}{d(V_g/V)} \tag{7a,b}$$

$$m_\rho = m_\rho(\rho \to \rho_g) = \frac{d \log_{10} \tau}{d(\rho/\rho_g)} \tag{7c}$$

However, there are serious fundamental problems associated with VFT-type equations discussed above, although hardly stated clearly. These equation should be able to transform into Arrhenius-type equations far away from the glass transition. For the VFT eq. (1) $\tau(T) = \tau_0 \exp[D_T T_0/(T - T_0)]$ and one obtains $\tau(T) = 1$ instead the Arrhenius equation $\tau(T) = \tau_0 \exp[E_a/T]$ for $T_0 = 0$.

This fundamental inconsistency was noted by Johari:[31] "… *this form* (i.e. eq. (1)) *of the VFT equation does not yield the Arrhenius equation for* $T=0$ …" Johari also questioned the physical sense of substitution $A_{VFT}=D_T T_0$ in eq. (1) and then the meaning of the fragility strength coefficient D_T.[31]

A similar inconsistency can be noted for the VFT-like equations associated with $\tau(P)$, $\tau(\rho)$ and $\tau(V)$ evolutions recalled above. Following the reasoning of Johari ([31]) no crossover to the Arrhenius-type equation when substituting $P_0=\infty$, $V_0=0$ or $\rho_0=\infty$ in eqs. (4), (5) and (6) seems to exists. The next basic problem of eqs. (1), (4), (5) and (6) is associated with τ_0 prefactors. In the basic VFT eq. (1) for $\tau(T)$ evolution the value $\tau_0=10^{-14}$ is often taken as a universal one for the high temperature limit in thermally activated systems.[1,19] However, in practice values $\tau_0=10^{-11}\div 10^{-16}s$ for various systems are obtained in experiments.[22,32,33] For eq. (4), associated with $\tau(P)$ behavior, the prefactor τ_0 is linked to $P=0$. However, in practice this value is taken from $\tau(T)$ studies under atmospheric pressure ($P=0.1MPa$). Consequently, the prefactor $\tau_0=\tau_0(P)_{T=const}$ can range from $10^{-11}s$ to $10s$, depending on the loci of the tested isotherm.[22,23,33] A similar fundamental inconsistency takes place for $\tau_0(V)$ in eq. (5) and $\tau_0(\rho)$ in eq. (6). For $\tau(\rho)=\tau_0\exp[D_\rho\rho/(\rho_0-\rho)]$ the prefactor is associated with $\tau_0=\tau_0(\rho=0)$, i.e. vacuum (!). However, in practice it is also linked to the atmospheric pressure state. Hence, an arbitrary value for this coefficient is possible. For $\tau(V)=\tau_0\exp[D_V V_0/(V-V_0)]$ the prefactor $\tau_0=\tau_0(V)$ is associated with the infinite volume, $V\to\infty$, what can be associated with the high temperature limit in the basic VFT eq. (1).[25-30] As shown in ref. ([13]) the analysis of the fragility based on VFT-type equations given above for $\tau(T)$ and $\tau(P)$ or $\eta(T)$ and $\eta(P)$ evolution lead to the nonequivalence of the temperature- and pressure-related fragilities (eqs. (2), (3) and (7a)). This issue was the focus of ref. ([20]) entitled "*Does fragility depends on pressure?...*", in which the following conclusion was given: (*i*) "*…m_P and D_T are equivalent measures of fragility if A is pressure invariant, and thus relaxation times for different isotherms will fall on a single master curve when plotted vs.* T_g/T" and (*ii*) "*since the pre-exponential factor in eq.(5) (which is just the ambient pressure value of τ) varies with temperature, m_P will always decrease with decreasing temperature, so that isotherms of τ(P) will not collapse onto a single master curve...*"

This paper presents a self-consistent set of VFT-type equations for portraying $\tau(T)$, $\tau(P)$, $\tau(V)$ and $\tau(\rho)$ behavior, associated with self-consistent, although modified, fragilities. These relations are free from "structural" problems encountered above.

2. Proposal of the self-consistent set *T, P, V* and ρ related VFT-like equations

When searching for the link between temperature- and pressure-related VFT-related VFT-type relations for the evolutions of dynamic properties one can consider a simple substitution $T \to A/P$, namely:

$$\boldsymbol{\tau(P)=\tau_0 exp\left(\frac{D_P P}{P_0-P}\right)} \to \tau_0 \exp\left(\frac{D_P(A/T)}{A/T_0 - A/T}\right) = \tau_0 \exp\left(\frac{D_P}{(T/T_0)-1}\right) = \tag{8}$$
$$= \tau_0 \exp\left(\frac{D_P(A/A_0)}{T/T_0 - 1}\right) \to \boldsymbol{\tau(T)=\tau_0 exp\left(\frac{D_T T_0}{T-T_0}\right)}$$

A similar reasoning can be used for VFT-like eqs. (5) and (6), $\rho = m/V$ and $\rho_0 = m/V_0$, namely:

$$\boldsymbol{\tau(\rho)=\tau_0 exp\left(\frac{D_\rho \rho}{\rho_0-\rho}\right)} \to \tau_0 \exp\left(\frac{D_\rho(m/V)}{m/V_0 - m/V}\right) = \tau_0 \exp\left(\frac{D_\rho}{V/V_0 - 1}\right) = \tag{9}$$
$$= \tau_0 \exp\left(\frac{D_\rho V_0}{V - V_0}\right) \to \boldsymbol{\tau_0 exp\left(\frac{D_V V_0}{V-V_0}\right) = \tau(V)}$$

Notwithstanding, the fundamental inconsistencies associated with τ_0 pre-factors remain. In the opinion of the authors these problems can be avoided if the existence of the absolute stability limit spinodal, partially hidden in the negative pressures domain, is taken into account.[33-35] In this respect noteworthy is the significant difference between the definition of temperature and pressure. Temperature is referenced to the absolute zero value (*T=0*), where thermodynamics of physical systems terminates. For the definition of pressure the reference is $P = 0$ (vacuum), but this value may be considered as the terminal one only for the gaseous state. For solids and liquids passing $P = 0$ does not yields any hallmark and the isotropic stretching ($P < 0$) down to absolute stability limit is possible. In this respect worth recalling is the statement of Lew D. Landau from his fundamental monograph "*Statistical Physics*" (1937):[36]

"*...There is a basic difference between negative pressures and negative temperatures. The latter are in a natural way unstable hence cannot exist in nature. Negative pressure states can exist in nature, although as metastable one...*".

The liquid state is limited by two absolute limit spinodals, located well above the boiling temperature and below $P < 0$. They are the loci of the absolute stability limit of the homogeneous nucleation. The significance of this issue showed studies on unusual properties of water where the estimation of the spinodal within the negative pressures domain is the fundamental checkpoint for theoretical models.[37] The significance of this issue taken into account in the novel pressure counterpart of the VFT equation recently proposed, namely:[33-35]

$$\tau(P) = \tau_0^P \exp\left[\frac{D_P \Delta P}{P_0 - P}\right] = \tau_0^P \exp\left[\frac{D_P P - D_P P_{SL}}{P_0 - P}\right], \qquad T = const \tag{10}$$

where $\Delta P = P - P_{SL}$, the prefactor τ_0^P is associated with the relaxation time at the liquid – gas stability limit at $P = P_{SL} < 0$.

For this equation the prefactor $\tau_0^P = \tau(P = P_{SL} < 0)$, i.e. it is associated with the absolute stability limit. Moreover it can describe experimental data in domains of negative and positive (hydrostatic) pressures, which was not possible for relations used so far. For $P_0 >> P$, where the latter is for pressures used in the given experiment, one can approximate $P_0 - P \approx P_0$ what yields the Arrhenius-type equation, namely:

$$\tau(P) \approx \tau_0^P \exp\left[\left(\frac{D_P}{P_0}\right)\Delta P\right] = \tau_0^P \exp[V_A \Delta P] = \tau_0^P \exp[V_A(P - P_{SL})] \tag{11}$$

where V_A is the measure of the constant activation volume.

Noteworthy is the difference between eq. (11) and the Arrhenius counterpart used so far: $\tau(P) = \tau_0 \exp(V_A P)$. Using eqs. (10) and (11) one can note the link between the "terminal" Arrhenius activation energy and the fragility: $V_a = \lim_{P \to P_{SL}} \left[D_P \Delta P / (P_0 - P)\right]$. As shown in refs. ([33-35]) basic parameters in eq. (10) can be estimated from the preliminary derivative-based analysis of $\tau(P)$ data via: (12)

$$\left[\frac{d\ln\tau}{dP}\right] = \left[\frac{V_a}{R}\right]^{-1/2} = \left[V_a'\right]^{-1/2} = \left[D_P(P_0 - P_{SL})\right]^{-1/2} P_0 - \left[D_P(P_0 - P_{SL})\right]^{-1/2} P = A + BP$$

where $V_a = V_a(P)$ is the apparent activation volume

The plot $\left(V_a'\right)^{-1/2}$ vs. P shows the range of validity of eq. (10) and yields optimal values of coefficients $P_0 = A/B$, but $1/AB = D_P P_0/(P_0 - P_{SL})$ prior to the final fitting of $\tau(P)$ data. The comparison of the "*old*" eqs. (4) and the "*new*" PVFT eq. (10) gives:[33-35]

$$\frac{D_P^{\text{“}old\text{”}}}{D_P^{\text{“}new\text{”}}} = \frac{P_0}{P_0 - P_{SL}} \qquad P_0^{\text{“}old\text{”}} = P_o^{\text{“}new\text{”}} = P_0 \tag{13}$$

It is visible that the "*old*" PVFT eq. (4) is associated with the unphysical loci of the gas – liquid stability limit, namely $P_{SL}(T)=0$ for arbitrary temperature an arbitrary liquid. Eq. (13) also shows that using the old PVFT eq. (4) the underestimated value of the fragility strength coefficient is obtained.

Basing on eq. (10) and assuming $P \approx A^{'}/T$ one can obtain the corrected form form of the "classical" VFT eq. (1), namely:

$$\tau(T) = \tau_0 \exp\left[\frac{D_P(T_{SL}-T)}{T-T_0}\frac{T_0}{T_{SL}}\right] \tag{14}$$

The dynamics of supercooled liquids is usually tested below the melting temperature, i.e. for $T_m < T < T_g$. In this domain the condition $T_{SL} >> T$ and then the condition $T_{SL} - T \approx T_{SL}$ may be assumed. This converts eq. (14) into:

$$\tau(T) = \tau_0 \exp\left[\frac{D_T T_0}{T-T_0}\right] \tag{15}$$

where the change $D_P \to D_T$ was introduced due to the difference in pressure and temperature related units.

The application of eqs. (1, 15) instead of eq. (14) can be a source of distortions of relevant parameters, since the comparison of eqs. (14) and (15) shows $D_T^{old} \to D_T^{New}\left[(T_{SL}-T)/T_{SL}\right]$. However, for molecular liquids the temperature T_{SL} is located well above the boiling temperature and the condition $T_{SL} >> T$ is well fulfilled for the typical experimental ranges of temperatures $T_m > T > T_g$. Consequently, the approximation of data via the old "classical" VFT eq. (1, 15) can introduce only a small shift of D_T values.

We would like to stress that, assuming that $T_0 \to 0$ yields $T - T_0 \approx T$. For the high temperature limit this yields:

$$\tau(T) = \tau_0 \exp\left[\frac{D_P(T_0/T_{SL})}{T/\Delta T}\right] \tag{16}$$

In the opinion of the authors this equation may be considered as the Arrhenius-parallel emerging at high temperatures. Noteworthy is the difference from the "classical" Arrhenius equation recalled in the introduction. There is an extensive evidence for the crossover from the VFT- to the Arrhenius-type behavior on heating above the melting temperature.[1-5,38] However, the analysis of high resolution experimental data for glass forming liquids indicate also the possibility of the non-Arrhenius dependence even up to the boiling temperature.[38,39] In the opinion of the authors both suggestions may be valid, within the limit of the experimental error. Notwithstanding, new tests focusing on this crossover and based on eqs. (14) and (16) may be advised. As mentioned in ref. ([22]) the preliminary derivative-based analysis of $\tau(T)$ data can show the domain of validity of the VFT description and estimate optimal values of relevant coefficients, namely:

$$\left[\frac{d\ln\tau}{d(1/T)}\right]^{-1/2} = \left[\frac{H_a(T)}{R}\right]^{-1/2} = \left(H_a'\right)^{-1/2} == \left[(D_T T_o)^{-1/2}\right] - \frac{\left[T_0(D_T T_0)^{-1/2}\right]}{T} = A - \frac{B}{T} \tag{17}$$

where $H_a(T)$ is for the apparent activation enthalpy and R denotes the gas constant. The linear regression analysis yields $T_0 = B/A$ and $D_T = 1/AB$.

The loci of the stability limit for the reference may be also important for $\tau(V)$ and $\tau(\rho)$ VFT-type evolutions. In an analogous way one can introduce (V_{SL}, ρ_{SL}) reference for $\tau(V)$ and $\tau(\rho)$ parameterizations. Then, the VFT-type eq. (5) can be transformed into:

$$\tau(\rho) = \tau_0 \exp\left(\frac{D_\rho \Delta\rho}{\rho_0 - \rho}\right) = \tau_0 \exp\left(\frac{D_\rho(\rho - \rho_{SL})}{\rho_0 - \rho}\right) \tag{18}$$

For this equation the prefactor $\tau_0 = \tau(\rho = \rho_{SL})$. The domain of its validity can be estimated from the linearized, derivative based, analysis of data:

$$\left[\frac{d\ln\tau(\rho)}{dP}\right] = \left[D_\rho(\rho_0 - \rho_{SL})\right]^{-1/2}\rho_0 - \left[D_\rho(\rho_0 - \rho_{SL})\right]^{-1/2} P = A + BP \tag{19}$$

The linear regression fit can yield optimal values of basic parameters: $P_0 = A/B$, and $1/AB = D_\rho\rho_0/(\rho_0 - \rho_{SL})$. In the case of "very strong" glass formers the condition $\rho_0 >> \rho$ is well fulfilled in the experimental domain. Then $\rho_0 - \rho \approx \rho_0$ and consequently eq. (18) can be approximated by the Arrhenius type relation:

$$\tau(\rho) = \tau_0 \exp\left(\frac{D_\rho \Delta\rho}{\rho_0 - \rho}\right) \to \tau(\rho) \approx \tau_0 \exp\left(\frac{D_\rho}{\rho_0}\Delta\rho\right) \tag{20}$$

Using eq. (19) one can compare parameters in the "old" ρVFT eq. (5) and the new one in eq. (18):

$$\frac{D_\rho^{\text{"old"}}}{D_\rho^{\text{"new"}}} = \frac{\rho_0}{\rho_0 - \rho_{SL}} \quad \text{and} \quad \rho_0^{old} = \rho_o^{new} = \rho_0 \tag{21}$$

A similar reasoning applied for $\tau(V)$ evolution yields the improved VFT-type equation in the form:

$$\tau(V)=\tau_0 \exp\left(\frac{D_V(V_{SL}-V)}{V-V_0}\frac{V_0}{V_{SL}}\right) \tag{22}$$

Also in this case the supplementary derivative-based analysis can facilitate the fitting of $\tau(V)$ data, namely:

$$\left[\frac{d\tau(V)}{dV}\right]^{-1/2}=(D_V\Delta V_0)^{-1/2}V-V_0(D_V\Delta V_0)^{-1/2}=BV-A \tag{23}$$

The plot based on eq. (23) indicates the domain of validity of eq. (22) and yields optimal values of parameters from the linear regression fit via: $V_0=A/B$ and $D_V\Delta V=(B)^{-1/2}$. For the low-volume limit, when $V\to V_0$, one can assume $V_{SL}-V\approx V_{SL}$ what leads to eq. (5). Far away from the glass transition one can assume $V-V_0\approx V$ and eq. (22) can be converted into:

$$\tau(V)=\tau_0 \exp\left(\frac{D_V(V_0/V_{SL})}{V/\Delta V}\right) \tag{24}$$

One can propose this dependence as the parallel of the Arrhenius equation, emerging in the high volume limit.

3. Conclusions

In this paper modified VFT-type equations for portraying $\tau(T)$ $\tau(P)$, $\tau(V)$ and $\tau(\rho)$ evolutions were proposed. In each case the prefactor are associated with the loci of the absolute stability limit T_{SL}, P_{SL}, V_{SL} or ρ_{SL} . The smooth transformations to the Arrhenius-like equations in the liquid state very far away from the glass transition was shown. Both VFT-type and Arrhenius-like equation contain the loci of the absolute stability limit as the reference. It was shown in ref. ([33]) that the nonequivalence of m_T and m_P fragilities if instead of the "traditional" eq. (4) the modified PVFT eq. (10) is used, namely:

$$m_T=\left[\frac{d\log_{10}\tau}{d\,\Delta P/\Delta P_g}\right]^{T=const.}_{P\to P_g}=\frac{1}{\ln 10}\frac{D_P(\Delta P_g/\Delta P_0)}{(1-\Delta P_g/\Delta P_0)^2}=B+\frac{B'^2}{D_P} \tag{25}$$

In the opinion of the authors the improved definitions should be used also for the volume-related and density-related fragility metrics, namely:

$$m_\rho = m_\rho\left(\rho \to \rho_g\right) = \left[\frac{d\log_{10}\tau}{d\,\Delta\rho/\Delta\rho_g}\right] \tag{26}$$

$$m_V = m_V\left(V \to V_g\right) = \left[\frac{d\log_{10}\tau}{d\,\Delta V_g/\Delta V}\right] \tag{27}$$

One can propose a similar correction for the temperature-related fragility:

$$m_P = m_P\left(T \to T_g\right) = \left[\frac{d\log_{10}\tau}{d\,\Delta T_g/\Delta T}\right] \tag{28}$$

However, in this case the change of the definition of the steepness index may be not important since for the typical range of experimental data the condition $T_{SL} >>> T_g$ is well fulfilled. Corrected definitions of steepness indexes are a clear consequence of corrected VFT-type equations for $\tau(T)$, $\tau(P)$, $\tau(V)$ and $\tau(\rho)$ dependences. The Angell plot, $\log_{10}\tau$ vs. T_g/T, is probably the most known hallmark of the glass transitions physics.[1,19,40] Recently, it was shown that its pressure-related counterpart is the plot $\log_{10}\tau$ vs. $\Delta P/\Delta P_g$.[33] Eqs. (26) and (27) suggest plots $\log_{10}\tau$ vs. $\Delta V_g/\Delta V$ for volume-related and $\log_{10}\tau$ vs. $\Delta\rho/\Delta\rho_g$ for density-related data. A similar correction for the isobaric, temperature related data may be important only very far away from the glass temperature, i.e. for $T_g/T \to 0$.

Recent discussions recalled the question of the general validity of the VFT equation for portraying dynamic data in glass forming liquids, at least for the temperature path.[4-6] Nevertheless, despite emerging objections it seems to remain a key tool. The above discussion shows that it is possible to construct a self-consistent set of VFT-type equation in respect to any path of approaching the glass transitions. The proposed relations are free from fatal problems of eqs. (1), (4), (5) and (6) used so far. The revision of fragility metrics may be also advised. The question of the existence of the crossover from the VFT to the Arrhenius type behavior remote from the critical point also re-appears due to new VFT counterparts. Finally we would like to stress that equations proposed in this paper made it possible to discuss the evolution of relaxation time, viscosity as well as fragility in negative pressures domain.

Acknowledgements

This research was carried out with the support of the CLG NATO Grant No. CBP NUKR.CLG 982312 and the research was also supported by the Ministry of Science and Higher Education (Poland) Grant No. N N202 231737, for years 2009-2012.

References

1. Donth, E. (2001) The Glass Transition: Relaxation Dynamics in Liquids and Disordered Materials (*Springer Verlag, Berlin*).
2. Rzoska, S. J., and Mazur, V. eds. (2006) Soft Matter Under Exogenic Impacts, *NATO Sci. Series II*, (Springer, Berlin), vol. **247**.
3. Castillo del, L. F., Goldstein, P., and Dagdug, L. (2002) Relaxation phenomena in the glass transition, *Revist. Mex. de Fisica* **48**, 174-181.
4. Kivelson, S. A., and Tarjus, G. (2008) In search of a theory of super-cooled liquids, *Nature Materials* **7**, 831-833.
5. McKenna, G. B. (2008) Diverging views on glass transition, *Nature Physics* **4**, 673-674.
6. Hecksher, T., Nielsen, A. I., Olsen, N. B., and Dyre, J. C. (2008) Little evidence for dynamic divergences in ultraviscous molecular liquids, *Nature Physics* **4**, 737-741.
7. Taniguchi, H. (1995) Universal viscosity-equation for silicate melts over wide temperature range, *J. Volc. and Geotherm. Res.* **66**, 1-8.
8. Vogel, H. (1921) Das temperature-abhängigketsgesetz der viskosität von flüssigkeiten, *Phys. Z.,* **22**, 645646.
9. Fulcher, G. S. (1925) Analysis of recent measurements of the viscosity of glasses. *J. Am. Ceram Soc.* **8**, 339-355.
10. Tammann, G., Hesse, W. (1926) Die Abhängigkeit der Viskosität von der Temperatur bei unterkühlten Flüssigkeiten, *Z. Anorg. Allg. Chem.* **156**, 245-257.
11. Greet, R. J., and Turnbull, D. (1967) Glass transition in o-terphenyl, *Journal of Chemical Physics* **4**, 1243-1251.
12. Turnbull, D., and Cohen, M. H. (1961) Free-volume model of the amorphous phase: glass transition, *Journal of Chemical Physics* **34**, 120-125.
13. Sciortino, F., Tartaglia, P., and Kob, W. (2002) Thermodynamics and aging in supercooled liquids: the energy landscape approach, *Physica A* **306**, 343-350.
14. Trachenko, K. (2008) The Vogel–Fulcher–Tammann law in the elastic theory of glass transition *J. Non-Cryst. Solids* **354**, 3903-3906.
15. Tamai, Y. (2003) Effects of pressure on the fragile nature of fluoro-zirconates studied by molecular dynamics simulations, *Phys. Rev. E* **67**, 031504.
16. Qin, Q., and Mckenna, G. B. (2006) Correlation between dynamic fragility and glass transition temperature for different classes of glass forming liquids, *J. Non-Cryst. Solids* **352**, 2977-2985.

17. Hodge, I. M. (1997) Adam-Gibbs Formulation of Enthalpy Relaxation Near the Glass Transition, *J. Res. Natl. Inst. Stand. Technol.* **102**, 195-205.
18. Xiaoyu Xia, and Peter G. Wolynes (2000) Fragilities of liquids predicted from the random First order transition theory of glasses, *PNAS* **97**, 2991-2994.
19. Böhmer, R., Ngai, K. L., Angell, C. A., and Plazek, D. J. (1993) Non-exponential relaxations in strong and fragile glass formers, *J. Chem. Phys.* **99**, 4201-4209.
20. Paluch, M., Gapiński, J., Patkowski, A., and Fischer, E. W. (2001) Does fragility depend on pressure? A dynamic light scattering study of a fragile glass-former, *J. Chem. Phys.* **114**, 8048-8055.
21. Novikov, V. N., and Sokolov, A. P. (2003) Universality of the dynamic crossover in glass-forming liquids: A "magic" relaxation time, *Phys. Rev. E* **67**, 031507.
22. Drozd-Rzoska, A., and Rzoska, S. J. (2006) Derivative-based analysis for temperature and pressure evolution of dielectric relaxation times in vitrifying liquids, *Phys. Rev. E* **73**, 041502.
23. Paluch, M., Rzoska, S. J., Habdas, P., and Ziolo, J. (1998) On the isothermal pressure behaviour of the relaxation times for supercooled glass-forming liquids, *J. Phys.: Condens. Matter* **10**, 4131-4135.
24. Johari, G. P., and Whalley, E. (1972) *Dielectric Properties of Glycerol in the Range 0.1-10^5 Hz, Faraday Symp. Chem. Soc.* **6**, 23.
25. Cangialosi, D., Wubbenhorst, M., Schut, Veen van, H. A., and Picken, S. J. (2004) Dynamics of polycarbonate far below the glass transition temperature: A positron annihilation lifetime study, *Phys. Rev. B* **69**, 134206.
26. Dlubek, G., Pointeck, J., Shaikh, M. Q., Hassan, E. M., and Krause-Rehberg, R. (2007) Free volume of an oligomeric epoxy resin and its relation to structural relaxation: Evidence from positron lifetime and pressure-volume-temperature experiments, *Phys. Rev. E* **75**, 021802.
27. Dlubek, G., Shaikh, M. Q., Raetzke, K., Faupel, F., Pionteck, J., and Paluch, M. (2009) The temperature dependence of free volume in phenyl salicylate and its relation to structural dynamics: A positron annihilation lifetime and pressure-volume-temperature study, *J. Chem. Phys.* **130**, 144906.
28. Roland, C. M., Hensel-Bielowka, S., Paluch, M., and Casalini, R. (2005) Supercooled dynamics of glass-forming liquids and polymers under hydrostatic pressure, *Rep. Prog. Phys.* **68**, 1405-1478.
29. Paluch, M., Patkowski, A., and Fischer, E. W. (2000) Temperature and Pressure Scaling of the a Relaxation Process in Fragile Glass Formers: A Dynamic Light Scattering Study *Phys. Rev. Lett.* **85**, 2140-3143.

30. Papadopoulos, P., Floudas, G., Schnell, I., Klok, H.-A., Aliferis, T., Iatrou, H., and Hadjichristidis, N. (2005) "Glass transition" in peptides: Temperature and pressure effects, *J. Chem. Phys.* **122**, 224906.
31. Johari, G. P. (2006) On Poisson's ratio of glass and liquid vitrification characteristics, *Phil. Mag.* **86**, 1567-1579.
32. Drozd-Rzoska, A., Rzoska, S. J., and Pawlus, S., Tamarit, J. Ll. (2006) Dynamics crossover and dynamic scaling description in vitrification of orientationally disordered crystal, *Phys. Rev. B* **73**, 224205.
33. Drozd-Rzoska, A., Rzoska, S. J., Roland, C. M., and Imre, A. R. (2008) On the pressure evolution of dynamic properties of supercooled liquids *J. Phys.: Condens. Matt.* **20**, 244103.
34. Drozd-Rzoska, A., Rzoska, S. J., and Imre A. R. (2007) On the pressure evolution of the melting temperature and the glass transition temperature, *J. Non-Cryst. Solids* **353**, 3915-3923.
35. Drozd-Rzoska, A., Rzoska, S. J., Paluch, M., Imre, A. R., and Roland, C. M. (2007) On the glass temperature under extreme pressures, *J. Chem. Phys.* **126**, 164504.
36. Landau, L. D. (1937) and Landau, L. D., and Lifshitz, E. M. (1976) Statistical Physics (*Nauka, Moscow*), in russian.
37. Imre, A. R., Maris, H. J., Williams, P. R. (2002) Liquids under Negative Pressures, *NATO Sci. Series II*, vol. **84** (Kluwer-Springer, Dordrecht).
38. Roessler, E., Hess, K.-U., and Novikov, V. N. (1998) Universal representation of viscosity in glass forming liquids, *J. Non-Cryst. Solids* **223**, 207-222.
39. Roessler, E. (2006) *Lecture and Discusion at Kia Ngai Fest 16th Sept.*, Pisa, Italy satellite event of *the IVth Workshop on Non-Equilibrium Phenomena in Supercooled Fluids, Glasses and Amorphous Materials*, 17-22, Pisa, Italy.
40. Martinez, L.-M., and Angell, C. A. (2001) A thermodynamic connection to the fragility of glass-forming liquids, *Nature* **410**, 663-667.

PART II

LIQUID CRYSTALS

STABILITY AND METASTABILITY IN NEMATIC GLASSES: A COMPUTATIONAL STUDY

MILAN AMBROZIC[1], TIMOTHY J. SLUCKIN[2], MATEJ CVETKO[3,4] AND SAMO KRALJ[1,4]
[1]Jožef Stefan Institute, Jamova 39, 1000 Ljubljana, Slovenia
[2]School of Mathematics, University of Southampton, Southampton SO17 1BJ, United Kingdom
[3]Regional Development Agency Mura Ltd, Lendavska 5a, 9000 Murska Sobota, Slovenia
[4]Laboratory of Physics of Complex Systems, Faculty of Sciences and Mathematic, University of Maribor, Koroška 160, 2000 Maribor, Slovenia

Abstract: The influence of randomly distributed impurities on liquid crystal (LC) orientational ordering is studied using a simple Lebwohl-Lasher type lattice model in two (d=2) and three (d=3) dimensions. The impurities of concentration p impose a random anisotropy field-type of disorder of strength w to the LC nematic phase. Orientational correlations can be well presented by a single coherence length ξ for a weak enough w. We show that the Imry-Ma scaling prediction $w\,\xi \propto w^{-2/(4-d)}$ holds true if the LC configuration is initially quenched from the isotropic phase. For other initial configurations the scaling is in general not obeyed.

Keywords: liquid crystals, metastability, symmetry breaking, weak disorder, Irmy-Ma scaling

1. Introduction

Condensed matter phases and structures are commonly reached via symmetry breaking transitions. In such systems, when the continuous symmetry is broken, temporary domain-type patterns are formed[1]. The domain structures eventually coarsen, and disappear in the long-time limit, leaving a uniform broken-symmetry state[2]. This state possesses so–called "long-range order" (LRO), in which the spatially dependent order parameter correlation function does not decay to zero in the limit of large distances.

However, the domain structures, temporary as they are in pure systems, can be stabilized by impurities. In the doped systems the long-time equilibrium or quasi-equilibrium structure can resemble an intermediate-time snapshot of the coarsening pure system. The domain formation in both pure and doped systems depends on a few basic properties of the system. The signature of these domain-dominated systems is a universal behavior (in the sense that statistical mechanicians use the word). The universality allows a mathematically wide variety of often apparently completely different systems to exhibit behavior which falls on the same curve. Despite their underlying simplicity, some important features of domain formation remain unresolved. This article will be concerned with domain formation in impure nematic liquid crystal glasses.

Many pure condensed matter systems can be quenched into a configuration susceptible to continuous symmetry breaking (CSB). Examples include magnets, liquid crystals and liquid helium. The basic characteristics of domain pattern kinetics of such a system following the quench are described by the Kibble-Zurek mechanism[1,3]. This model was originally introduced to explain the formation of topological defects in the early universe following the Big Bang.

To illustrate the Kibble-Zurek mechanism in a condensed matter physics context, we consider the coarsening dynamics of an isotropic (I) – nematic (N) phase transition of rod-like liquid crystal (LC) molecules[4]. In the isotropic phase, exhibiting continuous symmetry, the molecules stochastically fluctuate. In an equilibrium nematic phase, by contrast, the molecules on average orient along a single symmetry-breaking direction. Now, however, suppose that the isotropic phase is quenched very quickly into the nematic temperature regime. In different parts of the sample a randomly chosen configuration of the symmetry-breaking field is established. This choice is arbitrary and depends on the directions of local fluctuations. A domain structure then appears, which is well characterized by a single domain length $\xi_d(t)$ [1,2]. The order parameter spatial correlation functions are time dependent, but depend only on the single non-dimensional length scale $r/\xi_d(t)$. As time t increases, the domain growth eventually enters the so-called *dynamic scaling regime*, where the power law $\xi_d(t) \propto t^\gamma$ is obeyed[2]. The universal scaling coefficient γ depends on whether a conservation law for the order parameter exists or not.

However, impurities are almost unavoidably present in any system. These impurities can pin and stabilize the domains. The manner in which this occurs depends, among other factors, on the average separation between impurities, the coupling strength between the impurities and the dynamical variables whose symmetry is broken, and the size of proto-domains (i.e. the original size of the domains when they begin to form)[1].

structure? One key feature is the surface tension associated with the boundary between domains. Because the domain are formed from broken symmetry spins, this quantity is not a constant (as it is when the broken symmetry is discrete), but rather depends on the inverse power of the domain size. However, an extra feature limiting the growth of the domains is the existence of topological defects. Topological defects (points, lines or planes, but most likely to be lines) are regions over which the order parameter cannot relax smoothly to a uniform configuration. They appear as a consequence of local frustrations arising due to conflicting domain orientations. The relaxation process involves amalgamation or mutual annihilation of defect structures, which is a slower process than simple structural motion.

Domain structures in CSB systems experiencing a random-field type disorder stabilize in size. Many theoretical studies of such systems use approaches based on equilibrium statistical mechanics. Such systems are parameterized by the physical dimension of the system d, a disorder strength parameter w, the volume proportion of the impurities p, and the dimension n of the broken spin symmetry. There are two paradigms of the low temperature behavior of these systems.

One paradigm is due to Larkin, but rediscovered in the West by Imry and Ma[5]. The idea is that the uniform system is unstable with respect to break-up into domains of size ξ_d, whose size depends on balancing the energy associated with (a) disorder and (b) boundary energy. Roughly speaking, if the domains are too small, the system possesses a large number of boundaries, whose energy is unfavorable. But if they are too large, they cannot order locally to take advantage of the local random fields. The compromise is a universal domain pattern, which is characterized by so-called "short range ordering" (SRO). The ordering is short-range not because it is necessarily short on a molecular scale. Indeed it is not; the range of the correlations is long compared to molecular scales. The terminology arises because the correlations decay exponentially on length scale ξ_d, and do so only because of the presence of what can in principle be an infinitesimally weak local random ordering field w.

Detailed calculations predict the correlation length ξ_d to obey a universal scaling law $\xi_d \propto w^{-2/(4-d)}$. At least one detailed study seems to support this picture[6]. A weaker version of this result, the so-called Imry-Ma *theorem*, merely notes that this pictures *proves* quite conclusively that an arbitrarily small degree of disorder destroys a CSB ordered phase in a system of dimensionality less than four. If the physical dimension is greater than four, the ordered low temperature phase is robust with respect to the introduction of arbitrarily small disorder (though not, of course if the disorder parameter is sufficiently strong).

An alternative picture was first introduced by Aharony and Pytte[7] in the context of random magnets. In this picture the order parameter correlation function exhibits algebraic decay with distance instead. This situation, intermediate between SRO and LRO, has come to be known as quasi-long-range order (QLRO). The most well-known example of QLRO, due to Berezinsky and to Kosterlitz and Thouless[8] occurs in the low temperature phase of the two-dimensional XY model. A number of recent theoretical and computational studies have supported this point of view in random spin systems in a higher dimensionality[9].

Many recent studies have been carried in randomly perturbed LCs[6,9-13]. For these systems, the impact of weak quenched random disorder on CBS phases is relatively easily to observe experimentally. Disorder is in this case imposed by a porous matrix confining[11] a LC phase. Another related system consists of aerosil nanoparticles immersed in a liquid crystal matrix[6], and there have been speculations that this system too behaves in an immersed analogous fashion.

In this article we address the so-called *Random Anisotropy Nematic* (RAN)[11,14], in which interactions with arbitrarily oriented but quenched local spins can locally orient a nematic liquid crystal. We consider a slightly more generalized model than that discussed previously (see refs. ([6]) and ([10])), which allows for the density of impurity sites to be changed. This system belongs to the family of continuously broken spin systems, and is much amenable to experimental test than some of the magnetic systems used in the 1970s. Our study is computational and is therefore complementary to the high-powered theoretical approaches discussed elsewhere.

We are mainly concerned with domain properties well below the nematic–isotropic transition. We concentrate on the interaction between the *glass-like* properties of random nematics (specifically irreversibility and dependence of final behavior on initial conditions) and the *long-range order* properties of the final equilibrium or metastable state (specifically the question of whether the final state is LRO, QLRO or SRO). Specifically we are seeking to resolve the puzzle of when and how QLRO or SRO develops in these systems.

The literature exhibits a strong theoretical bias toward the existence of a QLRO state. But our numerical studies are carried out in the zero temperature limit. We shall find that, that if the system is started in a random configuration, in the long-time limit, the system usually exhibits SRO with Imry-Ma-like features.

By contrast, if it is started in a homogeneously ordered state, it may exhibit SRO, may exhibit QLRO, or it may exhibit LRO in the long-time limit, depending on the magnitude of the disorder parameter w.

The plan of the article is as follows. In the §2 we discuss the detailed model. In §3 we explain the numerical algorithm used to investigate the system. In §4 explain the parameters whose time-dependence we monitor in order to investigate the system. The results of our study are presented in §5. Finally in §6 we draw some brief conclusions and make suggestions for further work.

2. Model

Our simulations use a lattice-spin model of a liquid crystal, of the type pioneered by Lebwohl and Lasher. We use a simple Lebwohl-Lasher[15] pairwise interaction among rod-like lattice spins $\{\mathbf{S}_i\}$. The nature of the energy means that, as with all liquid crystal systems, there is never a distinction between $\mathbf{S}$ and $-\mathbf{S}$. This can simulate either a thermotropic or a lyotropic LC. The sites are arranged in a d-dimensional cubic lattice, of length L lattice constants, with total number of sites (i.e. particles) $N = L^d$, subject to periodic boundary conditions. In all subsequent work, distances are scaled with respect to the lattice constant.

We suppose that in addition the LC ordering is perturbed by local site random anisotropy disorder of strength w. This type of interaction was first introduced in magnets by Harris *et al* [14]. We have elsewhere labeled this model in a nematic context as the Random Anisotropy Nematic model (RAN)[11]. In this study the RAN is modified so that only spins at a random fraction p of sites are subject to random anisotropy, as discussed e.g. by Chakrabarti[10] and Bellini *et al* [6].

The interaction energy E of the system is:

$$E = -\frac{1}{2} J \sum_{i,j} (\mathbf{S}_i \cdot \mathbf{S}_j)^2 - w \sum_i \delta_i (\mathbf{S}_i \cdot \mathbf{e}_i)^2. \tag{1}$$

For liquid crystals in three dimensional space the quantities $\{\mathbf{S}_i\}$ are three-dimensional vector spins. It is tempting to identify the quantity $\mathbf{S}_i$ with the local nematic director which appears in continuum theories, but in fact the director only arises from time-averages of local spin directions. In principle, both the dimension of the space occupied by the spins n and the dimension of physical space d can be varied independently in a purely theoretical study.

Each of these quantities may affect the statistical mechanical properties of the model. In our studies, however, we shall restrict our study to cases for which $n = d$. We have examined two dimensional (2*D*) systems, $(d = n = 2)$ and three-dimensional (3*D*) systems $(d = n = 3)$.

In eq. (1) each site i is subject to a random anisotropy with probability p. The quantity δ_i, is a random variable, taking the values 0 or 1, defined formally as follows:

$\delta_i = 0$ if the site i *is not* subject to the local anisotropy field (2a)

$\delta_i = 1$ if the site i *is* subject to the local anisotropy field (2b)

subject to $p(\delta_i = 1) = p$.

The direction of the easy axis $\mathbf{e}_i$ is determined randomly at each site i for which $\delta_i = 1$, and is distributed uniformly on the surface of a d dimensional sphere.

3. Computational details

All simulations take place at zero temperature, and proceed by minimizing the energy. Different simulations range over different *p*, w and initial starting configuration, as well as for different system sizes. We consider two separate types of starting configuration:

(a) random (*r*), ($s = 1$), initial conditions, in which the spins in the starting configurations are distributed randomly over the allowed space of directions;
(b) homogeneous (*h*), ($s = 2$), initial conditions, in which the spins in the starting configuration are completely aligned along a single direction.

Each type of initial condition corresponds to known and common experimental situations. The *r* initial condition corresponds to a *zero-field-cooled* sample, whereas the *h* initial condition corresponds to a *field-cooled sample.* In the latter case, essentially perfect alignment is achieved for a high enough field and a slow enough quench. We note that the different starting configurations correspond to the *sample history.* It is known that history-dependent phenomena are very important in determining steady-state configurations in glassy systems.

Maximum simulation box sizes are $L = 400$ for $d = 2$ and $L = 140$ for $d = 3$. In order to diminish the influence of statistical variations, several simulations (typically $N_{rep} \sim 10$) have been carried out for a given set of parameters (i.e., *w*, *p* and *s*). In simulations all energies are measured with respect to the intersite coupling parameter J. In simulations we minimized the interaction energy *E* with respect to orientations of spins. The corresponding set of equations was solved using the Newton's method. Numerical and computational details are described in detail in[16].

After the simulation has been carried out, the spins reach a steady state configuration $\{\mathbf{S}_i = \sigma_i\}$. From this configuration, we now calculate the *orientational correlation function* $G(r)$. All our conclusions follow from analyses of $G(r)$. This measures the spin orientational correlation function as a function of their mutual separation $r = |\mathbf{r}_j - \mathbf{r}_i|$. $G(r)$ is defined as:

$$G(r) = \frac{1}{d-1}\left\langle d(\sigma_i \cdot \sigma_j)^2 - 1\right\rangle, \tag{3}$$

where *d* is the (spin) dimensionality of the system . The brackets $\langle ... \rangle$ denote the average over all lattice sites that are separated for a distance *r*. The peculiar factors of d are chosen so that if the spins are *completely correlated* (i.e, homogeneously aligned along a symmetry breaking direction), $(\sigma_i \cdot \sigma_j)^2 = 1 \Rightarrow G(r) = 1$, whereas if they are *uncorrelated* $(\sigma_i \cdot \sigma_j)^2 = d^{-1} \Rightarrow G(r) = 0$. We also note that for $d = 3$, $G(r_{ij}) = \langle P_2(\cos\vartheta_{ij})\rangle$, the second Legendre polynomial associated with the cosine of angle between spins at the sites in question. There is an analogous relation for $d = 2$: $G(r_{ij}) = \langle \cos 2\vartheta_{ij}\rangle$.

We can make some further comments about general properties of $G(r)$. Since each spin is necessarily parallel with itself, $G(0) = 1$. Furthermore, we normally expect the correlation function to be a decreasing function of distance *r*. In this model there is no coupling between directions in physical and spin space, and so it is possible to write down correlation functions as a function of scalar separation alone, i.e. $G(\mathbf{r}) = G(r)$. This is no longer true in models which include, for example, electrostatic, steric or dispersive forces. We have also checked this empirically. For the cases of SRO or QLRO, $G(r \to \infty) \to 0$, so that spins at far distant points are uncorrelated. But if there is LRO, on the other hand, we expect that $G(r \to \infty) = Q^2 \neq 0$ Q is the order parameter, and operationally it could be found from the largest eigenvalue of the mean order parameter tensor matrix.

For a truly *infinite* system $G(r)$ would be expected to take the form[17]:

$$G(r) = G_1(r) + Q^2, \tag{4}$$

where the long-range order is subsumed in Q, and all short-range or quasi-long-range order is included in. In general for LRO or SRO, the contribution $G_1(r)$ might be expected to decay exponentially, at least for large r, although we expect also a power law prefactor to reflect the fact that while for QLRO, $G_1(r) \sim r^{-(1+\eta)}$ and $Q = 0$ at large r. However, for finite systems we have found from experience that it is very difficult to analyse the data in such a way as to extract the exponent η from correlation function measurements.

In order to obtain structural details for *finite* systems from the observed properties of $G(r)$, we fit $G(r)$ to an empirical *ansatz* of the following form:

$$G(r) = (1 - s)e^{-(r/\xi)^m} + s, \tag{5}$$

where the coherence length ξ, the stretched exponential parameter *m*, and $s = Q^2(L)$ are adjustable parameters. The coherence length has an obvious interpretation. The stretched exponential parameter *m* is introduced by analogy with the stretched exponential temporal decay which occurs in many glassy systems[18]. For SRO $s = 0$. For QLRO, the exponent η can be extracted more reliably from an analysis of $Q(L) \sim L^{-(1+\eta)/2}$ than by analysing $G_1(r)$ for increasingly large systems. For LRO, $Q(L \to \infty) = Q \neq 0$. This fitting form has been chosen for empirical reasons; probably more sophisticated *ansätze* can be found. However this form is sufficient to distinguish the three basic regimes and to determine the dependence $\xi(w)$.

4. Results

Representative results for $G(r)$, for both *random* and *homogeneous* initial conditions, in 2*D* and 3*D* are shown Fig. 1. For the *random* case we obtain, $s \approx 0$. This holds true for all cases studied, although in the parameter régime $p < 0.1$, $w < 1$, convergence was extremely slow and definitive results could not be obtained. The orientational correlations vanish at long distances, which is a hallmark of SRO. By contrast $G(r)$ behaviour obtained in the presence of *homogeneous* initial conditions yield $s > 0$, so long as the anchoring strength *w* is not too large. In this case it was possible to carry out a finite size analysis of the quantity $s(L)$. Representative results are shown in Fig. 2. We find $s(L \to \infty) \to s(\infty)$, i.e. $s(L)$ seems to saturates at a finite value at high L. This is a signature of LRO. Simulations have been carried out on system sizes up to values $L = 400$ for $d = 2$ and $L = 140$ for $d = 3$.

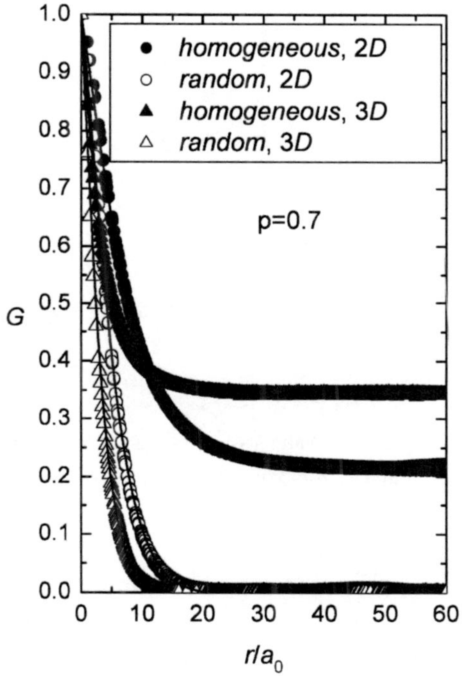

Figure 1. $G(r)$ for homogeneous and random initial configurations. $p = 0.7$; $d = 2$: $w = 1$, $L = 260$; $d = 3$: $w = 3$, $L = 80$. Note the difference between the results for r initial conditions, which decay to zero, and for h initial conditions, which do not.

However, for h initial conditions LRO may not always hold. For higher w, the LRO structure may be replaced by QLRO or even SRO. We give some indication of this in Fig. 3, where we plot $s(p)$ for different anchoring strengths w for d=2 and $L = 250$. The plot suggests that for each p there is a critical value $w_c(p)$, such that for $w > w_c(p)$, $s = 0$. If this is the case, SRO or QLRO holds at higher anchoring strengths, although the value of the critical cross-over anchoring strength changes with impurity concentration p. However, in order to verify this conclusion it will be necessary to carry out a complex finite size scaling analysis.

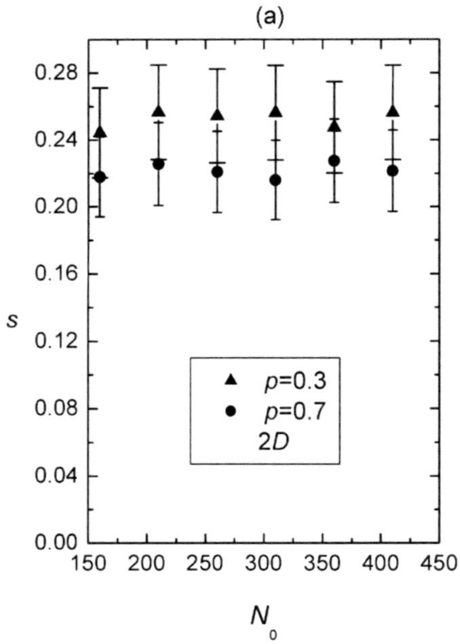

Figure 2. Behaviour of $s(L)$ for two values of p, for simulations with homogeneous initial configurations. a) $d = 2, w = 1$, and b) $d = 3, w = 3$ (The parameter N_0 coresponds to L).

In Figs. 4 we plot the stretched exponential parameter m as a function of p and w. We do not observe any systematic changes in behaviour of m below and above the percolation threshold on varying p. Indeed, values of m are strongly scattered because the structural details of $G(r)$ are relatively weakly m-dependent. The parameter m appears always to be essentially independent of the impurity concentration p. For h initial conditions it is also independent of the anchoring strength w. In all cases m is close to unity. Only for r initial conditions do we see any noticeable dependence on w. This dependence only occurs at low w, at increases m up to about 1.4.

Specifically, for the r initial configurations we obtain $m \sim 1.4$ for d=2, w=1, and $m \sim 1.2$ for d=3, w=3. On increasing w a value of m is decreasing in the weak anchoring regime and saturates at a constant values in the stronger anchoring regime. For h initial configurations we obtain $m \sim 1.18$ for d=2, and $m \sim 0.95$ for d=3.

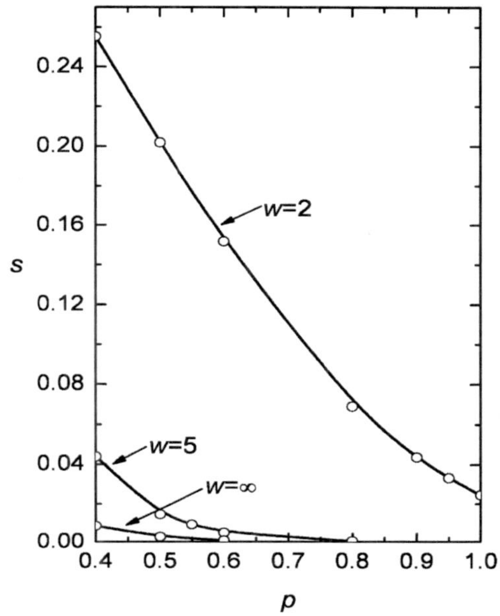

Figure 3. Simulations for d=2, $L = 250$, with h initial conditions, showing functional dependence $s(p)$ for three different values of w. Circles mark calculated points and the full line is fitted to these points. These results suggest the possibility of p-dependent critical anchorings $w_c(p)$ above which LRO no longer holds.

We now examine the $\xi(w)$ dependence. The Imry-Ma theorem makes a specific prediction that this obeys the universal scaling law

$$\xi \propto w^{-\frac{2}{4-d}}. \tag{6}$$

Eq. (6) predicts dimensionally dependent behavior: for $d = 2$, $\xi \propto w^{-1}$, whereas for $d = 3$, $\xi \propto w^{-2}$. We have analyzed results for p=0.3, p=0.5 and p=0.7, using both r and h initial configurations. Results are shown in Figs. 5. The figures use the *ansatz*:

$$\xi = \xi_0 w^{-\gamma} + \xi_\infty. \tag{7}$$

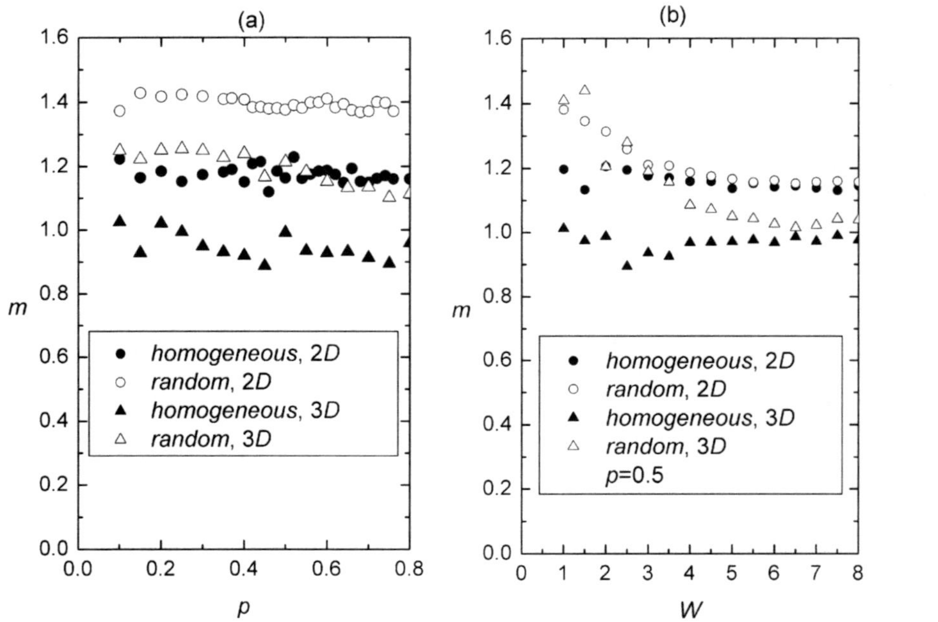

Figure 4. Changes in m on varying p and w. a) w=1, $N_0 = 260$ for $2D$, w=3, $N_0 = 80$ for $3D$. b) p=0.5; $N_0 = 260$ for $2D$, $N_0 = 80$ for $3D$.

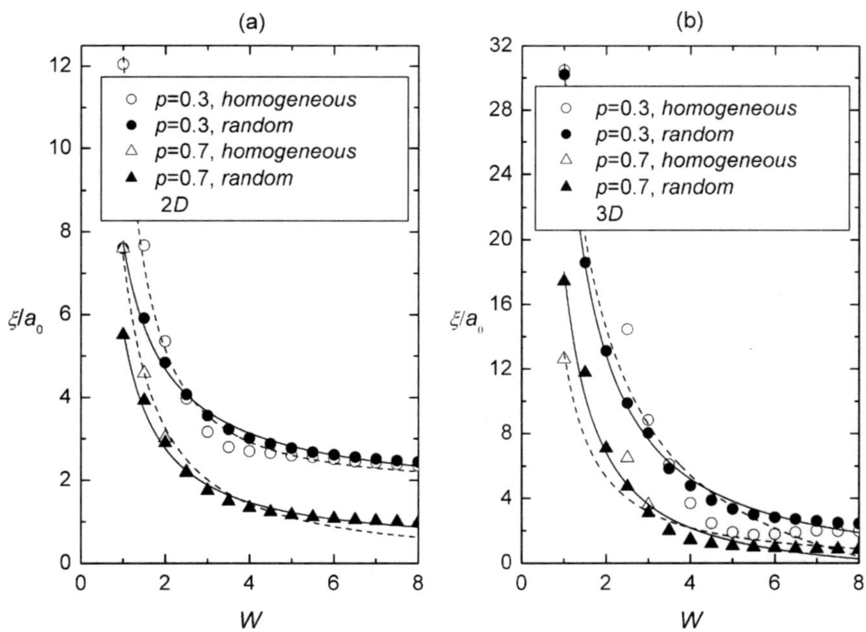

Figure 5. $\xi(w)$ variations for different initial configurations for a) 2D, $N_0 = 260$ and b) 3D, $N_0 = 80$. Only for the random initial configuration the Imry-Ma theorem is obeyed.

We expect that even in the strong anchoring limit, the finite size of the simulation cells will induce a non-zero coherence length. The fit of eq. (7) shows Imry-Ma behavior at low w, while still allowing finite-size correlated regions in the infinite anchoring limit. The fitting parameters for representative runs are exhibited in Table 1.

A summary of the conclusions from this fitting procedure is as follows. In the very strong anchoring limit ($w > 10$) the value of ξ does not depend on the history of the system. In the weak anchoring regime we find that $\xi_h > \xi_r$, where ξ_h and ξ_r represent the coherence lengths obtained from homogeneous and random initial configurations. For the r initial configurations, for which there is strong evidence of SRO, we obtain values close to the Imry-Ma prediction $\xi(2D) \propto w^{-1}$ and $\xi(3D) \propto w^{-2}$ for all p. But h initial configurations do not exhibit short range order. The value of ξ now comes from $G_1(r)$, that part of the correlation function which remains after the long range order has been subtracted. Here, Imry-Ma behavior is not expected, and indeed it is not observed. More interestingly however, we still find evidence of a scaling law, although the scaling parameters are now approximately $\xi \sim w^{-1.6}\ (d=2)$ and $\xi \sim w^{-3.2}\ (d=3)$. We do not, however, have any explanation of this result at this stage, and further investigation is required.

Table 1 Values of fitting parameters defined by Eq. (7) for representative simulation runs.

Initial condition	d	p	γ	ξ_0	ξ_∞
r	2	0.3	0.95±0.12	6.28±0.31	1.43±0.34
r	2	0.5	0.99±0.09	5.86±0.21	0.57±0.22
r	2	0.7	0.97±0.13	5.56±0.30	0.00±0.32
r	3	0.3	2.11±0.33	62±17	1.38±0.57
r	3	0.5	1.97±0.19	37±4	0.35±0.32
r	3	0.7	2.20±0.32	36±7	0.00±0.36
h	2	0.3	1.62±0.08	10.42±0.19	1.79±0.13
h	2	0.5	1.60±0.07	8.35±0.13	1.05±0.09
h	2	0.7	1.57±0.11	7.07±0.19	0.60±0.14
h	3	0.3	3.29±0.23	297±60	0.90±0.28
h	3	0.5	3.29±0.13	159±14	0.80±0.15
h	3	0.7	3.15±0.26	99±18	0.50±0.22

Legend: initial configuration, r: random initial configuration, h: homogeneous initial configuration.

5. Conclusions

We have studied the influence of randomly distributed impurities on orientational ordering of an ensemble of anisotropic rod-like objects. The impurities impose random-anisotropy disorder. We have used an impurity-modified version of the Lebwohl-Lasher interaction[6,10,15], on a d-dimensional cubic lattice with d=2, 3. The director configuration has been obtained by minimizing the total interaction energy of the system, where we have neglected the role of thermal fluctuations. This system represents the simplest toy model in which domain-type formation can be studied in phases or structures obtained via a continuous symmetry breaking transition. The pronounced universality in these systems is suggestive that detailed investigation of this simple model will reveal useful fundamental information. The system studied can be lyotropic (randomly perturbed nanorods dispersed in an isotropic liquid[19]) or thermotropic (liquid crystals in the nematic phase[6]). Examples of the source of the randomness could be either the random geometry due to substrate-confining anisotropic particles, or some kind of random network of pores acting as a matrix for the fluid. In the case related to the latter case, in LC-aerosol nanoparticle mixtures[20], the pore matrix is thought to form a network exhibiting fractal properties on large enough scale.

The simulations yielded configurations, for which we have calculated the orientational correlation function $G(r)$. This quantity enabled calculation of (a) the average coherence length ξ, and (b) the range of ordering. Within a volume $V_d \approx \xi^d$, which we refer to as a domain, the rod-like objects are relatively strongly correlated. In simulation we presented the objects as unit vectors exhibiting head-to-tail invariance. As initial configuration we either consider i) randomly distributed or ii) homogeneously aligned directors. The first case mimics experimental conditions where an isotropic phase is suddenly quenched (e.g., by sudden decrease in temperature or increase in pressure) into an ordered phase, or a zero-field cooled sample.

The second case corresponds to sudden switch-off of an ordering external electric or magnetic field in an orientationally ordered phase, or a field-cooled sample. We have studied domain characteristics as a function of impurity concentration p, coupling strength between impurities and directors w, and sample history.

Our results suggest that configurations reached by a quench from the isotropic phase always exhibit short range order. By contrast, configurations reached from a homogeneous initial condition can either exhibit long-range order or quasi- long-range order. This observation is the key result of our study. Different initial conditions can thus control material properties sensitive to domain patterning. We further show that structures with short-range order obey the Imry-Ma scaling law $\xi \propto w^{-2/(d-4)}$ in the weak and moderately weak anchoring regime[5]. Results indicate that $\xi(w)$ dependence obtained from random initial configurations show universal behavior. We have also discovered a scaling law for sizes of fluctuating domains in the LRO regime, but we do not know the cause of this interesting scaling law at this stage.

Acknowledgements

We are grateful to S. Rzoska and V. Mazur for the opportunity of presenting these results in Odessa. M. Cvetko acknowledge the support of the European Social Fund. S. Kralj acknowledges support of grant J1-0155 from ARRS, Slovenia. T.J. Sluckin acknowledges many useful conversations with P. Shukla (Shillong, India).

References

1. Zurek, W. H. (1985) *Nature* **317**, 505
2. Bray, A. J. (1994) *Adv. Phys* **43**, 357
3. Kibble, T. W. B. (1976) *J. Phys* A **9**, 1387
4. de Gennes, P. G., and Prost, J. (1993) *The Physics of Liquid Crystals* (Oxford University Press, 2nd edition : Oxford)
5. Larkin, A. I. (1970) *Sov. Phys. JETP* **31**, 784. (1970) *Zh. Eksp. Teor. Fiz.* **58**, 1466. See also Imry, Y. and Ma, S. (1975) *Phys. Rev. Lett.* **35**, 1399
6. Bellini, T., Buscagli, M., Chioccoli, C., Mantegazza, F., Pasini, P., and Zannoni, C. (2000) *Phys. Rev. Lett.* **85**, 1008
7. Aharony, A., and Pytte, E. (1980), *Phys. Rev. Lett.* **45**, 1583
8. Berezinskii, V. L. (1970) *Sov. Phys. JETP* **32**, 493. (1970) *Zh. Eksp. Teor. Fiz.* **59**, 907; Kosterlitz, J. M., and Thouless, D. J. (1973) *J. Phys.C* **6**, 1181.
9. Feldman, D. E. (2000) *Phys. Rev. Lett.* **85**, 4886
10. Chakrabarti, J. (1998) *Phys. Rev. Lett.* **81**, 385
11. Leaver, D. J., Kralj, S., Sluckin, T. J., and Allen, M. P. (1996) *Liquid Crystals in Complex Geometries Formed by Polymer and Porous Networks*, ed. Crawford G. P., and Zumer S. (Oxford University Press: London)
12. Radzihovsky, L., and Toner, J. (1997) *Phys. Rev. Lett.* **79**, 4214
13. Kralj, S., and Popa-Nita, V. (2004) *Eur. Phys. J.* E **14**, 115. Popa-Nita, V., and Kralj, S. (2006) *Phys. Rev.* E **73**, 041705
14. Harris, R., Plischke, M., and Zuckerman, M. J. (1973) *Phys. Rev. Lett.* **31**, 160
15. Lebwohl, P. A., and Lasher, G. (1972) *Phys. Rev.* A **6**, 426
16. Vetko, M., Ambrozic, M., and Kralj, S. (2009) accepted by *Liq. Cryst.*
17. Fabbri, U., and Zannoni, C. (1986) *Mol. Phys.* **58**, 763

18. See e.g. Phillips, C. (1996) *Rep. Prog. Phys.* **59**, 1133-1207
19. Lagerwall, J., Scalia, G., Haluska, M., Dettlaff-Weglikowska, U., Roth, S., and Giesselmann, F. (2007) *Adv. Mater.* **19**, 359
20. Haga, H., and Garland, C. W. (1997) *Phys. Rev.* E **56**, 3044

PHASE ORDERING IN MIXTURES OF LIQUID CRYSTALS AND NANOPARTICLES

BRIGITA ROŽIČ[1], MARKO JAGODIČ[2], SAŠO GYERGYEK[1], GOJMIR LAHAJNAR[1], VLAD POPA-NITA[3], ZVONKO JAGLIČIĆ[2], MIHAEL DROFENIK[1], ZDRAVKO KUTNJAK[1], SAMO KRALJ[1,4]

[1]*Jožef Stefan Institute, Jamova 39, 1000 Ljubljana, Slovenia*
[2] *Institute of Mathematics, Physics and Mechanics, Jadranska 19, 1000 Ljubljana, Slovenia*
[3]*Faculty of Physics, University of Bucharest, P.O.Box MG-11, Bucharest 077125, Romania*
[4]*Faculty of Natural Sciences and Mathematics, University of Maribor, Koroška 160, 2000 Maribor, Slovenia*

Abstract: We have studied the coupling interaction between liquid crystal (LC) molecules and nanoparticles (NPs) in LC+NPs mixtures. Using a simple phenomenological approach, possible structures of the coupling term are derived for strongly anisotropic NPs. The coupling terms include (i) an interaction term promoting the mutual ordering of the LC molecules and the NPs, and (ii) the Flory-Huggins-type term enforcing the phase separation. Both contributions exhibit the same scaling dependence on the diameter of the NPs. However, these terms only exist for a finite degree of nematic LC ordering. The magnetic response due to the LC-NPs coupling is probed experimentally for a mixture of weakly anisotropic magnetic NPs and a ferroelectric LC. A finite coupling effect was observed in the ferroelectric LC phase, suggesting such systems can be used as soft magnetoelectrics.

Keywords: liquid crystals, nanoparticles, mixtures, structural ordering, magnetoelectircs

1. Introduction

Nanoparticles (NPs) are expected to revolutionize many aspects of our lives. Consequently, intensive research has been devoted in recent years to producing new NPs exhibiting extraordinary properties. Among them, particular attention has been paid to carbon nanotubes (CNTs). These CNTs have a large aspect ratio and therefore exhibit most of their remarkable properties in a single direction, i.e., along the tube axis. However, the common CNT production methods[1,2,3] give rise to complex, entangled aggregates in which the anisotropic properties are drastically reduced. Standard CNT alignment techniques (field-assisted alignment[4], shearing[5], or molecular combing[6] processes) either result in poor CNT alignment over macroscopic scales or are too complicated for useful applications. Therefore, it is of great interest to develop efficient and relatively simple techniques appropriate for CNT alignment. In addition, mixtures of NPs with other conventional materials could exhibit exotic behaviours not found in either of the individual components [7].

Recently, it has been shown experimentally[8,9,10] that the spontaneous onset of liquid crystal (LC) ordering[11] could be a way to obtain extremely well aligned NPs. LCs are typical representatives of the class of soft materials, exhibiting a long-range orientational ordering under appropriate conditions. The most important property of soft materials is their extreme response to various perturbations (e.g., from the surface, from external electric or magnetic fields, or from impurities). LCs are optically transparent and consist of anisotropic molecules that become ordered, either over a given temperature interval (thermotropic LCs) or for appropriate concentrations of LC molecules (lyotropic LCs). For the purposes of an illustration the discussion will be limited to bulk (macroscopically large) samples of thermotropic LCs formed by rod-like molecules, thus neglecting the effects of sample boundary surfaces. For high enough temperatures LCs exist in the ordinary isotropic liquid phase. However, by decreasing the temperature T, various LC phases can appear before a solid phase is reached. A typical sequence is as follows. At $T=T_{IN}$, the nematic (N) phase is reached in which molecules tend to be aligned along a single symmetry-breaking direction $\vec{n}$. At $T=T_{NA}< T_{IN}$ a smectic A (SmA) ordering is established in which the molecules arrange in equidistant and parallel layers. Here, the average orientation of the molecules is along the normal $\vec{\nu}$ of layers. Consequently, in addition to the orientational order a quasi-long-range positional order is established. At $T=T_{AC}<T_{NA}$, a smectic C ordering is formed in which LC molecules become tilted with respect to $\vec{\nu}$. In the case of chiral molecules the structural ordering exhibits a helical structure and for symmetry reasons the electrical polarisation $\vec{P} \propto \vec{n}\times\vec{\nu}$ appears. The resulting ferroelectric phase is labelled SmC*.

We are also interested in magnetoelectrics[12], which form an important subgroup of multiferroics. Magnetoelectrics exhibit, simultaneously, ferromagnetic and ferroelectric properties, the combination of which is expected to revolutionize computers. A memory element combining both properties would permit an electric write operation combined with a magnetic read, eliminating the need for the destructive read (and reset) in present-day ferromagnetic RAMs, thus making possible fatigue-free memories. Furthermore, in such materials it would be possible to control the magnetic properties via the electric ones, and vice versa. Single-component magnetoelectrics, in which the electric and magnetic properties are directly coupled, are extremely rare. This is because the usual mechanisms that produce ferroelectricity and ferromagnetism are incompatible at the atomic level. In addition, the coupling of both properties is relatively weak, and such multiferroic states rarely exist at room temperature. For this reason, recently, interest has shifted to the magnetoelectric states in composites, in which the coupling between the magnetic and electric order is indirect, but nevertheless, much stronger than in the single-component case.

In this contribution mixtures of LCs and NPs are considered. First, the coupling between the components is theoretically analyzed using simple mesoscopic approaches for NPs possessing a large aspect ratio *L/D*. Here *D* and *L* stand for the diameter and length of the NP, respectively. For low concentrations of NPs, a weak and relatively strong coupling regime is considered, where the LC ordering is weakly and strongly perturbed by the NPs, respectively. Our modelling suggests that the effective coupling strength w_{eff} strongly depends on *D* and is independent on *L*. With a decrease in *D*, the w_{eff} value increases, favouring the parallel alignment of the NPs and the LC molecules. However, at the same time the phase-separation tendency increases. Next, we experimentally analyse mixtures of nearly spherical ferromagnetic NPs and ferroelectric LCs that could potentially form new magnetoelectric materials. Although a rough estimate suggests negligible coupling between the components, our magnetic measurements reveal a finite coupling strength. Based on our theoretical esimates the coupling strength could be further amplified by using strongly anisotropic NPs.

The paper is organised as follows. The theoretical background is presented in Sec. II. In Sec. III we theoretically analyse the coupling between the anisotropic NPs and the nematic LC ordering. We consider a weak and a relatively strong coupling regime. We estimate both the mutual ordering and the phase-separation effects. In Sec. IV we experimentally analyse the magnetic response of a mixture of ferromagnetic NPs and a ferroelectric LC across the SmA-SmC* phase transition. In the final section we summarize our results.

2. Theoretical background

In this section the continuum and order parameter fields of our simple mesoscopic approach is presented. The free energy of the mixtures and a review of the most important known facts relevant to our investigation are introduced.

2.1. Continuum and order parameter fields

The orientational ordering in the nematic phase is conventionally expressed in terms of the nematic director field $\vec{n}$. Here $\pm\vec{n}$ orientations are equivalent from the physical point of view and $|\vec{n}|=1$. The extent of fast (with respect to the characteristic time of collective excitations in $\vec{n}$) fluctuations about $\vec{n}$ is given by the nematic order parameter

$$S_{LC}(\vec{r})=\left\langle \tfrac{1}{2}\left(3\cos^2\vartheta-1\right)\right\rangle, \tag{1}$$

where ϑ is the angle between the average and the temporal orientation of a LC molecule at a mesoscopic point $\vec{r}$, and <...> stands for the local statistical averaging. The isotropic liquid-like phase is characterized by S_{LC} =0, and a rigidly aligned ordering along $\vec{n}$ yields S_{LC} =1.

In the same manner the orientational order parameter of the NPs is defined as

$$S_{NP}(\vec{r})=\left\langle \tfrac{1}{2}\left(3\cos^2\theta-1\right)\right\rangle, \tag{2}$$

where θ is the angle between the average and the temporal NP orientation at $\vec{r}$. The volume concentration of the NPs is labelled with ϕ. Therefore, the volume concentration of a LC phase is given by $1-\phi$.

2.2. Free energy of the mixture

The average free-energy density of the NP-LC mixture is expressed in terms of the spatially averaged orientational order parameters S_{LC} and S_{NP} of the mixture components LC and NP as

$$\overline{f}=\overline{f}_{mix}+\overline{f}_{NT}+\overline{f}_{LC}+\overline{f}_{c}, \tag{3}$$

where the over-bar denotes the spatial average.

The so-called mixing free-energy density term $\overline{f}_{mix}$ is expressed in the frame of the regular solution theory as [13]

$$\overline{f}_{mix}=\frac{k_BT}{V_{NP}}\phi\ln\phi+\frac{k_BT}{V_{LC}}(1-\phi)\ln(1-\phi)+\chi\phi(1-\phi). \tag{4}$$

Here, k_B stands for the Boltzmann constant, V_{NP} and V_{LC} represent the volumes of the NP and the LC molecule, respectively. The quantity χ is the mixing parameter, also called the Flory-Huggins parameter, which for a large enough positive value triggers the phase separation in the system. Its value is approximately given by $\chi \approx U_{LC-NP} - \frac{1}{2}(U_{LC-LC} + U_{NP-NP})$, where $U_{LC\text{-}NP}$, $U_{LC\text{-}LC}$, and $U_{NP\text{-}NP}$ stand for the representative LC-NP, LC-LC and NP-NP molecular interactions of neighbouring molecules, respectively. The term $\overline{f}_{mix}$ is essentially entropic in nature and roughly takes into account the number of ways of rearranging the particles before and after mixing. In the latter case a random mixing is assumed.

For the free-energy density of the NPs the Onsager-Flory-type model[14-16] is used, where

$$\overline{f}_{NP} = \frac{k_B T}{V_{NP}} \phi \left(\frac{1}{2}\left(1 - \frac{u}{3}\right) S_{NP}^2 - \frac{u}{9} S_{NP}^3 + \frac{u}{6} S_{NP}^4 \right). \tag{5}$$

The quantity u is a scaled volume fraction of the NPs. For rod-like NPs of diameter D and length L it is roughly given by $u \approx \phi L / D$. This term enforces a first-order structural transition from the isotropic (S_{NP}=0) to an orientationally liquid-crystal-like ordered phase (S_{NP}>0) at the critical scaled concentration $u = u_c \equiv 2.7$. For the LC free-energy density $\overline{f}_{LC} = \overline{f}_b + \overline{f}_e$ the simplified Landau-de Gennes-type approach [11] is used. The condensation bulk free-energy density is expressed as

$$\overline{f}_b = (1-\phi)\left(a(T - T_*) S_{LC}^2 - b S_{LC}^3 + c S_{LC}^4\right), \tag{6}$$

where the pre-factor $(1-\phi)$ takes into account the volume occupied by the LC (i.e., for $\phi = 1$ this term is absent). The material constants a, b and c are assumed to be independent of the temperature, and T_* is the spinoidal temperature of the isotropic LC phase. On decreasing the temperature the condensation term triggers a first-order phase transition into the nematic LC phase at $T_{IN} = T_* + b^2/(4ac)$.

The elastic term takes into account the elastic restoring force, tending to establish a spatially uniform LC ordering. It is expressed as

$$\overline{f}_e = (1-\phi)\left(\frac{k S_{LC}^2}{2} \overline{|\nabla \vec{n}|^2} + \frac{k_0}{2} \overline{|\nabla S_{LC}|^2} \right). \tag{7}$$

Only the most essential terms are taken into account, where k and k_0 are temperature-independent elastic constants. The quantity estimates the average nematic Frank elastic constant.[11]

$$K \approx k S_{LC}^2 \tag{8}$$

The term $\overline{f}_c \propto \phi(1-\phi)$ takes into account the coupling interaction between the LCs and the NPs. Its structure will be estimated in the next section.

2.3. LC-driven ordering of NPs

Recent experiments [8,9,10] have shown that a nematic LC phase could well align NPs along the preferential LC ordering and suggest the following relationship for the effective alignment free energy:

$$F(\psi) = F_{\perp} + (F_{II} - F_{\perp})\cos^2\psi\ . \tag{9}$$

Here, ψ is the angle between the average orientation of $\vec{n}$ in the sample and the NP orientation, and F_{II} and $F_{\perp}$ stand for the effective energies for $\psi = 0$ and $\psi = \pi/2$, respectively.

There have been several theoretical investigations analysing the nematic distortions of the LC surrounding anisotropic particles for different anchoring conditions at the LC-particle interface. In modelling the surface-anchoring free-energy density f_a it is usual to adopt the Rapini-Papular expressions[17], which can be written as

$$f_a = f_0 - \frac{W}{2}(\vec{n}\cdot\vec{e})^2 . \tag{10}$$

Here, f_0 stands for the isotropic part of the interfacial coupling, W measures the anchoring strength and the unit vector $\vec{e}$ points along the so-called easy direction. If the LC molecules are aligned along $\vec{e}$ at the interface, the anchoring free-energy penalty in minimized. The temperature dependence of W is, in most cases, approximated well by

$$W \approx wS_{LC}\ , \tag{11}$$

where w is a temperature-independent constant. An important parameter measuring the anchoring strength is the surface extrapolation length[11]

$$d_e \approx \frac{K}{W} . \tag{12}$$

A weak or strong anchoring regime is estimated from the value of the dimensionless ratio

$$\mu = \frac{D}{d_e} \approx \frac{WD}{K} . \tag{13}$$

Here, D stands for the characteristic geometrical length of the system. In the case of spherical particles or strongly anisotropic rod-like particles immersed in a nematic LC phase, D stands for the particle diameter or width, respectively. The strong anchoring regime, in which the surface-imposed tendency is strongly obeyed, corresponds to $\mu >> 1$. The weak anchoring regime is determined by $\mu \approx 1$.

For the case of anisotropic particles with a rod-like shape (of length L and diameter D) in a nematic LC phase it was shown that the typical free-energy costs ΔF due to the elastic distortions at the LC-NPs interface is roughly[18,19]

$$\Delta F \approx KL, \tag{14}$$

which further depends on the angle ψ between the particle's long axis and the average LC orientation far from the particle. For example, for strong enough anchoring and different anchoring conditions, where alignment either perpendicular or parallel to the interface is enforced, we have[19] $\Delta F_{\perp} \approx \pi KL \ln(D/d_e)$ and $\Delta F_{II} \approx \pi KL$ for $\psi = \pi/2$ and $\psi = 0$, respectively.

3. Orientational coupling between anisotropic nanoparticles and LC molecules

Next we will look at possible structures of the coupling term $\overline{f_c}$ in Eq. (3) for anisotropic particles. A more detailed analysis is presented in Ref.[[20]]. The NPs are treated as cylindrical objects with diameter D and length L, and $L/D>>1$. First, a dilute regime is considered, where the indirect interactions among the NPs play a secondary role. The very weak and strong anchoring regimes are analysed. Then an interaction between the NPs is considered in the case of apparently elastically distorted mixtures. Finally, the mutual ordering effects between the LC and the NPs are analysed, emphasising the qualitative change in the behaviour with respect to isolated components.

3.1. Weak anchoring regime

Consider a NP immersed in a nematic LC in the case of a weak anchoring regime, i.e., where $\mu = D/d_e \leq 1$. In this case $\vec{n}$ is negligibly affected by the presence of the NPs, although the reverse orientational effect is not negligible. Such a situation was suggested experimentally by Lynch and Patrick[8]. The simplest possible Rapini-Papoular-type ansatz for the anchoring consistent with Eq. (9) is

$$W = W_0 + W_a(\vec{n}\cdot\vec{e}_{II})^2 + W_p(\vec{n}\cdot\vec{e}_{\perp})^2. \tag{15}$$

Here, the quantities W_0, W_a and W_p stand for the isotropic, azimuthal, and polar anchorings, respectively. The unit vectors $\vec{e}_{II}$ and $\vec{e}_{\perp}$ point along the NP symmetry axis and perpendicular to it. W is integrated over the particle surface, neglecting its end parts. The resulting surface-anchoring free-energy penalty is

$$F_a = \pi LDW_i - \frac{\pi}{3} LD\Delta W P_2(\psi), \tag{16}$$

where $P_2(\psi) = (3\cos^2\psi - 1)/2$ is the second Legendre polynomial, ψ is the angle between the long axis of the NP and $\vec{n}$ (i.e., $\psi = \arccos(\vec{n}\cdot\vec{e}_{II})$), $W_i = W_0 + \frac{2}{3}W_\perp + \frac{1}{3}W_{II}$ is the net isotropic anchoring constant and $\Delta W = W_\perp - W_{II}$ is the net anisotropic anchoring constant. NPs tend to orient parallel to n for $\Delta W > 0$, and recent experiments[3-10] suggest this condition. Furthermore, taking into account Eq. (16) the distribution probability function $P(\psi)$ of the NPs within a homogeneously aligned nematic LC phase is approximately given by $P(\psi) = A\exp(-F_a / k_b T)$, where A is the normalization constant.

The average of F_a over the angles ψ yields $\overline{F}_a = \pi LDW_i - \pi LD\Delta W S_{NP}/3$, and the average coupling free-energy density term is expressed as

$$\overline{f}_c \approx N_{NP}\overline{F}_a / V. \tag{17}$$

Here, N_{NP} stands for the number of NPs within the volume V of the system. Furthermore, by taking into account that $\phi = N_{NP}V_{NP}/V$, $V_{NP} \approx \pi D^2 L/4$, $\Delta W \approx \Delta w S_{LC}$, and $W_i \approx w_i S_{LC}$ (see Eq. (11)), and that in the limit $\phi = 1$ the coupling term vanishes (i.e., the LC component is absent), it follows that

$$\overline{f}_c = f_{FH} + f_{\text{int}}, \tag{18}$$

where f_{FH} and f_{int} stand for the Flory-Huggins-type term and the interaction free-energy density term, respectively:

$$f_{FH} = \phi(1-\phi)\frac{4w_i}{D} S_{LC}, \tag{19a}$$

$$f_{\text{int}} = -\phi(1-\phi)\frac{4\Delta w}{3D} S_{LC} S_{NP}. \tag{19b}$$

Note that in this approximation both terms are independent of L. If the Flory-Huggins-type term is large enough it triggers a phase separation in the nematic LC phase, where $S_{LC}>0$. Its strength is inversely proportional to D. The interaction term promotes nematic ordering in both components, and $f_{\text{int}} \propto 1/D$. Therefore, with a decreasing diameter of NPs the interaction with the LC component is increasing, but at the same time the phase-separation tendency is also increasing.

3.2. The strong anchoring regime

Here, the case of an anisotropic NP immersed in a nematic LC is considered, where the anchoring strength at the NP-LC interface is relatively strong (i.e.,

$\mu = D/d_e > 1$). It is assumed that sufficiently far from a NP (a distance comparable to d_e) the nematic ordering is homogeneous along a single symmetry direction $\vec{n}_0$. Therefore, the validity of our estimates is limited to the diluted regime (i.e., $\phi << 1$). It is assumed that Eqs.(9) and (14) are roughly obeyed, and that we set $\Delta F_{II} \approx a_{II}\pi KL$, $\Delta F_{\perp} \approx a_{\perp}\pi KL$, where a_{II} and $a_{\perp}$ are positive constants of order one. It is further imposed that $a_{\perp} > a_{II}$, so that the LC molecules and the NPs tend to align parallel. After averaging Eq. (9) over the angle ψ it follows that

$$\overline{F} \approx \tfrac{2\pi}{3} KL\left(\frac{a_{II}}{2} + a_{\perp} - \frac{2}{3}(a_{\perp} - a_{II})S_{LC}\right). \tag{20}$$

Taking into account Eq. (17) and the relation $K \approx kS_{LC}^2$, and following the same steps as in the previous subsection, we obtain

$$f_{FH} = a_{FH}\phi(1-\phi)\frac{kS_{LC}^2}{D^2}, \tag{21a}$$

$$f_{\text{int}} = -a_{\text{int}}\phi(1-\phi)\frac{kS_{LC}^2}{D^2}S_{NP}. \tag{21b}$$

Here, the positive constants $a_{FH} = 4\left(\tfrac{1}{3}a_{II} + \tfrac{2}{3}a_{\perp}\right)$ and $a_{\text{int}} = \tfrac{8}{3}\left(a_{\perp} - a_{II}\right)$ are of order one. Therefore, also at the strong anchoring limit hawse have similar behaviour as in the weak anchoring regime. The essential differences are the proportionalities $\overline{f}_c \propto (S_{LC}/D)^2$ in the strong and $\overline{f}_c \propto S_{LC}/D$ in the weak anchoring regimes, respectively.

3.3. Interactions among NPs

In the previous subsections the structure of the coupling term in the diluted regime was estimated, where most of the LC molecules are aligned along a single direction. Here, the analysis is addressed to the question of which additional free-energy contributions could emerge if the LC-mediated interactions among nonhomogeneously aligned NPs are significant. For demonstration purposes it is assumed that the LC molecules tend to be oriented perpendicular to a NP surface area (the so-called homeotropic anchoring). The anchoring strength is either of moderate strength ($D/d_e \approx 1$) or it is strong ($D/d_e >> 1$). It is assumed that NPs significantly perturb the LC ordering, where the elastic free-energy penalties are estimated with Eq. (7).

We first consider the moderate anchoring strengths. The average distortions in the nematic director field are roughly given by $\overline{|\nabla \vec{n}|} \approx 1/L_a$, where L_a stands for the average separation between neighbouring NPs. The corresponding average free-energy density penalty is $\overline{f}_e \approx (1-\phi)kS_{LC}^2/(2L_a^2)$. The L_a value depends on the concentration of NPs. For homogeneously distributed NPs hawse have roughly $L_a \approx (V_{NP}/\phi)^{1/3}$, therefore

$$\overline{f}_e \approx (1-\phi)\phi^{2/3}\frac{kS_{LC}^2}{2V_{NP}^{2/3}}. \tag{22}$$

Thus, elastic distortions give rise to a term that is roughly of the Flory-Huggins type, enhancing the phase-separation tendency of the system.

Next, the strong anchoring regime is considered. The homeotropic anchoring condition gives rise to topological defects in the LC medium because each isolated NP introduces a topological charge of strength one (like a hedgehog defect) [21]. However, the overall topological charge of the system is conserved and stays zero (for the appropriate boundary conditions). Therefore, N_{NP} nanoparticles introduce the same number of defects (antihedgehogs) within the LC if the NPs are not in contact. For demonstration purposes it is instructive to restrict ourselves to such cases. Due to the existence of defects the elastic distortions in $\vec{n}$ are changed with respect to the case of the moderate anchoring strength. However, the essential qualitatively new feature with respect to the analysis just presented is that at the cores of topological defects the degree of nematic ordering is strongly suppressed. A typical elastic distortion within the core of volume $4\pi\xi^3/3$ is roughly given by $\overline{|\nabla S_{LC}|} \approx S_{LC}/\xi$, where ξ stands for the nematic order parameter's correlation length [11]. The resulting total elastic free-energy penalty from all the defects within the system is roughly given by $\overline{f_e} \approx (1-\phi)N_{NP}/V(2\pi/3)k_0\xi S_{LC}^2$. Therefore, again the Flory-Huggins-type term appears, which can be expressed as

$$\overline{f_e} \approx (1-\phi)\phi S_{LC}^2\frac{2\pi k_0\xi}{3V_{LC}}. \tag{23}$$

3.4. Effective ordering field

The existence of the LC ordering affects the anisotropic NPs as an effective ordering field. To emphasize this feature the average free energy of the system is expressed as

$$f = f_0 + f_{NP} - w_{eff}S_{NP}\ , \tag{24}$$

where f_0 contains terms independent of S_{NP} and w_{eff} stands for the effective field conjugated to the orientational order parameter of the NPs. In the weak and strong anchoring regimes (see Eq. (19a) and Eq. (21a)) it is expressed as

$$w_{eff} \approx \phi(1-\phi)\frac{k}{D^2}S_{LC}^2, \tag{25a}$$

$$w_{eff} = \phi(1-\phi)\frac{\Delta w}{D}S_{LC}, \tag{25b}$$

respectively. For $w_{eff} > 0$ we obtain $S_{NP} > 0$ (i.e., paranematic or nematic ordering) for any ϕ. Furthermore, on increasing ϕ, the orientational ordering of the NPs exhibits a gradual evolution for $w_{eff} \geq w_t$, where w_t stands for the tricritical value of the effective field. The tricritical point is defined via the condition $\frac{\partial f}{\partial S_{NP}} = \frac{\partial^2 f}{\partial S_{NP}^2} = \frac{\partial^3 f}{\partial S_{NP}^3} = 0$, corresponding to the conditions $S_{NP}^{(t)} = 1/6$, $u^{(t)} = 18/7$, $w_{eff}^{(t)} = 1/126$, where the superscript (t) refers to the tricritical state.

4. Magnetic behaviour in a mixture of LC and magnetic NPs

Next, a mixture of a ferroelectric LC and ferromagnetic NPs is considered, as it could potentially exhibit magnetoelectric properties. For this purpose magnetic measurements were performed to probe the coupling strength between the LC ordering and the magnetic properties of the NPs, where the NPs are weakly anisotropic (see Fig. 1). If a finite coupling exists, the theoretical framework described above suggests that strongly anisotropic NPs could further enhance the strength of this coupling. Note that the rough estimate given in Ref. [18] suggests a negligibly small coupling strength in mixtures of this type

Measurements were performed on a SCE9 liquid crystal, which contains the ferroelectric SmC* phase. The pure bulk SCE9 phase sequences with decreasing temperature from the isotropic (I) phase are as follows: the I-N, N-SmA, and SmA-SmC* phase transitions take place at $T_{IN} \approx 392K$, $T_{NA} \approx 360K$, and $T_{AC} \approx 334K$, respectively. For the magnetic NPs we used weakly anisotropic maghemite ($\gamma - Fe_2O_3$) particles of 17 nm diameter that were covered with oleic acid. The LC+NPs mixtures with concentrations $x = 0.005$ and 0.10 were investigated, where $x = m_{NP}/(m_{NP} + m_{LC})$, and m_{NP} and m_{LC} denote the masses of the NPs and the LC in the samples, respectively. A typical transmission electron microscopy (TEM) image of NPs dispersed in toluene is presented in Fig. 1.

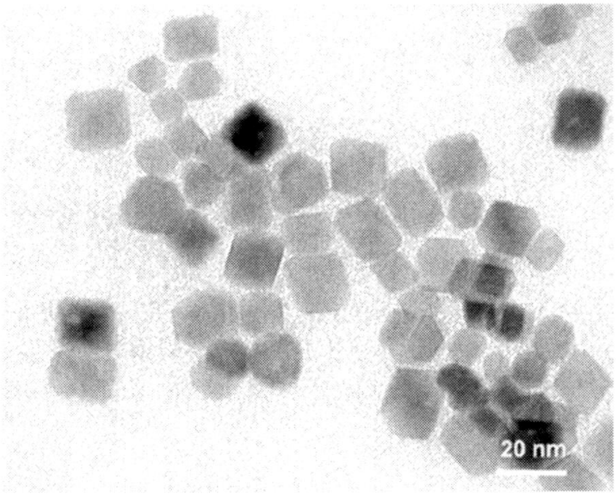

Figure 1. *TEM image of maghemite nanoparticles covered with the oleic acid dispersed in toluene.*

The preparation of the $\gamma - Fe_2O_3$ nanoparticles and mixtures was as follows. Oleic-acid-coated hydrophobic particles were synthesized by the coprecipitation of Fe(II) and Fe(III) cations using ammonia. The synthesized NPs were, on average, 11 nm in size. In order to promote NP growth, suspensions of NPs were treated hydrothermally at 200°C for 3 hours. The hydrophobic nanoparticles were precipitated by adding HNO_3. The particles were soaked in oleic acid and the excess oleic acid was removed by washing the nanoparticles in acetone. The NPs were then dispersed in toluene. The average linear size of the NPs was estimated to be 17 nm. The LC-NP mixtures were prepared by dissolving the LC in toluene and adding to this mixture the magnetic nanoparticles also dispersed in toluene. By thoroughly mixing these samples for approximately 2 hours at 393 K relatively homogeneous dispersions were obtained, and then all the solvent was allowed to evaporate. The samples obtained were inserted into thin glass tubes appropriate for the magnetic susceptibility measurements.

The magnetic properties of these mixtures were investigated on a commercial SQUID-based magnetometer with a 5 T magnet (Quantum Design MPMS XL-5). The measurements were performed in the temperature interval covering the pure bulk LC SmA-SmC* phase transition. The samples were first heated to about 350 K in a zero magnetic field. This temperature is above the phase-transition temperature of the liquid crystal (334 K). Next, an external magnetic field *H*=100 Oe was applied and the temperature dependence of the sample's magnetization was measured from 350 K to 320 K and back to 350 K with a cooling/heating rate of 0.15 K/min. In addition, another cooling experiment similar to this just described was performed. The sample was cooled from 350 K to 320 K in a magnetic field of 100 Oe but with a cooling rate of 1 K/min. Temperature dependencies of the excess magnetization $\Delta m = m - m_L$ show an apparent anomalous response due to the sufficiently strong coupling for x=0.1. A representative example is shown in Fig. 2, m_L represents the linear temperature dependence of the magnetization in the SmA phase, extrapolated to low temperatures. We attribute the departures of $m(T)$ from $m_L(T)$ below T_{AC} to the coupling between the magnetic moments and the LC director field. Therefore, our measurements confirm the finite coupling between the LC and magnetic NPs even for weakly anisotropic NPs.

5. Conclusions

To conclude, we have studied the coupling strength between NPs and LC molecules in LC+NP mixtures.

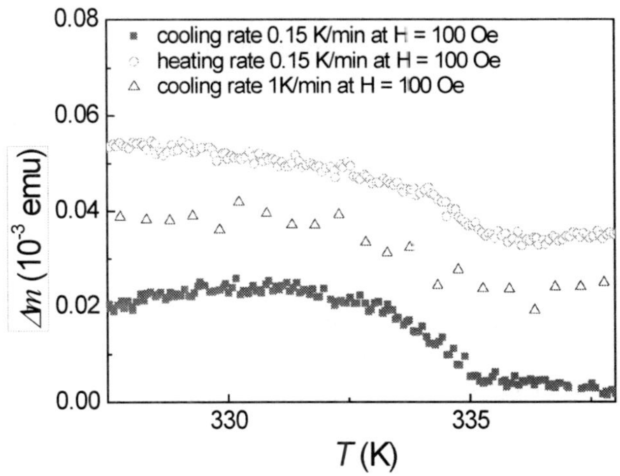

***Figure** 2. Excess magnetization* Δm *of the liquid-crystal compound mixed with maghemite nanoparticles of 17 nm. For clarity, the* Δm *data obtained during the heating (open circles) and cooling (open triangles) runs were shifted by* $0.035 \cdot 10^{-3}$ *emu and* $0.025 \cdot 10^{-3}$ *emu, respectively.* $x = 0.10$.

Theoretically, we have focused on the structure of the free-energy density coupling term in the nematic LC phase for strongly anisotropic particles. We used a simple phenomenological model covering both the weak and the strong anchoring regimes. Our investigation suggests that in all cases the coupling free-energy term $\overline{f}_c = f_{\text{int}} + f_{FH}$ strongly depends on the diameter D of the NPs. Both the interaction (f_{int}) and the Flory-Huggins-type (f_{FH}) contributions are proportional to S_{LC}/D and $(S_{LC}/D)^2$ in the weak and strong anchoring regimes, respectivelyWe have also studied the magnetic susceptibility for a mixture of the ferroelectric SCE9 liquid crystal and spherical maghemite NPs of average diameter 17 nm using a SQUID susceptometer. The anomaly in the excess magnetization observed at the SmA-to-SmC* phase transition reveals an apparent coupling strength between the liquid-crystal ordering and the magnetization of the nanoparticles. Such a coupling allows the possibility of an indirect interaction between the magnetic and ferroelectric order, thus making such mixtures candidates for indirectsoft magnetoelectrics.

References

1. Iijima, S. (1991) *Nature* **354**, 56
2. Endo, M., Takeuchi, K., Igarashi, S., Kobori, K., Shiraishi, M., and W. Kroto, H. (1993) *J. Phys. Chem. Solids* **54**, 1841
3. Colbert, D. T., Zhang, J., McClure, S. M., Nikolaev, P., Chen, Z., Hafner, J. H., Owens, D. W., Kotula, P. G., Carter, C. B., Weaver, J. H., Rinzler, A. G., and Smalley, R. E. (1994) *Science* **266**, 1218
4. Kamat, P. V., Thomas, K. G., Barazzouk, S., Girishkumar, G., Vinodgopal, K., and Meisel, D. (2004) *J. Am. Chem. Soc.* **126**, 10757
5. Wang, H., Christopherson, G., Xu, Z., Porcar, L., Ho, D., Fry, D., and Hobbie, E. (2005) *Chem. Phys. Lett.* **416**, 182
6. Gerdes, S., Ondarcuhu, T., Cholet, S., and Joachim, C. (1999) *Europhys. Lett.* **48**, 292
7. Balazs, A. C., et al. (2006) *Science* **314**, 1107
8. Lynch, M. D., and Patrick, D. L. (2002) *Nano. Lett.* **2**, 1197
9. Dierking, I., Scalia, G., and Morales, P. (2005) *J. of Appl. Phys.* **97**, 044309
10. Lagerwall, J., Scalia, G., Haluska, M., Dettlaff-Weglikowska, U., Roth, S., and Giesselmann, F. (2007) *Adv. Mater.* **19**, 359
11. de Gennes, P. G., and Prost, J. (1993) *The Physics of Liquid Crystals*, Oxford University Press, Oxford
12. Scott, J. F. (2007) *Science* **315**, 954
13. Doi, M., and Edwards, S. F. (1989) *Theory of Polymer Dynamics* (Clarendon, Oxford)

14. Onsager, L. (1949) *Ann. N. Y. Acad. Sci.* **51**, 727
15. Flory, P. J. (1956) *Proc. R. Soc. A* **243**, 73
16. Doi, M. J. (1981) *Polym. Sci., Part B: Polym. Phys.* **19**, 229
17. Rapini, A., and Papoular, M. (1969) *J. Phys. (Paris) Colloq.* **30**, C4-54
18. Brochard, F., and de Gennes, P. G. (1970) *J. Phys. (Paris)* **31**, 691
19. Burylov, S. V., and Raikher, Yu. L. (1994) *Phys. Rev. E* **50**, 358
20. van der Schoot, P., Popa Nita, V., and Kralj, S. (2008) *J. Phys. Chem. B* **112**, 4512
21. Lubensky, C., Pettey, D., Currier, N., and Stark, H. (1998) *Phys. Rev. E* **57**, 610

ANOMALOUS DECOUPLING OF THE DC CONDUCTIVITY AND THE STRUCTURAL RELAXATION TIME IN THE ISOTROPIC PHASE OF A ROD-LIKE LIQUID CRYSTALLINE COMPOUND

ALEKSANDRA DROZD-RZOSKA AND SYLWESTER J. RZOSKA
Institute of Physics, University of Silesia, ul. Uniwersytecka 4, 40-007 Katowice, Poland; e-mail: rzoska@us.edu.pl

Abstract: Recently, the isotropic phase of rod-like liquid crystalline compounds is advised as an experimental model system for studying complex glassy dynamics. One of unique phenomena occuring close to the glass temperature, for the time scale $10^{-7\pm1}s < \tau < \tau(T_g) \approx 100s$, is the fractional Debye-Stokes-Einstein (FDSE) behaviour $\sigma\tau^S = const$ with $S < 1$, i.e. the coupling of dc conductivity (σ, translational processes) and dielectric (structural) relaxation time (τ, orientational processes). It is shown that this relation may be found also in the isotropic phase of nematic liquid crystalline compounds *n*-pentylcyanobiphenyl (5CB), although surprisingly for time-scales $\tau < 10^{-8}s$. The application of the derivative based analysis revealed a change of S on cooling towards the isotropic – nematic transition. The optimal description of the evolution of relaxation time and conductivity by the mode-coupling theory (MCT) dependence in the isotropic phase is shown: $\tau(T) \propto (T - T_{MCT})^{-\phi}$ and $\sigma(T) \propto (T - T_{MCT})^{-\phi'}$, where $T_{MCT} \approx T_{I-N} - 33K$ and $T > T_{I-N}$ is shown. The link between this behavior and the FDSE is suggested, namely: $S = \phi'/\phi$. Finally the call for further pressure studies is formulated.

Keywords: glassy dynamics, *n*-pentylcyanobiphenyl (5CB), isotropic phase, dynamics dynamics, broad band dielectric spectroscopy, MCT

Broad band dielectric spectra of liquids enable an insight into relaxation processes associated both with the rotational and translational molecular motions.[1-3] The latter originates from small residual ionic dopants, always present in liquids.[1,2] The structural relaxation, called also the *alpha* relaxation, is often associated with reorientation of entire molecules coupled to the permanent dipole moment.[1-6] Recent studies gave strong evidence that supercooling can cause an enhancement of translational motions over reorientations. This decoupling manifest via empirical fractional Debye-Stokes-Einstein (FDSE) relation:[7-17]

$$\sigma\tau = 1 \quad \left(\tau < \tau(T_B) \sim 10^{-7\pm1} s\right) \quad \rightarrow \quad \sigma\tau^S = const \ \left(\tau > \tau(T_B) \sim 10^{-7\pm1} s\right) \tag{1}$$

where σ is for dc conductivity, τ denotes the relaxation times, the fractional exponent: $S<1$ and T_B denotes the dynamic crossover temperature associated with the change of parameters in the Vogel-Fulcher-Tammann (VFT) equation.[18-20]

Noteworthy is the strong evidence for the coincidence of T_B with the mode coupling theory (MCT) ergodic – non-ergodic crossover, "critical", temperature T_{MCT}. Moreover, the system-independent time-scale for the dynamic crossover: $\tau(T_B) \approx 10^{-7\pm1} s$ is suggested.[20] Despite many efforts no generally accepted explanation of the decoupling phenomenon has been proposed so far. A hypothetical dynamic phase transition underlying T_B, onset of cooperative molecular motions, the change in the free volume available for residual ions and dipoles or dynamic heterogeneities and hypothetical spatial heterogeneities are worth recalling here as a possible suggested artifacts linked to the FDSE behavior.[4, 7-17]

Hence, novel experimental facts may be of particular importance.

This contribution presents the evidence of the FDSE behavior in the isotropic phase of *n*-pentylcyanobiphenyl (5CB), a rod-like liquid crystalline compound with the isotropic (I) – nematic (N) – solid (S) phase sequence. Despite the fact that 5CB is probably the most "classical" nematic liquid crystal (NLC),[21] the experimental evidence for the clear glassy dynamics in the isotropic phase was obtained only recently. Particularly noteworthy is the strong influence of prenematic fluctuations on the dynamic, manifested even well above T_{IN}.[22-27] To the best of the authors knowledge there have been no discussion aiming on the violation of the DSE relation in 5CB or in other LC compounds up to now. In fact the lack of such investigation cannot be surprising because the time scale for the isotropic phase is in the range $\tau << 10^{-7} s$, whereas the FDSE behavior is expected for $\tau > 10^{-7} s$. Notwithstanding, a clear evidence for FDSE behavior in isotropic phase of 5CB is reported below. Experimental $\tau(T)$ data were taken from earlier authors' studies.[25] They are supplemented by $\sigma(T)$ data obtained during the same measurements but do not reported so far. Measurements were conducted via Novocontrol BDS 80 spectrometer as specified in refs. ([23-26]). The sample of 5CB was carefully degassed immediately prior to measurements.

Figure 1 shows $\tau(T)$ and $\sigma(T)$ dependences in the isotropic phase of 5CB. It is noteworthy that they cover an extraordinary broad range of temperatures. The "density" of experimental data, i.e. the number of tested temperatures per decade, strongly increases on approaching T_{IN}. The nonlinearity of obtained dependences proves the non-Arrhenius dynamics of both discussed magnitudes.

Figure 2 presents the log-log plot of $\sigma(T)$ versus $\tau(T)$ usually applied for showing the violation of the DSE relation.[9,8,15] The FDSE exponent $S = 0.77 \pm 0.02$ is determined from the linear regression analysis. The inset shows the derivative of data from the main part of the plot. This distortion-sensitive analysis of data, do not applied so far, revealed a secret feature of results in the main part of the Fig. 2: the value of the exponent S changes on approaching the isotropic – nematic (I-N) transition.

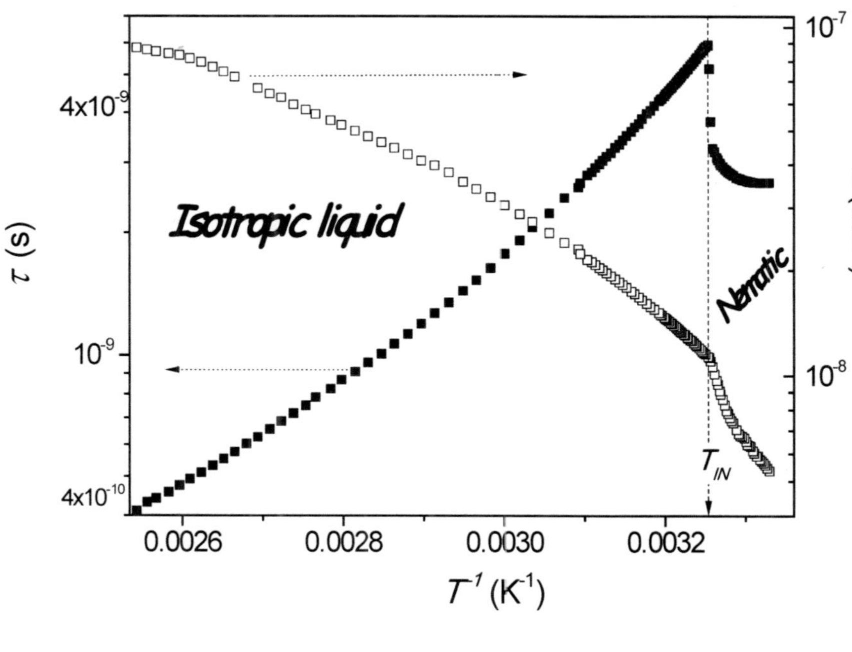

Figure 1. *The Arrhenius plot of temperature dependences of dielectric relaxation time (solid square) and dc-conductivity (open squares).*

It is noteworthy that the strong discrepancy from Eq. (1) occurs on approaching the I-N transition when the correlation length and the lifetime of quasi-nematic heterogeneities in the *fluidlike* surrounding boost, namely:[21, 25]

$$\tau_{fluct.} = \frac{\tau^0_{fluct.}}{T - T^*} \quad \text{and} \quad \xi_{fluct.} = \frac{\xi^0_{fluct.}}{\left(T - T^*\right)^{1/2}} \tag{2}$$

where $T > T_{IN} = T^* + \Delta T$, T^* is the temperature of the hypothetical continuous phase transition, ΔT is the measure of the discontinuity of the I-N transition. For 5CB: $\Delta T \approx 1.1K$.[25]

This is the case of nonlinear, i.e. nonlinear dielectric spectroscopy related, relaxation time.

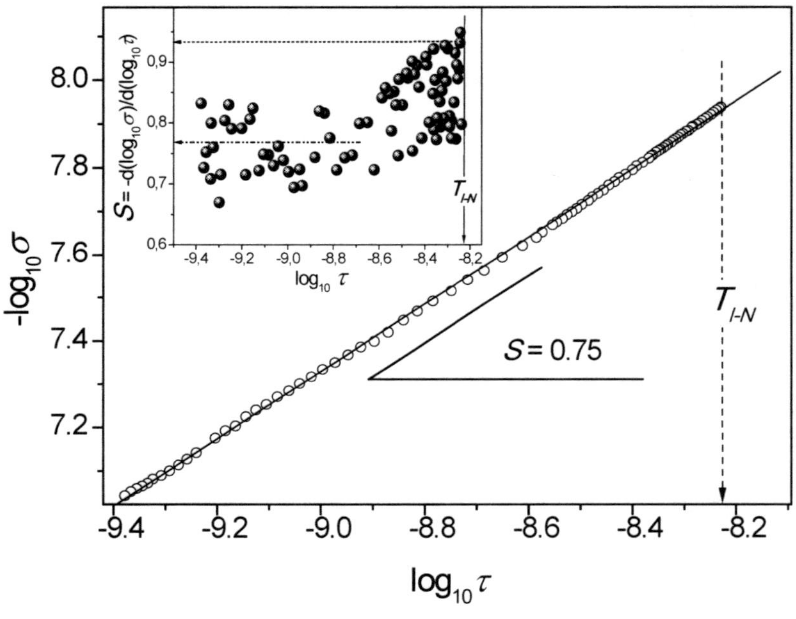

Figure 2. *The log-log plot of dc-conductivity versus dielectric relaxation time in the isotropic phase of 5CB. The line shows the validity of the FDSE relation (1). The inset shows results of the derivative analysis of data from the main plot. It shows the temperature evolution of the apparent value the exponent S on cooling towards the clearing temperature.*

The dielectric, structural, relaxation time is related to the "linear" regime. It reflects the evolution of the average permanent dipole moment, linked to the given molecule. The dielectric relaxation time detects heterogeneities indirectly, via changes of the average surrounding of a molecule. It was shown in ref. ([23]) that dielectric relaxation can be well portrayed, with small distortion only close to T_{IN}, by the MCT "critical-like" dependence:

$$\tau(T) = \tau_0^{MCT}\left(T - T_{MCT}\right)^{-\phi} \tag{3}$$

where T_{MCT} is the ergodic – non-ergodic crossover temperature, the exponent ϕ is a non-universal parameter which value depends on some coefficients describing the high frequency part of the BDS spectrum.

For non-mesogenic glassy liquids usually parameters T_C and γ instead of T_{MCT} and ϕ are used in relation (3). However, for liquids with the thermodynamic phase transition symbol T_C is reserved for the critical temperature and for liquid crystals for the clearing temperature, i.e. the temperature of the I-N weakly discontinuous phase transition.[21, 22] The exponent γ is for the universal description of the pretransitional anomaly of compressibility.[28] As shown in ref. ([18]) the application of the derivative based analysis enable unequivocal estimation of T_{MCT} and ϕ, using solely the linear regression and no "hidden", adjustable parameters. Fig. 3 shows that the analogous behavior takes place for the dc conductivity, namely:

$$\sigma(T) = \sigma_0\left(T - T_{MCT}\right)^{\phi'} \tag{4}$$

$$\left(\frac{d\ln\sigma}{d(1/T)}\right) = \frac{H_a^{\sigma}}{RT} = \frac{\phi' T^2}{T - T_{MCT}} \tag{5}$$

where $H_a^{\sigma}(T)$ is the apparent activation enthalpy related to transport processes which yields a linear dependence $T^2 / [d\ln\sigma / d(1/T)] = T^2 / H_a^{'} = A + BT$ with $T_{MCT} = B/A$ and $\phi' = A^{-1}$.

The validity of eq. (5) and hence also relation (4) for dc conductivity, is shown in Fig. 3. The slope of the solid line determines the exponent, i.e. $\phi' = A^{-1}$. the condition $RT^2 / H_a^{\sigma} = 0$ determines the "singular" temperature, i.e. $T_{MCT} = B/A$. Using MCT "critical-like" eqs. (3) and (4) and the FDSE eq. (1) one can obtain:

$$\sigma(T)[\tau(T)]^S = \frac{\sigma_0}{\tau_0} \frac{\left(T - T_{MCT}\right)^{\phi'}}{\left[\left(T - T_{MCT}\right)^{\phi}\right]^{S=\phi'/\phi}} = const \tag{6}$$

Then the FDSE exponent: $S = \phi'/\phi = 2.1/1.6 \approx 0.76$. The same value can be found in Fig. 2 basing on $\log_{10}\sigma$ vs. $\log_{10}\tau$ plot. Dividing further eq. (5) by its analog for dielectric relaxation time one can obtain the relation linking apparent activation enthalpies, MCT "critical-like" exponent and FDSE exponent:

$$\frac{H_a^{\sigma}(T)}{H_a^{\tau}(T)} = \frac{\phi'}{\phi} = S = const \tag{7}$$

The change of FDSE exponent visible in Fig. 2 reflects the fact that the MCT eq. (3) fails in the immediate vicinity of the I-N transition whereas such distortion is absent for $\sigma(T)$ behavior portrayed via eq. (4). This can induce the visible gradual change of the exponent ϕ.

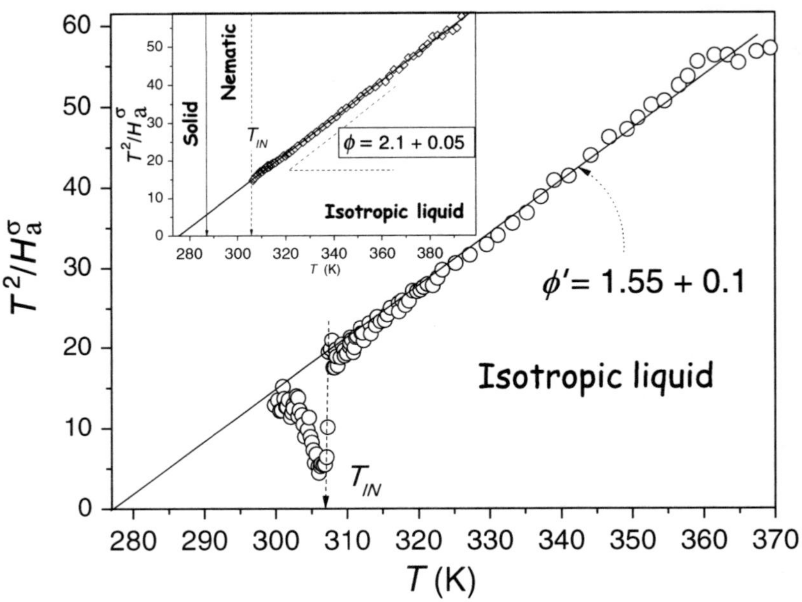

Figure 3. *The linearized derivative-based plot (see eq. (3)) showing the validity of the critical-like MCT behavior for dc conductivity (the main part of the plot) and dielectric relaxation time (the inset). Values of "critical" MCT exponents are given in the plot. For both magnitudes the singular temperature* $T_{MCT} = 275K \pm 3K$.

Concluding, results presented above showed a superior description of $\tau(T)$ and $\sigma(T)$ evolutions in the isotropic phase of 5CB using MCT eqs. (3) and (4), respectively. It is noteworthy that such description is qualitatively better that the VFT one, recommended so far. This report shows that the isotropic phase exhibit a fractional DSE behavior in the time domain where such phenomenon is inherently absent in "classical" molecular glass formers. The link between the MCT and FDSE behavior was also shown. Finally the conclusion from the theoretical analysis by Wang ([10]) is worth recalling: "*...coupling of the translation to rotation can also lead to a strong enhancement* *in a rotationally anisotropic molecular fluid. For a dynamically heterogeneous fluid, the probe size may play an important role. We show that if the diffusion probe is large so as to encompass several regions of high and low mobility, the rotation-translation coupling parameter needs to be modified to reflect the averaging heterogeneity effect. We show that, while the averaging effect arising from dynamic heterogeneity may alter the final enhancement, the coupling of translation to rotational degrees of freedom must be taken into account..."*

Although the above citation is related to "classical", non-mesogenic glass formers, it also clearly coincides with results for the isotropic 5CB presented above. This confirm that the isotropic phase of rod-like nematic liquid crystals may be considered as an important models system for glassy dynamics.[22-27] Particularly important may appear continuation of studies presented above for the pressure path of approaching the glass transition, due to the fact that pressure is linked to free volume changes and temperature to the shift in the activation energy.

Acknowledgements

This research was carried out with the support of the CLG NATO Grant No. CBP. NUKR.CLG 982312).

References

1. Donth, E. (2001) The Glass Transition: Relaxation Dynamics in Liquids and Disordered Materials (*Springer Verlag, Berlin*)
2. Kremer, F., and Shoenhals, A. (eds.) (2003) Broad Band Dielectric Spectroscopy (*Springer, Berlin*)
3. Rzoska, S. J., and Mazur, V. (eds.) (2006) Soft Matter Under Exogenic Impacts, *NATO Sci. Series II,* (*Springer, Berlin*), vol. **24**
4. Kivelson, S. A., and Tarjus, G. (2008) In search of a theory of supercooled liquids, *Nature Materials* **7**, 831-833
5. McKenna, G. B. (2008) Diverging views on glass transition, *Nature Physics* **4**, 673-674
6. Hecksher, T., Nielsen, A. I., Olsen, N. B., and Dyre, J. C. (2008) Little evidence for dynamic divergences in ultraviscous molecular liquids, *Nature Physics* **4**, 737-741
7. Douglas, J. F., Leporini, D. (1998) Obstruction model of the fractional Stokes-Einstein relation in glass-forming liquids, *J. Non-Cryst. Solids* **23-237**, 137-141
8. Corezzi, S., Lucchesi, M., Rolla P. A., Capaccioli, S., Gallone, G. (1999) Temperature and pressure dependences of the relaxation dynamics of supercooled systems explored by dielectric spectroscopy, *Phil. Mag. B* **79**, 1953-1963

9. Hensel Bielowka, S., Psurek, T., Ziolo, J., and Paluch, M. (2001) Test of the fractional Debye-Stokes-Einstein equation in low-molecular-weight glass-forming liquids under condition of high compression, *Phys. Rev. E* **63**, 062301
10. Wang, C. H. (2002) Enhancement of translational diffusion coefficient of a probe in a rotationally anisotropic fluid, *Phys. Rev. E* **66**, 021201
11. Funke, K., Banhatti, R. D., Brueckner, S., Cramer, C., and Wilmer, D. (2002) Dynamics of mobile ions in crystals, glasses and melts, described by the concept of mismatch and relaxation, *Solid State Ionics* **154-155**, 65-74
12. Cutroni, M., Mandanici, A., and De Francesco, L. (2002) Fragility, stretching parameters and decoupling effect on some supercooled liquids, *J. Non-Cryst. Solids* **307-310**, 449-454
13. Power, G., Johari, P., and Vij, J. K. (2002) Effects of ions on the dielectric permittivity and relaxation rate and the decoupling of ionic diffusion from dielectric relaxation in supercooled liquid and glassy 1-propanol, *J. Chem. Phys.* **116**, 419-4201
14. Bordat, P., Affouard, F., Descamps, M., and Mueller-Plathe, F. (2003) The breakdown of the Stokes–Einstein relation in supercooled binary liquids, *J. Phys.: Condens. Matt.* **15**, 5397-5407
15. Psurek, T., Ziolo, J., and Paluch, M. (2004) Analysis of decoupling of DC conductivity and structural relaxation time in epoxies with different molecular topology, *Physica A* **331**, 353-364
16. Richert, R. (2005) Dielectric responses in disordered systems: From molecules to materials, *J. Non-Cryst. Solids* **351**, 2716-2722
17. Becker, S. R., Poole, P. H., and Starr, F. W. (2006) Fractional Stokes-Einstein and Debye-Stokes-Einstein Relations in a Network-Forming Liquid, *Phys. Rev. Lett.* **97**, 055901
18. Drozd-Rzoska, A., and Rzoska, S. J. (2006) Derivative-based analysis for temperature and pressure evolution of dielectric relaxation times in vitrifying liquids, *Phys. Rev. E* **73**, 041502
19. Drozd-Rzoska, A., Rzoska. S. J., Roland, C. M., and Imre, A. R. (2008) On the pressure evolution of dynamic properties of supercooled liquids *J. Phys.: Condens. Matt.* **20**, 244103
20. Novikov, V. N., and Sokolov, A. P. (2003) Universality of the dynamic crossover in glass-forming liquids: a "magic" relaxation time, *Phys. Rev. E* **67**, 031507
21. Demus, D., Goodby, J., Gray, G. W., Spiess, H. W., and Vill, V. (eds.) (1998) *Handbook of Liquid Crystals, edited by vol. 1: Fundamentals (Springer, Berlin)*
22. Drozd-Rzoska, A., Rzoska, S. J., and Czupryński, K. (2000) Phase transitions from the isotropic liquid to liquid crystalline mesophases studied by "linear" and "nonlinear" static dielectric permittivity, *Phys. Rev. E* **61**, 5355-5360
23. Rzoska, S. J., Paluch, M., Drozd-Rzoska, A., Paluch, M., Janik, P., Ziolo J., and Czupryński, K. (2001) Glassy and fluidlike behavior of the isotropic phase of mesogens in broad-band dielectric, *Europ. Phys. J. E* **7**, 387

24. Rzoska, S. J., and Drozd-Rzoska, A. (2002) On the tricritical point of the isotropic – nematic transition in a rod-like mesogen hidden in the negative pressure region, in NATO Sci. Series II, vol. **84**, *Liquids under negative pressures*, eds.: Imre, A. R., Maris, H. J., and Williams P. R. (*Kluwer-Springer, Dordrecht*) p. 116
25. Drozd-Rzoska, A. (2006) Heterogeneity-related dynamics in isotropic n-pentylcyanobiphenyl, *Phys. Rev. E* **73**, 022501
26. Drozd-Rzoska, A. (2009) Glassy dynamics of liquid crystalline 4'-*n*-pentyl-4-cyanobiphenyl (5CB) in the isotropic and supercooled nematic phase, *J. Chem. Phys.* **130**, 234910
27. Cang, H., Li, J., Novikov, V. N., and Fayer, M. D. (2003) *J. Chem. Phys.* **119**, 10421
28. Anisimov, M. A. Critical Phenomena in Liquids and in Liquid Crystals (1992) (*Gordon and Breach, Reading*)

PART III

NEAR-CRITICAL MIXTURES

AN OPTICAL BRILLOUIN STUDY OF A RE-ENTRANT BINARY LIQUID MIXTURE

F. JAVIER BERMEJO *
CSIC, Instituto de Estructura de la Materia and Dept, Electricidad y Electrónica, Facultad de Ciencia y Tecnología, Universidad del Pais Vasco,P.O. Box 48080 Bilbao, Spain

LOUIS LETAMENDIA
CPMOH CNRS-Université Bordeaux1, 351 cours de la Libération 33405 Talence CEDEX France

Abstract: Optical spectroscopy studies on the system betapicoline-D_2O which forms a critical mixture showing a closed-loop phase diagram have been carried out along a large range of temperatures, comprising crossing of the two critical curves which define the closed loop of the coexistence curve. Data pertaining the elastic properties of the mixture either in one or two-phase states for the sound velocity and the sound absorption were obtained by means of analysis of the spectra of scattered radiation using Brillouin light scattering spectroscopy.

Keywords: reentrant phase transitions, Brillouin light scattering, sound velocity, sound absorption, critical phenomena.

1. Introduction

Our current knowledge on critical phenomena in binary liquids[1] concerning the critical exponents and amplitudes as well as corrections-to-scaling amplitude ratios can now be considered as complete. Such systems belong to the Ising universality class [2] where the factors determining critical behaviour are a scalar order parameter ζ, the presence of short-range interactions only, the isotropy as well as symmetry under inversion in the absence of an applied field ($\zeta \to -\zeta$). According to such a view, the behaviour of these systems under equilibrium conditions is fully specified by the factors listed above while other details such as the microscopic dynamics is deemed to be irrelevant. Most cases investigated so far involve the study of critical phenomena corresponding to changes from ordered to disordered phases or vice versa. There are however systems known since the pioneering study of Hudson[3,4] on the nicotine/ water mixture that show a closed-loop coexistence

*To whom correspondence should be addressed. JAVIER@LANGRAN.IEM.CSIC.ES

curve such as that shown on Fig. 1, i.e. the phase diagram of temperature versus composition shows a closed-loop outside of which the mixture becomes fully miscible but separates into two macroscopically distinguishable phases when the low- or high-temperature critical solution temperatures T_L and T_U are approached from lower or higher temperatures, respectively. Such behaviour, named as re-entrant since the system attains a state similar to the initial state after crossing two critical lines, is also shared by a wide variety of physical systems comprising ferroelectrics, liquid crystals, antiferromagnets or spin glasses[4]. While the presence of a miscibility line at T_U is easy to understand on energetic grounds, the fact that the mixture becomes miscible below T_L remains to be understood on quantitative grounds. The microscopic origin for this low temperature mixing phenomenon is most of the time interpreted on the basis of a conjecture brought forward by Hirschfelder a long time ago[5]. It attributes remixing below T_L to the presence of strong directional bonding interactions between the two unlike species. On the other hand, very recent reports on a novel phase behaviour where a tenuous solid-like structure appears at the liquid/liquid interface[6] make microscopic investigations of the forces driving such systems timely and well overdue. The advantage of studying mixtures of methyl pyridine and heavy water as physical realizations of re-entrant systems over other solid-state materials stems from the ease of controlling the width of the immiscibility loop. In fact, it is known that the loop size shows a extreme sensitivity to the isotopic composition of the hydrogen element (light or heavy water), the presence of salts [7], or the application of pressure.[8]

In previous papers we have reported on some detailed studies on the behaviour of the order parameter[7] ζ (T) which was assimilated to the correlation length for critical fluctuations accessible by means of neutron small-angle scattering as well some measurements on the molecular dynamics[8] across the two phase transitions as explored by neutron quasielastic scattering as well as computer simulations.

Further studies[8] were carried out in the quest for the double critcal point of the mixture 2-methyl-piridine(2MP)/D_2O. Within such studies we came across a remarkable anomaly appearing within 2MP neat liquid at applied pressures of about 200 bars. It manifests itself as a marked change of regime of the translation and rotational-diffusion coefficients versus density (pressure). To add more intrigue, the pressure range at which such an anomaly takes place basically coincides with that where the DCP was suspected to be located. In fact, the concurrent use of quasielastic neutron scattering and molecular dynamics simulations. evidenced a pronounced change of slope in the density dependence of both the translation and rotational diffusion coefficients for densities of ρ=0.975 g/cm^3. In turn, the description of the liquid structure carried out in terms of static pair distributions derived from computer simulations revealed indications of the presence of dynamical equilibria within the liquid as attested by clear isosbestic points.

Here we report on mesoscopic studies carried out on the mixture Beta-picoline (BP or 3-methylpyridine)/heavy water mixture (BP-D_2O) aiming to get detailed information on the elastic and viscous properties of the two 'external single-' phase and the 'internal two-phase' states in such mixtures by means of analysis of the optical Rayleigh-Brillouin spectra which have been measured over a wide range of temperatures.

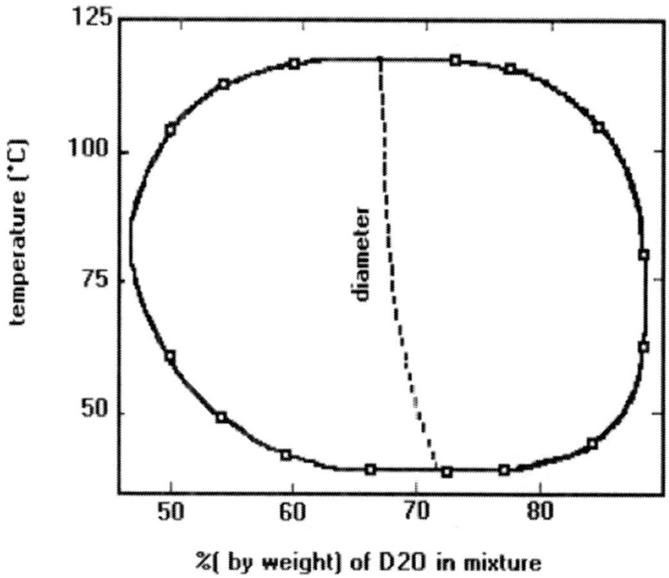

Figure 1. *Betapicoline/$D_2 0$ mixture after Cox (1952)T_L=38.5°C, T_U=117°C. When $\Delta T = T_U - T_L = 0$ the double critical point (DPC) is reached.*

The paper is structured as follows; the experimental setup and the obtained results are briefly described in the next section. The third section is then devoted to the description of the results in the single- and two-phase systems. Finally a discussion and comparison with other techniques and previous results is held in a fourth section.

2. Experiment

The experimental setup rests on a marble table, insulated from the mechanical vibrations of the floor. The laser, the Plan Fabry-Pérot (PFP), the sample and auxiliary optics are in the same table that insures the absence of relative movements between them. The source is a 2020 Spectra Physics laser working in monomode. The room temperature is controlled better than 0.1°C which insures good laser mode stability. The FP is insulated inside a plastic structure and covered by a black cloth that insures good protection from stray light and temperature fluctuations. The sample temperature is controlled by a thermostated circulating fluid bath. At temperatures below 90°C we use water as the control fluid whereas an adapted fluid is used for higher temperatures.

Two choppers allow us to divide each PFP scan into two parts. The first 200 points gives the apparatus function (the scattered spectra is stopped) and the 800 other points of the total scanned 1000 points are used to store the spectra. The FP is scanned using a highly linearized voltage ramp, built within our laboratory (CPMOH), which drives high quality Physics Instruments transducers. The piezoelectric plates work at high frequency in order to avoid coupling with parasitic frequencies. The working finesse of the FP is better than 30, during the accumulation time which usually takes about 15 minutes.

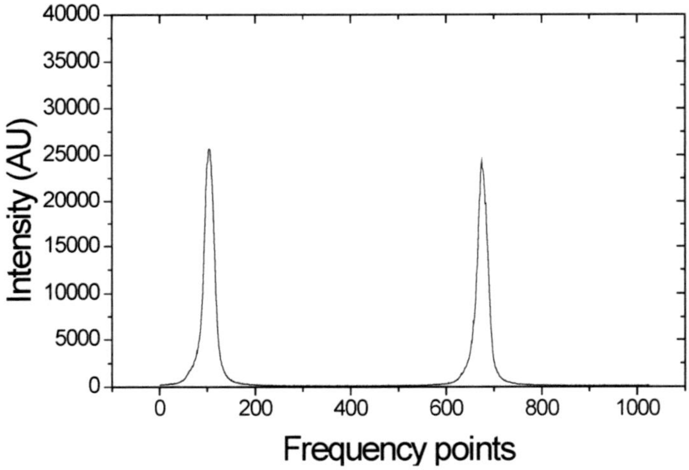

Figure 2. *Typical apparatus function of PFP.*

The scattered signal is collected by an optical fiber and detected by a diode working in the so-called 'Geiger' mode. The spectrum are collected and stored in a PC with a MCS National system. More details are given in reference[9]. The Free Spectral Range (FSR) of the PFP for this work was 12.712 GHz. As can be seen in Fig. 3 and 4, the Brillouin lines are far weaker than their Rayleigh counterparts, both inside and outside the coexistence curve (CC). The Rayleigh to Brillouin ratio is smaller outside the CC. With our current setup, the Rayleigh line provides us with the apparatus window function from where the FSR and the finesse for each spectrum are measured.

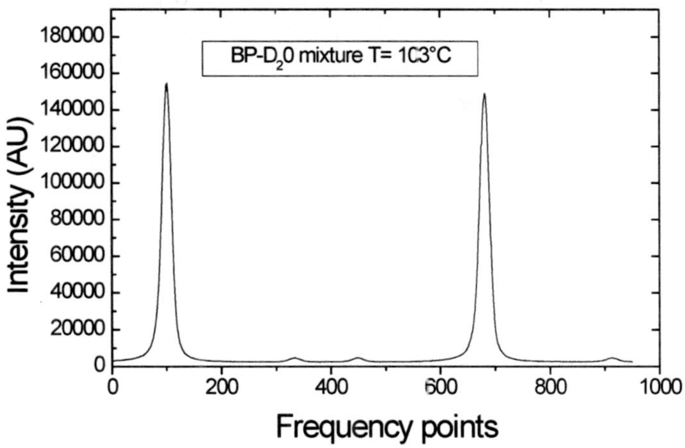

Figure 3. *Rayleigh-Brillouin spectra of the mixture BP-D_2O Inside de coexistence curve.*

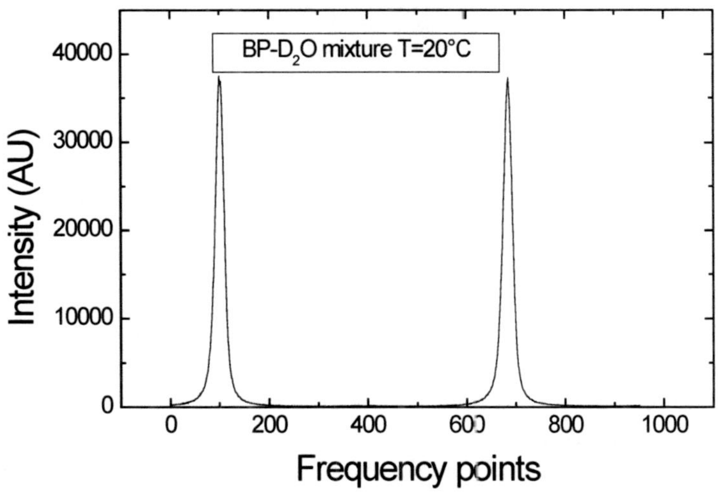

Figure 4. *Rayleigh Brillouin spectrum of BP-D2O mixture outside the coexistence curve.*

During a second step we carried out measurements of the Brillouin position and Brillouin line width which are related to the sound velocity and sound absorption. We use the method of Zamir, et al.[10] With our current setup, the Rayleigh line provides us with the apparatus window function from where the FSR and the finesse for each spectrum are measured. During a second step we carried out measurements of the Brillouin position and Brillouin line width which are related to the sound velocity and sound absorption. We use the method of Zamir, et al.[10]

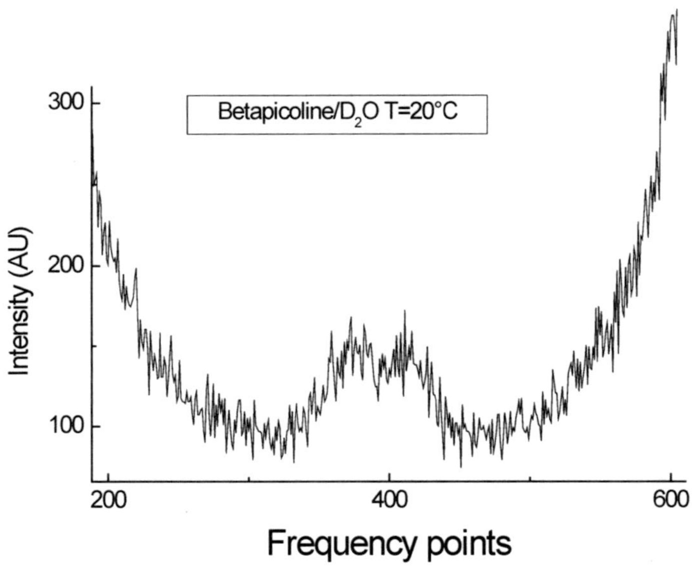

Figure 5. *Brillouin spectral 20°C. Notice the overlap between Brillouin lines.*

For the smaller Brillouin intensities, we extract the Brillouin lines from a typical spectra shown in Fig. 5. With our current setup, the Rayleigh line provides us with the apparatus window function from where the FSR and the finesse for each spectrum are measured. During a second step we carried out measurements of the Brillouin position and Brillouin line width which are related to the sound velocity and sound absorption. We use the method of Zamir, et al.[10]

3. Results

The Brillouin frequencies ω_B shown in Fig. 6 show three different regimes; first, from the lowest temperature up to T_L we get a decreasing behavior of frequency with increasing temperature (notice that we don't see clear signs of criticality since our point closest to T_L because about 1.5°C below this). Inside the two-phase region comprised for temperatures within the range T_L-T_U there are two values for ω_B, corresponding to the upper and lower phases respectively. The mixture within this range of temperatures separates into two phases, one in the upper side of the tube containing the sample and the other on the lower side, as can be witnessed with the naked eye. The ω_B values for both phases decrease with increasing temperature although the slopes signaling such changes are different.

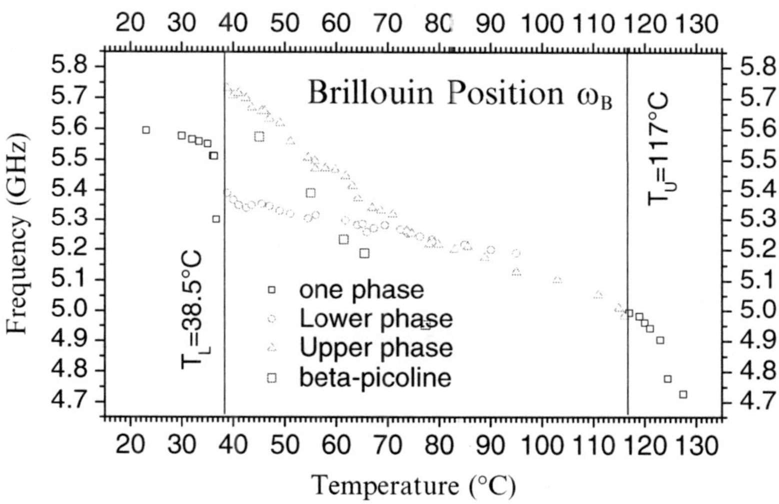

Figure 6. *Experimentally determined Brillouin frequencies, both inside and outside the co-existence curve.*

For comparison purposes, the Fig. 6 also shows data for pure BP which shows the stronger temperature dependence of the Brillouin frequency. A third temperature region is located above T_U, where we reenter into a single phase and a rather mild decrease of the Brillouin frequency with increasing temperature is observed. Data pertaining pure water or heavy water can be accessed from several sources.[11]

The half width at half height of the Brillouin line (Γ_B) are shown in Fig. 7. There we observe a strong decrease in linewitdth with increasing temperature below T_L, that is followed by a rather intriguing behaviour within the co-existence region. In fact, the figure shows that data for both the upper and the lower are basically superpossable. Furthermore, Γ_B decreases drastically the first 20 degrees above T_L and then continues decreasing at a lower pace.

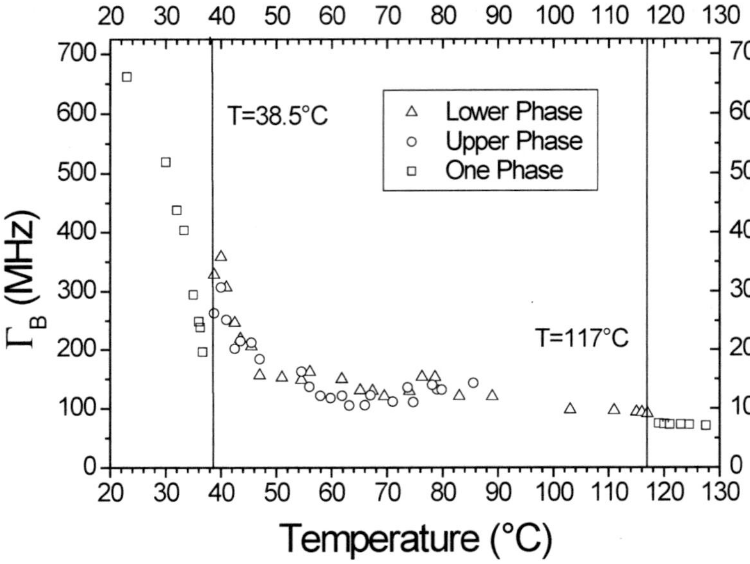

Figure 7. *Brillouin half width at half height inside and outside the coexistence curve.*

The Brillouin frequency can be connected to the sound velocity, V by the relationship:

$$\omega_B = V^*q \tag{1}$$

Γ_B is connected to the transport and thermal coefficient of the medium by

$$\Gamma_B = (q^2/2\rho_0)^*\{4/3\ \eta_S + \eta_B + [(\gamma-1)/c_p]\lambda\} \tag{2}$$

where q is the wave vector, η_S and η_B the shear and bulk viscosity coefficient; γ the ratio c_p/c_v and c_i the specific heath at i=v,p or constant volume or pressure. λ is the thermal conductivity and ρ_0 is the density of the medium.
the medium.

In doing so, the wave vector evaluation allows the determination of the sound velocity and the sound absorption within either single- or two-phase states.

We now use results of Fig. 1, giving the coexistence curve (CC) as well as the fact that the lower phase is the poorest in D_2O for the determination of the index of refraction n of both mixtures inside the CC. With the relation

$$q = (4\pi n/\lambda_0)\ \sin\theta/2 \tag{3}$$

where λ_0 is the wavelength of light)5145 A° and θ the scattering angle (90° here).

The polarisability of mixture is

$$\alpha_{mel}=\chi_{D2O}+(1-\chi_{D2O})\alpha_{B?} \quad (4)$$

and the polarisability of each component is given by Clausius-Mossoti relation

$$\alpha_i=(3M/N^*\rho_{D2O}(n_i^2-1/n_i^2+2) \quad (5)$$

where n_i is the index of refraction of component "i" and M the mass mixture. χ_{D2O} is deduced from Fig. 8. Here we use the CC as a usual mixture one.

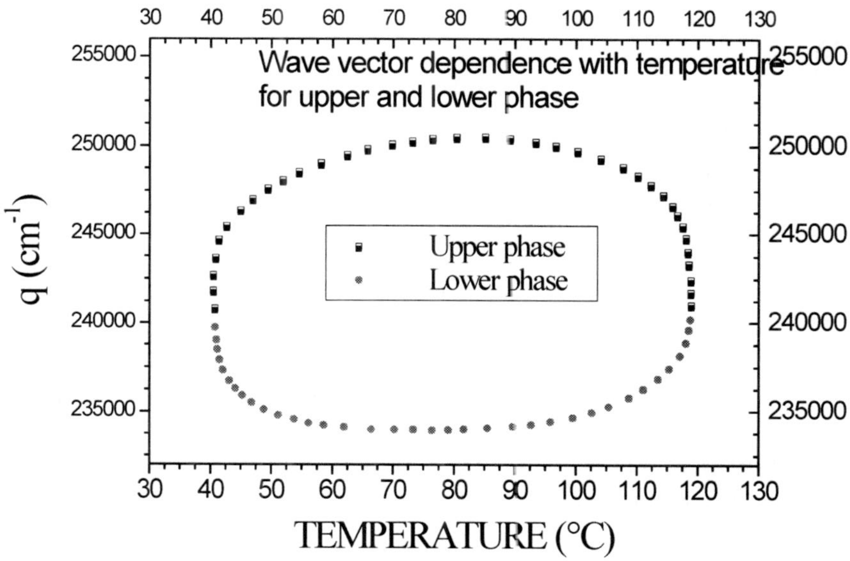

Figure 8. *Wave vector changes with D_2O concentration change of water in the mixtures inside de CC.*

The results of this calculation are given in Fig. 8. We see that the wave vector has a mean value of $2.41*10^5$ cm^{-1} and changes within values of $2.34*10^5$ cm^{-1} to $2.5*10^5$ cm^{-1} (+,- 3.5%). We see that it can have a linear effect in the sound velocity determination but a quadratic effect in sound absorption. From here, we calculate the sound velocity for all temperatures as shown in Fig. 9. The Fig. 9 provides data for the temperature dependence of the sound velocity once corrected by the refraction effects. There we see that the sound velocity of the lower phase smoothly decreases with temperature within the two-phase region, whereas that for the upper phase shows a stronger trend. In a previous work and in a different phase transition, K. V. Kovalenko et al.[12]

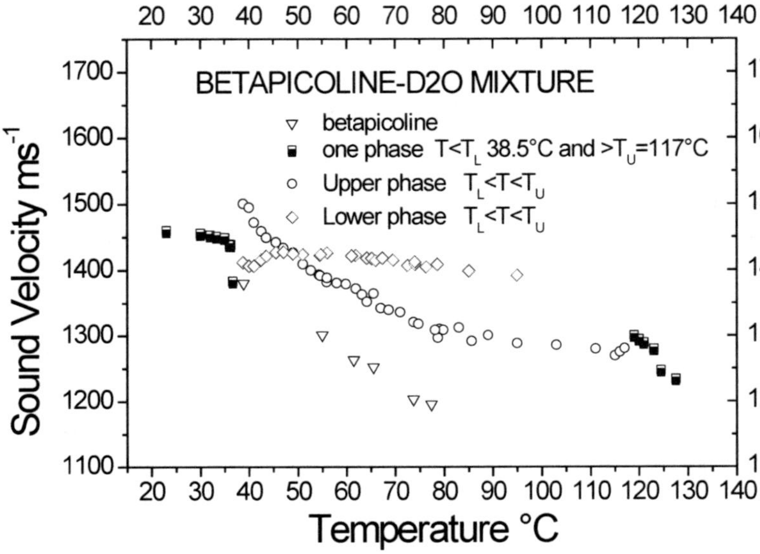

Figure 9. *Sound velocity of BP-D_2O mixture.*

found roughly a similar behavior for the sound velocity, and also the evolution of the sound velocity of the two phases has some similarity with temperature evolution of a mixture of water with alcohol.[11]

We could in principle connect the sound absorption, and the Brillouin frequency with frequency independent sound absorption by

$$\alpha/f^2 = 4\pi^2\Gamma_B / c_0\omega_B^2 \qquad (6)$$

but we do not have pursued such a route since there are not available results to compare with. In turn, we choose to evaluate the sound absorption in terms

$$\Gamma_B = (q^2/\rho_0)(4/3\ \eta_S + \eta_B) \qquad (7)$$

where we have neglected the last, thermal term. The quantity coming out from such an analysis is thus the linear viscosity term (LV), and the relevant results are shown in Fig. 10. The most striking behaviour concerning the LV concern the rather strong drop down to 60°C as well as the similitude of this transport property for the two phases inside the closed loop which by force have very different chemical compositions.

Data about 80°C show a small maximum, the physical soundness of which could not be ascertained. Also, the extremely mild dependence of the sound attenuation above 100 °C merits to be remarked.

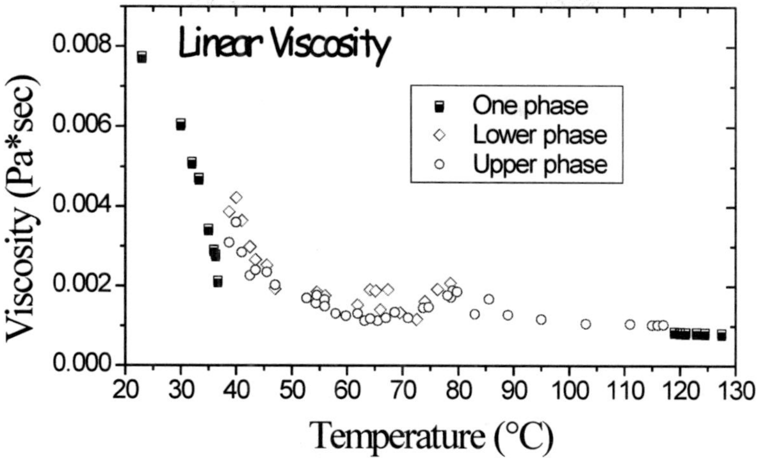

Figure 10. *Linear viscosity (LV) of BP-D_2O mixture.*

4. Discussion

We can now compare our results with previous measurements found in the literature. In Fig. 11 we plot measurements of shear viscosity of BP-D_2O critical mixture of Oleinikova[13], D_2O, and the non critical mixture BP-D_2O-04 containing a fraction of 0.4 of BP. Within the phase below T_L, our values are much more larger than the ones of Oleinikova et al.[6] derived from macroscopic measurements, a discrepancy understood from consideration of two facts appearing in formula (7). The first is the need to account for bulk viscosity, and also we have to add a term comprising one third of the value of the shear viscosity. Within the immiscibility loop we see that the viscosity presents a relaxation during the first 20°C, followed by a smooth decrease with temperature. Also for the sake of comparison we show the values for the mixture BP-D_2O 04 as reported by Oleinikova[13] that show that, as expected, the viscosity increases with the increase in molar fraction of BP. Finally, the viscosity of pure heavy water is also show as a background. As mentioned in previous sections, the purpose of the current set of experiments was to derive mesoscopic estimates for the longitudinal viscosity which could be contrasted versus information on the translational D_{trans} (T) and rotational $D_{rot}(T)$ diffusion coefficients derived from previous experiments by means of quasielastic neutron scattering.[7,8]

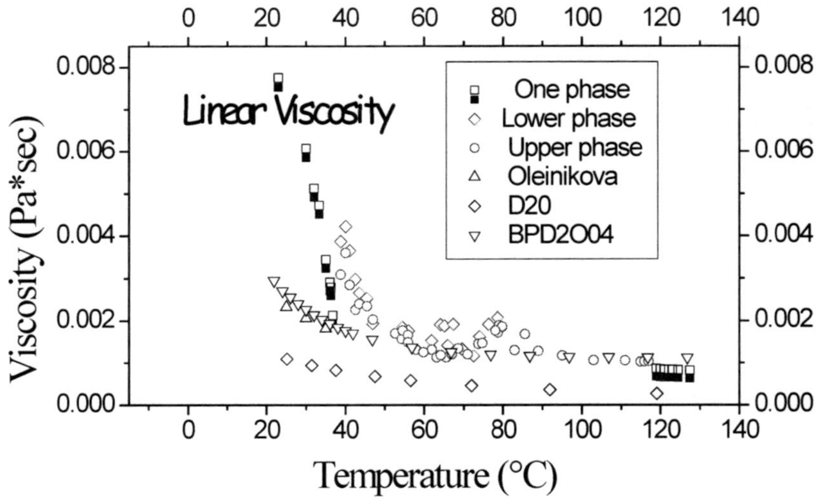

Figure 11. *Various plots of viscosities. See text for details.*

To such an avail, we have compared the measured transport coefficients to estimates derived from Stokes-Einstein (SE)

$$D_{trans}(T) = k_B T / 6\pi\eta r \tag{8}$$

and Stokes-Einstein-Debye (SED),

$$D_{rot}(T) = k_B T / 8\pi\eta r^3 \tag{9}$$

where η comprises the longitudinal viscosity term given above and r stands for an effective molecular radius, here set to the value used for the analysis of the neutron scattering data, that is r = 0.15 nm. The result of such a comparison is shown in Fig. 12, from where the following comments are in order. First, both coefficients as calculated from Eqns. 8 and 9 are reasonably close to the experimental values for the mixture provided that the temperature is comparable or larger than T_L. In striking contrast, data below T_L evidences a far higher mobility of the mixture which points towards to a decoupling from the behaviour followed by the viscosity. Second, data for both diffusion coefficients depict a lower mobility that those for the pure BP. Furthermore, data for the pure compound is seen to follow a simple thermally activated behaviour with activation energies E_{trans} = 140 meV and E_{rot} =73 meV. A fact which vividly illustrates the far larger freedom of relatively large molecules to rotate within the liquid than to execute translational motions. At this point it is worth comparing the results here reported on with recent data on the breakdown of SE and SED approximations within a hydrogen bonded liquid, that is a prototypical example of a system with strong directional interactions such is liquid HF[14].

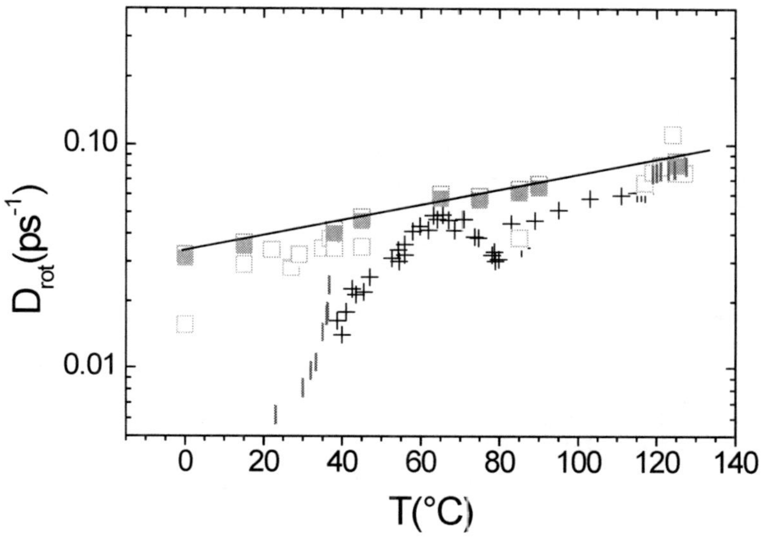

Figure 12. *Experimental estimates for the translational D_{trans} (T) and rotational D_{rot} (T) diffusion coefficients for pure BP (solid triangles and solid squares) as well as for the re-entrant mixture (circles with a dot and squares with a dot). Estimates of both transport coefficients as derived using Eqn. 8 and 9 are shown by vertical bars (miscible state) and crosses (phases within the immiscibility loop). The solid lines through 3MP data correspond to fits to a thermally activated process.*

As seen from Ref. 9, data for the rotational diffusion coefficient also shows a far milder dependence wit temperature than its translational counterpart. However the observed decoupling from the SE and SED behaviors is there seen to follow a far less drastic behaviour than that here found. On such a basis we deem that the presence of a strong directional interaction cannot account for the higher mobility observed by experiment if compared to the Brownian dynamics estimates of SE and SED. Finally it is worth to emphasize that the observed breakdown of both SE and SED approximations only appear for the miscible phase below T_L but not after the re-entrance above T_U into the high temperature, miscible phase. Such fact is thus suggestive of the existence of phenomena additional to those responsible of the re-entrant behavior being particularly important at low temperatures. Detailed studies conducted under pressure will hopefully shed important light into the origin of the anomaly here reported on.

References

1. Sengers, J. V., and Levelt-Sengers, J. M. H. (1986) *Ann. Rev. Phys. Chem.* **37**, 189; Kumar, A., Krishnamurthy, H.R., and Gopal, E. S. R. (1983) *Phys. Rep.* **98**, 57
2. Ma, S. K. (1976) *Modern Theory of Critical Phenomena*, Benjamin, M. A.
3. Hudson, C. S. (1904) *Z. Phys. Chem.* **47**, 113
4. Narayanan, T., and Kumar, A. (1994) *Phys. Rep.* **249**, 135
5. Hirschfelder, J., Stevenson, D., and Eyring, H. (1937) *J. Chem. Phys.* **5**, 896
6. Jacob, J., Kumar, A., Asokan, S., Sen, D., Chitra, R., and Mazumder, S. (1999) *Chem. Phys. Lett.* **304**, 180
7. Maira-Vidal, A., Gonzalez, M. A., Cabrillo, C., Bermejo, F. J., Jimenez-Ruiz, M., Saboungi, M. L., Otomo, T., Fayon, F., Enciso, E., and Price, D. L. (2003) *Chem. Phys.* **292**, 273
8. Maira-Vidal, A., González, M. A., Jimenez-Ruiz, M., Bermejo, F. J., Price, D. L., Enciso, E., Saboungi, M. L., Fernández-Perea, R., and Cabrillo, C. (2004) *Phys.Rev. E* **70**, 021501
9. Letamendia, L., Pru-Lestret, E., Panizza, P., Rouch, J., Sciortino, F., Tartaglia, P., and Hashimoto, C., Ushiki, H., and Risso, D. (2001) *Physica A* **300**, 53
10. Zamir, E., Gershon, N. D., and Ben Reuven, A. (1971) *J. Chem. Phys.* **77**, 3397
11. Conde, O., Teixeira, J., and Papon, P. (1982) J. *Chem. Phys.* **76**, 3747
12. Kovalenko, K.V., Krivokhizha, S.V., Fabelinkii, I. L., and Chaikov L. L., (1993) *Pisma Zh. Eksp. Teor. Fiz.* **58**, 395
13. Oleinikova, O., Bulanin, L., and Pipich, V. (1997) *Chem. Phys. Lett.* **278**, 121
14. Fernandez-Alonso, F., Bermejo, F. J., McLain, S. E., Turner, J. F., Molaison, J.J., and Herwig, K. W. (2007) *Phys. Rev. Lett.* **98**, 077801

NEW PROPOSALS FOR SUPERCRITICAL FLUIDS APPLICATIONS

SYLWESTER J. RZOSKA AND ALEKSANDRA DROZD-RZOSKA
Institute of Physics, University of Silesia, ul. Uniwersytecka 4, 40-007 Katowice, Poland

Abstract: Supercritical fluids (SCF) constitute a promising platform for future technological and environmental applications. The basic advantage of this method is their enormous selectiveness matched with a broad-range tuning of useful properties. So far, SCF technologies have made use of the vicinity of the gas – liquid critical point in a one component system, such as CO_2. For specific applications an additional component may be used to reach the desired technological target. In this report novel possibilities associated with supercriticality near the liquid – liquid critical point in binary solutions of limited miscibility and in a one component fluid are presented. The latter is discussed for a weakly supercooled nitrobenzene and *trans*-1,2-dichlorethylene. For the critical consolute point the possibility of obtaining arbitrary values and sings of the pressure shift of the critical temperature dT_C/dP are stressed.

Keywords: supercritical fluids technology, liquid-liquid transitions, high pressures, critical phenomena

1. Introduction

The interest in supercritical fluids (SCF) technologies is associated with their efficient and environment friendly application.[1-6] Generally, this technology is linked to the gas – liquid (GL) critical point in a one component system.[3,5] The supercritical fluid domain is defined by values of temperatures and pressures slightly above the critical temperature T_C and the critical pressure P_C. Technologists call this region a "dense gas phase".[1-6] Unique properties of SCF technology are due to the anomalous increase of some physical magnitudes on approaching the critical point (T_C, P_C). The divergence of compressibility (susceptibility) and the solvating ability, matched with broad-range tuning, is the basic artifact of technological relevance.[1-6] For SCF technology materials with values of T_C and P_C close to ambient values are desired. Moreover they should be cheap, non-toxic and exhibit a good solvation power in relation to possible application targets. In practice CO_2 ($P_C = 7.29MPa$ and $T_C = 31.3°\text{C}$) is the most often used system. Other common choices are (i) N_2O ($P_C = 7.25MPa$ and $T_C = 36.6°\text{C}$) which can be explosive, (ii) SF_6 ($P_C = 3.71 MPa$ and $T_C = 45.5°C$), linked to global warming problems and (iii) Xe ($P_C = 5.84 MPa$ and $T_C = 16.6°C$) which is expensive.

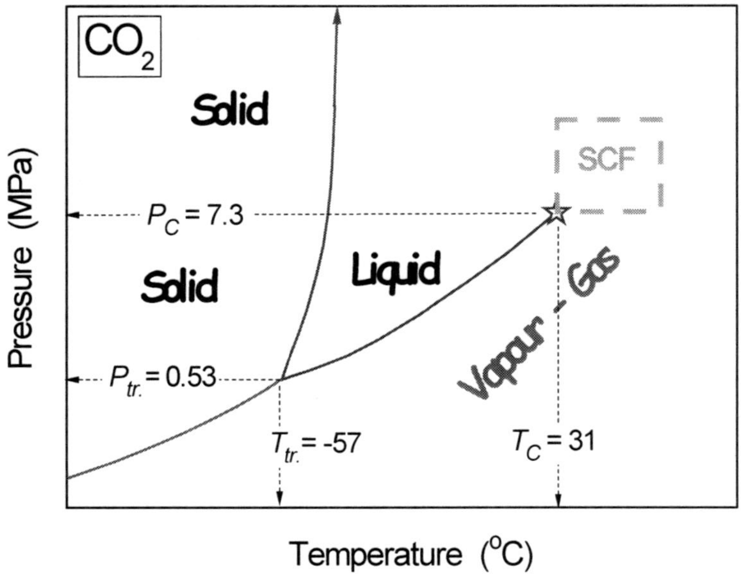

Figure 1. *Phase diagram for CO_2 with coordinates of the triple point ($T_{tr.}$, $P_{tr.}$) and the liquid-gas critical point T_C, P_C. The loci of the supercritical fluid (SCF) domain is also shown.*

Consequently, the vast majority of SCF applications are based on CO_2 near the GL critical point, with a possible admixture to support the ability for solvating dipolar components. The extraction of carcinogenic aromatic hydrocarbons and their nitro derivatives from diesel particulates by CO_2 + toluene or methanol SCF can serve as an example. CO_2 based SCF also helps in cleaning polyethylene from undesired polymer additives. In a similar way one can consider technologies focused on so called hyper-coal, an extremely pure and environment friendly fuel for turbines in power plants. Recently, the first power plants based on this idea are being constructed in China. The removal of pesticides from meat, decaffeinated coffee and denicotinized cigarettes are the next society-relevant applications. Noteworthy is the hyper-oxidation with supercritical water and bitumens extraction based on supercritical toluene. The latter system is also used for the liquefaction of coal.[1-6]

It seems that SCF technologies may be considered as the future platform for society relevant material engineering as well as for dealing with waste and pollutions. This can be of particular importance in decomposition of aggressive agents, where the currently used extreme temperature technologies can only lessen the undesired influence on the environment. Supercritical fluids constitute a unique and powerful medium due to their enormous possibilities of selective controlling of solvent properties and reaction rates. This is associated with anomalous changes on approaching the critical point of such properties as dielectric permittivity, density, diffusion, coefficient, solubility parameter, ionic strength or compressibility (susceptibility). The greatest advantage of SCF is that it can replace some toxic technologies used so far. Nevertheless, traditional technologies are still less expensive and often conceptually simpler for users. It is noteworthy that for industrial application pressurization, which is inherently associated with SCF–GL technology, constitutes a particularly aggravating factor.

It seems plausible that the following conditions may boost SCF technological and environmental applications:

(i) a qualitative increase of the number of systems used as the basic platform
(ii) location of the SCF domain close to the ambient pressure and temperature
(iii improvement of the efficiency and controlling of the desired processes

Some emerging possibilities of reaching these targets are discussed below.

2. Critical mixtures of limited miscibility

In 1926 Kohnstamm extended the Gibbs phase rule[7] to encompass the appearance of critical points in one- and multicomponent systems[8]. He assumed that the critical point may be considered as a specific, additional phase. If "*p*" phases coexist, and next become critical, $p-1$ meniscuses disappear in a solution consisting of "*c*" components. Hence, the Gibbs phase rule supplemented by the "critical phase" has the following form $f_{crit} = c-(p+p-1)+2 = c-2p+3$.[8,9] Consequently, for *p*-critical point at least $c = 2p-1$ component system is required. This yields for $f_{crit} = 0$, i.e. a single "critical" point, following conditions:

For $p = 2 \quad \rightarrow \quad c = 1$ and a gas-liquid critical point in 1-component system.

For $p = 3 \quad \rightarrow \quad c = 3$ tricritical point in a 3-component mixture.

This short analysis shows that in one component fluids only a single, "isolated" liquid – gas critical point should exist. This is associated with the selected values of temperatures and pressures (T_C, P_C). For binary mixtures of limited miscibility with a critical consolute point (CP) a continuous line of critical points, in addition to the gas-liquid critical point, should appear, namely" for

$f_{crit} = 1$ one obtains $p = 2$ and $c = 2$. Then in the a two component system a line of critical points, associated with liquid – liquid immiscibility, should exists. Consequently, for such mixtures a single, isolated critical point under arbitrary pressure, including the atmospheric one, may be expected. For GL and CP the transition is associated with the loss of the translational degree of freedom, i.e. with the transition from the one-phase (gas or homogeneous liquid) to the two-phase (gas-liquid or liquid-liquid) system of coexisting phases. It is noteworthy that GL and CP critical points belong to the same universality class in the modern theory of critical phenomena. They are described by the same universal values of critical exponents and ratios of critical amplitudes.[9]

Consequently, one may consider the application of binary mixtures near the critical consolute point for SCF technologies instead of one component fluids near the gas – liquid critical point. It is also noteworthy that for one component fluids the presence of the second liquid – liquid (L-L) near-critical point has been found.[10-13] Then, the question arises whether the latter kind of L-L critical points can be useful for SCF technology?

The basic advantages of using binary mixtures with the critical CP instead of the GL critical point can be found in the history of critical phenomena, namely establishing the basic universal parameters appeared to be much simpler for binary mixtures with CP than for GL systems. Firstly, CP investigations can be carried out under atmospheric pressure. Secondly, one can select a binary mixture with CP close to room temperature.[14] Finally, it is possible to select a mixture which emphasizes the desired specific feature, for instance: (a) methanol – cyclohexane mixture can simulate weightless conditions since densities of both components are almost equal (b) there are almost no critical opalescence for isooctane – cyclohexane mixture since their refractive indices are almost the same (c) one can considerably change the concentration of the dipole component of the mixture. The latter feature can strongly influence both dielectric properties and solvency.

Worth recalling is the case of nitrobenzene – *n*-alkane mixtures.[14,15] Their technological importance arises from the large dielectric permittivity and dipole moment of nitrobenzene and the ability to solvate some environmentally hazardous agents.[12 and refs therein] Basic properties of such mixtures are presented below, with the use of the authors' earlier studies and some novel results.

Figure 2 shows values of the critical temperature and critical concentration in nitrobenzene – *n*-alkanes homologous series. Noteworthy is the power-type evolution of the critical concentration, described by the power exponent 1/2. The pressure dependence of the critical consolute temperature for these critical mixtures are shown in Fig. 3. Noteworthy is the change of the sign from $dT_C/dP < 0$ to $dT_C/dP > 0$ when increasing the length of *n*-alkane.

Figure 4 indicates that this behavior can be associated with the evolution of the excess volume (V^E), since the excess enthalpy (H^E) is always positive for mixtures with an upper critical consolute point.

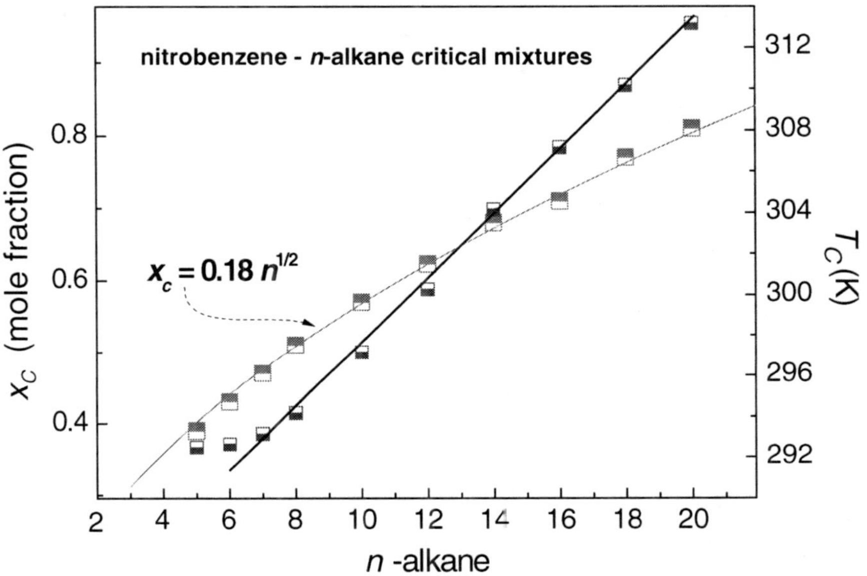

Figure 2. *The evolution of critical temperatures (in blue) and concentrations (in red) for the homologous series of nitrobenzene – n-alkanes mixtures. It is based on authors' studies from refs.* ([15]) *supplemented by novel results for n-alkanes with* $n > 12$*. They were obtained using visual – cathetometric method described in ref.* ([14])*. Results are for the atmospheric pressure* $P = 0.1 MPa$.

Particularly noteworthy is the possibility of reaching the domain where $dT_C/dP \to 0$, i.e. $V^E \to 0$. In such a case using the isothermal, pressure-related path one can easily reach the very immediate vicinity of the critical point. On approaching the critical point as a function of temperature reaching the $T - T_C < 0.01K$ domain is an enormously difficult experimental task. For the isothermal pressurization in binary mixtures with $dT_C/dP \to 0$ one can reach the immediate vicinity of the critical point, even equivalent to $T - T_C << 0.01K$. Moreover, on depressurizing the distance from the critical point (the line of critical points) increases extremely slowly. Hence, the precise and technically simple control of the distance from the critical point, matched with precise tuning, is possible.

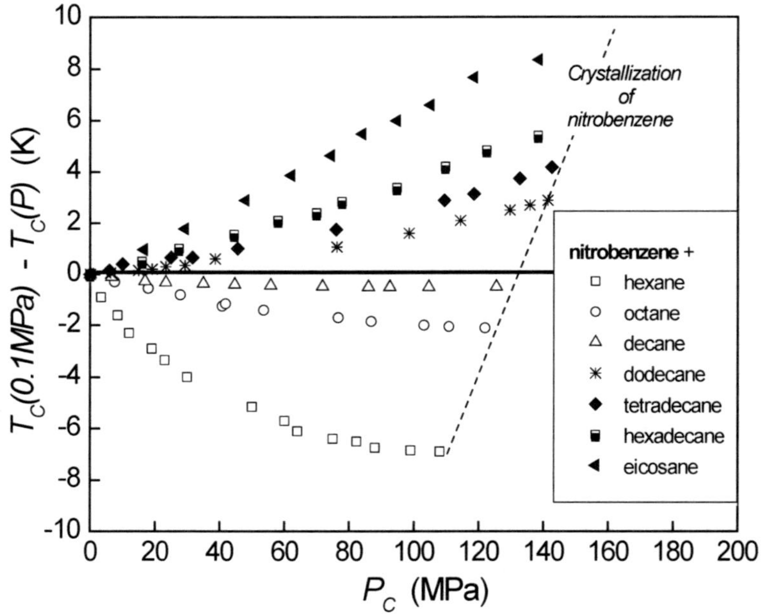

Figure 3. *The pressure evolution of critical temperature in the series of nitrobenzene – n-alkanes mixtures. It is based on authors' studies from ref.* ([16]) *supplemented by novel results for n-alkanes with n >12. The set-up used for studies is the same as in refs.* ([16, 17]).

Such conditions are difficult to obtain for the gas – liquid critical point. Studies of pretransitional anomalies via dielectric permittivity measurements can illustrate the unique possibilities of the isothermal pressurization towards the critical consolute point.[16] These anomalies are much stronger than for the temperature path and their reliable description without additional power terms, so called corrections-to-scaling, is possible. The change of the sign of dT_C/dP on pressuring within a homologous series is characteristic not only for nitrobenzene – *n*-alkanes series of critical mixtures. The same occurs for oligostyrene – *n*-alkanes[17] or *o*-nitrotoluene – *n*-alkanes[15] mixtures and for a series of primary alcohols and water, i.e. $C_nH_{2n+1}OH + H_2O$.[18]

Figure 5 is related to the latter series of near critical mixtures, namely for $n = 4$, where the crossover from $dT_C/dP < 0$ to $dT_C/dP > 0$ on compressing occurs. It was shown that in nitrobenzene – hexane mixture such $T_C(P)$ evolution can be portrayed via:[19]

$$T_C(P) = T_C^0\left(1 + \frac{\Delta P}{\pi + P_C^0}\right)^{1/b} \exp\left(-\frac{\Delta P}{c}\right) \tag{1}$$

where $\Delta P = P - P_C^0$ and $\Pi = \pi + P_C^0$, $-\pi$ is the negative pressure asymptote for $T \to 0$, P_m^0 and T_m^0 are the reference pressure and temperature, c denotes the damping pressure coefficient; $-\pi$ is the negative pressure asymptote for $T \to 0$.

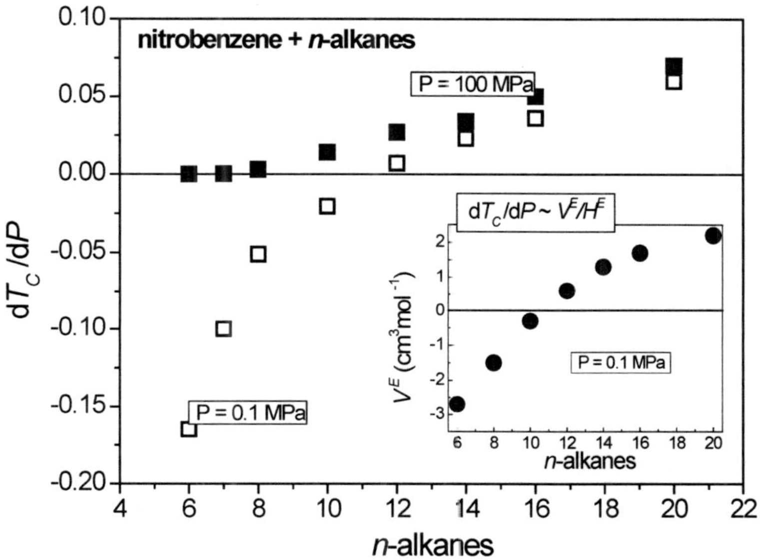

Figure 4. *The evolution of dT_C/dP in the series of nitrobenzene – n-alkanes mixtures. It is based on authors' studies from ref. ([15]) supplemented by novel results for n-alkanes with n >12. The set-up used for studies is the same as in ref. ([15]). The inset shows the evolution of the excess volume, indicated the validity of the thermodynamic relation ([15]) given in the Figure.*

Basic parameters for this dependence can be estimated using the preliminary derivative based analysis:

$$\left[\frac{d(\ln T_C)}{dP} + c^{-1}\right]^{-1} = b\pi + bP \tag{2}$$

For the optimal selection of the damping coefficient "c" the transformed $T_C(P)$ experimental data should exhibit a linear behavior. Subsequently, the linear regression analysis can yield optimal values of 'b' and 'π' parameters. The latter is located in the negative pressures domain. Since values of 'b' and 'π' and 'c' are pressure invariant, the resulting $T_C(P)$ evolution can be extended even well beyond the domain determined by experimental data.

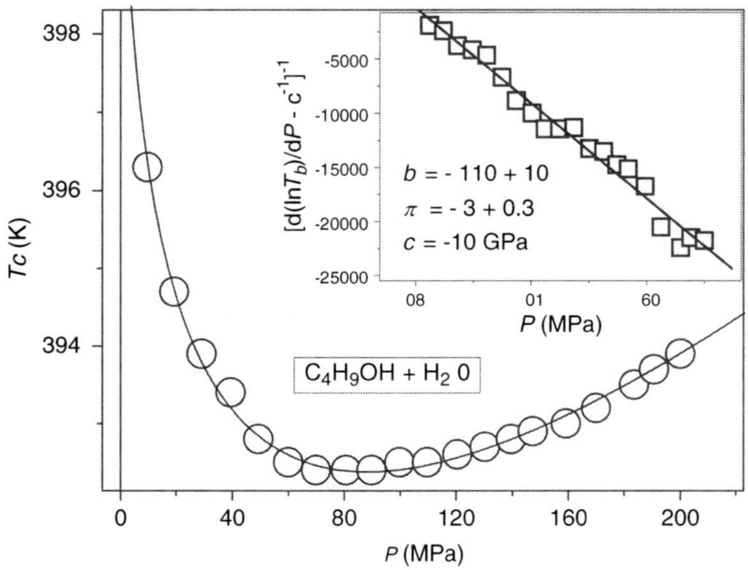

Figure 5. *The pressure evolution of the critical consolute temperature in n-butanol – water mixture. The solid curve presents the parameterization via eq. (1) with basic parameters derived from the derivative based analysis via eq. (2), which results are shown in the inset.*

In ref. ([20]) the applications of the analogous dependences (eq. (1) and (2)) for the pressure evolution of the glass temperature and the melting temperature in supercooled liquids were shown. It is noteworthy that both alcohols and water are important technological agents, also used as additives to the CO_2 basic critical system. For the discussed case of binary mixtures of limited miscibility the critical behavior is the inherent feature of the system containing water and alcohol or nitrobenzene or nitrotoluene and alkanes, even under atmospheric pressure. When critical binary mixtures are considered as the base for the SCF technologies, no additional component is needed.

The influence of pressure on critical concentration in binary mixtures is not discussed here. However, studies presented in ref. ([17]) showed that it can be smaller than 0.01 mole fraction when pressurized by 1 kbar (100 MPa).[21]

3. The liquid – liquid critical point in a one component liquid

The last decade has given clear evidence for the existence of a second critical or near–critical liquid-liquid (L-L) transition in a one component fluid.[10,11] This suggests that the Kohnstamm-Gibbs phase rule[8] has to be supplemented since it suggests the existence of a single gas – liquid critical point in a one component fluid. So far, there is no response to the basic question: "*Is this phenomenon general for any liquid or is it restricted to a specific group of liquids?*"

It is often suggested that L-L transition in one component liquids is always "*secret*", for instance hidden below the glass temperature, and then the evidence for its existence is non-direct. The most "classical" examples for this phenomenon are water[10] or and triphenyl phosphite (TPP).[11] Recently clear evidence for the L-L near critical transition in the experimentally available domain for two novel liquids of vital technological and environmental significance have been given. They are: nitrobenzene[12] and *trans*-1,2-dichlorethylene.[13]

The most pronounced evidence of the critical-like behavior in these compounds can be found by using the nonlinear dielectric effect (NDE).[12] *NDE* is coupled via the 4-point correlation to multimolecular heterogeneities – fluctuations. It was shown in refs. ([22]) that:

$$NDE \sim \left\langle \Delta M^2 \right\rangle \chi \sim \frac{\left\langle \Delta M^2 \right\rangle}{\left(T - T^*\right)^{\gamma}} \quad , \gamma = 1 \tag{3}$$

where $\left\langle \Delta M^2 \right\rangle$ is the average of mean square of the local order parameter fluctuations and $\chi \sim \left(T - T^*\right)^{-\gamma}$ denotes the susceptibility linked to the order parameter. T_C is temperature of the continuous phase transition. In experiment this magnitude is defined by $NDE = \varepsilon\left(^{E} - \varepsilon\right)/E^2$, where ε^E and ε are dielectric permittivities in a weak and strong electric field of intensity E.

The susceptibility-related exponent $\gamma = 1$, i.e. it has the mean field value.[22] Such a value is characteristic for $d \geq 4$ dimensionality.[9] This is equivalent to the situation when the number of neighbors of a given molecule or an assembly of molecules is significantly larger than the number for near spherical molecules/fluctuations surrounding a molecule/assembly in a three dimensional space.

This can be also illustrated as follows: a fatty man can be surrounded by 4 – 5 "closely-packed" fat men. But the same man can be surrounded even by 10 slim men. The latter can parallel the increasing dimensionality of space for the fat man. Consequently, the elongation of fluctuations due to the action of the strong electric field[22] in the homogeneous phase of binary mixtures of limited miscibility is equivalent to increased dimensionality d = 4. The uniaxial symmetry is natural also for the isotropic phase nematic liquid crystals.[23] It may be considered for supercooled nitrobenzene due to intermolecular interactions (Fig. 5).[12]

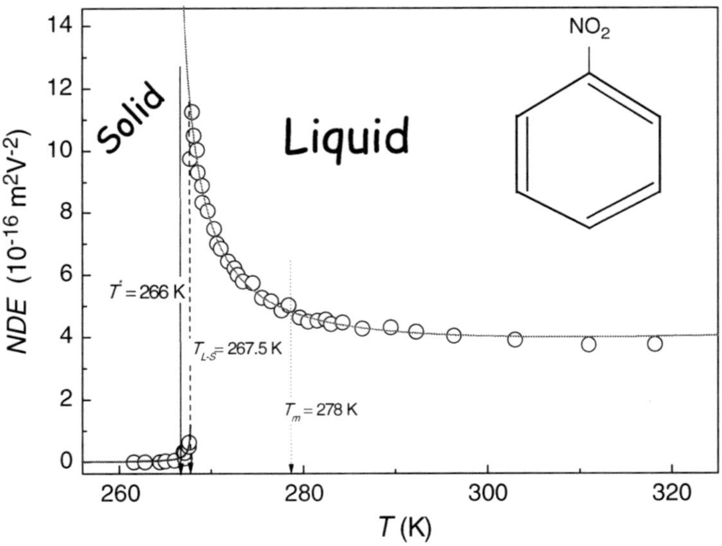

Figure 6. *The temperature evolution of nonlinear dielectric effect in nitrobenzene. The red solid curve is for eq. (3).*

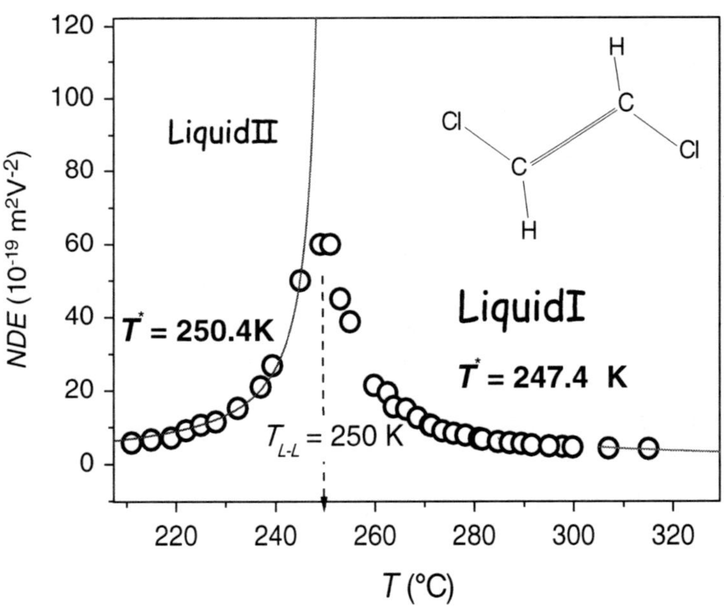

Figure 7. *The temperature evolution of nonlinear dielectric effect in trans-1,2-dichlorethylene. The red solid curve is for eq. (3).*

Trans -1,2-dichlorethylene shown in Fig. 7 also exhibits the uniaxial form. It is noteworthy that for non-classical critical systems, such as binary mixtures of limited miscibility, $\langle \Delta M^2 \rangle \sim (T - T^*)^{2\beta}$, where $\beta \approx 0.33$ is the order parameter critical exponent.[22] For naturally rod-like molecules $\langle \Delta M^2 \rangle \sim \Delta\alpha\Delta\alpha'$,[22,23] where $\Delta\alpha$ and $\Delta\alpha'$ are anisotropies of polarizability for the frequency of the strong electric field and for the measurement frequency, respectively. This occurs for the isotropic phase of rod-like nematic liquid crystals,[23] nitrobenzene[12] and *trans*-1,2-dichlorethylene[13]. Results for the latter are shown in Figs. 6 and 7. It should be stressed here that for *trans*-1,2-dichlorethylenethe L-L near-critical point is located at $T_{L-L} \approx 25\,°C$, i.e. well above the melting temperature.

This material is a fundamental solvent in polymer material engineering hence its application as the L-L SCF may facilitate the production processes and the degradation of plastic wastes.[14 and refs therein]

Nitrobenzene is considered as a basic model system for studying molecular interactions and fundamental properties of nitro-aromatic compounds. They are essential compounds for manufacturing of explosives, pesticides and other chemicals.[12 and refs. therin]

4. Conclusions

Supercritical fluids belong to the most promising platforms for solving technological and environmental challenges facing the XXI century. They enable precise, enormously efficient and selective extraction processing. The efficiency of SCF processes can be easily changed by decades only by temperature and/or pressure shift. It seems that these conditions are fulfilled also for the liquid – liquid transition associated with the consolute point in binary mixtures or with the L-L second critical point in a one component fluid. For these systems one can reach the SCF conditions without an additional component, which is often a prerequisite for the SCF–GL technology. Consequently, the description of systems with the L-L critical point is possible solely via the basic principles of the modern theory of critical phenomena. This is of particular importance due to the fact that during any extraction process an addition, 'parasitic', solvent is introduced to the near-critical system. Hence, the distance from the critical point strongly increases causing the qualitative decrease of extraction efficiency. For L-L systems one can apply the basics of the theory of critical phenomena as the universality concept,[9] Fisher's renormalization[24] and the isomorphism (smoothness) postulate.[9, 16] Then, for the L-L based SCF technology monitoring the distance from the critical point, e.g.

via NDE or dielectric permittivity measurements, can yield on-line estimations of the critical point position. Consequently, a simple feedback procedure maintaining the distance from the critical point and the constant efficiency of the SCF technology is possible. The emerging advantages of using L-L transitions for SCF technologies are additionally boosted by the possibility of reaching arbitrary values and signs of the dT_C/dP coefficient.

Acknowledgements

This research was carried out with the support of the CLG NATO Grant No. CBP. NUKR.CLG 982312 and of the Ministry of Science and Higher Education (Poland) Grant No. N202 147 32/4240).

References

1. Wu, B. C., Klein, M. T., and Sandler, S. I. (1991) Solvent effects on reactions in supercritical fluids, *Ind. Eng. Chem. Res.* **30**, 822-82
2. Hawthorne, S. B., et al. (1992) Extraction of polycyclic aromatic hydrocarbons from diesel particulates, *J. Chromatography*, **609**, 333-340
3. Cansell, F., and Rey, S. (1998) Thermodynamic aspects of supercritical fluids processing: application to polymer and wastes, *Rev.. de L'Institute Frances du Petrolse*, vol. **53**, 71-98
4. Engelhardt, H., and Haas, P. (1993) *J.Chromatographic Sci.* **31**, 13-19
5. Hauthal, W. H. (2001) Advances with supercritical fluids (review), *Chemo-sphere* **43**, 123-135
6. Nalawade, S. P., Pocchioni, F., and Janssen, L. P. B. M. (2006) Supercritical carbon dioxide as a green solvent for processing polymer melts: processing aspects and applications, *Prog. Polym. Sci.* **31**, 19-43
7. Gibbs, J. W. (1982) Collected Papers: Thermodynamics and Statistical mechanics (*Nauka, Moscow*) in russian
8. Kohnstamm, Ph. (1926) in: H. Geiger and K. Scheel, Editors, *Handbuch der Physik* vol. **10**, Springer, Berlin p. 223
9. Rzoska, S. J. (1979) Dielectric permittivity near the gas – liquid critical point in diethyl ether, *MSc Thesis* (*Silesian University*)
10. Anisimov, M. A. (1994) Critical Phenomena in Liquids and Liquid Crystals (*Gordon and Breach, Reading*)
11. Kumar, P., Buldyrev, S. V., and Stanley, H. E. (2007) Water liquid-liquid dynamic crossover and liquid-liquid critical point in the TIP5P model of water, in S. J. Rzoska and V. Mazur (eds.) "*Soft Matter under Exogenic Impacts*", NATO Sci. Series II, vol. **242** (Springer, Berlin)
12. Tanaka, H., Kurita R., and Mataki, H. (2004) Liquid-liquid transition in the molecular liquid triphenyl phosphite, *Phys. Rev. Lett.* **92**, 025701
13. Drozd-Rzoska, A., Rzoska, S. J., and Ziolo, J. (2008) Anomalous temperature behavior of nonlinear dielectric effect in supercooled nitrobenzene, *Phys. Rev. E* **77**, 041501

14. Rzoska, S. J., Zioło, J., Drozd-Rzoska A., Tamarit, J. Ll., and Veglio, N. (2008) New evidence for a liquid – liquid transition in a one component liquid, *J. Phys.: Condens. Matt.* **20**, 244124
15. Rzoska, S. J. (1990) Visual methods for determining of coexistence curves in liquid mixtures, *Phase Transitions* **27**, 1-13
16. Urbanowicz, P., Rzoska, S. J., Paluch, M., Sawicki, B., Szulc, A., and Ziolo, J. (1995) Influence of intermolecular interactions on the sign of dT_C/dP in critical solutions, *Chem. Phys.* **201**, 575-582
17. Rzoska, S. J., Urbanowicz, P., Drozd-Rzoska, A., Paluch, M., and Habdas, P. (1999) Pressure behaviour of dielectric permittivity on approaching the critical consolute point, *Europhys. Lett.* **45**, 334-340
18. Imre, A. R., Melnichenko, G., van Hook, W. A., and Wolf, B. A. (2001) *Phys. Chem. Chem. Phys.* **3**, 1063
19. Schneider, G. M. (1993) Phase equilibrium of fluid system at high pressures, *Pure & Appl. Chem.* vol. **65**, 173
20. Drozd-Rzoska, A., Rzoska, S. J., and Imre, A. R. (2004), Liquid-liquid equilibria in nitrobenzene – hexane mixture under negative pressure, *Fluid Phase Equilibria* **6**, 2291-2294
21. Drozd-Rzoska, A., Rzoska, S. J., and Imre, A. R. (2007) On the pressure evolution of the melting temperature and the glass transition temperature, *J. Non-Cryst. Solids* **353**, 3915-3923.
22. Urbanowicz, P., and Rzoska, S. J. (1996) Influence of high hydrostatic pressure on a nitrobenzene - dodecane critical solution, *Phase Transitions* **56**, 239-244
23. Rzoska, S. J. (1993) Kerr effect and nonlinear dielectric effect on approaching the critical consolute point, *Phys. Rev. E* **48**, 1136-1143
24. Drozd-Rzoska, A., and Rzoska, S. J. (2002) Complex relaxation in the isotropic phase of *n*-pentylcyanobiphenyl in linear and nonlinear dielectic studies, *Phys. Rev. E* **65**, 041701
25. Rzoska, S. J., Chrapeć, J., and Zioło J. (1987) Fisher's renormalization for the nonlinear dielectric effect from isothermal measurements, *Phys. Rev. A* **36**, 2885-2889

2D AND 3D QUANTUM ROTORS IN A CRYSTAL FIELD: CRITICAL POINTS, METASTABILITY, AND REENTRANCE

YURI A. FREIMAN
B. Verkin Institute of Low Temperature Physics and Engineering, 47 Lenin Avenue, Kharkov, UA-61103, Ukraine

BALÁZS HETÉNYI
Institut für Theoretische Physik, Technische Universität Graz, Petersgasse 16, Graz, A-8010, Austria

SERGEI M. TRETYAK
B. Verkin Institute of Low Temperature Physics and Engineering, 47 Lenin Avenue, Kharkov, UA-61103, Ukraine

Abstract: An overview of results of models of coupled quantum rotors is presented. We focus on rotors with dipolar and quadrupolar potentials in two and three dimensions, potentials which correspond to approximate descriptions of real molecules adsorbed on surfaces and in the solid phase. Particular emphasis is placed on the anomalous reentrant phase transition which occurs in both two and three-dimensional systems. The anomalous behaviour of the entropy, which accompanies the reentrant phase transition, is also analyzed and is shown to be present regardless if a phase transition is present or not. Finally, the effects of the crystal field on the phase diagrams are also investigated. In two-dimensions the crystal field causes the disappearance of the phase transition, and ordering takes place via a continuous increase in the value of the order parameter. This is also true in three dimensions for the dipolar potential. For the quadrupolar potential in three dimensions turning on the crystal field leads to the appearance of critical points where the phase transition ceases, and ordering occurs via a continuous increase in the order parameter. As the crystal field is increased the range of the coupling constant over which metastable states are found decreases.

Keywords: quantum rotors, phase transition, mean-field theory, solid hydrogen

1. Introduction

In molecular solids the energy scales of translation, rotation, and vibration can be expected to be of different orders of magnitude. In the solid hydrogens,[1-3] the rotational lines are clearly distinct from the spectral signatures of the translations and rotations. For a large pressure interval in such systems models of coupled rigid rotors are sufficient to understand the general features of phase transitions in particular those of the orientational kind. In this work we refer to such systems as orientational crystals.

There exist systems in both two and three dimensions which can be thought of as orientational crystals. Two dimensional examples are physisorbed molecules on inert surfaces, such as N_2 or H_2 and its isotopes on graphite or boron-nitride. The former can be approximated by a model of planar rotors, known as the anisotropic planar rotor (APR) model.[4-7] This model exhibits an orientational order-disorder phase transition from an orientationally disordered state to the orientationally ordered herringbone structure. In N_2 on a graphite surface this transition takes place at 30K,[8,9] well below the liquid-solid ordering temperature of 47K.[10] While the classical APR model accounts for the orientational ordering, a more quantitative description of the system necessitates the inclusion of quantum effects.[11] Also, models of coupled quantum planar rotors are useful in describing other systems, such as granular superconductors[12-17] and more recently the bosonic Hubbard model.[18,19]

Three-dimensional examples are the solid phases of the hydrogens and different isotopes. The behavior of the hydrogens is generally made more complex by the fact that ortho-para conversion times are slow on the time-scale of rotations (in the pressure ranges considered here $\leq$100GPa),[20] hence it is a reasonable approximation to take the ortho-para ratio to be a fixed parameter.[21] The existence of ortho and para species is due to the coupling of nuclear spins and the rotational quantum numbers characterizing a particular molecule. For the H_2 molecule a rotation of angle π corresponds to an exchange of the constituent atoms, hence the wave function has to be anti-symmetric in such a rotation. Since the H atoms are of spin ½, the possible spin states of the molecule as a whole are three symmetric and one anti-symmetric spin state. To preserve the overall antisymmetry of the wave function the symmetric spin states couple with anti-symmetric spatial states (odd angular momentum or odd-J) and the anti-symmetric spin states couple with symmetric spatial states (even-J). In HD, where the atoms are indistinguishable all angular momentum states are allowed (all-J). In D_2 the constituent atoms are bosons, hence the wavefunction has to be symmetric. However this leads to a qualitatively similar situation: here symmetric(anti-symmetric) spin-states are even-J(odd-J).

The orientational ordering properties of odd-J, even-J, and all-J systems show striking differences. Odd-J systems show orientational ordering in the ground state, whereas even-J systems order at higher pressures. At low pressures and low temperatures the even-J systems can be thought of as spheres. An interesting anomalous feature that was first predicted for all-J systems is the *reentrant* phase diagram. Upon cooling, in certain pressure ranges, the system orders orientationally due to a decrease in thermal fluctuations. In the reentrant region, the orientationally ordered phase is destroyed by quantum fluctuations (also know as quantum melting). This effect was first predicted in the mean-field phase diagram of the all-J hydrogen system (HD),[22,23] and experimentally verified thereafter.[24]

For the quantum generalization of the APR model (QAPR) the system corresponding to the all-J case also shows reentrance. This was first predicted by mean-field theory,[25] and then verified via quantum Monte Carlo calculations[26] as well as quantum Monte Carlo calculations analyzed via finite size scaling.[27,28] Reentrance was also found in the corresponding model of granular superconductors.[16,17]

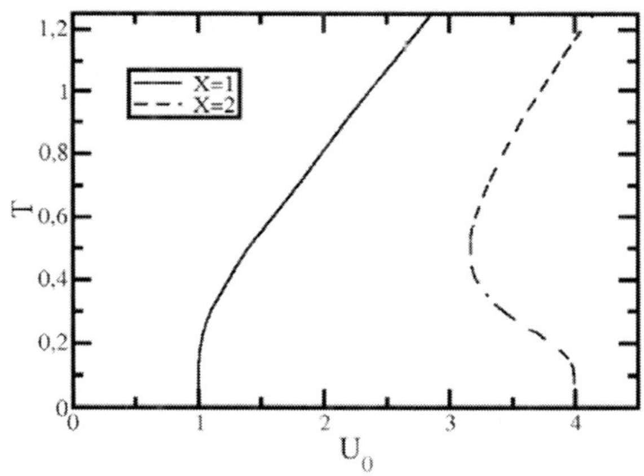

Figure 1. *Phase diagrams for the systems without crystal field for* X=1,2.

Recently a set of studies[29-31] have suggested that if the thermal equilibrium distribution of the ortho-para ratio is reached then reentrance can occur in the homonuclear systems H_2 and D_2. This conclusion is supported by experimental evidence.[32]

In this paper we give an overview of the mean-field theory of phase transitions in coupled rotors with particular attention to the issues of reentrance, other quantum anomalies, and meta-stability. We comparatively analyze coupled planar rotors (two-dimensional model) and coupled linear rotors (three-dimensional). We show that the dipolar potential does not exhibit the reentrance anomaly, whereas the quadrupolar one does. The phase transition turns out to be second order in all cases except for the linear rotors in a quadrupolar potential where it is first order. We also investigate the effects of the crystal field: in the case of the linear rotor model with quadrupolar potentials the crystal field causes the appearance of critical points which separate lines of the phase diagram where the transition is first order from regions where there is no phase transition, but simply a continuous change of the order parameter.[33] We show that the range over which meta-stable states (which accompany a first-order phase transition) depends on the crystal field: as it is increased this region becomes smaller, and disappears when the phase transition itself disappears. We also analyze the behaviour of the entropy in all cases.

2. Coupled rotors in two-dimensions

The model we study in this section is described by the Hamiltonian

$$H = -B\sum_{i=1}^{N}\frac{\partial^2}{\partial\phi_i^2} - \frac{U}{2}\sum_{\langle i,j\rangle}\cos(X\phi_i)\cos(X\phi_j) - U_1\sum_{i=1}^{N}\cos(X\phi_i), \quad (1)$$

where B, U, and U_1 denote the rotational constant, the coupling constant, and the strength of the crystal field respectively, and where the sum $\langle i,j\rangle$ runs over nearest neighbors. The parameter X specifies the periodicity of the potential. In this work we will investigate the cases $X=1,2$, which show qualitatively different behaviour. As a unit of energy and temperature we choose the rotational constant B in all of the subsequent cases.

Applying the mean-field approximation to this Hamiltonian results in

$$H_{MF} = -B\sum_{i=1}^{N}\frac{\partial^2}{\partial\phi_i^2} - (U_0\gamma + U_1)\sum_{i=1}^{N}\cos(X\phi_i) + \frac{U_0 N}{2}\gamma^2, \quad (2)$$

where γ denotes the order parameter, and $U_0=Uz$ with z denoting the coordination number. We note that had we used the dipole-dipole (quadrupole-quadrupole) potential in Eq. (1) the resulting approximate Hamiltonian can be shown to be the same as the one in Eq. (2) with $X=1(X=2)$ with a modified coupling constant.

The mean-field phase diagrams without crystal field for $X=1,2$ are shown in Figure 1. The phase diagrams separate the orientationally disordered phase (at lower values of the coupling constant) from the orientationally ordered phase. The two striking differences between the two curves are the quantitative difference between the onset of order and the shape of the phase diagram. The former can be attributed to the width of the barrier through which the quantum systems tunnel. The $X=1$ system has a wider barrier than the $X=2$ system. The reentrance has been found in the related QAPR model via quantum Monte Carlo[26-28] and is known to be due to the ordering tendency of higher energy states (the states with angular momentum zero are disordered as they are of polar symmetry, the first odd angular momentum states are ordered). In both cases we have found the transition to be of second order.

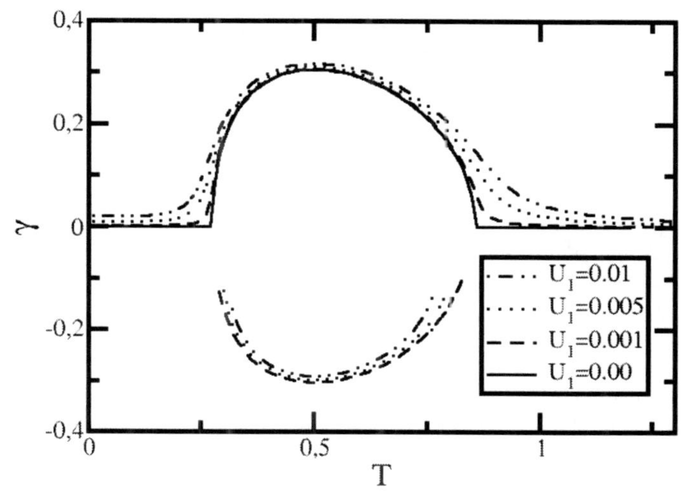

Figure 2. *Order parameter as a function of temperature for the X=2 system at U_0=3.50. The curves with negative values indicate a metastable state in the case of finite crystal field.*

Calculations for the order parameter are presented in Fig. 2 for a system with *X=2* (*U_0=3.50* reentrant region). In the case of no crystal field both transitions are manifestly second-order. As the temperature is increased the order parameter is zero until *T~ 0.27*, it increases up to *T~ 0.5*, then the slope switches sign and decreases until *T~0.86*. Subsequently the order parameter is zero. The effect of the crystal field is also shown in Fig. 2. The order parameter for the system with crystal field shows no discontinous change in the slope of the order parameter, however a change in sign of the slope occurs at *T~0.5* as in the case of no crystal field. Another feature of the crystal field is the appearance of a metastable state with negative order parameter as shown in Fig. 2. The behaviour of the entropy for the system with *X=2* without crystal field is shown in Figure 3. As has been shown for the three dimensional case,[30] the entropy displays an anomaly in the case of the reentrant phase diagram.

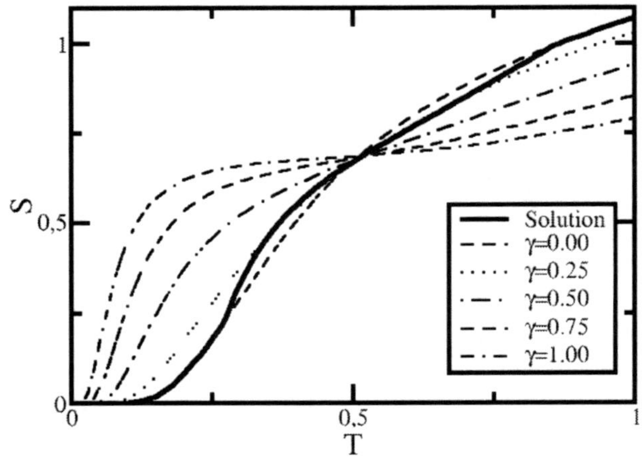

Figure 3. *Entropy calculations for a system without crystal field. The comparison is for the entropy of the actual system (solution of the mean-field equations) and for fixed order parameter* (γ*=0.00,0.25,0.50,0.75,1.00),* U_0*=3.50.*

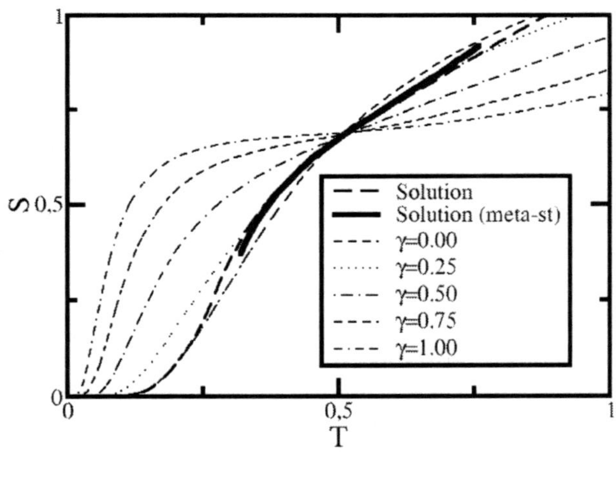

Figure 4. *Entropy calculations for a system with crystal field (*U_1*=0.01). The conparison is for the entropy of the actual system (solution of the mean-field equations), the meta-stable solution, and for fixed order parameter (*γ*=0.00,0.25,0.50,0.75,1.00). The coupling constant is* U_0*=3.5.*

The entropy curves for the fixed order parameter show qualitatively different behavior above and below $T{\sim}0.5$, where the slope of the order parameter switches sign (Fig. 2). The entropy of the disordered state (γ=0) is the lowest below the temperature $T{\sim}0.5$, and the entropy increases as the system orders. This behaviour is unexpected from a classical point of view. Above $T{\sim}0.5$ the entropy of the ordered state is the lowest, and it increases upon disordering, as expected based on the classical view. This unusual feature can be understood from considering the expression of the entropy for the quantum mechanical system, $S=\sum P_i \ln P_i$ where P_i denote the probability for a particular state. In the quantum mechanical system the states are obtained after diagonalizing the Hamiltonian (in the corresponding classical system the sum in the expression for the entropy is an integral over the angles, and the probability is a function of the angles as well). As the lowest state, which dominates the behaviour of the system at low temperatures (i.e. has the highest probability), corresponds to a disordered state, it is not surprising that the entropy decreases and that simultanously the system disorders. In the state-space to which the probabilities in the entropy expression refer the number of possible states does in fact decrease (i.e. in that sense the system orders), however the states themselves are disordered in real space.

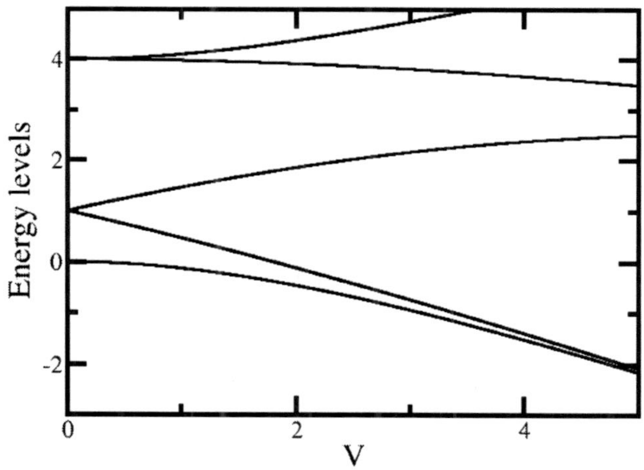

Figure 5. *Energy levels for a system with potential $V\cos(2\varphi)$.*

In some sense this picture is similar to Bose-Einstein condensation,[34] where the single lowest state becomes populated (and which corresponds to a state that is spatially disordered), with the important difference that here the state is not a collective state.

We also note that the entropy of the solution of the mean-field equations corresponds to the disordered case below and above the phase transition points. At the phase transition points the slope of the entropy is discontinuous.

The effect of the crystal field on the entropy is shown in Fig. 4 (U_1=*0.01*). The same behavior is observed with regard to the ordering pattern as in Fig. 3. Below the turning point of the slope of the order parameter (Fig. 2) the entropy of the ordered state is higher than that of the disordered state. Here the slope of the entropy does not change discontinously as a function of temperature, as no phase transitions are experienced.

The entropy anomaly can also be understood in terms of the local energy spectrum. The eigenvalues of a planar rotor with potential $V\cos(2\varphi)$ are shown in Fig. 5 as a function of V. At $V=0$ (disordered state) the ground state is a singlet and the first excited state is doubly degenerate. As V is increased the degeneracy of the first excited state is split, and the lower energy state becomes degenerate with the ground state adding a factor of $R\ln 2$ to the entropy at low temperatures.

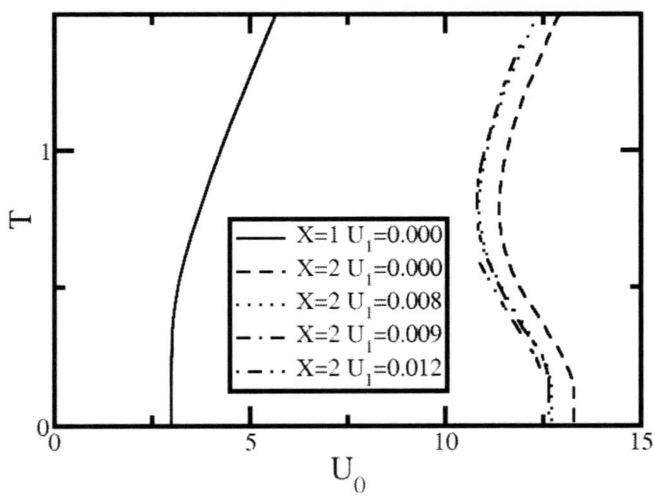

Figure 6. *Phase diagram for linear rotors, X=1 and X=2, with several values of the crystal field in the case of the latter.*

3. Coupled rotors in three dimensions

In this section we calculate the mean-field phase diagram of a system of coupled three-dimensional rotors under a crystal field.

$$H = B\sum_{i=1}^{N} \hat{L}_i^2 - \frac{U}{2}\sum_{\langle i,j\rangle} \mathrm{P}_{Xm}(\Omega_i)\mathrm{P}_{Xm}(\Omega_j) - U_1\sum_{i=1}^{N} \mathrm{P}_{X0}(\Omega_i). \qquad (3)$$

where

$$\mathrm{P}_{X0}(\Omega_i) = \sqrt{4\pi/(2X+1)}\mathrm{Y}_{X0}(\Omega_i)$$

The mean-field approximation to the Hamiltonian in Eq. (3) results in

$$H_{MF} = B\sum_{i=1}^{N} \hat{L}_i^2 - (U_0\gamma + U_1)\sum_{i=1}^{N} \mathrm{P}_{X0}(\Omega_i) + \frac{U_0 N}{2}\gamma^2. \qquad (4)$$

The phase diagrams for the two cases *X=1* and *X=2*, and for several crystal fields in the case of the latter, are shown in Figure 6.

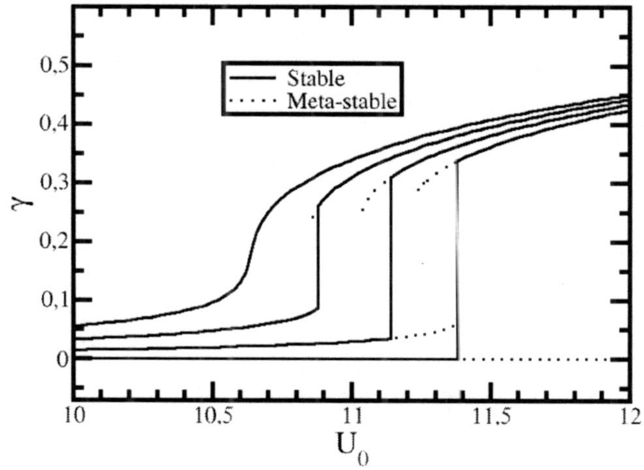

Figure 7. *Order parameter for different values of the crystal field as a function of the coupling constant U_0 at a temperature of T=0.75 The values of the crystal field from left to fight the crystal field from left to right are U_1 = 0.012,0.008,0.004,0.000. The dotted lines indicate the value of the order parameter for metastable states.*

For the systems without crystal field, the two features identified in the previous section in the case of the planar rotors, namely the stronger ordering tendency in the *X=1* case, and the reentrant phase transition in the *X=2* case are present in the case of linear rotors as well.

An important difference is that the *X=2* case exhibits a first order phase transition. An unusual feature develops upon turning on the crystal field. As shown before[33] the crystal field gives rise to critical points which separate regions in the phase diagram where the transition is first order from regions where no phase transition occurs, rather a continuous increase in the order parameter (the exact quantitative features of the phase diagram are explained in Ref. 33). In Figure 7 we show the order parameter as a function of the coupling constant at a temperature of $T=0.75$ (approximately where the reentrant turning point occurs) for the $X=2$ system. The calculations are presented for different values of the crystal field, $U_1=0.012$ $0.008, 0.004, 0.000$. The dotted lines indicate the meta-stable states. As usual in first-order phase transitions, as the parameter U_0 is varied a meta-stable state develops before the phase transition, which becomes the stable state upon crossing the phase transition point. Simultaneously the stable state becomes meta-stable. When no crystal field is present we found that as U_0 is increased from the left, the ordered meta-stable phase first appears at $U_0 \sim 11.2$ and becomes the stable state at $U_0=11.38$. Subsequently the disordered phase $\gamma=0$ becomes metastable. As the crystal field is turned on the range where metastability is encountered decreases. For $U_1=0.004$, as U_0 is increased from the left we find evidence for a meta-stable phase at $U_0 \sim 11.0$, the phase transition is encountered at $U_0 \sim 11.14$, but the less ordered phase (which was stable at $U_0 \leq 11.14$ persists as a meta-stable phase until $U_0 \sim 11.4$.

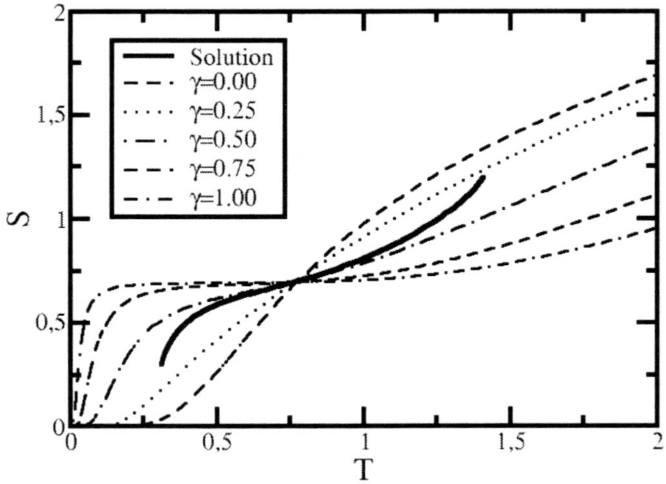

Figure 8. *Entropy of the ordered state and at fixed values of the order parameter for the system of linear trotors with X=2 at no crystal field, U =12.50.*

For $U_l = 0.008$ the phase transition is found at $U_0 = 10.88$ and metastability is encountered only in a range ~*0.04* around the phase-transition point. For $U_1 = 0.012$ no phase transition is encountered, only a continuous increase in the order parameter.

The entropy curves verify the general tendency shown in the case of linear rotors in the previous section. In Figures 8 and 9 the value of the entropy corresponding to the solution are shown as well as the value of the entropy at fixed order parameter for the case without crystal field ($U_0 = 12.50$) and with a crystal field of $U_1 = 0.018$ ($U_0 = 12.00$). The inset shows the value of the order parameter at $U_1 = 0.018$ as a function of temperature: as the temperature is decreased the order parameter increases, it experiences a turning point at $T = 0.75$ and then begins to decrease. This happens continuously, without any phase transition. The entropy of the ordered state, as was the case for the planar rotors, is higher at low temperature ($T \leq 0.75$) than that of the disordered state. Thus the reversal of ordering as the temperature is cooled appears to be correlated with the entropy anomaly, however, whether the disordering occurs as a result of a phase transition is not.

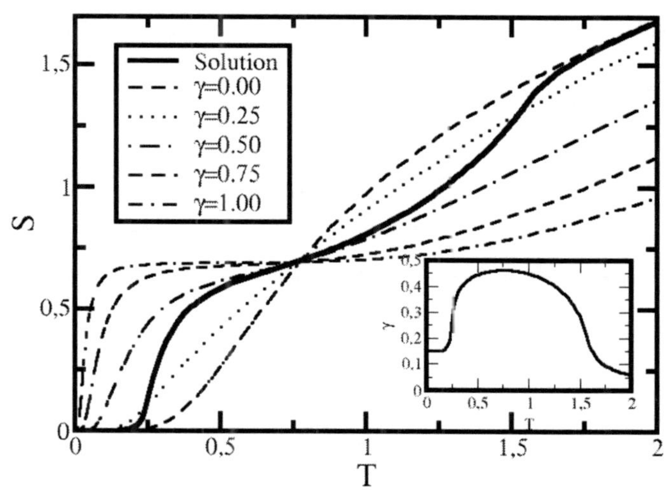

Figure 9. *Entropy of the ordered state and at fixed values of the order parameter for the system of linear rotors with X=2 at a crystal field of* $U_l = 0.018$, $U_0 = 12.00$. *The inset shows the order parameter.*

In the absence of the crystal field the quantum melting phase transition is second order for planar rotors, first order for linear rotors. When a crystal field is turned on the phase transition is absent for planar rotors, whereas a more complicated situation develops for linear rotors (see Figures 6 and 7 and Ref. 30), but if the crystal field is large enough the ordering and disordering also happens continuously. The entropy anomaly accompanies all of these ordering patterns. The energy levels for the linear rotors in an external potential of $VY_{20}(\Omega)$ are shown in Fig. 10. As in the case of the planar rotors increase of V from zero causes one state to move down and approach the ground state causing an increase of $\sim R\ln 2$ in the entropy.

4. Conclusions

We have presented a comparative review of the mean-field theory of different types of coupled rotors. We have considered planar and linear rotors in dipolar and quadrupolar potentials.

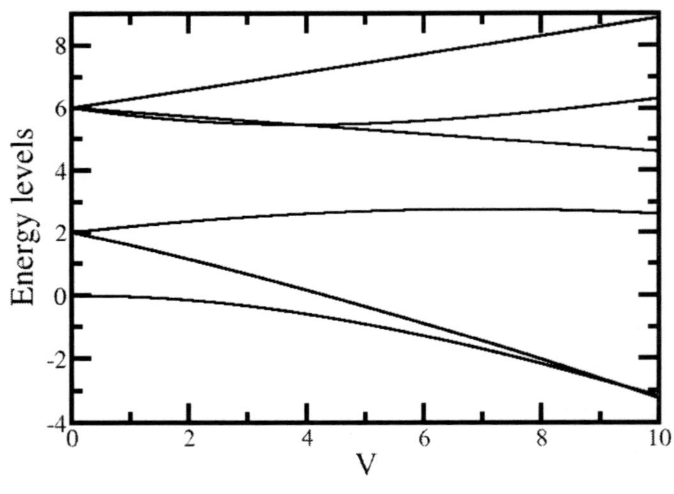

Figure 10. *Energy levels for a system with potential $VY_{20}(\Omega)$.*

These models have corresponding physical realizations: diatomic molecules (heteronuclear in the dipolar case, homonuclear in the quadrupolar case) physisorbed on surfaces (two dimensional system) or in the solid phase (three dimensional system). The dipolar potentials in both cases lead to a usual phase diagram where above a particular value of the coupling constant the temperature vs. coupling constant phase diagram increases with coupling constant. The quadrupolar potentials lead to reentrant phase diagrams in both cases: at low temperatures, for some values of the coupling constant, quantum melting takes place. The phase transition for the planar rotors is always second order. For the linear rotors the dipolar potential leads to a second-order phase transition, in the quadrupolar potential the phase transition is first order.

We have also shown the different effects found when the systems are subjected to a crystal field. For the dipolar potentials the crystal field causes a disappearance of the phase transition line, as temperature is decreased, and as the coupling constant is increased only a continuous increase in the order parameter is found. As the ordering increases a metastable state is also found with a negative order parameter. We have also found this for the planar rotors coupled via a quadrupolar potential. For the linear rotors the situation is more complicated. As previously found[30] increasing the crystal field causes the appearance of critical points which separate the phase diagram into lines where the phase transition is first order from regions where no phase transition, but a continuous change in the order parameter occurs. An interesting accompanying feature is that where there is a phase transition, the range in which a metastable state is found decreases with the strength of the crystal field.

The reentrance in the case of the quadrupolar systems is accompanied by an entropy anomaly: if the order parameter is held fixed the entropy of the ordered state is higher at low temperatures than that of the disordered state. The situation reverses when the temperature is increased. This entropy anomaly is present in all the systems which exhibit quantum melting, irrespective whether the melting takes place via a phase transition (either first or second order), or via a continuous change in the order parameter. Calculation of the spectrum of the mean-field potentials shows that the entropy anomaly can be explained in terms of the change in the degeneracies of states as a function of the coupling constant, as the ground state becomes doubly degenerate. It can also be argued that the entropy anomaly is a natural consequence of quantum mechanics: the entropy decreases with temperature, as a single state begins to dominate, but this single state is a delocalized one (zero angular momentum state), hence it is disordered.

References

1. Silvera, I. (1980) *Rev. Mod. Phys.* **52**, 393
2. Van Kranendonk, J. (1983) *Solid Hydrogen: Theory of the Properties of solid H_2, HD, and D_2* (Plenum Press, New York)
3. Mao, H. -K., and Hemley, R. J. (1994) *Rev. Mod. Phys.* **66**, 671
4. O'Shea, S. F., and Klein, M. L. (1979) *Chem. Phys. Lett.* **66**, 381
5. O'Shea, S. F., and Klein, M. L. (1982) *Phys. Rev. B* **25**, 5882
6. Mouritsen, O. G., and Berlinsky, A. J. (1982) *Phys. Rev. Lett.* **48**, 181
7. Marx, D., and Wiechert, H. (1996) *Adv. Chem. Phys.* **95**,181
8. Chung, T. T., and Dash, J. D. (1977) *Surf. Sci.* **66**, 559

9. Eckert, J., Ellenson, W. D., Hastings, J. B., and Passell, L. (1977) *Phys. Rev. Lett.* **43**, 1329
10. Kjems, J. K., Passell, L., Taub, H., and Dash, J. D. (1977) *Phys. Rev. Lett.* **32**,724
11. Presber, M., Löding, D., Martoňák, R., and Nielaba, P. *Phys. Rev. B* **58**, 11937 (1998)
12. McLean, W. L., and Stephen, M. J. (1979) *Phys. Rev. B* **19**, 5925
13. Šimánek, E. (1981) *Phys. Rev. B* **22**, 459
14. Maekawa, S., Fukuyama, H., and Kobayashi, S. (1981) *Solid State Comm.* **37**, 45
15. Doniach, S. (1981) *Phys. Rev. B* **24**, 5063
16. Šimánek, E. (1985) *Phys. Rev. B* **32**, 500
17. Simkin, M. V. (1991) *Phys. Rev. B* **44**, 7074
18. Polak, P., and Kopeć, T. K. (2008) *Acta Physica Polonica A* **114**, 29
19. Kopeć, T. K. (2004) *Phys. Rev. B* **70**, 054518
20. Strzhemechny, M. A., and Hemley, R. J. (2000) *Phys. Rev. Lett.* **85**, 5595
21. Harris, A. B., and Meyer, H. (1985) *Can. J. Phys.* **63**, 3
22. Freiman, Y. A., Sumarokov, V. V., Brodyanskii, A. P., and Jezowski, A. (1991) *J. Phys. Condens. Matter* **3**, 3855
23. Brodyanskii, A. P., Sumarokov, V. V., Freiman, Y. A., and Jezowski, A. (1993) *Sov. J. Low Temp. Phys.* **19**, 520
24. Moshary, F. N., Chen, H., and Silvera, I. (1993) *Phys. Rev. Lett.* **71**, 3814
25. Martoňák, R., Marx, D., and Nielaba, P. (1997) *Phys. Rev. E* **55**, 2184
26. Müser, M. H., and Ankerhold, J. (1997) *Europhys. Lett.* **44**, 216
27. Hetényi, B., Müser, M. H., and Berne, B. J. (1999) *Phys. Rev. Lett.* **83**, 4606
28. Hetényi, B., and Berne, B. J. (2001) *J. Chem. Phys.* **114**, 3674
29. Hetényi, B., Scandolo, S., and Tosatti, E. (2005) *Phys. Rev. Lett.* **94**, 125503
30. Freiman, Y. A., Tretyak, S. M., Mao, H. -K., and Hemley, R. J. (2005) *J. Low Temp. Phys.* **139**, 765
31. Hetényi, A., Scandolo, S., and Tosatti, E. (2005) *J. Low Temp. Phys.* **139**, 753
32. Goncharenko, I., and Loubeyre, P. (2005) *Nature* **435**, 1206
33. Freiman, Y. A., Tretyak, S., Antsygina, T., and Hemley, R. J. (2003) *J. Low Temp. Phys.* **133**, 151
34. Leggett, A. J. (2001) *Rev. Mod. Phys.* **73**, 307

PART IV

WATER AND LIQUID-LIQUID TRANSITONS

METASTABLE WATER UNDER PRESSURE

[1]KEVIN STOKELY, [1] MARCO G. MAZZA, [1]H. EUGENE STANLEY, AND[2] GIANCARLO FRANZESE
[1]*Center for Polymer Studies and Department of Physics, Boston University – Boston, MA 02215 USA*
[2]*Departament de Fısica Fonamental – Universitat de Barcelona,Diagonal 647, Barcelona 08028, Spain*

Abstract: We have summarized some of the recent results, including studies for bulk, confined and interfacial water. By analyzing a cell model within a mean field approximation and with Monte Carlo simulations, we have showed that all the scenarios proposed for water's P–T phase diagram may be viewed as special cases of a more general scheme. In particular, our study shows that it is the relationship between H bond strength and H bond cooperativity that governs which scenario is valid. The investigation of the properties of metastable liquid water under pressure could provide essential information that could allow us to understand the mechanisms ruling the anomalous behavior of water. This understanding could, ultimately, lead us to the explanation of the reasons why water is such an essential liquid for life.

Keywords: water, anomalous behavior, simulations

1. Introduction

Water's phase diagram is rich and complex: more than sixtee crystalline phases[1], and two or more glasses[2]. The liquid state also displays intersting behavior. In the stable liquid regime water's thermodynamic response functions behave qualitatively differently than a typical liquid. The isothermal compressibility K_T and isobaric specific heat C_P each display a minimum as a function of temperature (at 46°C and 36°C for 1 atm, respectively) while for a typical liquid these quantities monotonically decrease upon cooling. Water's anomalies become even more pronounced as the system is cooled below the melting point and enters the metastable supercooled regime[3]. Here K_T and C_P increase rapidly upon cooling, with an apparent divergence for 1 atm at −45°C.[4] A precise understanding of the physico–chemical properties of liquid water is important to provide accurate predictions of the behavior of biological molecules[5,6], geophysical structures[7], and

S. Rzoska et al. (eds.), *Metastable Systems under Pressure*,

nanomaterials[8] to mention just a few subjects of interest. Microscopically, the anomalous liquid behavior is understood as resulting from the tendency of water molecules to form hydrogen (H) bonds upon cooling, with a decrease of potential energy, decrease of entropy, and increase of distance between the bonded molecules. The low temperature phase behavior which results from these interactions, however, remains unknown because experiments on bulk water below the crystal homogenous nucleation temperature T_H (−38°C at 1 atm) are unfeasible. Four different scenarios for the pressure–temperature ($P - T$) phase diagram have been debated:

(i) The stability limit (SL) scenario[9] hypothesizes that the superheated liquid spinodal at negative pressure re-enters the positive P region below $T_H(P)$ leading to a divergence of the response functions.
(ii) The singularity–free (SF) scenario[10] hypothesizes that the low-T anticorrelation between volume and entropy gives rise to response functions that increase upon cooling and display maxima at non–zero T, but do not display singular behavior.
(iii) The liquid–liquid critical point (LLCP) scenario[11] hypothesizes a first–order phase transition line with negative slope in the P − T plan, separating a low density liquid (LDL) from a high density liquid (HDL), which terminates at a critical point C′. Below the critical pressure PC′ the response functions increase on approaching the Widom line (the locus of correlation length maxima emanating from C′ into the one–phase region), and for $P > P_{C'}$ by approaching the spinodal line. Evidence suggests[11–13] that $P_{C'} > 0$, but the possibility $P_{C'} < 0$ has been proposed.[14]
(iv) The critical–point free (CPF) scenario[15] hypothesizes a first–order phase transition line separating two liquid phases and extending to $P < 0$ down to the (superheated) limit of stability of liquid water. No critical point is present in this scenario.

Though experiments on bulk water are currently unfeasible, freezing in the temperature range of interest can be avoided for water in confined geometries[16–18] or on the surface of macromolecules.[19–25] Since experiments in the supercooled region are difficult to perform, an intense activity of numerical simulations has been developed in recent years to help interpret of the data[26, 27]. However, simulations at very low temperature T are hampered by the glassy dynamics of the empirical models of water.[28,29] It is therefore important to study simple models, which are able to capture the fundamental physics ofwater while being less computationally expensive. We analyze a microscopic cell model[30] of water that has been shown to exhibit any of the proposed scenarios, depending on choice of parameters.[10,13, 31] The model, whose dynamics behavior compares well with that of supercooled water,[29,32] is here studied using both mean-field (MF) analysis and Monte Carlo (MC) simulations.

2. The cell model

The model consists of dividing the fluid into N cells with index $i \in [1, \ldots, N]$, each with volume v_0, and occupation variable $n_i = 0$ (for a cell with gas–like density) or $n_i = 1$ (for a cell with liquid–like density). Each cell is assumed in contact with 4 nearest neighbor (n.n.) cells, mimicking the first shell of liquid water, in the simplified assumption of no interstitial molecules.

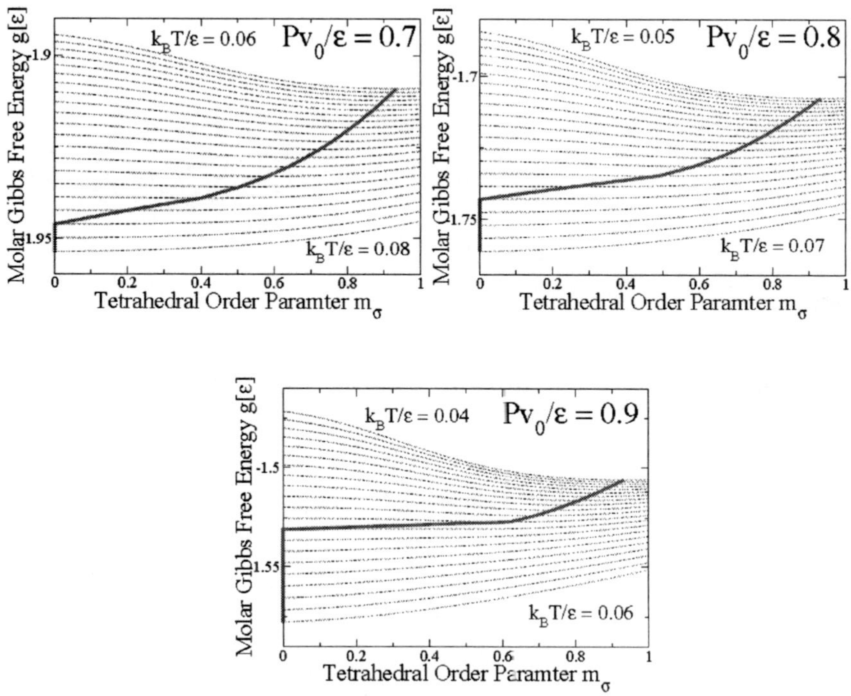

Figure 1. *Numerical minimization of the molar Gibbs free energy g in the mean field approach. The model's parameters are $J/\varepsilon = 0.5$, $J\sigma/\varepsilon = 0.05$, $v_{HB}/v_0 = 0.5$ and $q = 6$. In each panel we present g (dashed lines) calculated at constant P and different values of T. The thick line crossing the dashed lines connects the minima $m_\sigma^{(eq)}$ of g at different T. Upper panel: $Pv_0/\varepsilon = 0.7$, for T going from $k_BT/\varepsilon = 0.06$ (top) to $k_BT/\varepsilon = 0.08$ (bottom). Middle panel: $Pv0/\varrho = 0.8$, for T going from $k_BT/\varepsilon = 0.05$ (top) to $k_BT/\varepsilon = 0.07$ (bottom). Lower panel: $Pv_0/\varepsilon = 0.9$, for T going from $k_BT/\varepsilon = 0.04$ (top) to $k_BT/\varepsilon = 0.06$ (bottom). In each panel dashed lines are separated by $k_B\delta T/\varepsilon = 0.001$. In all the panels $m_\sigma^{(eq)}$ increases when T decreases, being 0 (marking the absence of tetrahedral order) at the higher temperatures and ≈ 0.9 (high tetrahedral order) at the lowest temperature. By changing T, $m_\sigma^{(eq)}$ changes in a continuous way for $Pv_0/\varepsilon = 0.7$ and 0.8, but discontinuous for $Pv_0/\varepsilon = 0.9$ and higher P.*

The system is described by the Hamiltonian [30]:

$$\mathscr{H} = -\epsilon \sum_{\langle i,j \rangle} n_i n_j - J \sum_{\langle i,j \rangle} n_i n_j \delta_{\sigma_{ij},\sigma_{ji}} - J_\sigma \sum_i n_i \sum_{(k,\ell)_i} \delta_{\sigma_{ik},\sigma_{i\ell}}. \tag{1}$$

The first term with $\varepsilon > 0$ accounts for the van der Waals attraction and hard-core volume exclusion, such that neighboring liquid cells are energetically favorable. This term is due to the long–range attraction and short–range repulsion of the electron clouds [33]. The sum is over all n.n. cells h_i, j_i.

The second term with $J > 0$ accounts for the directional H bond interaction between neighboring liquid cells, which must be correctly oriented in order to form a bond.

Figure 2. *Three snapshots of the system, for $N = 100\times100$, showing the Wolff's clusters of correlated water molecules. For each molecule we show the states of the four arms and associate different colors to different arm's states. The state points are at pressure close to the critical value P_C ($Pv_0/\varepsilon = 0.72 \approx P_Cv_0/\varepsilon$) and $T > T_C$ (top panel, $k_BT/\varepsilon = 0.053$), $T \approx T_C$ (middle panel, $k_BT/\varepsilon = 0.0528$), $T < T_C$ (bottom panel, $k_BT/\varepsilon = 0.052$), showing the onset of the percolation at $T \approx T_C$. At $T \approx T_C$ (middle panel) there is one large cluster, in red on the right, with a linear size comparable to the system linear extension and spanning in the vertical direction.*

This term is associated with the covalent nature of the bond [34]. Bond variables σ_{ij} represent the orientation of the molecules in cell i with respect to the n.n.

molecule in the cell j, and $\delta_{a,b} = 1$ if $a = b$ and $\delta_{a,b} = 0$ otherwise. We choose $q = 6$, giving rise to $6^4 = 1296$ possible orientational states per molecule.

Experiments show that the formation of a H bond leads to a local volume expansion[2], so the total volume is given as

$$V = Nv_0 + N_{HB}v_{HB} \tag{2}$$

$$N_{HB} \equiv \sum_{\langle i,j \rangle} n_i n_j \delta_{\sigma_{ij},\sigma_{ji}} \tag{3}$$

is the total number of H bonds, and v_{HB} is the specific volume increase due to H bond formation.[10] The third term in Eq. (1) with $J_\sigma \geq 0$ represents the many–body interaction among H bonds, related to the T-dependent O–O–O correlation[35], driving the molecules toward a local tetrahedral configuration.[36–39] Here $(k, \ell)_i$ indicates one of the six different pairs of the four bond variables of molecule i. This interaction introduces a cooperative behavior among bonds, which may be fine tuned by changing J_σ. Choosing $J_\sigma = 0$ leads to fully independent H bonds, while $J_\sigma \to \infty$ leads to fully dependent bonds.

3. The mean field analysis

In the MF analysis the macrostate of the system in equilibrium at constant P and T is determined by a minimization of the Gibbs free energy per molecule, $g \equiv \left(\langle \mathcal{H} \rangle - PV + TS\right)/N_W$

$$N_w = \sum_i n_i \tag{4}$$

the total number of liquid-like cells, and $S = S_n + S_\sigma$ is the sum of the entropy Sn over the variables n_i and the entropy S_σ over the variables σ_{ij}. A MF approach consists of writing g explicitly using the approximations

$$\sum_{<ij>} n_i n_j \longrightarrow 2Nn^2 \qquad \sum_{<ij>} n_i n_j \delta_{\sigma_{ij},\sigma_{ji}} \longrightarrow 2Nn^2 p_\sigma$$
$$\sum_i n_i \sum_{(k,l)_i} \delta_{\sigma_{ik},\sigma_{il}} \longrightarrow 6Nnp_\sigma \tag{5-7}$$

where $n = N_w/N$ is the average of n_i, and p_σ is the probability that two adjacent bond indices σ_{ij} are in the same state. Therefore, in this approximation we can write

$$V = Nv_0 + 2Nn^2 p_\sigma v_{HB}$$
$$\langle \mathcal{H} \rangle = -2Nn\left[\epsilon n + (Jn + 3J_\sigma)\, p_\sigma \right. \tag{8,9}$$

The probability p_σ that two adjacent bond variables form a bond is properly defined as the thermodynamic average of $\delta_{\sigma ij,\sigma ji}$ over the entire system. It is here approximated as the average over two neighboring molecules, under the effect of the mean-field h of the surrounding molecules,

$$p_\sigma = \langle \delta_{\sigma_{ij},\sigma_{ji}} \rangle_h \tag{10}$$

The ground state of the system consists of all N variables $n_i = 1$, and all σ_{ij} in the same state. At low temperatures the symmetry will remain broken, with the majority of the σ_{ij} in a preferred state. We associate this preferred state with the space-filling tetrahedral network of H bonds formed by liquid water, and define n_σ as the density of bond indices in this tetrahedral state, with $1/q \le n_\sigma \le 1$. An appropriate form for h is[30]

$$h = 3J_\sigma \frac{1 + (q-1)m_\sigma}{q} \tag{11}$$

where $0 \le m_\sigma \le 1$ is an order parameter associated with the number of bond variables in the preferred state.
Equating the MF relation

$$p_\sigma \equiv n_\sigma^2 + \frac{(1-n_\sigma)^2}{q-1} \tag{12}$$

with the approximate expression in Eq. (10) allows us to express n_σ in terms of T, P, and m_σ, which may be substituted into the MF expression for g. The MF approximations for the entropies S_n of the N variables n_i, and S_σ of the $4Nn$ variables σ_{ij}, are[40]

$$S_n = -k_B N(n\log(n) + (1-n)\log(1-n))$$
$$S_\sigma = -k_B 4Nn[n_\sigma \log(n_\sigma) + (1-n_\sigma)\log(1-n_\sigma) + \log(q-1)] \tag{13,14}$$

where k_B is the Boltzmann constant.

Minimizing numerically g with respect to n and m_σ, we find the equilibrium values $n^{(eq)}$ and $m_\sigma^{(eq)}$. By substitution into Eqs. (4) and (2), we calculate the density ρ at any (T, P), the full equation of state. An example of the minimization of g is presented in Fig. 1 where, for the model parameters $J/\varepsilon = 0.5$, $J_\sigma/\varepsilon = 0.05$, $v_{HB}/v_0 = 0.5$ and $q = 6$, a discontinuity in $m_\sigma^{(eq)}$ is observed for $Pv_0/\varepsilon > 0.8$. As discussed in Refs. [13, 30] this discontinuity corresponds to a first order phase transition between two liquid phases with different degree of tetrahedral order and, as a consequence, different density. The P at which the change in $m_\sigma^{(eq)}$ becomes continuous corresponds to the pressure of a LLCP. The occurrence of the LLCP is consistent with one of the possible interpretations of the anomalies of water, as discussed in Ref. [40]. However, for different choices of parameters, the model reproduces also the other proposed scenarios.[31]

4. The Monte Carlo simulations

To perform MC simulations in the NPT ensemble, we consider a modified version of the model in which we allow for continuous volume fluctuations. To this goal, (i) we assume that the system is homogeneous with all the variables n_i set to 1 and all cells have volume $v = V/N$; (ii) we consider that $V \equiv V_{MC} + N_{HB}v_{HB}$, where $V_{MC} > N v_0$ is a dynamical variable allowed to fluctuate in the simulations; (iii) we replace the first (van der Waals) term of the Hamiltonian in Eq. (1) with a Lennard-Jones potential with attractive energy $\varepsilon > J$ plus a hard-core interaction

$$U_W(r) \equiv \begin{cases} \infty & \text{if } r \leqslant r_0, \\ \epsilon\left[\left(\frac{r_0}{r}\right)^{12} - \left(\frac{r_0}{r}\right)^{6}\right] & \text{if } r > r_0. \end{cases} \tag{15}$$

where $r_0 \equiv (v_0)^{1/d}$;[13] the distance between two n.n. molecules is $(V/N)^{1/d}$, and the distance r between two generic molecules is the Cartesian distance between the center of the cells in which they are included. The simplification (i) could be removed, allowing the cells to assume different volumes v_i and keeping fixed the number of possible n.n. cells. However, results of the model under the simplification (i) compare well with experiments.[40] Furthermore, the simplification (i) allows to drastically reduce the computational cost of the evaluation of the $U_{W(r)}$ term from $N(N-1)$ to $N-1$ operations.

MC simulations are performed with $N = 10^4$ molecules, each with four n.n. molecules on a 2d square lattice, at constant P and T, and with the same model parameters as for the MF analysis. To each molecules we associate a cell on a square lattice. The Wolff's algorithm is based on the definition of a cluster of variables chosen in such a way to be thermodynamically correlated.[41, 42] To define the Wolff's cluster, a bond index (arm) of a molecule is randomly selected; this is the initial element of a stack. The cluster is grown by first checking the remaining arms of the same initial molecule: if they are in the same Potts state, then they are added to the stack with *probability* $p_{same} \equiv min[1, 1-exp(-\beta J_\sigma)]$,[43] where $\beta \equiv (k_BT)^{-1}$. This choice for the probability p_{same} depends on the interaction J_σ between two arms on the same molecule and guarantees that the connected arms are thermodynamically correlated.[41] Next, the arm of a new molecule, facing the initially chosen arm, is considered. To guarantee that connected facing arms correspond to thermodynamically correlated variables, is necessary[42] to link them with the probability $p_{facing} \equiv min[1, 1-exp(-\beta J')]$ where $J' \equiv J - Pv_{HB}$ is the P–dependent effective coupling between two facing arms as results from the enthalpy H +PV of the system. It is important to note that J' can be positive or negative depending on P. If $J' > 0$ and the two facing arms are in the same state, then the new arm is added to the stack with probability p_{facing}; if $J' < 0$ and the two facing arms are in different states, then the new arm is added with probability p_{facing}.[44] Only after every possible direction of growth for the cluster has been considered the values of the arms are changed in a stochastic way; again we need to consider two cases: (i) if $J' > 0$, all arms are set to the same new value

$$\sigma^{\text{new}} = \left(\sigma^{\text{old}} + \phi\right) \mod q \tag{16}$$

where φ is a random number between 1 and q; (ii) if $J' < 0$, the state of every single arm is changed (rotated) by the same random constant $\varphi \in [1, \ldots q]$

$$\sigma_i^{\text{new}} = \left(\sigma_i^{\text{old}} + \phi\right) \mod q. \tag{17}$$

In order to implement a constant P ensemble we let the volume fluctuate. A small increment $\Delta r/r_0 = 0.01$ is chosen with uniform random probability and added to the current radius of a cell. The change in volume $\Delta V \equiv V^{new} - V^{old}$ and van der Waals energy ΔE_W is computed and the move is accepted with probability $min(1, exp[-\beta(\Delta E_W + P\Delta V - T\Delta S)])$, where $\Delta S \equiv -Nk_B ln(V^{new}/V^{old})$ is the entropic contribution. The cluster MC algorithm turms out to be hundreds of time faster, in generating uncorrelated configurations, than a Metropolics MC dynamics when the system has P and T in the vicinity of the liquid critical point. The efficiency of the Wolff's cluster algorithm is a consequence of the exact relation between the average size of the finite clusters and the average the size of the regions of thermodynamically correlated molecules. The proof of this relation at any T derives straightforward from the proof for the case of Potts variables[41] This relation allows to identify the clusters built during the MC dynamics with the correlated regions and emphasizes (i) the appearance of heterogeneities in the sturctural correlations,[45] and (ii) the onset of percolation of the clusters of tetrahedrally ordered molecules at the liquid–liquid critical point,[46] as shown in Fig. 2.

5. Effects of the hydrogen bond strength and cooperativity

From the MF analysis, when $J_\sigma = 0$ the model coincides with the one proposed in[10] which gives rise to the SF scenario (Fig. 3a). When $J_\sigma > 0$ the model displays a phase diagram with a LLCP (Fig. 3b) [13]. For $J_\sigma \to 0$, keeping J and the other parameters constant, we find that $T_{C'} \to 0$, and the power–law behavior of K_T and the isobaric thermal expansion coefficient α_P is preserved. Further, we find for the entropy S that, for any value of J_σ, $(\partial S/\partial T)_P \sim |T - T_{C'}|^{-1}$.

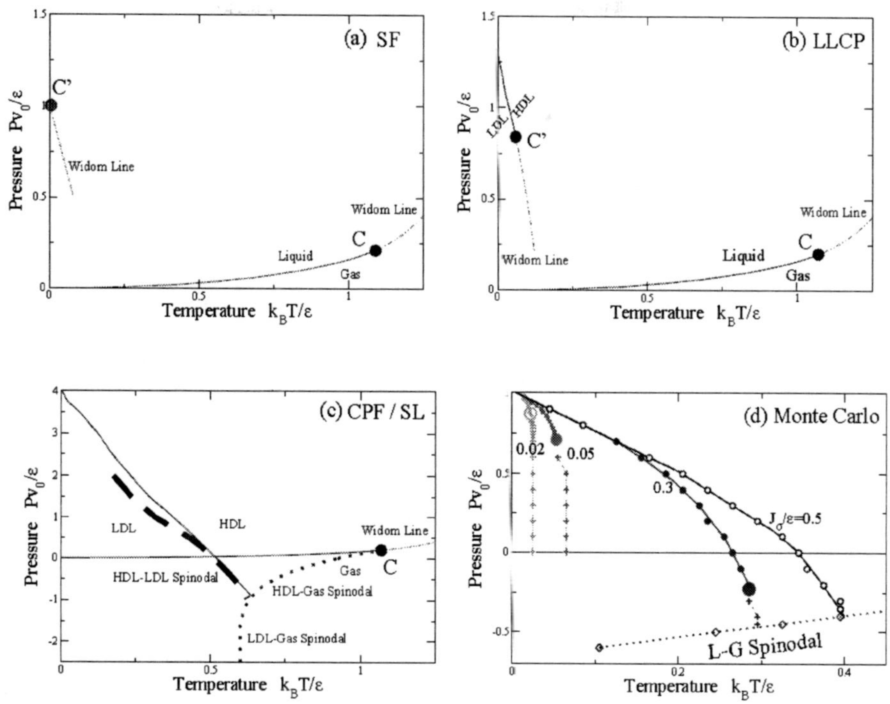

Figure 3. *Phase diagram predicted from our calculations for the cell model with fixed H bond strength ($J/\varepsilon = 0.5$), fixed H bond volume increase ($v_{HB}/v_0 = 0.5$), and different values of the H bond cooperativity strength J_σ. (a) Singularity-free scenario ($J_\sigma = 0$) from MF calculations. At high T, liquid (L) and gas (G) phases are separated by a first order transition line (thick line)ending at a critical point C, from which a L–G Widom line (double–dot–dashed line) emanates. In the liquid phase, the α_P maxima and the K_T maxima increase along lines that converge to a locus (dot–dashed line). In C' both α_P and K_T have diverging maxima. The locus of the maxima is related to the L-L Widom line for $T_{C'} \to 0$ (see text). (b) Liquid–liquid critical point scenario (for$J_\sigma/\varepsilon = 0.05$) from MF calculations. At low T and high P, a high density liquid (HDL) and a low density liquid (LDL) are separated by a first order transition line (thick line) ending in a critica lpoint C', from which the L-L Widom line emanates. Other symbols are as in the previous panel.(c) Critical–point free scenario ($J_\sigma/\varepsilon = 0.5$) from MF calculations. The HDL–LDL coexistence line extends to the superheated liquid region at $P < 0$, merging with the liquid spinodal (dotted line) hat bends toward negative P. The stability limit (SL) of water at ambient conditions (HDL) is limited by the superheated liquid–to–gas spinodal and the supercooled HDL–to–LDL spinodal (long–dashed thick line), giving a re-entrant behavior as hypothesized in the SL scenario. Other symbols are as in the previous panels. (d) Phase diagram from MC simulations, for $J_\sigma/\varepsilon = 0.02, 0.05, 0.3, 0.5$ (thick lines with symbols and labels). For $J_\sigma/\varepsilon = 0.5$, we find the CPF scenario, as in panel (c). For $J_\sigma/\varepsilon = 0.3$, we find C' (large circle) at $P < 0$ [14], with the L-L Widom line (crosses). For J_σ/ε= 0.05, we find the LLCP scenario with C' at $P > 0$, as in panel (b). For $J_\sigma/\varepsilon = 0.02$, C' approaches $T = 0$ as in the SF scenario in panel (a). Errors are of the order of the symbol sizes. Lines are guides for the eyes. In all panels, k_B is the Boltzmann constant.*

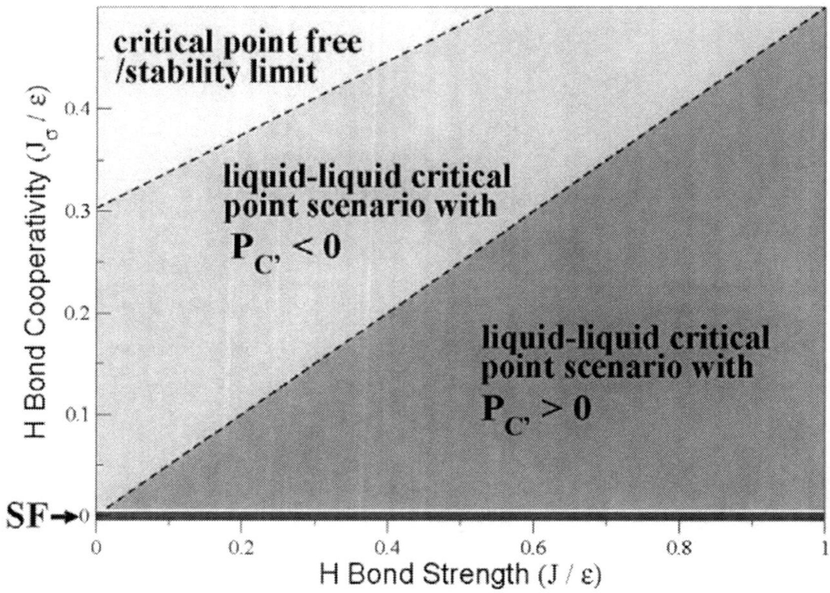

Figure 4. *Possible scenarios for water for different values of J, the H bond strength, and J_σ, the strength of the cooperative interaction, both in units of the van der Waals energy ε. The ratio v_{HB}/v_0 is kept constant. (i) If$_\sigma J = 0$ (red line), water would display the singularity free (SF) scenario, independent of J. (ii) For large enough J_σ, water would possess a first–order liquid–liquid phase transition line terminating at the liquid–gas spinodal—the critical point free (CPF) scenario; the liquid spinodal would retrace at negative pressure, as in the stability limit (SL) scenario (yellow region). (iii) For other combinations of J and J_σ, water would be described by the liquid–liquid critical point (LLCP) scenario. For large J_σ, the LLCP is at negative pressure (ochre region). For small J_σ, the LLCP is at positive pressure (orange region). Dashed lines separating the three different regions correspond to mean field results of the microscopic cell model. The $P - T$ phase diagram evolves continuously as J and J_σ change.*

This critical behavior of the derivative of S implies that $C_P \equiv T(\partial S/\partial T)_P$ diverges when is non–zero ($J_\sigma > 0$), but C_P is constant for the case $T_{C'} = 0$ ($J_\sigma = 0$), which corresponds to the SF scenario.[10] Therefore, the SF scenario coincides with the LLCP scenario in the limiting case of $T_{C'} \to 0$ for $J_\sigma \to 0$ (Fig. 4). Next, we increase J_σ/J, keeping J constant, and observe that C' moves to larger T and lower P. For $J_\sigma > J/2$, we observe that $P_{C'} < 0$ as in.[14] By further increasing J_σ, we observe that the liquid–liquid coexistence line intersects the liquid–gas spinodal, which is precisely the CPF scenario (Fig. 3c).[15,47] As in Ref. [12], we find that the superheated liquid spinodal merges with the supercooled liquid spinodal, giving rise to a retracing spinodal as in the SL scenario. Hence, the CPF scenario and the SL scenario (i) coincide and (ii) correspond to the case in which the cooperative behavior is very strong. In

Fig 4. we summarize our results in the J/ε vs. J_σ/ε parameter space. The MC simulations confirm the MF results (Fig. 3d). For large values of J_σ ($J_\sigma = J = 0.5\varepsilon$), we find a HDL–LDL first–order phase transition that merges with the superheated liquid spinodal as in the CPF scenario. At lower Jσ ($J_\sigma = 0.6J = 0.3\varepsilon$), a HDL–LDL critical point appears at $P < 0$,[14] with the liquid–liquid Widom line intersecting the superheated liquid spinodal. By further decreasing J_σ ($J_\sigma = J/10 = \varepsilon/20$), the HDL–LDL critical point occurs at $P > 0$ as in the LLCP scenario, with the liquid–liquid Widom line intersecting the $P = 0$ axis. By approaching $J_\sigma = 0$ ($J_\sigma = J/25 = \varepsilon/50$), we find that the temperature of the HDL–LDL critical point approaches zero and the critical pressure increases toward the value $P = \varepsilon/v_0$ independent of J_σ. The liquid–liquid Widom line approaches the $T = 0$ axis, consistent with our MF results for $\mathbf{J_\sigma \to 0}$. Thus, we offer a relation linking the four proposed scenarios, showing that (i) all can be included in one general scheme and (ii) the balance between the energies of two components of the H bond interaction determines which scenario is valid.

6. Changes with pressure of the specific heat

Our MF calculations and MC simulations of the cell model allow us to offer also an intringuing interpretation[51] of a phenomenon recently observed. Recent experiments on water confined in cylindrical silica gel pores with diameters of 1.2–1.8 nanometers allow to probe extremely low temperatures that are inaccessible to bulk water. Under these conditions, two maxima in C_P have been observed as the temperature decreases.[48–50] A prominent peak at low T is accompanied by a smaller and broader peak at higher T. These experiments have been interpreted in terms of non-equilibrium dynamics [50]. Our analysis, instead, provides a thermodynamic interpretation, supported by very recent experiments .[52, 53] From simulations for the model parameters J/ε = 0.5, $J_\sigma/\varepsilon = 0.05$, $v_{HB}/v_0 = 0.5$ and $q = 6$, we calculate $C_P \equiv (\partial H/\partial T)_P$, where $H = \langle E\rangle + P\langle V\rangle$ is the enthalpy, and $\langle\ \rangle$ denotes the thermodynamic average. For low pressure isobars, such as $Pv_0/\varepsilon = 0.001$, we observe the presence of two C_P maxima: one, at higher T, and the second, at lower T, sharper [Fig. 5(a)]. The less sharp maximum moves to lower T and eventually merges with the sharper maximum as P is raised toward P_c. The temperature of the sharper maximum does not change much with P at low P; its value slowly increases, reaching the largest values at the critical pressure P_c .[54] Approaching P_c from below the two maxima merge. For $P > P_c$ this maximum occurs at the temperature of the first-order liquid-liquid (LL) phase transition. For $P >> P_c$ the two maxima split: C_P for the sharper maximum decreases in value and shifts to lower T along the LL phase transition line, while C_P for the less sharp maximum is independent of P [Fig. 5(b)], as has been noted.[55, 56]

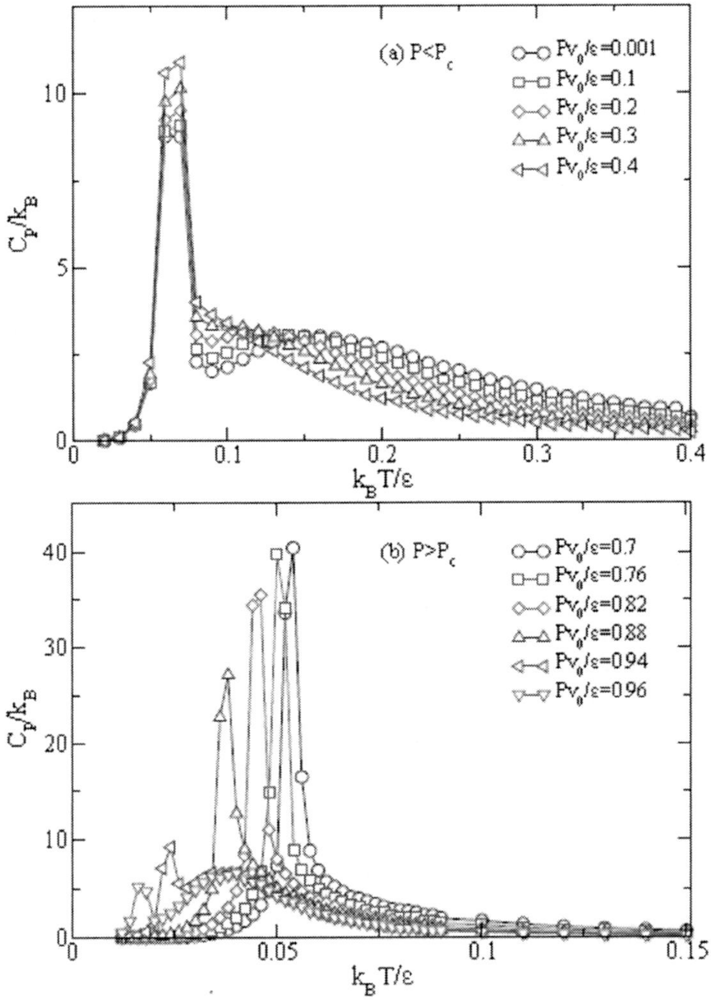

Figure 5. *(a) Temperature dependence of the specific heat CP from MC simulations, for the parameters in the text, along low pressure isobars with* $P < P_C$. *A broad maximum is visible along with a more pronounced one at lower T. The first maximum moves to lower T as the pressure is raised and it merges with the low–T maximum at* $Pv/\varepsilon \approx 0.4$. *Upon approaching* $P_C v_0/\varepsilon = 0.70 \pm 0.02$ *the sharp maximum increases in value. (b) Same for* $P \geq P_C$: *the two maxima are separated only for* $Pv_0/\varepsilon > 0.88$; *the sharp maximum decreases as P increases. In both panels errors are smaller han symbol size.*

We also calculate C_P in the MF approximation.[40] We find that the two maxima are distinct only well below P_c [Fig. 6(a)]. Both maxima move to lower T as P increases, though the less sharp maximum at higher T has a more pronounced

P–dependence. Above $P_{v0}/$ *0.3*, the two maxima merge into a single maximum.

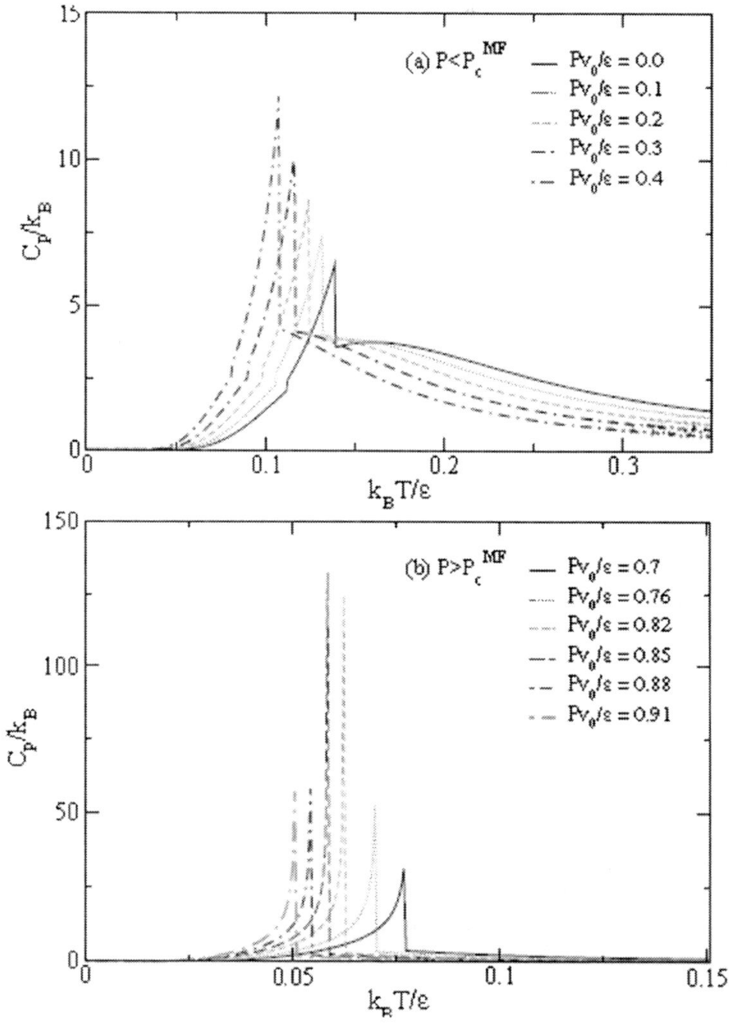

Figure 6. *Same as in Fig. 5 but from mean-field calculations (a) at P<PMFc and (b) at* $P > P_c^{MF}$*. The mean-field critical pressure is = 0.81 0.04.*

We also find that for higher P [Fig. 6(b)] the maximum of C_P increases on approaching the MF critical pressure $P_c^{MF} v_0/$ $= 0.81 \pm 0.04$ and that the single maximum for $P > P_c^{MF}$ marks the LL phase transition line.[54, 57]

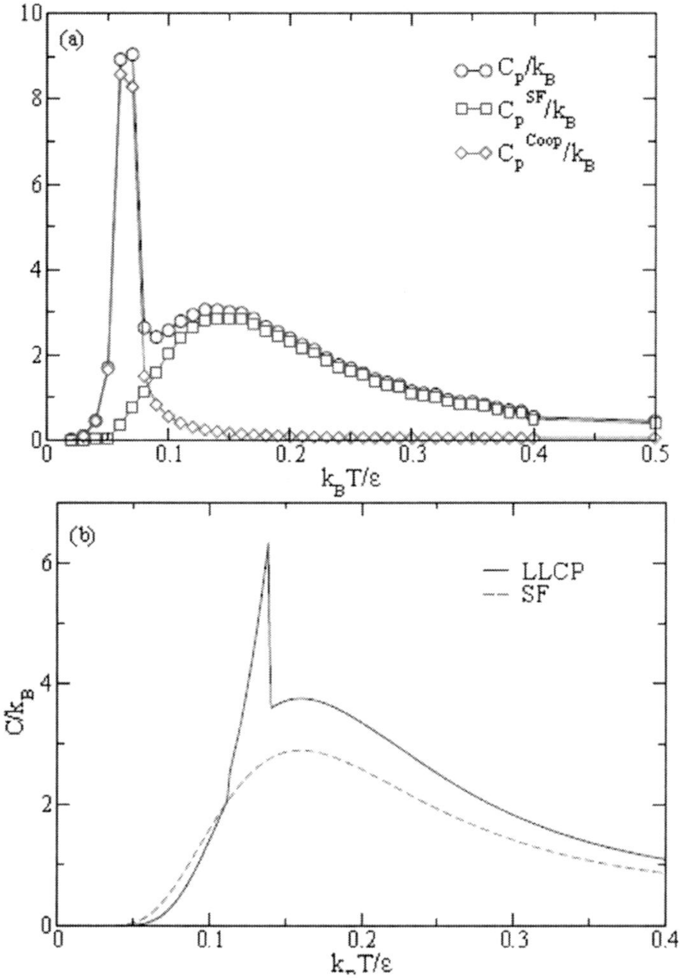

Figure 7. *(a) Decomposition of C_P from MC simulations [Fig. 5] for Pv0/ϱ = 0.1 into the cooperative component C_P^{Coop} and the SF component C_P^{SF} . (b) Comparison of MF calculations for the LLCP scenario case (J_σ/ε = 0.05) and the SF case (J_σ = 0). The low-T maximum is present only in the LLCP case. Both lines are calculated at Pv_0/ε= 0.1.*

To understand the origin of the two C_P maxima, we write the enthalpy as the sum of two terms

$$H = H^{SF} + H^{Coop} \tag{18}$$

where $H^{SF} \equiv \langle -JN_{HB} + P(V_{MC} + NH_{B}v_{HB})\rangle$ and $H_{Coop} \equiv H - H^{SF}$. Hence, we consider $C_P = C_P^{SF} + C_p^{Coop}$, where we define the SF component $C_P^{SF} \equiv (\partial H^{SF}/\partial T)_P$ and the cooperative component $C_P^{Coop} \equiv \partial H^{Coop}/\partial T)_P$ [Fig. 7(a)].

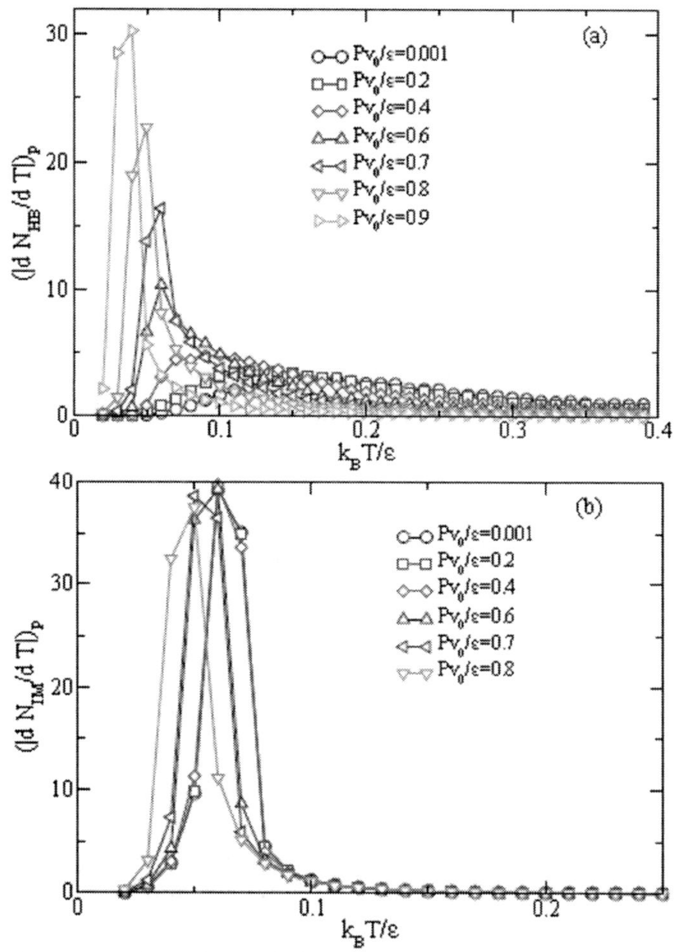

Figure 8. *(a) Temperature dependence of* $(|dN_{HB}/dT|)_P$ *for different isobars. (b) Temperature dependence of* $(|dN_{IN}/dT|)_P$ *for different isobars.*

C^P_{SF} is responsible for the broad maximum at higher T. C_P^{SF} captures the enthalpy fluctuations due to the hydrogen bond formation given by the terms proportional to the hydrogen bond number N_{HB}. This term is present also in the SF model[10]. To show that this maximum is due to the fluctuations of hydrogen bond formation, we calculate the locus of maximum fluctuation of N_{HB}, related to the maximum of $|dN_{HB}/dT|_P$ [Fig. 8(a)], and find that the temperatures of these maxima correlate very well with the locus of maxima of C_P^{SF} [Fig. 9].We find in Fig. 7(a) that the maximum of C_P at lower T is given by the maximum of C_P^{Coop}. To show that C_P^{Coop} corresponds to the enthalpy fluctuations due to the IM term in Eq. 1 proportional to J_σ,[58] we calculate $|dN_{IM}/dT|_P$, where N_{IM} is the number of molecules with complete tetrahedral order. We find that the locus of maxima of $|dN_{IM}/dT|_P$ [Fig. 8(b)] overlaps with the locus of maxima of C_P^{Coop} [Fig. 9].

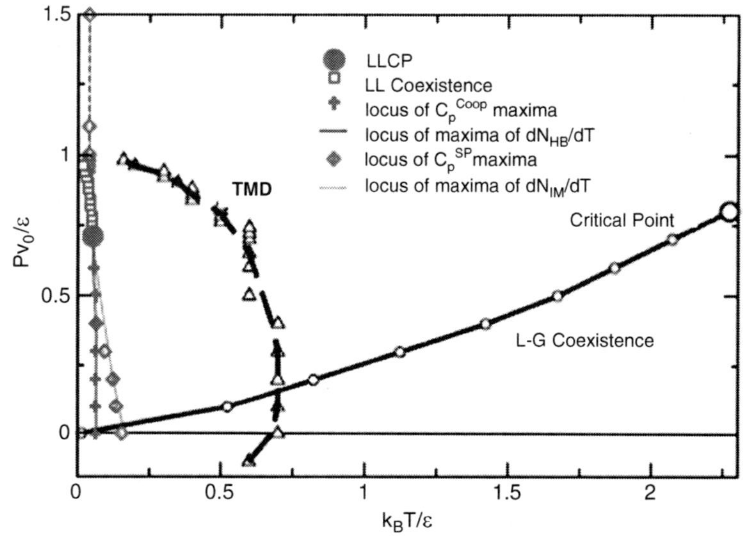

Figure 9. *Phase diagram from MC simulations showing the liquid–gas transition (thick line), the liquid–liquid transition (squares) and the temperature of maximum density (TMD). Emanating from the LLCP (full circle) is the locus of maxima of C_P^{Coop} (crosses), the locus of maxima of C_P^{SF} (diamonds), the locus of maxima of $|dN_{HB}/dT|$ (dark line) and the locus of maxima of $|dN_{IM}/dT|$ (light line). At pressure above the LLCP, a dashed line connects as a guide for the eyes the locus of maxima of C_P^{SF}.*

Therefore, the maximum of C_P^{Coop} occurs where the correlation length associated with the tetrahedral order is maximum, i.e. along the Widom line associated with the LL phase transition.[40] In MF we may compare C_P calculated for the LLCP scenario ($J_\sigma > 0$) with C_P calculated for the SF scenario ($J_\sigma = 0$) [Fig. 7(b)]. We see that the sharper maximum is present only in the LLCP scenario, while the less sharp maximum occurs at the same T in both scenarios. We conclude that the sharper maximum is due to the fluctuations of the tetrahedral order, critical at the LLCP, while the less sharp maximum is due to fluctuations in bond formation. The similarity of our results with the experiments in nanopores is striking.[50] Data in ref. [[50]] show two maxima in C_P. They have been interpreted as an out–of–equilibrium dynamic effect in [[15,50]], but more recent experiments [52,53] show that they are a feature of equilibrated confined water. Therefore, our interpretation of the two maxima is of considerable interest.

7. Conclusion

The behavior of metastable water under pressure is the object of an intense experimental and theoretical investigation. Here we have summarized some of the recent results, including studies for bulk, confined and interfacial water. By analyzing a cell model within a mean field approximation and with Monte Carlo simulations, we have showed that all the scenarios proposed for water's P–T phase diagram may be viewed as special cases of a more general scheme. In particular, our study shows that it is the relationship between H bond strength and H bond cooperativity that governs which scenario is valid. We have also considered recent experiments on confined water at low temperatures that display two maxima in the specific heat. Our analysis of metastable water at very low T and for increasing P, provides an intriguing interpretation of the phenomenon, based exclusively on the thermodynamic properties of water. In conclusion, the investigation of the properties of metastable liquid water under pressure could provide essential information that could allow us to understand the mechanisms ruling the anomalous behavior of water. This understanding could, ultimately, lead us to the explanation of the reasons why water is such an essential liquid for life.

References

1. Zheligovskaya, E. A., and Malenkov, G. G. (2006) *Russ. Chem. Rev.* **75**, 57
2. Debenedetti, P. G., (2003) *J. Phys.: Condens. Matter* **15**, R1669
3. Debenedetti, P. G. and Stanley, H. E. (2003) The Physics of Supercooled and Glassy Water, *Physics Today* **56**, 40
4. Angell, C. A. (1982) in: *Water: A Comprehensive Treatise* vol. 7, edited by F. Franks (Plenum, NewYork)
5. Ball, P. (2008) *Chem. Rev.* **108**, 74
6. Franzese, G., and Rubi, eds. M. (2008) *Aspects of Physical Biology: Biological Water, Protein Solutions, Transport and Replication*, (Springer, Berlin)
7. Anisimov, M. A., Sengers, J. V., and Levelt-Sengers, J. M. H. (2004) in *Aqueous System at Elevated Temperatures and Pressures: Physical Chemistry in Water, Stream and Hydrothmaler Solutions*, edited by D.A. Palmer, R. Fernandez-Prini, A. H. Harvey (Elsevier, Amsterdam)
8. Granick and S., and Bae, S. C. (2008) *Science* **322**, 1477
9. Speedy, R. J. (1982) *J. Phys. Chem.* **86**, 3002
10. Sastry, S., Debenedetti, P. G., Sciortino, F., and Stanley, H. E. (1996) *Phys. Rev. E* **53**, 6144

11. Poole, P. H., Sciortino, F., Essmann, U., and Stanley, H. E. (1992) *Nature* **360**, 324
12. Poole, P. H., Sciortino, F., Grande, T., Stanley, H. E., and Angell, C. A. (1994) *Phys. Rev. Lett.* **73**, 1632
13. Franzese, G., Marques, M., and Stanley, H. E. (2003) *Phys. Rev. E.* **67**, 011103
14. Tanaka, H. (1996) *Nature* **380**, 328
15. Angell, C. A. (2008) *Science* **319**, 582
16. Faraone, A., Liu, L., Mou, C. Y., Yen, C. W., and Chen, S. H. (2004) *J. Chem. Phys.* **121**, 10843
17. Liu, L., Chen, S. H., Faraone, A., Yen, C. W., and Mou, C. Y. (2005) *Phys. Rev. Lett.* **95**, 117802
18. Mallamace, F., Broccio, M., Corsaro, C., Faraone, A., Wanderlingh, U., Liu, L., Mou, C. Y., and Chen, S. H. (2006) *J. Chem. Phys.* **124**, 161102
19. Chen, S.-H., Liu, L., Fratini, E., Baglioni, P., Faraone, A., and Mamontov, E. (2006) *Proc. Natl. Acad. Sci.* USA **103**, 9012
20. Mamontov, E. (2005) *J. Chem. Phys.* **123**, 171101
21. Jansson, H., Howells, W. S., and Swenson, J. (2006) *J. Phys. Chem. B* **110**, 13786
22. Chen, S.-H. et al., (2006) J. Chem. Phys. **125**, 171103
23. Chu, X., Fratini, E., Baglioni, P., Faraone, A., and Chen, S.-H. (2008) *Phys. Rev. E* **77**, 011908
24. Franzese, G., Stokely, K., Chu, X.-Q., Kumar, P., Mazza, M. G., Chen. S.-H., and Stanley, H. E. (2008) *J. Phys.: Cond. Matt.* **20**, 494210
25. Stanley, H. E., Kumar, P., Franzese, G., Xu, L. M., Yan, Z. Y., Mazza, M.G., Chen, S.-H., Mallamace, F., Buldyrev, S. V. (2008) "Liquid polyamorphism: Some unsolved puzzles of water in bulk, nano-confined, and biological environments", in Complex Systems, M. Tokuyama, I. Oppenheim, H. Nishiyama, H, eds. *AIP Conference Proceedings*, **982**, 251
26. Xu, L., Kumar, P., Buldyrev, S. V., Chen, S.-H., Poole, P. H., Sciortino, F., and Stanley, H. E. (2005) *Proc. Natl. Acad. Sci.* **102**, 16558
27. Kumar, P., Yan, Z., Xu, L., Mazza, M. G., Buldyrev, S. V., Chen. S.-H., Sastry, S., and Stanley, H. E. (2006) Phys. Rev. Lett. **97**, 177802
28. Stanley, H. E., Buldyrev, S. V., Franzese, G., Giovambattista, N. F., Starr, W. (2005) *Phil. Trans. Royal Soc.* **363**, 509; Kumar, P., Franzese, G., Buldyrev, S. V., and Stanley, H. E. (2006) *Phys. Rev. E* **73**, 041505
29. Kumar, P., Franzese, G., and Stanley, H. E. (2008) *Phys. Rev. Lett.* **100**, 105701

30. Franzese, G., and Stanley, H. E. (2002) *J. Phys. Cond. Matter* **14**, 2201 (2002); *Physica A* **314**, 508
31. Stokely, K., Mazza, M. G., Stanley, H. E., and Franzese, G. (2008) arXiv: 0805.3468v3
32. Kumar, P., Franzese, G., and Stanley, H. E. (2008) *J. Phys.: Cond. Matt.* **20**, 244114
33. Pendas, A. M., Blanco, M. A., and Francisco, E. (2006) *J. Chem. Phys.* **125**, 184112
34. Isaacs, E. D., Shukla, A., Platzman, P. M., Hamann, D. R., Barbiellini, B., and Tulk, C. A. (2000) *J. Phys. Chem. Solids* **61**, 403
35. Ricci, M. A., Bruni, F., Giuliani, A. (2009) Similarities between confined and supercooled water, to appear on *Faraday Discussion*, in press
36. Ohno, K., Okimura, M., Akai, N., and Katsumoto, Y. (2005) *Phys. Chem. Chem. Phys.* **7**, 3005
37. Cruzan, J. D., Braly, L. B., Liu, K., Brown, M. G., Loeser, J. G., and Saykally, R. J. (1996) *Science* **271**, 59
38. Schmidt, D. A., and Miki, K. (2007) *J. Phys. Chem. A* **111**, 10119
39. Chaplin, M. (2007) "Water's Hydrogen Bond Strength", cond-mat/ 0706.1355
40. Franzese, G., and Stanley, H. E. (2007) *J. Phys.: Condens. Matter* **19**, 205126
41. Coniglio, A., and Peruggi, F. (1982) *J. Phys. A* **15**, 1873
42. Cataudella, V., Franzese, G., Nicodemi, M., Scala, A., and Coniglio, A. (1996) *Phys. Rev. E* **54**, 175; Franzese, G. (1996) *J. Phys. A* **29** 7367
43. Wolff, U. (1989) *Phys. Rev. Lett.* **62**, 361
44. The results of [41, 42] guarantee that the cluster algorithm described here satisfies the detailedbalance and is ergodic. Therefore, it is a valid Monte Carlo dynamics
45. Mazza M.G. et al. (2006) *Phys. Rev. Lett.* **96**, 057803; N. Giovambattista et al., (2004) *J. Phys. Chem. B* 1086655; M.G. Mazza et al. (2007) *Phys. Rev. E* **76**, 031203
46. Oleinikova, A., Brovchenko, I., (2006) *J. Phys.: Condens. Matter* **18**, S2247
47. We fit the boundary of the CPF scenario with the functional form $J_\sigma = a + bJ$, with $a = 0.30 \pm 0.01$ and $b = 0.36 \pm 0.01$
48. Maruyama, S., Wakabayashi, K., and Oguni, M. (2004) *AIP Conf. Proc. Proc.* **708**, 675
49. Oguni, M., Maruyama, S., Wakabayashi, K., and Nagoe, A. (2007) *Chem. Asian* 2, 514

50. Oguni, M., Kanke, Y., and Namba, S. (2008) *AIP Conference Proceedings* **982**, 34
51. Mazza, M. G., Stokely, K., Stanley, H. E., and Franzese, G. (2008) *arXiv:0810.4688*
52. Mallamace, F. (2008) preprint
53. Bruni, F. (2008) private communication
54. *Our resolution in T does not allow us to observe the expected divergence of C_P upon approaching the critical point.*
55. Marques, M. I. (2007) *Phys. Rev. E* **76**, 021503
56. *The difference of our results with those in [55], i.e. the presence of two maxima also at $P < P_C$ and $T > T_C$ is due to the different choice of parameters for the model: here $J_\sigma < J < \varepsilon$ as in [13, 30, 31], while in [55] is $\varepsilon < J_\sigma < J$ which gives rise to a different phase diagram.*
57. *The non-zero value of C_P at low T is reminiscent of the appearance of the broad maximum. However the MF approximation is not able to reproduce the splitting of the maxima seen in MC at $P \gg P_C$*
58. *In the range of T of interest here the contribution to H of the UW term is negligible*

CRITICAL LINES IN BINARY MIXTURES OF COMPONENTS WITH MULTIPLE CRITICAL POINTS

SERGEY ARTEMENKO, TARAS LOZOVSKY, VICTOR MAZUR
Departemnt of Thermodynamics, Academy of Refrigeration, 1/3 Dvoryanskaya Str., 65082 Odessa, Ukraine

Abstract: The principal aim of this work is a comprehensive analysis of the fluid phase behavior of binary fluid mixtures via the van der Waals like equation of state (EoS) which has a multiplicity of critical points in metastable region. We test the modified van der Waals equation of state (MVDW) proposed by Skibinski et al. (2004) which displays a complex phase behavior including three critical points and identifies four fluid phases (gas, low density liquid (LDL), high density liquid (HDL), and very high density liquid (VHDL)). An improvement of repulsive part doesn't change a topological picture of phase behavior in the wide range of thermodynamic variables. The van der Waals attractive interaction and excluded volume for mixture are calculated from classical mixing rules. Critical lines in binary mixtures of type III of phase behavior in which the components exhibit polyamorphism are calculated and a continuity of fluid-fluid critical line at high pressure is observed.

Keywords: critical lines, equation of state, multiple critical points, binary mixtures, one fluid mixture model

1. Introduction

Knowledge fluid phase behavior is of immense interest to decode the puzzle phenomena associated with novel and emergent technologies exploiting high pressures. Several fluids have been reported to exist in different density states under extremes of temperature and pressure. Experimental data about liquid-liquid phase transitions published over the last decade confirmed a surprising behavior for diversity of single-component systems such as carbon[1,2], phosphorous[3-5,] triphenyl phosphite[6,7], silica [8], nitrogen[9], Y_2O_3-Al_2O_3 glasses[10]. Water is one of vivid examples of molecular systems where quite different structures are formed in vitro by computer simulation but needs experimental verification[11]. At the moment the complete phase diagram of water is still missing and experimental existence proof of second and third critical points is a subject of debates. Detailed discussions of different *pro et contra* exploratory scenarios of water behavior were published in the thorough reviews [11-13]. It should be noted that anomalous behavior of thermodynamic variables and their derivatives is not only prerogative of water. Liquid helium isotopes also exhibit non-conventional properties at very low temperature (maximum density, the Pomeranchuk's effect in liquid He^3, temperature decreasing under adiabatic compression, and etc.). The main mechanism of unusual from daily experience but thermodynamically correct behavior of different substances is a competition of entropic measures among inherent clandestine structures at given state parameters.

The appearance of polyamorphism phenomenon and the related phase transitions between disordered states exaggerate greatly the variety of mixture phase behavior. The classical van Konynenburg and Scott classification[14] based on the one-fluid van der Waals model of binary mixture reproduces qualitatively the main topologically different phase diagrams at moderate pressures. At high pressures the traits of the van der Waals equation of state lead to discontinuity of critical lines for II, III, and IV phase behaviour types for binary mixtures. A forecast of real phase changes for materials with open and less dense structures that pack tightly under extreme conditions where hard matter becomes "soft" is a challenge for more general classification of fluid phase behavior of mixtures. Disregarding the reasonable doubts associated with true thermodynamic description of the fluid-fluid phase transitions and experimental observations their ending in critical point, we suggest here the virtual reality of multiple metastable liquid-liquid transitions to study phase behavior of mixtures with components which can exhibit polyamorphism.

The first step in quantitative description of pure polyamorphic fluid is a selection of the model that can qualitatively describe a possible multiplicity of critical points in wide range of temperatures and pressures. A great many of explanations of multicriticality in monocomponent fluids (perturbation theory models [15,16], semiempirical models [17-20], lattice models [24-26], two-state models [27-29], field theoretic models [30], two-order-parameter models [31-35], and parametric crossover model [36] has been disseminated after the pioneering work by Hemmer and Stell [37]. Here we test more extensively the modified van der Waals equation of state (MVDW) proposed in work [20] and refine this model by introducing instead of the classical van der Waals repulsive term a very accurate hard sphere equation of state over the entire stable and metastable regions [38].

This paper is organized as follows. In Sec. 2 we review the MVDW model proposed by Skibinsky et al.[20] and take into account more exact hard sphere term from Liu's paper [38]. It is demonstrated that improvement of repulsive part doesn't change a topological portrait of phase behavior of polymorphic fluid in the wide range of thermodynamic variables. Section 3 displays the picture of the phase behavior for different parameters of MVDW model and third critical point which didn't observed earlier for this model is clearly established. It allows to interpret four fluid phases as gas, low density liquid (LDL), high density liquid (HDL), and very high density liquid (VHDL). In Section 4 thermodynamic model of binary mixture and critical line calculation methods are discussed. Section 5 presents the results of critical lines calculations for the one-fluid van der Waals like model of binary mixtures with polyamorphic pure components. The paper ends with some conclusions and an outlook to further work.

2. The van der Waals-like equation of state with multiplicity of critical points

A mean field EoS is a major tool for the description of general thermodynamic behavior in the existence domain of state variables. The various physical approximations don't change a topological structure of thermodynamic surface which is generated by mean field theories. From these reason the simplest models of the van der Waals like EoS demonstrating the great variety of features of thermodynamic and phase behavior for mono- and multicomponent fluids have been chosen.

The total compressibility factor is expressed as the sum of repulsive and attractive parts

$$Z = Z_{rep} + Z_{attr} \tag{1}$$

To compare a very accurate and very rough approximations for repulsive term the classical van der Waals expression

$$Z_{rep} = \frac{1}{1-4\eta} \tag{2}$$

and the wide range hard sphere EoS for stable and metastable regions from [21]

$$Z_{rep} = 1 + \sum_{i=1}^{12} a_{i+1}\eta^{i} + \frac{c_0\eta}{1-\alpha\eta} + c_1\eta^{40} + c_2\eta^{42} + c_3\eta^{44} \tag{3}$$

are considered.

Here $Z = PV, NkT$ is the compressibility factor, and P, the pressure, V, the total volume, T, the temperature, N, the total number of particles, k, the Boltzmann constant; □η□, the packing fraction, defined as $\eta = \pi\rho d^3/6$, $\rho = N, V$, the number density and d, the hard sphere diameter. The coefficients a_i (i = 1,2,...12) and c_i (i = 0...3) were taken from[38] and reproduce the virial coefficients up to the 12th. The most important parameter for the metastable region is $\alpha = 1/0.635584$. The inverse value gives the maximally random jammed packing and places a limit of EoS applicability that is very close to computer simulation result $\eta_0 = 0.6418$.

The attractive term has the same form as the classical van der Waals EoS expression

$$Z_{attr} = -\frac{a\eta}{NkT} \tag{4}$$

where a is the interaction constant.

The conventional van der Waals approach where model parameters d and a are the constants cannot describe more than one first order phase transition and one critical point. Therefore a key question is a formulation of temperature - density dependency for EoS parameters generating more than one critical point in the mono-component matter. There are several approaches of the effective hard sphere determination from spherical interaction potential models that have a region of negative curvature in their repulsive core (the so-called core softened potentials). To avoid the sophistication of EoS and study a qualitative picture of phase behavior we adopt an approach Skibinsky et al.[20] for one-dimensional system of particles interacting via pair potential

$$U(R) = \begin{cases} \infty, & R \le d_h \\ U_R, & d_h < R \le d_s \\ 0, & R > d_s \end{cases} \tag{5}$$

where d_h is a diameter associated with hard core, d_s is a diameter associated with impossibility of particle to penetrate into soft core at low densities and low temperatures.

The potential has three dimensionless parameters: d_h/d_0, d_s/d_0, and U_R/U_A where $d_0 = 1$ and $U_A = 1$ have been chosen as units of length and energy, respectively. This potential generates three critical points in metastable region with respect to a solid phase. The algorithm of excluded volume calculation $b_i(\rho,T) = \frac{2}{3}\pi d_i^3$, $i = h,s$ is given in [20]. The behavior of the excluded volume in the entire range of densities and temperatures is illustrated in Figure1.

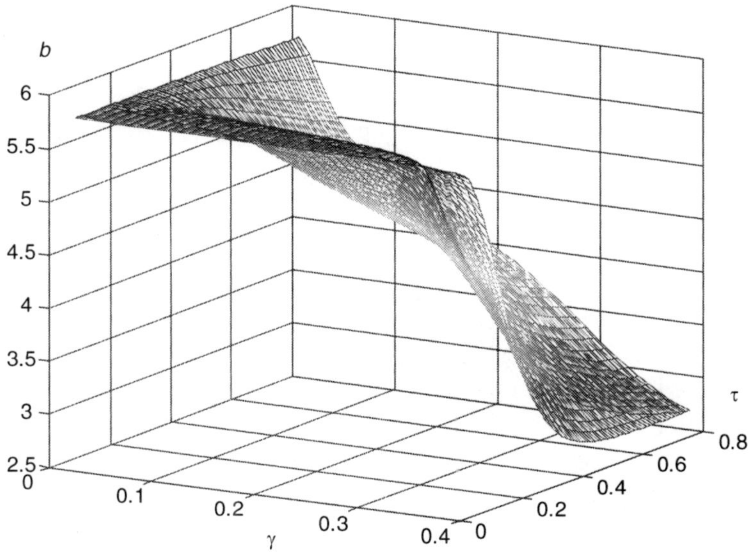

Figure 1. The mapping of the core softened potential (5) on the hard sphere diameter. The temperature ($\tau = kT/U_R$) - density ($\gamma = b_h\rho$) dependence of the excluded volume (b) for model parameters set: $d_h = 2.27$, $U_R/U_A = 2$, $d_s = 10.29$

3. Phase behavior of pure component with multiple critical points

Topology of the fluid-fluid phase transitions depends on the concurrence between the repulsive and the attractive parts of EoS. The binodal location at given temperature, T, and pressure, P is a solution of the set equations:

$$\mu(\rho', T) - \mu(\rho'', T) = 0$$

$$p(\rho', T) - p(\rho'', T) = 0 \tag{6}$$

where ρ' and ρ'' are the densities of the coexisting phases, the pressure, p, is calculated from the EoS described, the expression for the chemical potential, μ, can be derived from an equation of state using standard thermodynamic relations. Spinodals are determined via the following thermodynamic condition:

$$\left(\frac{\partial p}{\partial V}\right)_T = 0 \qquad (7)$$

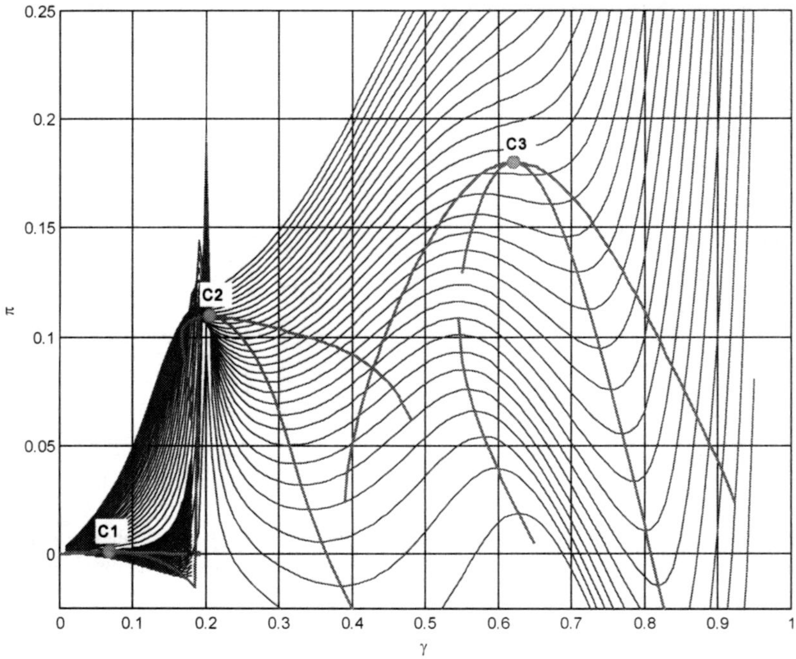

Figure 2. Evolution of isotherms in the $P - \rho$ phase diagram for the core softened potential with three critical points. C1 - gas + liquid, C2 - LDL + HDL, and C3 - HDL + VHDL critical points. Red curves (online) are coexistence curves; green curves (online) are spinodals. Critical point location: π_{C1} =0.832e-3, τ_{C1} = 0.0327, γ_{C1} = 0.0678; π_{C2} =0.1096, τ_{C2} = 0.2297, γ_{C2} = 0.2058; π_{C3} =0.1799 , τ_{C3} =0.1746, γ_{C3} =0.6209. Model parameter set: a = 2.272; b_h =2.27, U_R/U_A =2, b_s=10.29.

Figures 2–7 show the phase behavior for the van der Waals like EoS where hard sphere diameter depends on the state variables. It was detected an appearance of third critical point with repulsive term (2) that surprisingly broadens the possibilities of very simple EoS model. It allows considering the liquid state as a mixture of the two corresponding fluid phases, LDL and HDL.

Figure 5 illustrates a possible scenario of the isotherms behavior in the P - T phase diagram for the core softened potential with third critical point in the metastable region. This result confirms a suggestion[39] that HDL is not stable but rather is highly metastable structure, relaxing to VHDA as glasses generated with hyperquenched methods relax on slow heating to glasses generated with conventional cooling rates.

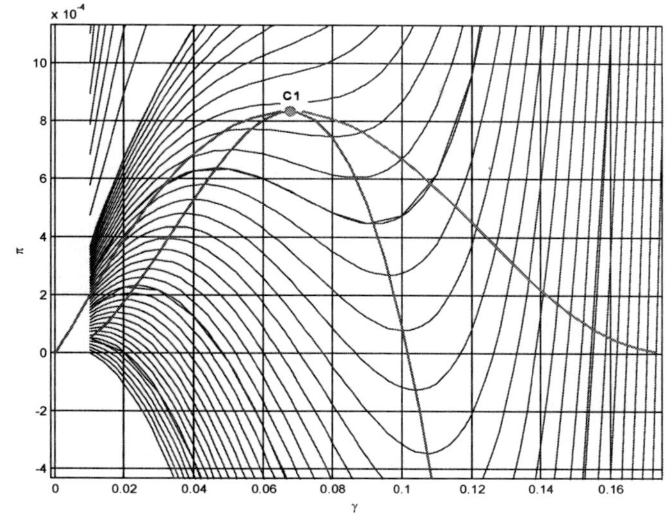

Figure 3. Evolution of isotherms in the $P - \rho$ phase diagram near gas + liquid critical point. C1 - gas + liquid. Red lines (online) are coexistence curves; green lines (online) are spinodals. Critical point location: π_{C1} =0.7949e-3, τ_{C1} = 0.0284, γ_{C1} = 0.0678. Model parameter set: B_h =2.27, U_R, U_A =2, D_s=10.29.

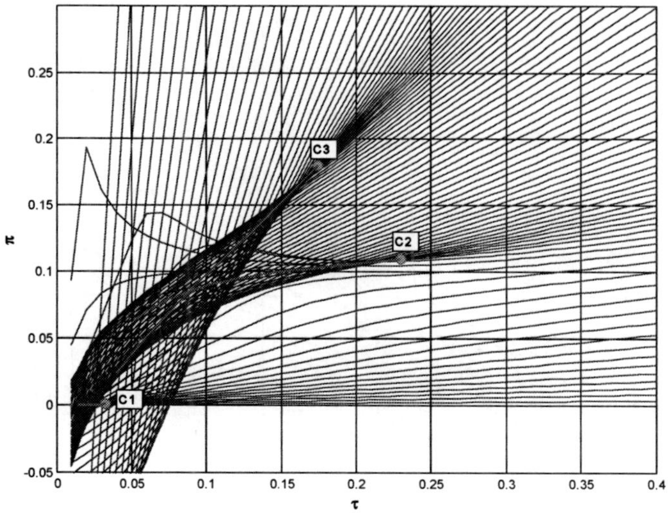

Figure 4. Evolution of isochors in the *P - T* phase diagram for the core softened potential with three critical points. C1 - gas + liquid, C2 - LDL + HDL, and C3 - HDL + VHDL critical points. Red lines (online) are coexistence curves. Critical point location: π_{C1} =0.832e-3, τ_{C1} = 0.0327, γ_{C1} = 0.0678; π_{C2} =0.1096, τ_{C2} = 0.2297, γ_{C2} = 0.2058; π_{C3} =0.1799 , τ_{C3} =0.1746, γ_{C3} =0.6209. Model parameter set: *a* = 2.272; b_h =2.27, U_R/U_A =2, b_s=10.29.

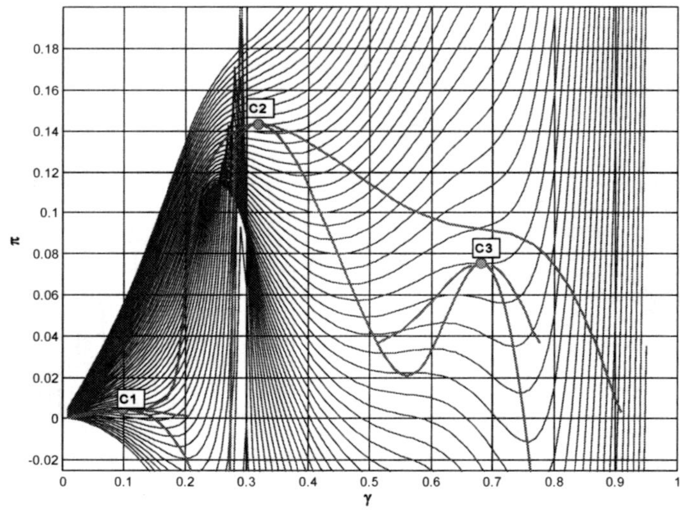

Figure 5. Evolution of isotherms in the $P-\rho$ phase diagram for the core softened potential with third critical point in metastable region. C1 - gas + liquid, C2 - LDL + HDL, and C3 - HDL + VHDL critical points. Red lines (online) are coexistence curves; green lines (online) are spinodals. Critical point location: π_{C1} = *0.0064*, τ_{C1} = *0.1189*, γ_{C1} =*0.0998*; π_{C2} = *0.1423*, τ_{C2} = *0.3856*, γ_{C2} = *0.33*; π_{C3} = *0.07487*, τ_{C3} = *0.2398*, γ_{C3} =*0.6856*. Model parameter set: a = 6.962, b_h =*2.094*, U_R/U_A=*3*, b_s=*7.0686.*

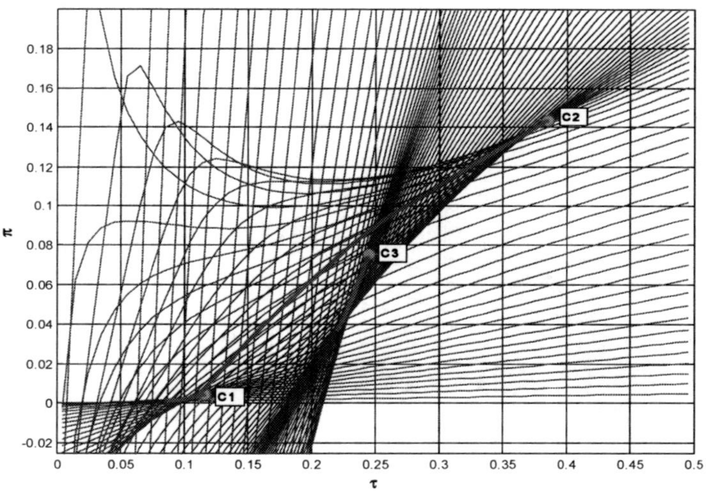

Figure 6. Evolution of isochors in the $P-T$ phase diagram for the core softened potential with third critical point in metastable region. C1 - gas + liquid, C2 - LDL + HDL, and C3 - HDL + VHDL critical points. Red lines (online) are coexistence curves. Blue curves (online) are isochors. Critical point location: π_{C1} = *0.0064*, τ_{C1} = *0.1189*, γ_{C1} =*0.0998*; π_{C2} = *0.1423*, τ_{C2} = *0.3856*, γ_{C2} = *0.33*; π_{C3} = *0.07487*, τ_{C3} = *0.2398*, γ_{C3} = *0.6856*. Model parameter set: a = 6.962, b_h =*2.094*, U_R/U_A=*3*, b_s=*7.0686.*

An improvement of a classical repulsive expression (2) for one dimensional system of hard sphere by the very accurate presentation of the Liu's EoS[38] (Figure 7) doesn't change a topologic picture of phase diagram in comparison with classical van der Waals expression for repulsive term. It seems that an improvement of repulsive term makes more plausible of isotherm behavior near second critical point. To analyze a qualitative behavior of thermodynamic surface anomalies in whole via simpler model is preferable due to a topological equivalence of models under consideration.

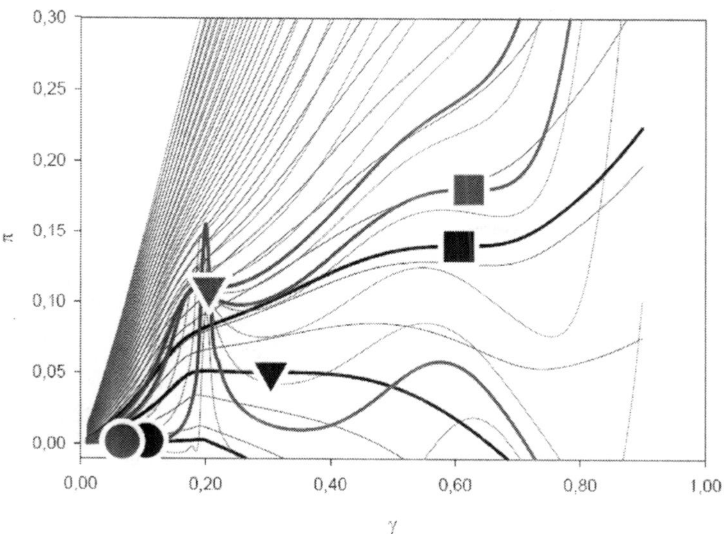

Figure 7. Evolution of isotherms in the $P-\rho$ phase diagram from the core softened potential with three critical points. The filled circles are C1 - gas + liquid critical point, the triangles correspond to C2 - LDL + HDL second critical point, and squares are C3 - HDL + VHDL critical points. Blue curves (online) are isotherms according to the van der Waals like model with Liu's repulsive term[38]. Critical point location: $\pi_{C1}=1.5824e\text{-}3$, $\tau_{C1}=0.0416$, $\gamma_{C1}=0.1059$; $\pi_{C2}=0.0501$, $\tau_{C2}=0.1597$, $\gamma_{C2}=0.3049$; $\pi_{C3}=0.1389$, $\tau_{C3}=0.2708$, $\gamma_{C3}=0.6055$. Red curves (online) are isotherms according to the van der Waals model. Critical point location: $\pi_{C1}=8.3242e\text{-}4$, $\tau_{C1}=0.0327$, $\gamma_{C1}=0.0678$; $\pi_{C2}=0.1096$, $\tau_{C2}=0.2297$, $\gamma_{C2}=0.2060$; $\pi_{C3}=0.1799$, $\tau_{C3}=0.1746$, $\gamma_{C3}=0.6214$. Model parameter set: $a=2.272$, $b_h=2.27$, $U_R/U_A=2$, $b_s=10.29$.

4. Thermodynamic model of binary mixture

We consider here the one-fluid van der Waals model where the EoS parameters a and b of a mixture depends on the mole fractions x_i and x_j of the components i and j and on the corresponding parameters a_{ij} and b_{ij} for different pairs of interacting molecules:

$$a = \sum_{i=1}^{2}\sum_{j=1}^{2} x_i x_j a_{ij}\left(1 - k_{ij}\right),$$
$$b = \sum_{i=1}^{2}\sum_{j=1}^{2} x_i x_j b_{ij}. \tag{8}$$

where k_{ij} is a binary interaction parameter for long range attraction.

Qualitatively the main phase diagram types for binary mixtures with single critical point components are shown in Fig. 8.

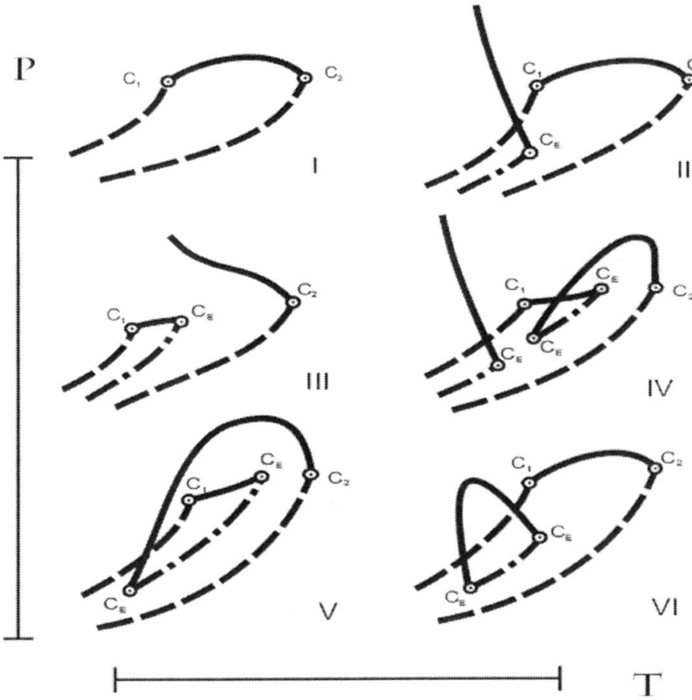

Figure 8. *P-T* projections of main phase diagram types. The roman numbers correspond to the classification introduced by Scott and van Konynenburg[14] : the solid lines are critical curves; the dashed lines are vapor pressure curves of pure components with critical points C_1 and C_2; the dash dotted lines are three phase lines; C_E is critical end point.

For a normal critical point when two fluid phases are becoming identical critical conditions are expressed in terms derivatives of the molar Gibbs energy in the following way:

$$\left(\frac{\partial^2 G}{\partial x^2}\right)_{p,T} = \left(\frac{\partial^3 G}{\partial x^3}\right)_{p,T} = 0\ . \tag{9}$$

Corresponding critical conditions for the composition - temperature – volume variables are:

$$\begin{aligned} &A_{xx} - WA_{xV} = 0; \\ &A_{xxx} - 3WA_{xxV} + 3W^2 A_{xVV} - 3W^3 A_{VVV} = 0; \end{aligned} \tag{10}$$

where A is the molar Helmholtz energy,

$W = \frac{A_{xx}}{A_{VV}}$, $A_{mVnx} = \left(\frac{\partial^{n+m} A}{\partial x^n \partial V^m}\right)_T$ is a contracted notation for differentiation operation which can be solved for V_C and T_C at given x.

The calculation of critical lines attends numerical instability due to the extremely large changes of pressure derivatives in the vicinity of critical point and uncertainty in the definition of initial conditions for iterative search of excluded volume from algorithm[20]. The derivatives appearing in Eq. (10) are calculated numerically because an analytical derivation is impossible for this model. The lack of a priori information concerning the solution structure makes the numerical analysis for non-analytical models of equation of state considerably more complicated. The constructive approach to selection of a root-finding algorithm which combines generalization and reliability is based on thermodynamic model (10) formulation as the problem of multiextrema nonlinear programming[40].

5. Results and discussion

An occurrence of several critical points for monocomponent fluid leads to complication of binary mixture phase behavior. Following Varchenko's approach[41], generic phenomena encountered in binary mixtures when the pressure p and the temperature T change, correspond to singularities of the convex envelope (with respect to the x variable) of the "front" (a multifunction of the variable x) representing the Gibbs potential $G(p,T,x)$. Pressure p and temperature T play the role of external model parameters like k_{12}. A total amount of 26 singularities and 56 scenarios of evolution of the p - T diagram were found[42]. The increasing of critical point number should involve an enhancement of singularity amount and evolution scenarios.

It is most likely the critical line asymptotes observed for the conventional types II, III, and IV (Figure 8) should end not at the infinite pressure but at the pure component second (or third) critical point. To study the possibility of continuous critical line path from stable critical point of one component to metastable critical point of other component the type III of phase behavior was chosen. The selection criterion of thermodynamic model parameters for type III was extracted from global phase diagram for the binary van der Waals mixture[14].

$$\xi = \frac{b_{11} - b_{22}}{b_{11} + b_{22}} \approx 0.5,$$

$$\lambda = \frac{\dfrac{a_{22}}{b_{22}^2} - \dfrac{2a_{12}}{b_{11}b_{22}} + \dfrac{a_{11}}{b_{11}^2}}{\dfrac{a_{22}}{b_{22}^2} + \dfrac{a_{11}}{b_{11}^2}} \approx 0.5. \tag{11}$$

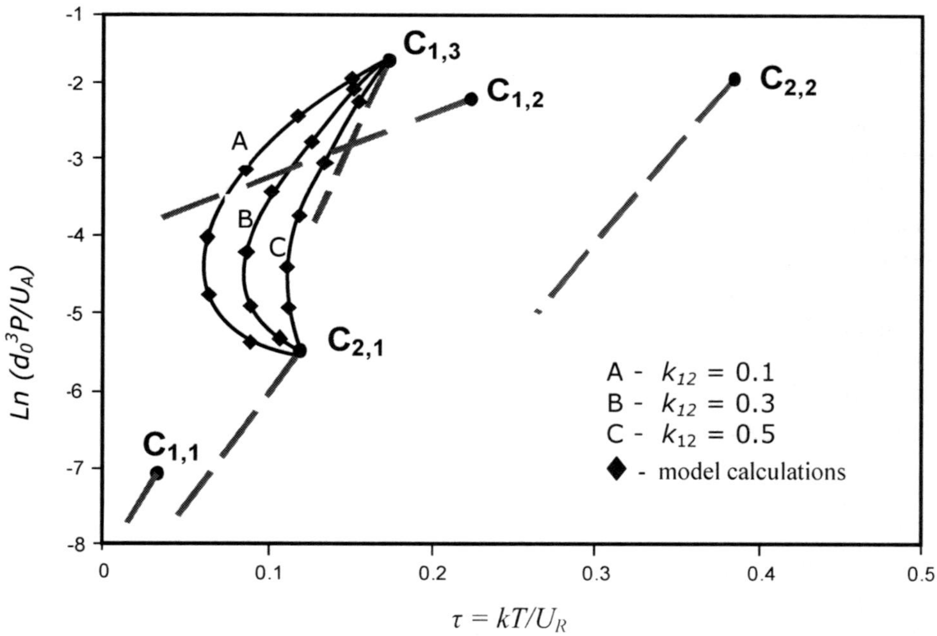

Figure 9. Critical lines for a binary mixture of components with several critical points. Solid lines (A, B, C) indicate binary mixture critical lines; dashed lines are phase existence curve of pure components; $C_{n,m}$ are the mth critical point ($m \geq 1$) for the nth pure component ($n = 1,2$); $m = 1$ identifies the vapor-liquid critical point; $m > 1$ corresponds to the fluid-fluid critical points.

To reproduce type III of phase behavior the model parameters for pure components have been chosen for first and second components from data presented in Figures 4 and 6, correspondingly. Figure 9 shows the $P - T$ projection of critical lines of binary mixture with pure components having several critical points. The initial approximation for critical line calculations was chosen in vicinity of second component critical point to check the hypothesis of critical line continuity. Interaction coefficient k_{12} has been varied within interval [0…0.5]. We have not detected the traces of additional critical lines which were found in the Truskett-Ashbaugh model[43,44] at the analysis of fluid phase behavior of binary mixture with polyamorphic component[45]. One of possible explanations of these distinctions is the opposite slopes of liquid-liquid curves in the models under comparison. For interaction potential (5) we could not find any parameter set to reproduce a negative slope of liquid-liquid curve as it is displayed by the Truskett-Ashbaugh model. The molecular dynamics study of water-like solvation thermodynamics in a spherically symmetric solvent model with two characteristic lengths[46] also doesn't confirm the appearance of new singularities in comparison with the classical Scott – van Konynenburg picture.

6. Conclusions

We have studied one–fluid model of binary fluids with polyamorphic components and found that multicritical point scenario gives opportunity to consider the continuous critical lines as the pathways linking isolated critical points of components on the global equilibria surface of binary mixture. It enhances considerably the landscape of mixture phase behavior in a stable region at the account of hidden allocation of other critical points in metastable region.

This study suggests realizing a future research program including a study of the boundaries of global phase diagram (tricritical points, double critical end points, and etc) for binary mixture with polyamorphic components.

Acknowledgements

We thank Professor G. Franzese for helpful discussions and assistance.

References

1. van Thiel M., Ree F. (1993) High-pressure liquid-liquid phase change in carbon, *Phys. Rev. B* **48(6)**, 3591-3599.
2. Togaya M. (1997) Pressure Dependences of the Melting Temperature of Graphite and the Electrical Resistivity of Liquid Carbon, *Phys. Rev. Lett.* **79**(13), 2474-2477.

3. Katayama Y., Mizutani T., Utsumi W., Shimomura O., Yamakata M., Funakoshi K. (2000) A first-order liquid–liquid phase transition in phosphorus, *Nature* **403**, 170-173.
4. Katayama Y., Inamura Y., Mizutani T., Yamakata M., Utsumi W., Shimomura O. (2004) Macroscopic Separation of Dense Fluid Phase and Liquid Phase of Phosphorus, *Science* **306** (5697), 848 - 851.
5. Monaco G., Falconi S., Crichton W., Mezouar M. (2003) Nature of the First-Order Phase Transition in Fluid Phosphorus at High Temperature and Pressure, *Phys. Rev. Lett.* **90**(25), 255701-255705.
6. Tanaka H., Kurita R., Mataki H. (2004) Liquid-Liquid Transition in the Molecular Liquid Triphenyl Phosphite, *Phys. Rev. Lett.* **92**(2), 025701-025705.
7. Kurita R., Tanaka H. (2004) Critical-Like Phenomena Associated with Liquid-Liquid Transition in a Molecular Liquid, *Science* **306**(5697), 845-848.
8. Angell C., Borick S., Grabow M. (1996) Glass transitions and first order liquid-metal-to-semiconductor transitions in 4-5-6 covalent systems, *Journal of Non-Crystalline Solids* **205-207**, 463-471.
9. Mukherjee G., Boehler R. (2007) High-Pressure Melting Curve of Nitrogen and the Liquid-Liquid Phase Transition, *Phys. Rev. Lett.* **99**(22), 225701-2250705.
10. Wilding M.C., Mcmillan P.F., Navrotsky A. (2002) Calorimetric study of glasses and liquids in the polyamorphic system Y_2O_3-Al_2O_3, *Physics and Chemistry of Glasses* **43**, 6, 306-312.
11. Mishima O., Stanley H. (1998) The relationship between liquid, supercooled and glassy water, *Nature* **396**, 329-335.
12. Angell C.A. (2008) Insights into Phases of Liquid Water from Study of its Unusual Glass-Forming Properties, *Science* **319**, 582-587
13. Debenedetti P.G. (2003) Supercooled and glassy water, *J. Phys.: Condens. Matter* **15**, R1669-R1726.
14. van Konynenburg P.H., Scott R. (1980) Critical lines and phase equilibria in binary van der Waals mixtures, *Phil. Trans. Roy. Soc. London* **298**, 495-540.
15. Fomin Yu. D., Ryzhov V. N., Tareyeva E. E. (2006) Generalized van der Waals theory of liquid-liquid phase transitions, *Phys. Rev. E* **74**, 041201.
16. Cervantes L.A., Benavides A.L., delRío F.(2007)Theoretical prediction prediction of multiple fluid-fluid transitions in monocomponent fluids, *J. Chem. Phys.* **126**, 084507.
17. Poole P.H., Sciortino F., Grande T., Stanley H., Angell C. (1994) (1994) Effect of Hydrogen Bonds on the Thermodynamic Behavior of Liquid Water, *Phys. Rev. Lett.* **73**, 1632 - 1635.

18. Truskett T.M., Debenedetti P.G., Sastry S, Torquato S. (1999) A single-bond approach to orientation-dependent interactions and its implications for liquid water. *J. Chem. Phys.* **111**, 2647.
19. Jeffery C.A., Austin P.H. (1999) A new analytic equation of state for liquid water *J. Chem. Phys.* **110**, 484.
20. Skibinsky A., Buldyrev S.V., Franzese G., Malescio G., Stanley H.E. (2004) Liquid-liquid phase transitions for soft-core attractive potentials, *Phys. Rev. E* **69**, 061206.
21. Franzese G., Malescio G., Skibinsky A., Buldyrev S., Stanley H. (2001) Generic mechanism for generating a liquid–liquid phase transition, *Nature* **409**, 692-695.
22. Franzese G., Malescio G., Skibinsky A., Buldyrev S., Stanley H. (2002) Metastable liquid-liquid phase transition in a single-component system with only one crystal phase and no density anomaly, *Phys. Rev. E*, **66**(5), 051206-051220.
23. Malescio G., Franzese G., Skibinsky A., Buldyrev S., Stanley H. (2005) Liquid-liquid phase transition for an attractive isotropic potential with wide repulsive range, *Phys. Rev. E* , **71**(6), 061504-061512.
24. Borick S., Debenedetti P., Sastry S. (1995) A Lattice Model of Network-Forming Fluids with Orientation-Dependent Bonding: Equilibrium, Stability, and Implications for the Phase Behavior of Supercooled Water, *J. Phys. Chem.* **99**(11), 3781-3792.
25. Roberts C., Debenedetti P. (1996) Polyamorphism and density anomalies in network-forming fluids: Zeroth- and first-order approximations, *J. Chem. Phys.* **105**(2), 658.
26. Franzese G., Stanley H. (2002) Liquid-liquid critical point in a Hamiltonian model for water: analytic solution, *J. Phys.: Condens. Matter* **14**, 2201-2209.
27. Franzese G., Malescio G., Skibinsky A., Buldyrev S., Stanley H. (2002) Metastable liquid-liquid phase transition in a single-component system with only one crystal phase and no density anomaly, *Phys. Rev. E* **66**, 051206-051220.
28. Malescio G., Franzese G., Skibinsky A., Buldyrev S., Stanley H. (2005) Liquid-liquid phase transition for an attractive isotropic potential with wide repulsive range, *Phys. Rev.* E **71**(6), 061504-061512.
29. Ponyatovsky E.G., Sinitsyn V.V. (1999) Thermodynamics of stable and metastable equilibria in water in the *T–P* region, *Physica B*: *Condensed Matter* **265**, Issues 1-4, 121-127.
30. Sasai M. (1990) Instabilities of hydrogen bond network in liquid water, *J. Chem. Phys.* **93**, 7329.

31. Tanaka H. (1998) Simple Physical Explanation of the Unusual Thermodynamic Behavior of Liquid Water, *Phys. Rev.* Lett. **80**(26), 5750-5753.
32. Tanaka H. (1999) Two-order-parameter description of liquids: critical phenomena and phase separation of supercooled liquids, *J. Phys.: Condens. Matter* **11,** L159-L168.
33. Tanaka H. (2000) Thermodynamic anomaly and polyamorphism of water, *Europhys. Lett.* **50,** 340-346.
34. Tanaka H. (2000) Simple physical model of liquid water, *J. Chem. Phys.* **112**(2), 799.
35. Tanaka H. (2000) General view of a liquid-liquid phase transition, *Phys. Rev.* E **62**(5), 6968-6976.
36. Kiselev S., Ely J. (2002) Parametric crossover model and physical limit of stability in supercooled water, *J. Chem. Phys*. **116** (3), 5657.
37. Hemmer P.C., Stell G. (1970) Fluids with Several Phase Transitions *Phys. Rev. Lett*. **24**, 1284-1287.
38. Liu H. (2006) A very accurate hard sphere equation of state over the entire stable and metstable region. ArXiv.org:cond-mat 0605392, 26 p.
39. Giovambattista N., Stanley H., Sciortino F. (2005) Relation between the High Density Phase and the Very-High Density Phase of Amorphous Solid Water, *Phys. Rev. Lett.* **94**(10), 107803-107807.
40. Mazur V., Boshkov L., Murakhovsky V. (1984) Global Phase Behaviour of Binary Mixtures of Lennard-Jones Molecules, *Phys. Lett.* **104A**, 8, 415-418.
41. Varchenko A.N., (1990). Evolution of convex hulls and phase transition in thermodynamics, *J. Sov. Math.* **52**(4):3305-3325.
42. Aicardi F., Valentin P., Ferrand E. (2002). On the classification of generic phenomena in one-parameter families of thermodynamic binary mixtures. *Phys. Chem. Chem. Phys*., **4**, 884-895.
43. Ashbaugh H., Truskett T., Debenedetti P.G. (2002) A simple molecular thermodynamic theory of hydrophobic hydration, *J. Chem. Phys.* **116**, 2907-2921.
44. Chatterjee S., Ashbaugh H., Debenedetti (2005) Effects of non-Polar Solutes on the Thermodynamic Response Functions of Aqueous Mixtures, *J. Chem. Phys*. **123**, 164503.
45. Chatterjee S., Debenedetti P. (2006) Fluid-phase behavior of binary mixtures in which one component can have two critical points, *J. Chem. Phys.* **124**, 154503.
46. Buldyrev S., Kumar P., Debenedetti P., Stanley H.E. (2007) Water-like solvation thermodynamics in a spherically symmetric solvent model with two characteristic lengths, *Proc. Natl. Acad. Sci.USA* **104,** 20177- 20181.

ABOUT THE SHAPE OF THE MELTING LINE AS A POSSIBLE PRECURSOR OF A LIQUID-LIQUID PHASE TRANSITION

ATTILA R. IMRE*
KFKI Atomic Energy Research Institute, H-1525 POB 49, Budapest, Hungary (imre@aeki.kfki.hu)

SYLWESTER J. RZOSKA
Institute of Physics, Silesian University, ul. Uniwersytecha 4, 40-007, Katowice, Poland

Abstract: Several simple, non-mesogenic liquids can exists in two or more different liquid forms. When the liquid-liquid line, separating two liquid forms, meets the melting line, one can expect some kind of break on the melting line, caused by the different freezing/melting behaviour of the two liquid forms. Unfortunately recently several researchers are using this vein of thinking in reverse; seeing some irregularity on the melting line, they will expect a break and the appearance of a liquid-liquid line. In this short paper, we are going to show, that in the case of the high-pressure nitrogen studied recently by Mukherjee and Boehler, the high-pressure data can be easily described by a smooth, break-free function, the modified Simon-Glatzel equation. In this way, the break, suggested by them and consequently the suggested appearance of a new liquid phase of the nitrogen might be artefacts.

Keywords: liquid-liquid transition, melting line, high pressures

While solids can exists in several solid forms, regular, non-mesogenic fluids mainly exists only two forms, namely liquid and gas ones. Obviously, liquids (liquid crystals) with special molecular structure (like disc- or banana-shaped molecules) can form different phases, but for regular fluids, most people would not expect more than one liquid form. Surprisingly there are a few liquids (like cesium, selenium, phosphorus), which can exist in more than one liquid forms, like normal ones and dense ones.[1,2] Being both phases disordered, the transition between the two liquid phases should be very similar to the transition between liquid and vapour phases, i.e. there should be a phase transition line (liquid-liquid line) terminated by a critical point. One of the recent candidate for the membership of the "Club" for liquids with more than one form is the water [3,4,5]; in this case the dense liquid phase is hidden in the deeply undercooled region, masked by freezing. The search for the second form of liquid water is hot and it revitalized the study of other materials which have the potential to have more than two liquid forms. Some of the studies are indirect, i.e. instead of trying to find the second liquid phase, researchers are trying to find something else which might be caused by the existence of the dense liquid. Break on the melting line in the pressure-temperature space is handled sometimes as one of the hallmark of the hidden dense liquid.[6] Here we would like to show that these "virtual" breaks are actually not real evidences for any liquid-liquid phase transition.

When the liquid-liquid line, separating two liquid forms, meets the melting line, one can expect some kind of break on the melting line, caused by the different freezing/melting behaviour of the two liquid forms. Several researchers are using this vein of thinking in an upside-down form; seeing some irregularity on the melting line, they will expect a break and the appearance of a liquid-liquid line. Using formal logic, when B is caused by A, A is not necessarily caused by B (here A is the existence of a liquid-liquid phase transition and B is a break on the melting line). In this way, we can say that seeing a break on the melting line is not an evidence for a hidden liquid-liquid phase transition, although it makes the system suspicious.

As a further step, having a set of (p,T) data for the melting line, one cannot decide very easily whether the line is smooth or broken. Here we will demonstrate that the high-pressure melting data of the nitrogen[6] can be interpreted in two different ways: showing a break (and being a suspect to have two different liquid forms as it has been predicted earlier[7]) or being smooth (remaining a candidate to be "normal", one-form liquid). Further examples of these kinds of systems will be published soon.[8]

We are using a modified for of the Simon-Glatzel relation[9-12], introduced by Drozd-Rzoska et al.[10,12]:

$$T_m(P)=T_m^0\left(1+\frac{\Delta P}{\pi+P_m^0}\right)^{1/b}\exp\left(-\frac{\Delta P}{c}\right) \tag{1}$$

where $\Delta P = P - P_m^0$ and $\Pi = \pi + P_m^0$, $-\pi$ is the negative pressure asymptote for $T \to 0$, P_m^0 and T_m^0 are the reference pressure and temperature, c denotes the damping pressure coefficient. Comparing to the original Simon-Glatzel relation, this form has two advantages, namely it can reproduce the maximum on the melting curve in (P, T) space and it can yield a negative pressure asymptote. This later one might be related to the so-called crystal spinodal[13], which is the existence limit of a solid under negative pressure (i.e. in isotropically stretched state)[14], although the relation of these two quantities requires further studies.

On Figure 1 one can see the high-pressure melting data of nitrogen.[6]

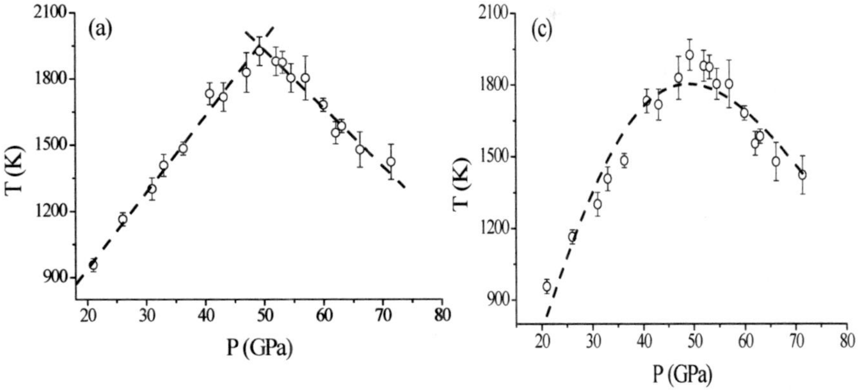

Figure 1. *(a) Double-linear fitting of the high-pressure melting data of nitrogen, suggested by Mukherjee and Boehler [6], forcing us to see a break on the melting line as a possible precursor of a liquid-liquid phase transition- (b) The same dataset, fitted by the modified Simon-Glatzel equation (Eq.1), showing smooth, break-free melting line. Fitting parameters are:* $T_0=3.1$, $P_0=2.77$, $\Pi=1.61$, $b=0.38$ *and* $c=18.2$.

Here we have to mention that the low pressure data can be fitted properly neither by Eq. 1, nor by fitting two linear parts, therefore those data are not shown here. On Figure 1/a one can see the data fitted by two linear parts, proposed by Mukherjee and Boehler.[6] Using this fitting, we are forcing our eyes to see a sharp break in the melting line, which might be the sign of a liquid-liquid line joining into the melting line. On the other hand, using the modified Simon-Glatzel curve with the parameters shown in the figure legend (Figure 1/b), one can see that the break will disappear and the melting line will be a smooth curve.

The existence of this proposed break can be clarified by further measurements with better accuracy (smaller error), but the break itself will not be a proper precursor for the existence of a second liquid phase of the nitrogen. To prove the existence of than new form, other kinds of experiments are needed.

Acknowledgements

The authors would like to acknowledge the helpful advices of Prof. G. Franzese (Barcelona). The financial support of NATO under grant No. CLG 982312 is also acknowledged. A. R. Imre was supported by Hungarian Research Fund (OTKA) under contract No. K67930, and by the Bolyai Research Grant of the Hungarian Academy of Science.

References

1. Franzese, G. (2007) Differences between discontinuous and continuous soft-core attractive potentials: The appearance of density anomaly, *J. Mol. Liq.*, **136**, 267-273
2. Oliveira, A.B., Franzese, G., Netz, P.A. and Barbosa, M.C. (2008) Waterlike hierarchy of anomalies in a continuous spherical shouldered potential, *J. Chem. Phys.*, **128**, 064901
3. Mishima, O. and Stanley, H.E. (1998) The relationship between liquid, supercooled and glassy water, *Nature* **392**, 329-335
4. Debenedetti, P.G. (1998) Condensed matter - One substance, two liquids? *Nature* **392**, 127-128
5. Mishima, O. (2000) Liquid-liquid critical point in heavy water, *Phys. Rev. Lett.* **95**, 334-336
6. Mukherjee, G.D. and Boehler, R. (2007) High-pressure melting curve of nitrogen and the liquid-liquid phase transition, *Phys. Rev. Lett.* **99**, 225701
7. Ross, M. and Rogers, F. (2006) Polymerization, shock cooling, and the high-pressure phase diagram of nitrogen, *Phys. Rev. B* **74**, 024103
8. Imre, A.R. and Rzoska, S.J. (2009) High pressure melting curves and liquid-liquid phase transition, *Int. J. Liq. State Sci.*, submitted
9. Drozd-Rzoska, A. (2005) Pressure dependence of the glass temperature in supercooled liquids, *Phys. Rev. E.* **72**, 041505
10. Drozd-Rzoska, A., Rzoska, S.J. and Imre, A.R. (2007) On the pressure evolution of the melting temperature and the glass transition temperature, *J. Non-Cryst. Solids* **353**, 3915-3923
11. Drozd-Rzoska, A., Rzoska, S.J., Paluch, M., Imre, A.R. and Roland, C. M. (2007) On the glass temperature under extreme pressures, *J. Chem. Phys.* **126**, 164504
12. Drozd-Rzoska, A., Rzoska, S.J., Roland, C.M. and Imre, A.R. (2008) On the pressure evolution of dynamic properties of supercooled liquids, *J. Phys.: Condens. Matter.* **20**, 244103
13. McMillan, P.F. (2002) New materials from high-pressure experiments, *Nature Materials* **1**, 19-25
14. Imre, A. R. (2007) On the existence of the negative pressure states, *Phys. Stat. Sol. B* **244**, 893-899

DISORDER PARAMETER, ASYMMETRY AND QUASIBINODAL OF WATER AT NEGATIVE PRESSURES

VITALY B. ROGANKOV
Odessa State Academy of Refrigeration, Dvoryanskaya str. 1/3, 65082 Odessa, Ukraine

Abstract: The virtual terms "binodal" and "spinodal" are equivalent to the experimental terms "coexistence curve" (CXC) and "metastability limit" (ML), respectively, within an inherent accuracy of any semi-empirical EOS at the description of a real fluid behavior. Any predicted location of mechanical spinodal at positive pressures $P_{sp}(T)\geq 0$ merits verification because the Maxwell rule is a model (EOS)-dependent method based on the non-measurable values of chemical potential for both phases. It is not a reliable tool of CXC- and ML-prediction especially at low temperatures between the triple and normal boiling ones $[T_t, T_b]$ where the actual vapor pressures $P_s(T) > 0$ are quite small while the spinodal pressures $P^l_{sp}(T) < 0$ are huge and negative for a superheated liquid. In this paper the alternative substance (non-model)-dependent method of binodal/spinodal formalism consistent with the actual CXC-data: ρ_l, ρ_g, $P_s(T)$ is proposed. The crucial distinction from the conventional results is the novel virtual curve of the negative vapor pressures: $P^-_s(T) < 0$ *symmetrical to the actual pressures of saturation*: $P_s(T) + P^-_s(T) \approx 0$. For water this curve crosses the liquid spinodal branch $P^l_{sp}(T)$ at the point: $P^0_{sp} \approx -13{,}16\ MPa; T^0_{sp} \approx 605K$ which is the top of a novel quasibinodal. The respective branches of it (taken at the negative pressures of "saturation": $P^-_s < 0$ are formed by the liquid-like: $\rho^+(T)$ and gas-like: $\rho^-(T)$ densities of a "coexistence" at the (P^-_s, T)-conditions. The predicted gas-like branch $\rho^-(T)$ is localized completely within the spinodal but the liquid-like-branch $\rho^+(T)$ has the "near-critical" metastable and even the low-temperature stable parts where $\rho^+(T) \approx \rho_l(T)$.

The predicted quasibinodal has the *practically T-independent rectilinear diameter*: $\rho^0_d = (\rho^+ + \rho^-)/2 \approx \rho^0_{sp}(P^0_{sp}, T^0_{sp})$ as it is in the discrete lattice-gas gas model. Simultaneously, the proposed continuum model of a real fluid is consistent with the *possible asymptotic singularity of the actual T-dependent rectilinear diameter*: $\rho_d(T) = (\rho_l + \rho_g)/2$ *without any appealing to the scaling formalism*. The prediction of ML- and quasibinodal – parameters is based on the fluctuation EOS (FEOS) proposed by author. The reduced slope $A_s(T)$ of the vapor pressure $P_s(T)$ is used as the factor of asymmetry to control the realistic interrelation between the entropy-disorder (s_g-s_l)- and density-order $(\rho_l-\rho_g)$-parameters along the CXC.

Keywords: disorder parameter, order parameter, metastability limit, asymmetry, quasi-binodal, particle-hole-type symmetry

1. Introduction

Certainly the most important models for the development of modern scaling theory of critical phenomena have been the discrete Ising model of ferromagnetism and its antipode – the continuum van der Waals model of fluid. The widespread belief is that real fluids and the lattice-gas 3D-model belong to the same universality class but the absence of any particle-hole-type symmetry in fluids requires the revised scaling EOS. The mixed variables were introduced to modify the original Widom EOS and account the possible singularity of the rectilinear diameter.

One may assume the existence of coupling entropy-disorder (s_g-s_l) and density-order $(\rho_l-\rho_g)$ parameters at any state-point of saturation controlled by the thermodynamic Clapeyron's differential equation. It is shown in Section 2 that incorporation of the undimensional slope-parameter $A_s(T)$ along the vapor-pressure curve $P_s(T)$ is the crucial step to provide the adequate representation of CXC-properties between the triple T_t and critical T_c temperatures by the fluctuational EOS(FEOS). The consistent description of CXC for a real fluid and the respective spinodal is considered, firstly, at low temperatures of water in Section 3 and, then, for the whole fluid range in Section 4. The novel quasibinodal curve is predicted at negative pressures and the relevant hypothetical phase diagram (HPD) is discussed in the frameworks of FEOS. The impressed result of the proposed continuum model is its consistency with the possible singularity of the rectilinear diameter without any appealing to the scaling formalism.

2. Fluctuation equation of state for water

The normal water has the long range $[T_t, T_c]$ of gas-liquid transition and the complicated molecular structure. It is a good object to demonstrate the thermodynamical universality of proposed HPD-concept. The first problem is the evaluation of T-dependent FEOS-coefficients [1-3]:

$$P = \frac{\rho RT[1-c(T)]}{1-b(T)\rho} - a(T)\rho^2, \tag{1}$$

$$a(T) = \frac{P_S(A_S - 1)}{\rho_l \rho_g}, \tag{2}$$

$$b(T) = \frac{A_s - 2}{(\rho_l + \rho_g)(A_s - 1)}, \tag{3}$$

$$1 - c(T) = Z_l \left[1 + \rho_l(A_s - 1)/\rho_g\right](1 - b\rho_l), \tag{4}$$

where $Z_l = P_s/(\rho_l RT)$ and the reduced slope of $P_s(T)$-curve is:

$$A_s(T) = \frac{T}{P_s} \cdot \frac{dP_s}{dT} = \frac{T(s_g - s_l)\rho_l \rho_g}{P_s(\rho_l - \rho_g)}. \tag{5}$$

The undimensional T-dependent parameter A_s is the main measurable quantity to control the real interrelation between the order $(\rho_l - \rho_g)$- and disorder $(s_g - s_l)$-parameters at the coexistence of stable phases.

The former parameter is the conventional factor of asymmetry in the expansions truncated after linear terms[4]. It can be used to introduce the presumed scaling relations at subcritical temperatures $\overline{T} = 1 - T/T_c \geq 0$ as well as to obtain the consistent description of stable phases, in which the asymptotic power laws are used. Unfortunately, the conventional analysis of the scaling consistency fails, often, even in the asymptotic range of temperatures: $\overline{T} \leq 10^{-3}$ because the adjustable system-dependent amplitudes of the power laws are rather inaccurate. Besides, the implicit assumption of scaling, the parameter $(\rho_l - \rho_g)$ to be the single factor of asymmetry, must be corroborated especially in the extended critical region.

It was found[1-3] for the original van der Waals – Maxwell – Gibbs (WMG) model of CXC:

$$\rho_g = \left\{b\left[1 + y(x)e^x\right]\right\}^{-1}, \quad \rho_l = \left\{b\left[1 + y(x)e^{-x}\right]\right\}^{-1}, \tag{6}$$

$$y(x) = \frac{shxchx - x}{xchx - shx}, \quad shx = \frac{e^x - e^{-x}}{2}, \quad chx = \frac{e^x + e^{-x}}{2}, \tag{7}$$

where x is the reduced (classical) disorder parameter:

$$x = (s_g - s_l)/2R, \quad b = 1/(3\rho_c), \tag{8}$$

that the symmetrical linear dependencies $\rho_l(x)$ and $\rho_g(x)$ exist only within the extremely small asymptotic range of x [0;0,5]. This fact was confirmed also[1] for the set of real fluids: Ar, C_2H_4, CO_2, H_2O (Fig.1) where the respective slopes $d\rho_i/dx$ $at \leq x\ 0,5$ are close to the system-dependent values $\pm\ (A_c Z_c)^{-1}$: Ar(0.5764), C_2H_4(0.5597), CO_2(0.5172), H_2O(0.5551), van der Waals fluid (2/3).

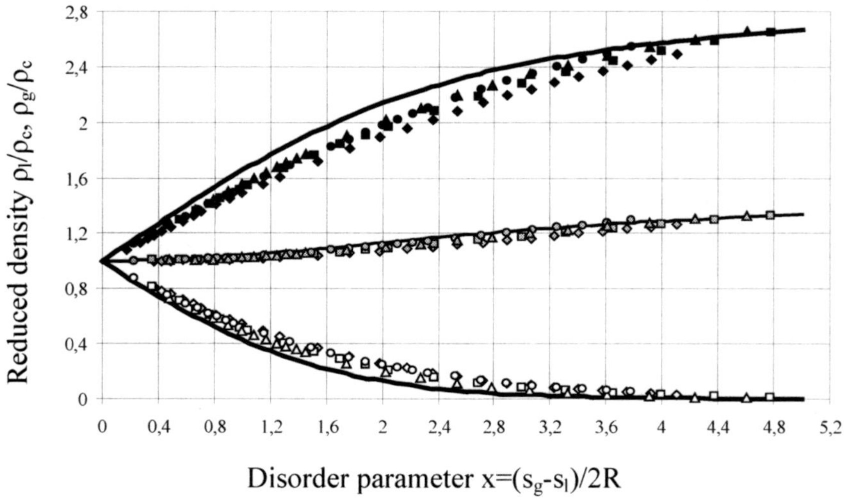

Figure 1. *Comparison of the reduced CXC-densities $\rho_{l,g}/\rho_c$ for real substances: Ar (▲/△), C_2H_4 (■/□), CO_2 (◆/◇), H_2O (●/○) with the van der Waals-Maxwell-Gibbs model's predictions (▬) based on the disorder parameter x; the respective rectilinear diameter $(\rho_l+\rho_g)/2\rho_c$ as a function of x (—).*

The sharp intersection of the liquid $\rho_l(x)$- and gas $\rho_g(x)$-branches at the critical point shown in Fig. 1 implies that the molar internal energy $e(v,s)$ is not a continuous differentiable function of v and s along the CXC for both: real and WMG-model fluids. It is the direct confirmation of the singular concept introduced [1-3] to represent the actual CXC-data of any real fluid by FEOS (1-5). Put in thermodynamic terms, any state-point P_s,T of CXC including the critical point P_c,T_c is, simultaneously, the one phase:

$$P_s = T_i\left(\frac{\partial P}{\partial T}\right)_{\rho_i} + \rho_i^2\left(\frac{\partial e}{\partial \rho}\right)_{T_i} = T_i\left(\frac{\partial s}{\partial v}\right)_{T_i} + \rho_i^2\left(\frac{\partial e}{\partial \rho}\right)_{T_i} \tag{9}$$

($i = l$ or g) and two-phase one:

$$P_s = T\frac{dP_s}{dT} + \rho_l\rho_g\frac{e_l - e_g}{\rho_l - \rho_g} = T\frac{s_g - s_l}{v_g - v_l} + \rho_l\rho_g\frac{e_l - e_g}{\rho_l - \rho_g}. \tag{10}$$

This statement is, of course, in a contradiction with the conventional mean-field assumption that the orthobaric curve forms a line along which e,s,v all increase monotonically on passing from liquid, through the critical point, to the gas.

The important consequences of the above consideration are:

1) the mean-field expansions along the CXC are not applicable for real fluids as well as for the original van der Waals EOS;
2) the reduced slope $A_s(T)$ is the most appropriate factor of asymmetry to control the realistic interrelation between the disorder $(s_g–s_l)$- and order $(\rho_l–\rho_g)$-parameters;
3) the system of FEOS-eqs.(1-5,9,10) is completely consistent in the whole range $[T_t, T_c]$ of a fluid phase transition.

One may note from eq. (3) that the rectilinear diameter ρ_d and its derivative $d\rho_d/dT$ can be explicitly represented by equations in terms of A_s:

$$\rho_d = \frac{(A_s - 2)}{2b(T)(A_s - 1)}, \tag{11}$$

$$\frac{d\rho_d}{dT} = -\frac{1}{2b^2}\cdot\frac{db}{dT} + \frac{1}{2b(A_s - 1)^2}\cdot\frac{dA_s}{dT}, \tag{12}$$

$$\frac{dA_s}{dT} = -\frac{A_s(A_s - 1)}{T} + \frac{T}{P_s}\cdot\frac{d^2P_s}{dT^2}. \tag{13}$$

There are two possible reasons for the derivative $d\rho_d/dT$ to be divergent in eq. (12) – possible divergences of db/dT and/or dA_s/dT.

To study the problem and compare the non-mean-field and classical CXC-description, we propose to generalize the original WMG-model by incorporating into eq. (6) a T-dependent coefficient $b(T)$. Then one may use the ratio of the experimental or tabular densities ρ_g/ρ_l in the whole range $[T_t, T_c]$ to solve the transcendent equation for certain classical $x(T)$-value of disorder parameter from eqs. (7,8):

$$\frac{\rho_g}{\rho_l} = \frac{1 + y(x)e^{-x}}{1 + y(x)e^{x}} . \tag{14}$$

Let me remind here that the actual non-mean-field disorder parameter $(s_g–s_l)$ (guaranty of the actual chemical potential evaluation) can be calculated by the Clapeyron eq. (5) or determined from the measured latent-heat: $r_s = T(s_g–s_l)$-values.

The next step of the generalized WMG-model is the calculation of the second coefficient $a(T)$ from eqs. (2,3):

$$a(T) = \frac{P_s}{\rho_l\rho_g\left[1 - b(T)(\rho_g + \rho_l)\right]}, \tag{15}$$

based on the coefficient $b(T)$ found from eq. (6) and on the input CXC-properties ρ_g, ρ_l, P_s. The well-known power CXC-functions of water proposed by Saul and Wagner [5] have been used to evaluate the actual as well as the classical FEOS-coefficients represented in Figs. 2-5. The both sets of coefficients describe the $P_s(T)$, $\rho_g(T)$, $\rho_l(T)$-correlations within the experimental uncertainties.

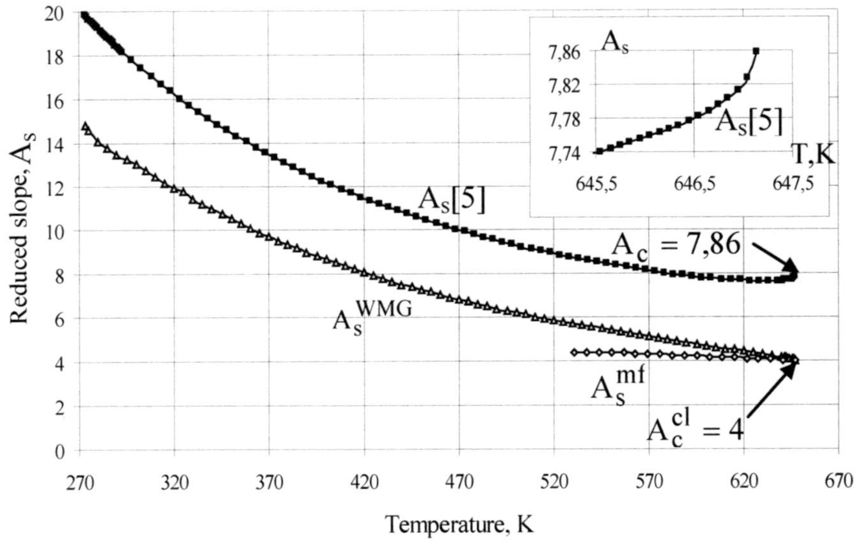

Figure 2. *Reduced slope $A_s(T)$ of the vapor pressure curve $P_s(T)$ for water predicted:* –△– *- by the generalized WMG-model and* –◇– *- by the analytic expansion along CXC in comparison with* –■– *the actual data[5]; the possible reason of near-critical singularities at subcritical temperatures is also shown for* $\overline{T} \leq 2{,}5 \cdot 10^{-3}$.

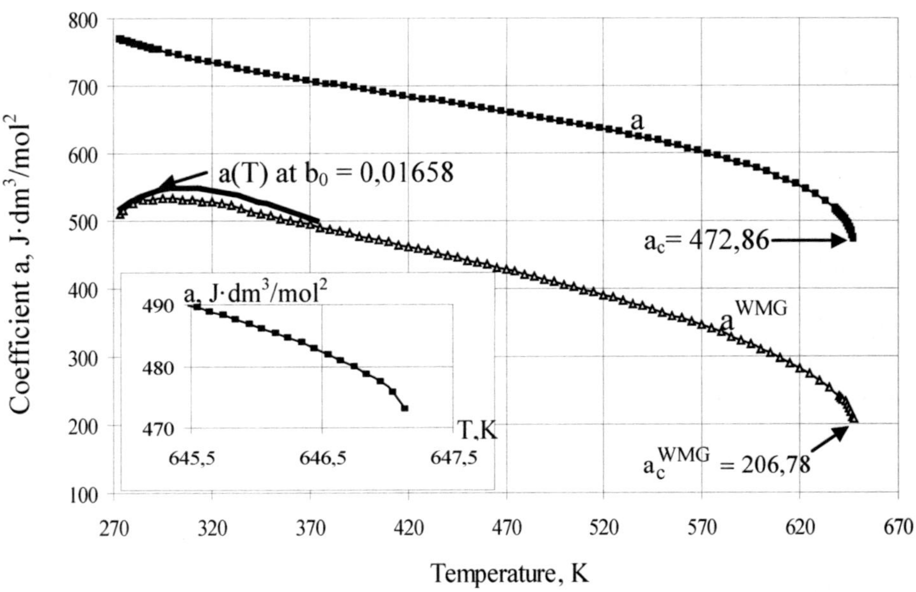

Figure 3. *Variants of the FEOS-coefficient $a(T)$ for water predicted:* –■– *a - by the actual values $A_s(T)^5$;* –△– *a^{WMG} - by the WMG- model; $a(T)$ at b_0 – by the low-temperature variant of FEOS; the near-critical behavior of $a(T)$ is also shown for* $\overline{T} \leq 2{,}5 \cdot 10^{-3}$.

These Saul-Wagner nonanalytic equations for water[5] have been transformed into the FEOS-coefficients by one-to-one map without any adjustable coefficients. The distinction is that the transformation into the set: $a(T)$, $b(T)$, $c(T)$ by eqs. (2-4) is based on the actual $A_s(T)$-values from eq. (5) while the predicted by the generalized WMG-model $A_s^{WMG}(T)$-values in Fig. 2 are classical. In other words, the consistent description of the order parameter is not a guaranty of the correct disorder parameter prediction.

The actual $A_s(T)$-function has a minimum at $T/T_c \approx 0{,}98$ in opposite to the predicted monotonic decreasing of $A_s^{WMG}(T)$ down to $A_c^{cl} = 4$. This difference is crucial to provide the non-mean-field description of the CXC. It is obvious, also, the essential distinction in the fluctuational coefficient $c(T)$ and classical coefficient $c^{WMG}(T)$ are represented in Fig. 5.

By contrast to the scaling formalism based on the concepts and results for the discrete 3D-Ising model, the above-discussed phenomenological crossover model of CXC is based on the continuum van der Waals – type FEOS (1-5). Much effort has been devoted toward applying the various semiempirical EOSs to real near-critical fluids by assumption the coefficients be T- or/and ρ- dependent. One may find the relevant comparative review and analysis of these attempts in the works on the crossover problem formulated by Sengers and coauthors [6,7]. In opposite to the conventional crossover approach, the proposed phenomenological model does not incorporate any analytical or nonanalytical truncated expansions with the fitted asymmetrical terms and mixed scaling variables. The developed here formalism is much simpler than the alternative methods but provides the reliable quantitative results for the whole subcritical range 273.16…647.14K of water.

2. Low-temperature behavior of water

The known low-temperature anomaly of water is interesting by itself and is connected with the, so-called, "reentrant spinodal" form of liquid branch $P_{sp}^{l}(T)$ in the range $[T_t, T_b]$. This behavior has been investigated by Speedy [8] in the context of a truncated analytic expansion of the pressure $P(\rho, T)$ about the limit of stability $P_{st}(\rho_{st}, T)$ along each isotherm:

$$\frac{P}{P_{st}} = B\left\{\left[2+\left(\frac{1}{B}-1\right)\frac{\rho_{st}}{\rho}\right]\left(\frac{\rho}{\rho_{st}}\right)-\left(\frac{\rho}{\rho_{st}}\right)^{2}\right\}. \tag{16}$$

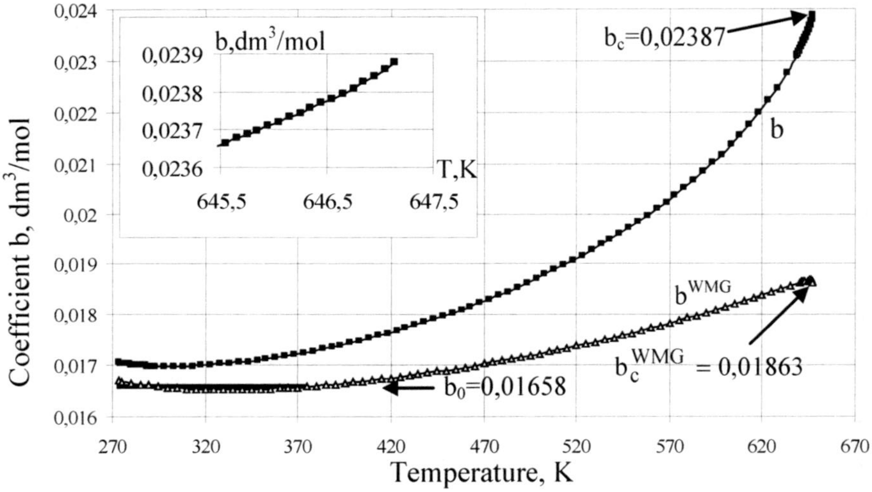

Figure 4. *Variants of the FEOS-coefficient b(T) for water predicted:* —■— *b - by the actual values* $A_s(T)^5$*;* —△— b^{WMG} *- by the WMG-model;* b_0 *– by the low-temperature variant of FEOS; the near-critical behavior of b(T) is also shown for* $\overline{T} \le 2{,}5 \cdot 10^{-3}$.

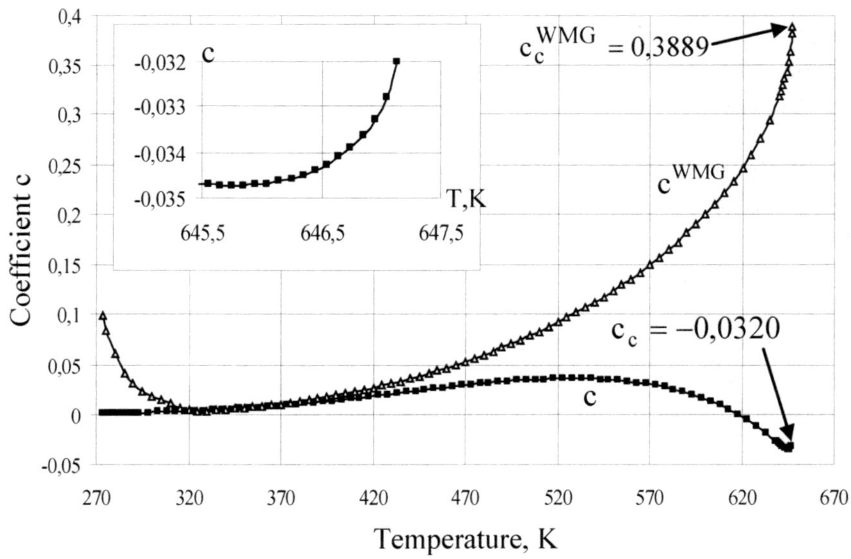

Figure 5. *Variants of the FEOS-coefficient c(T) for water predicted:* —■— *c - by the actual values* $A_s(T)^5$*;* —△— c^{WMG} *- by the WMG-model; the near-critical behavior of c(T) is also shown for* $\overline{T} \le 2{,}5 \cdot 10^{-3}$.

It is evident that the adjustable functions $P_{st}(T)$, $\rho_{st}(T)$ and T-dependent coefficient $B(T)$ are built into EOS (16) only for a liquid (stable or metastable) phase. They are consistent [8] with the power-low divergences of the response functions: χ_T, α_P, C_P and the common pseudospinodal exponent $\gamma=1/2$ along a subcritical isobar:

$$\chi_T \sim \alpha_P \sim C_P \sim \left[T_{st}(P)-T\right]^{-1/2} . \tag{17}$$

It should be noted that author [8] distinguishes the spinodal itself $P^l_{sp}(T)$ from the virtual line of stability limits $P_{st}(T)$. There are two main reasons of a such caution. Firstly, it is the difference between the mean-field critical exponent $\gamma=1$ and mean-field pseudospinodal exponent $\gamma=1/2$ from eq. (17). Secondly, Speedy[8] has determined the $P_{st}(T)$- and $\rho_{st}(T)$-lines by extrapolation and adjustment of the stable-liquid properties for water in the ranges 0–100°C and 0–100MPa to the EOS-form (16).

I do not believe this difference to have general meaning, so have used the consistent spinodal equations obtained from FEOS (1):

$$P_{sp} = a(T)\rho^2_{sp}\left[1-2b(T)\rho_{sp}\right], \tag{18}$$

$$RT\left[1-c(T)\right] = 2a(T)\rho_{sp}\left[1-b(T)\rho_{sp}\right]^2 . \tag{19}$$

It follows straightforwardly from eqs. (1,18,19) that the similar FEOS-form exists:

$$\frac{P}{P_{sp}} = \overline{B}\left[2\frac{\left(1-b\rho_{sp}\right)^2}{\left(1-b\rho\right)}\left(\frac{\rho}{\rho_{sp}}\right)-\left(\frac{\rho}{\rho_{sp}}\right)^2\right] \tag{20}$$

with the well-defined coefficient $\overline{B}(T)$:

$$\overline{B}(T) = \frac{1}{1-2b\rho_{sp}} \tag{21}$$

for the whole subcritical range. By contrast to the Speedy EOS (16), the obtained FEOS-form (20,21) does not contain any adjustable function of T. The former approach is based on the extrapolated one-phase properties while the latter one uses only the measurable CXC-data and, as a result, predicts the actual (i.e. consistent with the actual CXC) spinodal.

For all substances in the low-temperature range $[T_t, T_b]$ the asymptotic form of FEOS with constant excluded volume b_0, zero fluctuation coefficient $c=0$ and T-dependent interaction coefficient $a(T)$ is adequate[3] for a liquid phase:

$$P = \frac{\rho RT}{1-b_0\rho} - a(T)\rho^2 . \tag{22}$$

The mostly measurable input data on the density $\rho_0(T)$ at ambient pressure have been used below to develop the predictable model. For the vapor pressure at low temperatures it is proposed to use the perfect-gas equation:

$$P_s(T) = \rho_g(T)RT \tag{23}$$

with the original WMG-expressions (6,7) at the changed coefficient b_0 (instead of the vdW-value b from eq. (8)).

If any CXC-state-point is unknown except the normal boiling point (T_b, P_0, ρ_{0l}) one needs to estimate, firstly, the disorder parameter $x(T)$ and the constant excluded volume b_0. The FEOS-parameter A_s is expressed at low temperatures as:

$$A_s = 2x(T) = 1 + \left(T / \rho_g\right)\left(d\rho_g / dT\right) \tag{24}$$

in the framework of the Clapeyron-Clausius approximation $\rho_l >> \rho_g$. By eliminating the coefficient $a(T)$ from two variants of eq. (22) written for $P_s(T)$ (unknown) and $P_0(T)$ (given) the important solution can be obtained at the same assumption $\rho_l >> \rho_g$:

$$\rho^{\pm} = \frac{1}{2b_0}\left[1 \pm \sqrt{1 - \frac{4b_0\rho_0(1-b_0\rho_0)}{1-Z_0(1-b_0\rho_0)}}\right] \tag{25}$$

based on the known data $\rho_0(T)$ and $Z_0 = P_0/(\rho_0 RT)$ along the initial atmospheric isobar P_0. The coefficient $b_0 = 0{,}01658 dm^3/mol$ has been estimated by eqs. (14,23) at the single normal-boiling CXC-point for water.

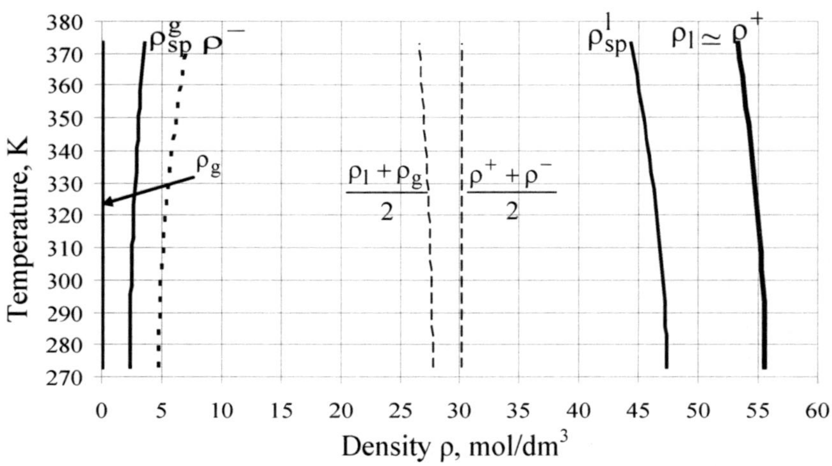

Figure 6. *Hypothetical low-temperature phase ρ,T-diagram for water as super-position of symmetric $(\rho^+ + \rho^-)/2$ and asymmetric $(\rho_l + \rho_g)/2 = \rho_d(T)$ behavior.*

It is remarkably that the exact particle-hole-type symmetry of a novel quasibinodal (25) exists at any low-temperature:

$$\rho^{\pm}=\frac{1}{2b_0}\left[1\pm\sqrt{1-\frac{4b_0RT}{a(T)}}\right] \tag{26}$$

where *a(T)* was calculated by eq.(15) at $b_0=0{,}01658 dm^3/mol$. The "liquid branch" $\rho^+(T)$ coincides at these conditions with the stable saturated liquid[5]: $\rho^+\approx\rho_l(T)$ as it is shown for water in Fig. 6. It can be used for the reliable calculation of *x(T)*:

$$x=\left(1-b_0\rho^+/2\right)/\left(1-b_0\rho^+\right) \tag{27}$$

and the prediction of latent heat $r_s(T)$ by eq. (24).

The most unwonted result is shown in Fig. 7 for low-temperature water. The substitution of the quasibinodal data $\rho^{\pm}(T)$ predicted by eq. (25) into the respective FEOS-form (22) gives the novel branch of negative "vapor pressures" $P_s^-(T)$. *It provides practically symmetrical map of the stable vapor pressures:* $P_s(T)+P_s^-(T)\approx 0$. The predicted low-temperature spinodal form is represented in Fig. 8. It is obvious that the reentrant behavior of superheated liquid in P_{sp}^l, T-plane is connected with the anomaly of a low-temperature liquid phase.

4. Hypothetical phase diagram of water

Figures.9 and 10 represent the most expressive confirmation of the latent symmetry in the HPD for water obtained by the generalized WMG-model. It was suggested the existence of the quasicritical point to extrapolate the low-temperature results of Figs. 6,7 and predict the whole quasibinodal. It is the point (Fig. 9) in which the branch of negative pressures $P_s^-(T)$ intersects the liquid branch of the WMG-spinodal $P_{sp}^l(T)$ at the parameters: $T_{sp}^0\approx 605K$ and $P_{sp}^0\approx -13{,}16\,MPa$. Substitution of pressures $P_s^-(T)$ into FEOS (1) with the respective coefficients a^{WMG}, b^{WMG}, c^{WMG} gives the almost symmetrical "liquid and gas" branches $\rho^{\pm}(T)$ of a quasibinodal shown in Fig. 10. It is interesting that the locus of unstable solutions (the third root of FEOS) for the WMG-model represented also in Fig. 10 is a symmetrical map of the actual CXC-diameter $\rho_d(T)$.[5] Strong resemblance of the above results with the magnetic transition is important since the presence of the non-ordered saturated gas-phase $\rho_g(T)$ is the evident reason of asymmetry observed in a real fluid. In terms of ferromagnetic system, presence of the paramagnetic (non-ordered) component may destroy the ideal symmetry of spontaneous magnetization at zero field $h=0$.

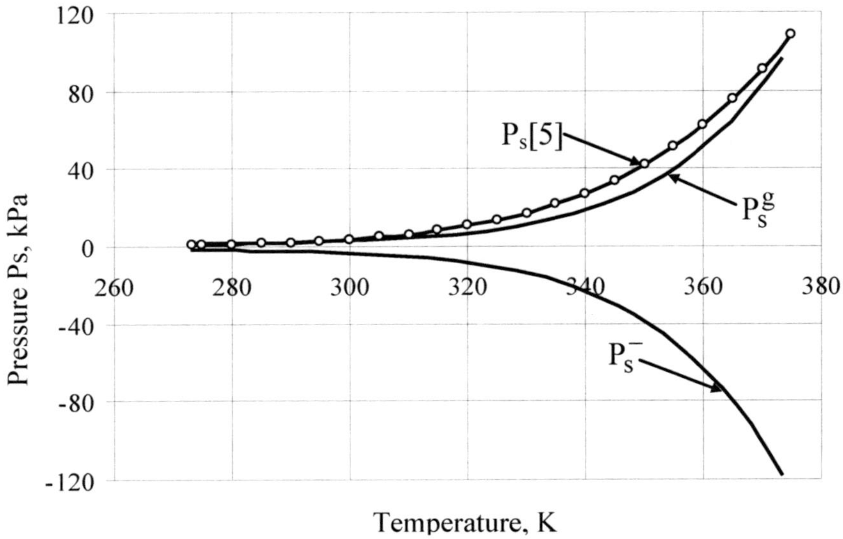

Figure 7. *Comparison of the predicted two-valued vapor-pressures* $\pm P_s^g$ $(-P_s^g \simeq P_s^-)$ *with the tabular* $P_s(T)$*-data for water*[5].

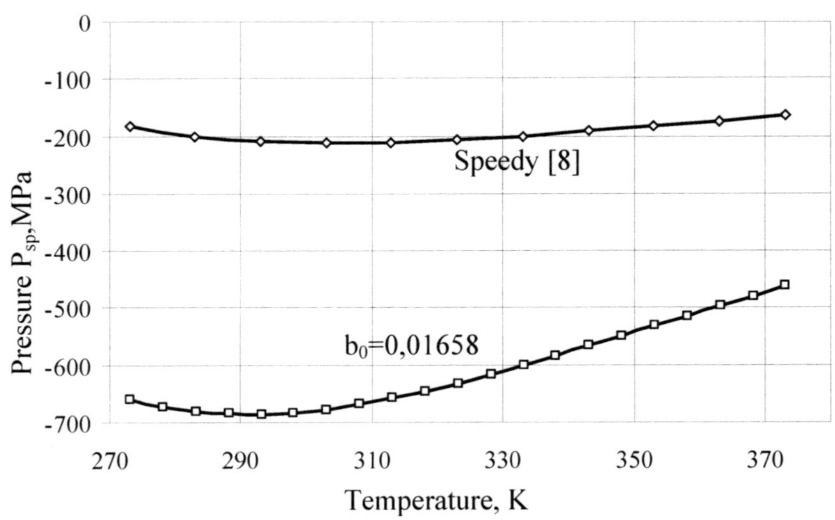

Figure 8*. Predicted form of spinodal for superheated liquid in water at the selected value of the effective excluded volume* b *[dm/mol].*$_0$ ^{3}

Strong resemblance of the above results with the magnetic transition is important since the presence of the non-ordered saturated gas-phase $\rho_g(T)$ is the evident reason of asymmetry observed in a real fluid. In terms of ferromagnetic system, presence of the paramagnetic (non-ordered) component may destroy the ideal symmetry of spontaneous magnetization at zero field $h=0$. The relevant feature follows from the up-down symmetry of the Ising spin model when the magnetic field h is replaced by $-h$. It implies [9] that any phase transition that occurs at nonzero field h must occurs at both $\pm h$. However, the ferromagnetic transition has to be occurred just at zero field because the Gibbs free energy: $g(h,T)=f(m,T)-mh$ is everywhere analytic in h except at $h=0$. It is preferably to consider the Helmholtz free energy derivatives: $P=-(\partial f/\partial v)_T$, $s=-(\partial f/\partial T)_v$ for a fluid and the respective analogies:[9] $P \leftrightarrow h, v \leftrightarrow -m$ or $P \leftrightarrow -h, v \leftrightarrow m$ if the EOS-form is discussed. In these terms the phase transitions at $h=0$ and $P=0$ are formally similar.

By contrast to the lattice-gas model, the CXC-diameter $\rho_d(T)$ and non-zero vapor pressure $P_s(T)$ are T-dependent for the whole subcritical range. One may conclude, some paradoxically, that the symmetrical map of $P_s(T)$ into the negative pressures i.e.: $-P_s(T)$ must exist if the latent particle-hole-type symmetry is in a real fluid. From what has been said above, the FEOS-model confirms this rather unusual possibility.

The Ising spin model does not consider the coexistence of the ordered (ferromagnetic) and non-ordered (paramagnetic) phases at subcritical temperatures. As a result, there is no latent heat $r_s(T)$ and disorder parameter associated with the ferromagnetic transition. The condition $dh/dT = 0$ must be added to $h=0$. The known CXC-dependence of the lattice-gas chemical potential [9]:

$$\mu_s = -\left\{kT\,ln\left[v_0 / \lambda_B^3(T)\right] + z\varepsilon / 2\right\} \tag{28}$$

($\lambda_B = h_P /(2\pi mkT)^{1/2}$, ε - well depth) provides the T-dependent interpretation of the condition $h=0$ in terms of coupling constant J and μ_s:

$$h = (kT/2)\,ln\left[v_0 / \lambda_B^3(T)\right] + zJ + \mu_s / 2 = 0 \tag{29}$$

where the condition $dh/dT=0$ can be transformed into equality:

$$d\mu_s / dT = -d/dT\left\{kT\,ln\left[v_0 / \lambda_B^3(T)\right]\right\}. \tag{30}$$

It denotes that the lattice-gas CXC coincides with the critical (passing through the critical temperature T_c) and, simultaneously, the ideal-gas isoentrop $s_c(T)$ at all subcritical temperatures:

$$s_c / k = ln\left[v_0 / \lambda_B^3(T)\right] - 3\left[T / \lambda_B(T)\right] d\lambda_B / dT\,. \tag{31}$$

To confirm the coincidence one may use the Clapeyron-type equation at the special lattice-gas condition: $s_g=s_l=s_c$:

$$\frac{d\mu_s}{dT} = \frac{\rho_g s_g - \rho_l s_l}{\rho_l - \rho_g} = -s_c \,. \tag{32}$$

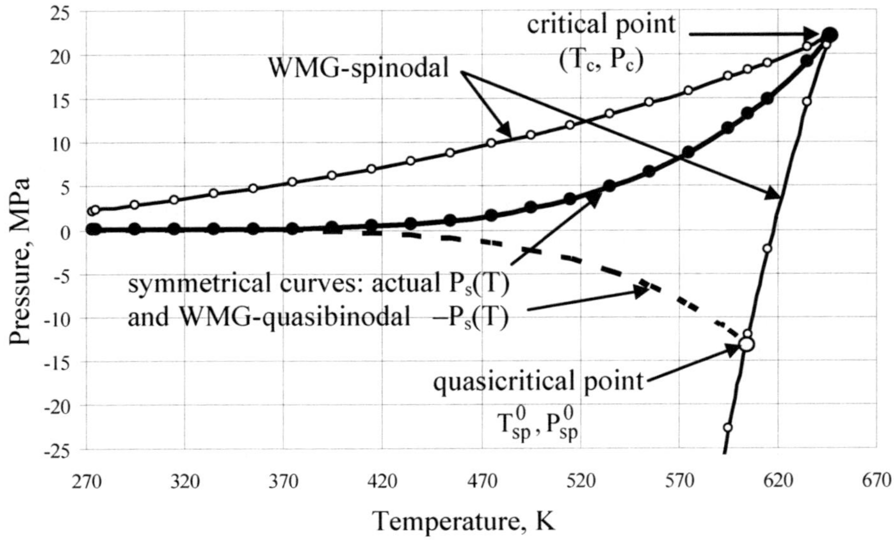

Figure 9. *Hypothetical phase diagram of water predicted by the generalized WMG-model.*

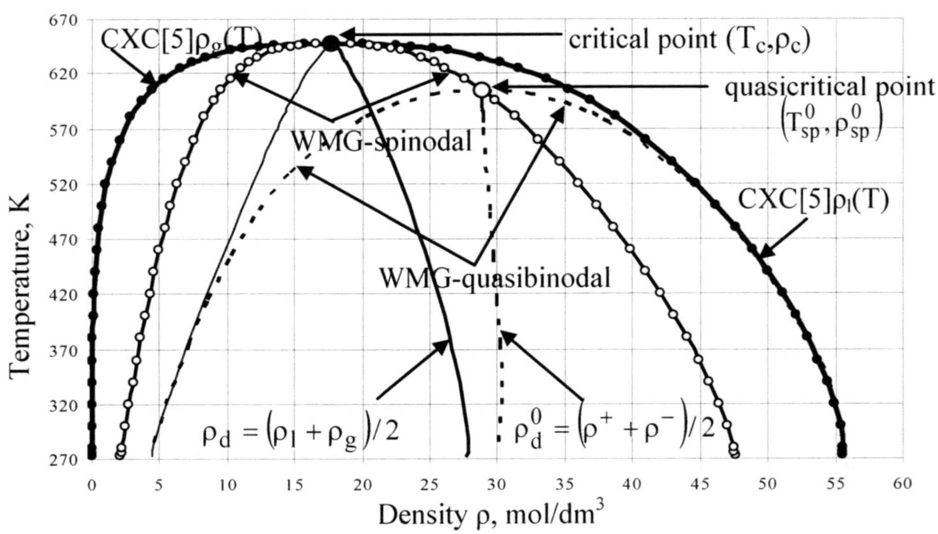

Figure 10. *Hypothetical phase diagram of water predicted by the generalized WMG- model (see also Fig. 9) as a superposition of asymmetry (CXC-WMG-spinodal with the T-dependent diameter $\rho_d(T)$) and symmetry (WMG-quasibinodal with the weakly T-dependent diameter $\rho_d^0(T)$ at $-P_s(T)$).*

Absence of the disorder parameter *(s_g-s_l)* is the serious restriction of the lattice-gas model. This restricted concept is used also in the above-discussed scaling expansions adopted in the study of asymmetry for a real fluid[4].

5. Conclusions

The phenomenological FEOS-model and its consequence – the HPD with the formal particle-hole-type symmetry of a quasibinodal has some specific distinctions from the relevant conventional approaches.[6,7] First of all, it is based on the continuum, exactly solvable WMG-model of a phase transition without any adjustable parameters. Besides, the study of the novel substances and mixtures can be carried out within the framework of the common FEOS which is applicable to any low-molecular and high-molecular compounds. This property can be quite useful in many applications such as the supercritical extraction or the low-temperature phase transition in the complex mixtures.

References

1. Rogankov, V. B., and Boshkov L. Z. (2002) Gibbs Solution of the van der Waals–Maxwell Problem and Universality of the Liquid-gas Coexistence Curve, *Phys. Chem. Chem. Phys.* **4,** 873
2. Mazur, V. A., and Rogankov, V. B. (2003) A Novel Concept of Symmetry in the Model of Fluctuational Thermodynamics, *J. Molec Liq.* **105/2-3,** 165-177
3. Rogankov, V. B., Byutner, O. G., Bedrova, T. A., and Vasiltsova, T. V. (2006) Local Phase Diagram of Binary Mixtures in the Near-Critical Region of Solvent, *J. Molec. Liq.* **127,** 53
4. Stephenson, J. (1976) On thr Continuity of Isohore Slopes and the Divergence of the Curvature of the Vaporization Curve at the Critical Point of a Simple Fluid, *Phys. Chem. Liq.* **6,** 55-69
5. Saul, A., and Wagner, W. (1987) International Equations for the Saturation Properties of Ordinary Water Substance, *J. Phys. Chem. Ref. Data,* **16,** 893
6. Chen, Z. Y., Albright, P. C., and Sengers, J. V. (1990) Crossover from Singular to Regular Classical Thermodynamic Behavior of Fluids, *Phys. Rev.* **A41,** 3161-3177
7. Kostrowicka, A., Wyczalkowska, A. K., Sengers, J. V., and Anisimov, M. A. (2004) Critical Fluctuations and the Equation of State of van der Waals, *Physica A* **334,** 482-512

8. Speedy, R. J. (1982) Limiting Forms of the Thermodynamic Divergencies at at the Conjectured Stability Limits in Superheated and Super-cooled Water, *J. Phys. Chem.* **86,** 3002
9. Wheeler, J. C. (1977) Decorated Lattice-gas Models of Critical Phenomena in Fluids and Fluid Mixtures, *Ann. Rev. Phys. Chem.* **28,** 411-443

EXPERIMENTAL INVESTIGATIONS OF SUPERHEATED AND SUPERCOOLED WATER (REVIEW OF PAPERS OF THE SCHOOL OF THE ACADEMICIAN V. P. SKRIPOV)

VLADIMIR G. BAIDAKOV
Institute of Thermal Physics, Ural Branch of Russian Academy of Sciences, Amundsen St., 620016, Yekaterinburg, Russia

Abstract: The review presents the results of experimental investigations of nucleation in superheated light and heavy water in the range of nucleation rates from 10^4 to 10^{29} $s^{-1}m^{-3}$. A study is performed of the kinetics of crystallization of droplets of superheated water and amorphous water layers. Measurements have been made of the density, sound velocity, dielectric constant of light and heavy water in the vicinity of the phase equilibrium line with deep entry into the region of metastable (superheated) states. The local and integral characteristics of streams of boiling-up water flowing out into the atmosphere through a short channel have been investigated. One can see the determining role of a vapor phase in such a process at a temperature above $0.9T_c$, where T_c is the temper-temperature at the critical point.

Keywords: metastable state, water, heavy water, superheating, supercooling, nucleation rate, explosive boiling up, density, sound velocity

1. Introduction

Water is the most widespread liquid on our planet. Phase transitions in water are often accompanied by a considerable deviation from equilibrium conditions, with one of the phases being in the metastable state. Examples of metastable states of water are superheated and supercooled water. In nature high water superheats are observed in geysers and active volcanoes and supercoolings in atmospheric phenomena.

The development of new technologies is almost always connected with the intensification of processes. If a process is accompanied by a phase transition, an inevitable concomitant of intensification is the metastability of one or several phases. The control of such a process presupposes a sufficient knowledge of the phenomenon of phase metastability. In developing powerful steam turbines it is necessary to reckon with the initiation of a supersaturation surge in the running-water part of the low-pressure stages, and in development of pumps with the phenomenon of liquid cavitation. Steam generators of atomic power plants should have a sufficient margin of stability with respect to boiling crisis, but at the same time operate in a regime of very intense heat exchange.

A supercooled liquid poses its own problems. The interest in nucleation under considerable supercoolings has appreciably increased in connection with problems of obtaining new noncrystalline materials by the methods of rapid melt cooling. Thus, the problems of the kinetics of nucleation under considerable supercoolings border with the technological problems of production and thermal stability of new materials (amorphous films, ultradisperse systems).

The existence of metastable states is caused by the activation character of the initial stage of a first-order phase transition. Homogeneous nucleation determines the upper boundary of the liquid superheat and supercooling. The appearance of a viable new-phase nucleus in a metastable liquid is connected with the performance of the work W_* determined by the height of the thermodynamic potential barrier, which is to be overcome for the subsequent irreversible growth of a new phase. The dimensionless complex $W_* / k_B T$, where k_B is the Boltzmann constant and T is the temperature, is the stability measure of the metastable phase.[1]

In homogeneous nucleation the work W_* is performed at the cost of fluctuations. In this sense homogeneous nucleation is a fluctuation process. For a nucleus that consists of $n \sim 10^2 - 10^3$ molecules the magnitude of $W_* / k_B T$ is equal to several tens of unities, and the spontaneous process of nucleation at an appropriate supersaturation proceeds with an appreciable rate J.

The work W_* may be related to the probability of fluctuation nucleation and the nucleation rate. For the stationary nucleation rate we have[2]

$$J = \rho B \exp\left(-W_* / k_B T\right). \tag{1}$$

Here ρ is the number of molecules in a unit volume of the metastable phase, B is the kinetic factor determining the rate of the nucleus transition through the critical size.

A check of the validity of the main result of the homogeneous nucleation theory (Eq. (1)) presupposes an experimental study of the behavior of $J(T, p)$ in a wide range of temperature T and pressure p. It can be done by using different techniques. As shown below, for water a range of J from 10^4 to 10^{29} $s^{-1}m^{-3}$ can be spanned in such a way. The investigation of liquids in the metastable state is not limited by homogeneous nucleation. The allowance for metastability in engineering practice requires extending the existing tables (banks) of data on thermophysical properties of liquids to the region of metastable states. For solving this problem it was first of all necessary to ascertain the very possibility of measuring quasi-statically (irrespective of the system's history) the properties of a metastable system. On this way V. P. Skripov[3] formulated the notion of a well-defined metastability. This term presupposes a system with "unremoved" metastability, but relaxed with respect to all other motions. The condition of a well-defined metastable state is fulfillment of the requirement

$$\tau_i \approx l^3 / D_i t_x \tag{2}$$

where l is the characteristic linear dimension of the system, D_i is the kinetic kinetic coefficient for relaxation of the i-th type, t_x is the characteristic time of expectation of decay of the metastable phase.

Choosing the characteristic time of experiment $t_{\mathrm{exp}} < t_x$, one can study thermodynamic and kinetic properties of a metastable system in the "pure" state.

The results of investigating thermodynamic properties make it possible to approximate a spinodal, which is determined by the conditions

$$\left(\frac{\partial p}{\partial v}\right)_T = 0, \qquad \left(\frac{\partial T}{\partial s}\right)_p = 0. \tag{3}$$

Here v is volume, s is entropy.

The spinodal is not connected directly with nucleation and is the limit of stability of the metastable phase against infinitesimal changes in the state variables.

The fact of a very abrupt increase in the nucleation rate J under changes of temperature and pressure established in homogeneous nucleation theory and confirmed by experiment makes it possible to realize the shock regime of phase transition, when boiling or crystallization on heterogeneous centres gives a weak and blurred signal against the background of a powerful burst of evaporation or crystallization caused by homogeneous nucleation.

The conception of a system strong response to the boiling-up of a liquid in the shock regime proves to be useful in solving problems characterized by high rates of changes in the liquid state (laser heating, depressurization of a hot liquid, rapid melt cooling, etc.). As any limiting case, this approach has its field of application; in particular, it proves to be very efficient in describing high-speed flows of boiling-up liquids.[4]

The paper presents the results of experimental investigations of nucleation, thermophysical properties and processes in superheated and supercooled water. This work was initiated and performed for a number of years under the guidance of the academician V. P. Skripov at first at the Department of Molecular Physics of the Ural Polytechnical Institute, and then at the Institute of Thermal Physics of the Ural Branch of the Russian Academy of Sciences.

2. Nucleation in superheated water

Experimental investigations of the kinetics of stationary nucleation presuppose the determination of the rate J as a function of temperature and pressure. Information on the nucleation rate may be obtained from data on distribution functions and moments of appearance of the first critical nucleus. It requires repeated experiments with one sample or measurements with a system of equivalent samples.

In studying the kinetics of spontaneous boiling-up of superheated water use was made of the method of measuring the lifetime (quasi-static method) and the method of pulse superheat of a liquid on a thin platinum wire (dynamic method).

In the method of measuring the lifetime the liquid under investigation was contained in a thermostatted tube (with volumes V ~ 50 – 150 mm^3) and transferred to the metastable state by a pressure release to a given value of p.[5]

Measurements were made of temperature, pressure and the time τ of the liquid stay in the superheated state. The results of 30–100 measurements of τ were used to determine the mean lifetime $\bar{\tau}$ related to nucleation by the relation $J = (\bar{\tau}V)^{-1}$. The method of measuring the lifetime covers a range of J from 10^4 to 10^9 $s^{-1}m^{-3}$.

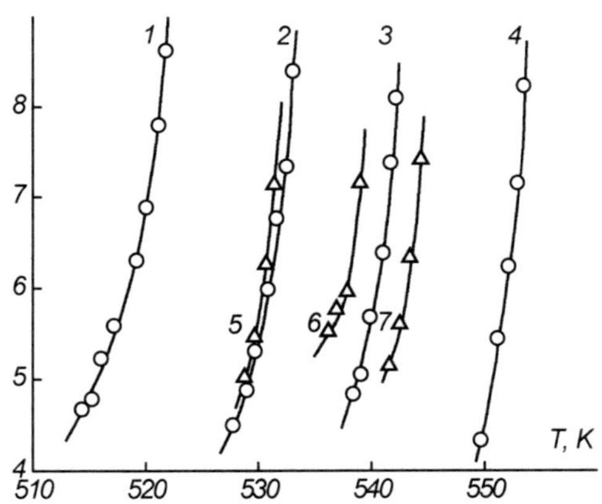

Figure 1. *Temperature dependence of the nucleation rate in superheated light (1 – $p = 0.1$ MPa, 2 – 1.0, 3 – 2.0, 4 – 3.3) and heavy (5 – $p = 0.1$ MPa, 6 – 1.1, 7 – 2.1) water.*[5, 6]

Investigation of nucleation in superheated light and heavy water in quasi-static conditions[5,16] has revealed their anomalous behavior, which is manifested in the fact that superheat temperatures achieved in experiments at different pressures have proved to be much lower than theoretical ones, and the character of the dependence $J(p,T)$ is different from other, for instance cryogenic, liquids.[7]

Light water was superheated in tubes of optical quartz or pyrex glass. Experiments were made in the pressure ranges 0.1–3.3 MPa (H_2O) and 0.1–2.1 MPa (D_2O) at nucleation rates $3\cdot10^4-5\cdot10^8$ $s^{-1}m^{-3}$. On experimental isobars (Fig. 1) there are no flattened sections and sections with curvature of different sign characteristic of other liquids.[2,7] The maximum value of an experimental temperature of superheat in light water at atmospheric pressure is 521.4 K, which is 55 K lower than the theoretical value. For heavy water these values are 531.4 K and 44 K, respectively. With increasing pressure, discrepancies between theory and experiment decrease, which is mainly connected with a decrease in the dimensions of the metastable region.

For the elucidation of the reasons for the anomalous water behavior a study was made of nucleation in superheated hydrogen-bonded liquids with different energies of hydrogen bonds and different bond characters (ammonia, Freons F-11, F-21, F-113).[8] The energy of hydrogen bonds in ammonia is comparable with the energy of hydrogen bonds in water. Freon F-21 forms considerably weaker hydrogen bonds. Freons F-11 and F-113 do not form such bonds. It has been found that in all the liquids listed the kind of kinetic curves $J = J(T)$ does not depend on the degree of associativity of a substance and is close to those observed for ordinary liquid.[2,7] The sections of the kinetic curves corresponding to spontaneous boiling-up within 0.2–1.5 K coincide with those calculated by the homogeneous nucleation theory. Thus, the ability of a substance to form hydrogen bonds is not a sufficient condition for the anomalous behavior of the stability of a superheated liquid, as it is in the case of water.

Water is very aggressive and destroys the surfaces of practically all glasses, including Pyrex and quartz. Surface defects may be variously shaped and, accordingly, may variously reduce the work of nucleus formation. In a capillary of molybdenum glass, which is the least tolerant of water, n-hexane was superheated at atmospheric pressure. A superheat temperature $T_s = 453.4$ K was obtained. The theoretical value of T_s is equal to 453.9 K. Then the kinetics of spontaneous boiling-up of superheated water was studied in this capillary. In the course of an experiment the capillary surface was destroyed. The experiment was stopped when the capillary surface became mat. The average size of defects was $5\cdot10^{-7}$ m. Nevertheless, for water the superheat temperature achieved was $T_s = 521$ K, the same as in quartz and pyrex capillaries with a smooth surface. The radius of a water critical bubble at this temperature is equal to $1.4\cdot10^{-8}$ m. At last n-hexane was superheated in the capillary again and, as before, the temperature obtained was $T_s = 453.4$ K, to which corresponds a critical-bubble radius of $5.8\cdot10^{-9}$ m. If the reason for the premature boiling-up of water were surface defects, it would be impossible to superheat n-hexane above 428 K. Therefore, it is not defects that cause the premature boiling-up of water.

To achieve high nucleation rates in superheated water, use was made of pulse methods [9]. In heating a liquid at a rate of 10^5–10^8 K/s a thin platinum wire was used as a heater and temperature-sensitive element. The wire was included in a special metering circuit and heated by a current pulse. When the liquid in the wall boundary layer was heated to a certain temperature T_s, its explosive boiling-up was observed. An electric signal of explosive boiling-up, based on the solution of the corresponding thermophysical problem, made it possible to determine the nucleation rate. The temperature of the wire heater surface was determined in synchrony with it. This method allowed one measuring the water superheat temperature in the range of nucleation rates from 10^{19}–10^{29} $s^{-1}m^{-3}$.

In experiments on pulse water superheat on a thin platinum wire it has been established that the shock boiling-up regime, when the determining contribution to evaporation is made by centres of fluctuation nature, is realized at heating rates above 10^7 K/s.[10-12] In experiments with other liquids a heating rate of 10^5 K/s will suffice to achieve the shock regim.[9]

In the case of pulse water superheat the agreement between theory and experiment improves with increasing pressure in the liquid and recorded nucleation rate (Fig. 2).

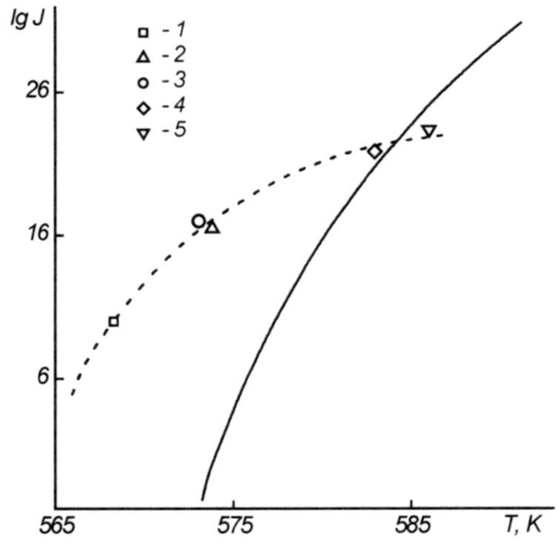

Figure 2. *Temperature dependence of the nucleation rate in superheated water at atmospheric pressure. Data of dynamic experiments: 1 – [12], 2 – [9], 3 – [9], 4 – [10], 5 – [11]. The solid line shows calculation by homogeneous nucleation theory.*

In the region of negative pressures water also behaves anomalously with respect to superheat.[13] A negative pressure in water was created when a short compression wave (□ 3 μs) was reflected from a free liquid surface. The compression wave was formed by a duralumin membrane during a discharge of a low-inductance capacitor onto a flat coil pressed to the membrane. A platinum wire heated by a current pulse was immersed in the liquid. The pressure pulse and the heating pulse were reconciled in time in such a way as to make the moment of the liquid boiling-up on the wire coincide with the passage through it of the maximum negative-pressure pulse. Nucleation rates of 10^{24}–10^{26} $s^{-1}m^{-3}$ were realized in the experiments.

Experimental investigations of limiting superheats of organic liquids in the region of positive and negative pressures have shown that the limiting superheat boundary passes continuously from the region of positive into the region of negative pressures and is close to that calculated by homogeneous nucleation theory.[14] In the case of water one can observe a change in the slope of the dependence $T_s(p)$ ($J = \mathrm{const}$) in passing from positive to negative pressures[13] (Fig. 3). It is shown that additions of ethanol to water smooth out the dependence $T_s(p)$, increasing the cavitation strength of water at high negative pressures. On dissolving acetone in water the temperature of the water superheat decreases. So does the slope of the curves $T_s(p)$ in the region of negative pressures.

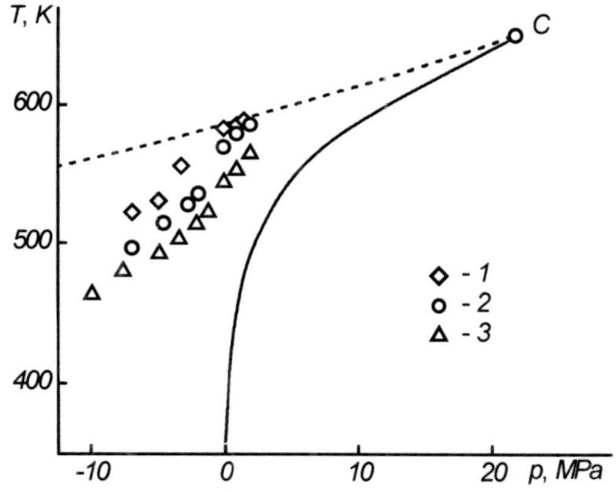

Figure 3. *Boundary of limiting superheats of water and solutions of acetone with water: 1 – water, 2 – water + 5 % acetone, 3 – water + 15 % acetone. Solid line – line of liquid–vapor phase equilibrium, C – critical point, dashed line -- calculation by homogeneous nucleation theory for* $J = 10^{24}$ $s^{-1}m^{-3}$ *(water) [15].*

For pure acetone a smooth extension of the curve $T_s(p)$ from the region of positive into the region of negative pressures is observed (Fig. 4). However, at stretches exceeding –4.0 MPa the experimental curve deviates from the theoretical line. Additions of water into acetone flatten the dependence $T_s(p)$, increasing the temperature of the acetone limiting superheat at high negative pressures. The dependence $T_s(p)$ becomes similar to the dependence $T_s(p)$ for normal liquids and agrees well with homogeneous nucleation theory. The courses of the curves $T_s(p)$ described are observed up to a volume concentration of water of about 60%, which is close to the azeotropic composition of a solution.[15]

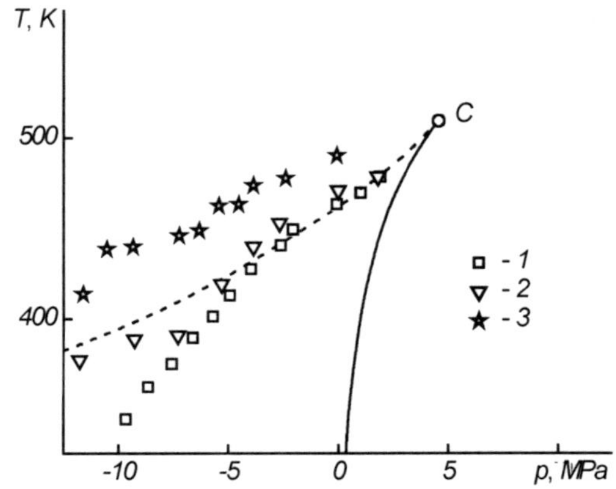

Figure 4. Boundary of limiting superheats of acetone and water solutions in acetone: 1 – acetone, 2 – acetone + 10 % water, 3 – acetone + 30 % water. Solid line – line of liquid–acetone vapor phase equilibrium, C – critical point, dashed line – calculation by homogeneous nucleation theory for $J = 10^{24}$ $s^{-1}m^{-3}$ (acetone).[15]

3. Spontaneous crystallization of supercooled water

Considerable supercoolings are realized in small liquid drops.[16] Water drops from 500 to 20 μm in diameter in oil were located on the junction of a differential thermocouple. Every drop was melted down and crystallized several tens of times. Measurements at the same temperature were made on 5-10 drops similar in size. The distribution of crystallization events of isolated drops was studied in repeated experiments under isothermal conditions and continuous supercooling.[16]

Experimental results obtained in isothermal conditions for light and heavy water are presented in Fig. 5. For light water they cover a range of nucleation rates of 5 orders in the interval of supercoolings from 33.9 to 37.8 K.[17] In this interval the nucleation rate increased 10 times when the temperature decreased by 0.8 K. The effective value of the surface tension calculated from experimental data is $\sigma_l = 28.7$ mN/m^2, and the value of the preexponential factor $\rho B = 10^{37\pm1}$ s^{-1}m^{-3} is close to the theoretical evaluation by (1) for homogeneous nucleation ($\rho B = 10^{36}$ s^{-1}m^{-3}).

Experiments in the regime of continuous cooling [17] give additional arguments in favor of the homogeneous mechanism of crystal nucleation of supercooled water. The half-width of the temperature distribution of crystallization events has proved to be equal to 0.85 ± 0.10 K, which is in good agreement with the expected value $\delta T_{1/2} = 0.8$ K for homogeneous nucleation.

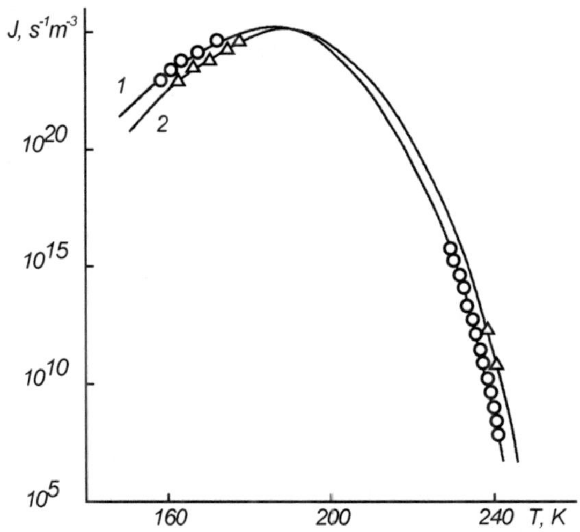

Figure 5. *Temperature dependence of the nucleation rate of crystals in light (1) and heavy (2) water. Dots on the low-temperature branch of the curve – data on crystallization of amorphous layers ,*[18,19] *on the high temperature-branch – crystallization of droplets.*[17] *Solid line – calculation by homogeneous nucleation theory.*

In conditions of high viscosity of the metastable phase the time of establishment of a stationary concentration of nuclei becomes longer. The process of non-stationary nucleation may be characterized by the stationary nucleation rate J and the lag time τ_0. Non-stationary nucleation shows up in the crystallization of amorphous layers of water. The crystallization of such layers proceeds during continuous heating or an isothermal allowance after a stepwise rise in the temperature.

Amorphous layers of light and heavy water 50-500 μm thick were obtained by condensation of vapor in vacuum on a copper substrate cooled by liquid nitrogen.[18] The condensation rate was 50-500 μm/hour. Crystallization was detected by the method of differential-thermal analysis. In experiments with amorphous layer of light and heavy water for the same heating rate the position of the abrupt temperature jump pointing to the sample crystallization was independent within 0.5 K of both the thickness of the sample and the condensation rate in the process of its preparation. In heating amorphous layers of water at a rate of 0.25 K/s crystallization took place at $T_* \sim 166$ K.[18]

Experimental data on the sample heating rate $\dot{T}$ and the crystallization temperature corresponding to it T_*, and also the fraction $X = 0.1$ of the crystallized substance referred to it, made it possible to evaluate the activation energy, E. On calculating it one can evaluate the stationary nucleation rate J, the lag time τ_0 at different crystallization temperatures of amorphous layers.[16,19]

Fig. 5 gives temperature dependences of the stationary nucleation rate for supercooled and heavy water under crystallization of amorphous layers (dots on the low-temperature branch of the dome of $J(T)$).

4. Thermophysical properties and the spinodal of superheated water

The thermodynamic interpretation of first-order phase transitions assumes that the thermodynamic potentials of each of the phases exist on either side of the phase-equilibrium line, and this line is in no way distinguished for the potentials of each of the phases. At the same time the appearance of a "growth channel" for pre-critical nuclei in the metastable phase makes the analyticity of a thermodynamic potential on the phase-equilibrium line nonobvious. The uncertainty arises because the system is essentially relaxing. Evaluations show that at $W_* / k_B T > 18$ the uncertainty is small as compared with the level of thermal fluctuations, which allows one to speak about the uniqueness of extension of the substance properties deep beyond the phase equilibrium line into the metastable region.[20]

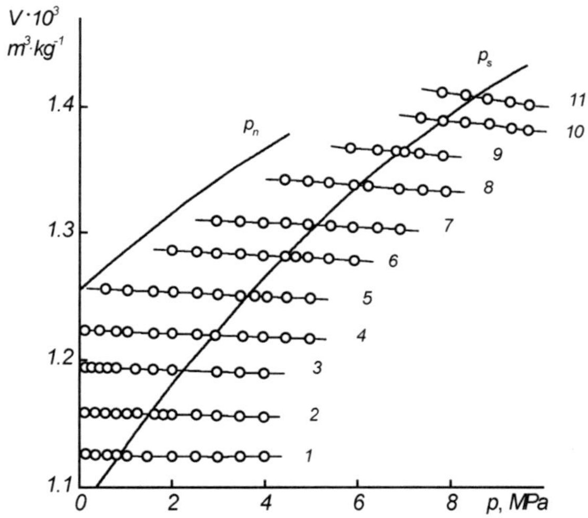

Figure 6. *Water isotherms: 1 – $T = 452.2$ K, 2 – 474.2, 3 – 493.6, 4 – 507.8, 5 – 520.8, 6 – 533.4, 7 – 542.7, 8 – 552.6, 9 – 560.6, 10 – 567.9, 11 – 572.5. p_s – saturation line, p_n – line of attainable superheats.*[21]

The method of a piezometer of variable volume in glass cells has been used to measure the density of superheated light and heavy water.[21,22] Experimental data have been obtained in the range of temperatures $(0.7-0.95)T_c$ and pressures from the saturation line to those close to the boundary of spontaneous boiling-up.

Figure. 6 presents water isotherms. From experimental data follows the smoothness of extension of isotherms, isochores, isobars from the stable into the metastable region and the absence of singularities, at least for the first two derivatives of the thermodynamic potential, on the phase-equilibrium line. As distinct from isotherms and isobars, which are essentially nonlinear, isochores are close to straight lines in the metastable region up to the critical point.

The sound velocity ($f = 1-3$ MHz) in superheated ordinary and heavy water was measured by the pulse method.[23] A liquid was superheated in a glass acoustic cell of volume □ 3 cm^3. Measurements were made along isotherms. The entry into a metastable region was realized by a pressure release.

The depth of the entry into a metastable region was limited by the action of the radiation background and easily activated boiling sites. A water superheat was accompanied by a decrease in the sound velocity (an increase in the adiabatic compressibility). Values of the sound velocity on the binodal and the line of attainable superheat ($J = 10^5$ s^{-1}m^{-3}, $T = \text{const}$) differ on average by 8-12 %.

The static dielectric constant of superheated water was measured by the relative noncontact bridge method.[24] The glass measuring cell was relieved of pressure. Measurements were made in the range from 423 to 573 K along isotherms with an interval of 10 K. Within the measurement error the static dielectric constant of superheated water remains unchanged along the isotherms.

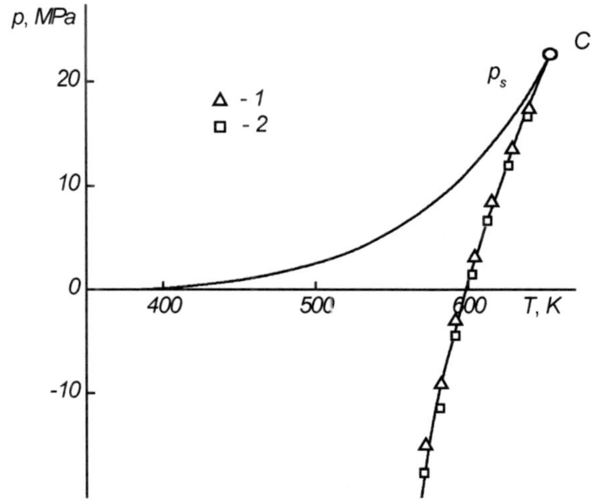

Figure 7. *Spinodal of superheated liquid water: solid line – by the empirical equation of state*[21]*, 1 – by Fürth equation*[20]*, 2 – by Gimpan equation*[20]*. p_s – saturation line, C – critical point.*

Experimental data on thermodynamic properties of water in the stable and the metastable states make it possible to approximate the spinodal. An empirical equation of state is set up with the use of p, ρ, T – data and data on the sound velocity or isochoric heat capacity. The spinodal is found by its simultaneous solution with Equation (3).The validity of the international equation of state for water has been confirmed[21, 22] (in an accessible region of metastable states) for the subregion 1 .[25] The results of several means of finding the water spinodal are compared in Fig. 7. Data obtained by the empirical and by the international equation of state are closely analogous and on the scale of Fig. 7 coincide.

5. Flows of boiling-up water

Jets of a boiling-up liquid may originate in an emergency in various thermal-power and chemical machines. The consequences of an accident with local depressurization of a high-pressure pipeline (vessel) are affected by diverse factors, for instance, the flow rate of a heat-transfer agent, the jet form and its dynamic reaction to the construction elements.

An experimental study of jets of boiling-up water has been made on a setup of short-term action, which ensures a stationary regime of an outflow from a high-pressure chamber into the atmosphere for 5–10 s.[26] The initial state (p_0, T_0) of water in the chamber varied along the saturation line from $T = 200\,^0\mathrm{C}$ to temperatures close to T_c and isobars. Considerable water superheats in a flow were ensured by the use of short channels $d/l \approx 1$ (d is the diameter of a cylindrical channel, l is its length), in which high rates of pressure decrease are realized (of the order of 10^6 MPa/s).

The main results of these investigations are reduced to the following. A thermodynamically non-equilibrium flow of a boiling-up liquid is realized in a short channel. Owing to the delay in boiling-up and the short time of the liquid stay within the channel the mass flow rate during the outflow into the atmosphere may be twice the equilibrium flow rate.[27] For a wide range of superheated states $T \leq 0.9T_c$, $0 < p < p_s$ the liquid in the channel remains practically in a one-phase state. Here for describing the outflow use may be made of the approximation of the ideal incompressible liquid, and the flow rate may be calculated by the Bernoulli formula. Homogeneous nucleation theory predicts quite low values of the nucleation rate at low and moderate superheats ($J = 1$) and an extremely high intensity and rate of increase in J for positive pressures at temperatures $T > 0.9T_c$. The high intensity, and above all, the extremely strong dependence of J on p and T lead to an abrupt increase in the local vapor content in the flow and a rapid decrease in the velocity of propagation of small perturbations upstream to values of the order of the outflow velocity. This results in a crisis of the outflow regime. It leads to a channel choking and an abrupt decrease in the liquid flow rate at $T > 0.9T_c$ (flow-rate crisis).[28] The jet shape beyond the channel changes essentially depending on the value of the liquid superheat. The bar form of the jet gives way to the conic, parabolic (with a large angle of opening α at the outlet section), gas one (at $T/T_c > 1$). Water at $T/T_c < 0.9$ is characterized by anomalously high nucleation rates as compared with most organic liquids. This affects the jet form. In experiments with water a complete jet opening (α □ 180^0) $\alpha = 180^o$ was observed even at T/T_c □ 0.75, whereas in n-pentane it happened only at $T/T_C = 0.9$.

The flow instability shows up at high liquid superheats. A jet may be "entrapped" by the wall of the channel superimposed flange and spread out in the plane perpendicular to the direction of its motion[29] (the Coanda effect[30]). The force of the jet recoil R acting on a chamber with a liquid increases with saturation pressure in the chamber, but on attaining conditions of explosive boiling-up and a jet collapse (spread along the surface of the operating chamber) the value of R decreases (Fig. 8).

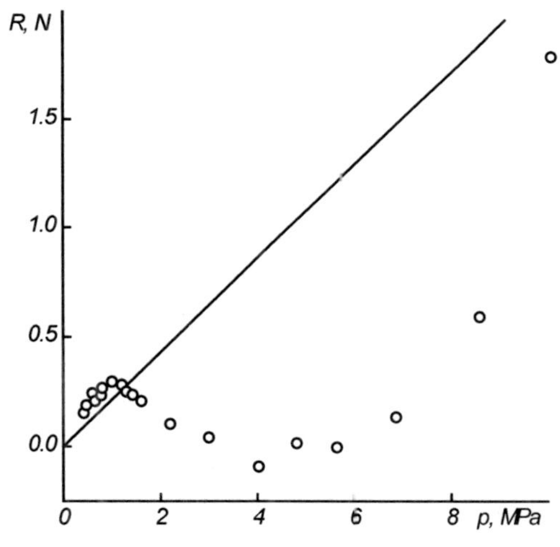

Figure 8. *Dependence of the reactive force of a superheated-water jet on the initial pressure corresponding to the saturation line.*[29] *Solid line – calculation for the hydraulic regime of outflow of a one-phase (non-boiling-up) liquid.*

In the process of observing jets of boiling-up water not only were characteristic jet shapes under certain superheats established, but considerable fluctuations of the flow parameters were noted as well. In particular, noticeable fluctuations were observed in the angle of opening of the jet cone and the local density of the outgoing two-phase medium. The method of photometry of laser radiation was used to study spectral characteristics of the fluctuation phenomena in different regimes of boiling-up of water jets. For the bar shape of a jet (with boiling-up on isolated centres in the flow) the frequency distribution of the intensity of fluctuations corresponded to white noise. For the conic jet shape (intense volume boiling-up on heterogeneous centres) in a region of low frequencies the spectral density of the power of fluctuations had a dependence inversely proportional to frequency (flicker or $1/f$ noise). When the homogeneous mechanism of evaporation was realized in a water jet ($T/T_c \geq 0.9$), the frequency interval of flicker noise widened.[29]

6. Conclusion

Water is a peculiar liquid in many respects. Metastable water is not an exception. Unlike all liquids investigated at present, water cannot be superheated in quasi-static experiments to the point of spontaneous boiling-up, in dynamic experiments the value of the water superheat decreases abruptly in passing from the region of positive into the region of negative pressures. At the same time in experiments on water supercooling in drops and warming of amorphous layers crystallization proceeds quite analogously to other molecular liquids. It may be suggested that in superheated water there are some specific centres which initiate water boiling-up, but they do not affect its stability against crystallization.

Experimental investigations of thermophysical properties of superheated water do not reveal any peculiarities in their behavior. However, it should be borne in mind that by now measurements of properties of metastable water have been made in a very narrow range of state variables. In quasi-static experiments a deep entry into the metastable region of water was hindered by a great number of easily activated boiling sites. Dynamic experiments cannot as yet ensure an acceptable accuracy of measurement of thermophysical properties of metastable liquids.

For slightly metastable states of superheated water no problems arise in describing its thermophysical properties. They differ little from properties on the saturation line. But a problem will arise at the approach of the spinodal, when isothermal compressibility, thermal expansion and isobaric heat capacity tend to infinity.

Water in the supercooled state has been studied less thoroughly than in the superheated one. Experimental data mainly refer to pressures close to atmospheric. Ice exists in different crystalline forms, and the water phase diagram has an elaborate form if one does not restrict oneself to the region of low pressures ($p < 200$ MPa). The polymorphism of ice may manifest itself at a low pressure too. It has been found that during crystallization of amorphous layers of light and heavy water there forms a mixture of hexagonal and cubic ice.

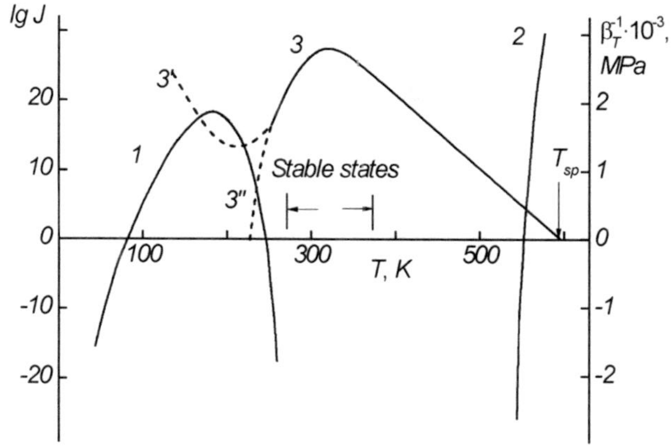

Figure 9. *Temperature ranges of states of stable, superheated and supercooled water at atmospheric pressure. Stationary homogeneous nucleation rate during crystallization (1) and boiling-up (2). Inverse isothermal compressibility for stable and metastable states of water (3) in the absence of the spinodal in a supercooled liquid (3′) and in the case of its presence according to [33] (3″), T_{sp} – the temperature of the spinodal of a superheated liquid.*

Indicated in Fig. 9 are temperature ranges of supercooled, stable and superheated water at atmospheric pressure.[31] Ibidem one can see curves representing the temperature dependence of the logarithm of the homogeneous nucleation rate for crystallization (curve 1) and boiling-up (curve 2). The maximum rate of formation of vapor nuclei is attained at the approach of the spinodal determined by condition (3). Fig. 9 also shows how the inverse isothermal compressibility $\beta_T^{-1} = -v(\partial p / \partial v)$ changes with temperature (curve 3). An arrow shows the temperature of the spinodal of superheated water.

In considering the kinetics of crystallization of supercooled water and representing the domelike curve 1 for the crystallization rate we left aside the question of the spinodal of supercooled liquid. If such a spinodal exists, it means that, at least, a part of curve 1 (on the left) does not conform to the actual possibility of nucleation in a homogeneous system. The decrease of the inverse isothermal compressibility of water with a temperature decrease below 319 K is interpreted by the authors [32,33] as a trace of thermodynamic singularity at 228 K (curve 3″). However, it does not agree with the liquid capacity for much greater supercoolings established by experiment.

There is another viewpoint on the stability of a supercooled liquid,[34] according to which the region of metastable states of a one-component liquid does not pass into a labile region with decreasing temperature. A supercooled liquid has no spinodal determined by condition (3). V. P. Skripov thought that at $T = 228$ K there was no divergence of β_T, c_p of supercooled water, but there was a sufficiently blurred and small "normal" maximum. The dashed line 3′ (Fig. 9) corresponds to this point of view.

The work has been done with a financial support of a project by the Programme of integrated investigations of the Ural and Far Eastern Branches of the Russian Academy of Sciences, grant of the President of Russia "Leading Scientific Schools" НШ – 2999.2008.8.

References

1. Gibbs, W. (1928) The Collected Works, Vol. 1. *Thermodynamics. Longmans and Green,* (*New York, London, Toronto*)
2. Skripov, V. P. (1974) Metastable Liquids. *Wiley,* (*New York*)
3. Skripov, V. P. (1989) Metastable Phases as Relaxing Systems. *In Termodinamika metastabilnykh system. Ural Branch of the USSR Academy of Sciences* (*Sverdlovsk*)
4. Skripov, V. P., Shuravenko, N. A., Isaev, O. A. (1978) Flow Choking in Short Channels in Shock Boiling-Up of Liquids. *Teplofizika vysokikh temperatur* **16**, 563-568
5. Chukanov, V. N., Evstefeev, V. N., (1976) Attainable Water Superheat. *In Atomnaya i molekularnaya fizika. UPI* (*Sverdlovsk*)
6. Skryabin, A. N., Chukanov, V. N., Shipitsyn, V. F. (1976) Experimental Investigation of Boiling-Up Kinetics of Superheated Heavy Water. *Zhurn. Fiz. Khim.* **53**, 1622-1623
7. Baidakov, V. G., (2007) Explosive Boiling of Superheated Cryogenic Liquids. *WILEY–VCH* (*Weinheim*)
8. Skryabin, A. N., Chukanov, V. N., Drokin, V. N. (1978) Kinetics of Boiling-Up of Superheated Liquid Ammonia. *Teplofizika vysokikh temperatur* **16**, 1107-1109
9. Pavlov, P. A. (1988) Dynamics of Boiling-Up of Highly Superheated Liquids. *Ural Branch of the USSR Academy of Sciences* (*Sverdlovsk*).
10. Skripov, V. P., Pavlov, P. A., Sinitsyn, E. N. (1965) Liquid Boiling-Up under Pulse Heating. 2. Experiments with Water, Alcohols, n-Hexane and Propane. *Teplofizika vysokikh temperatur* **3**, 722-726
11. Pavlov, P. A., Nikitin, E. D. (1980) Kinetics of Nucleation in Superheated Water. *Teplofizika vysokikh temperatur* **18**, 354-358
12. Smolyak, B. M., Pavlov, P. A. (1986) Investigations of Volume Water Superheat. *Teplofizika vysokikh temperatur* **24**, 396-398
13. Vinogradov, V. E., Pavlov, P. A. (2000) Boundary of Limiting Superheats of n-Heptane, Ethanol, Benzene and Toluene in a Region of Negative Pressures. *Teplofizika vysokikh temperatur* **38**, 402-406
14. Vinogradov, V. E., Pavlov, P. A. (2000) Extension of the Boundary of Limiting Superheats of Liquids into a Region of Negative Pressures. *Trudi 4 Mezhdunarodnogo Minskogo Foruma. V. 5* (*Minsk*)

15. Vinogradov, V. E., Pavlov, P. A. (2002) Limiting Superheat in a Region of Negative Pressures. *Trudi 3 Rossiyskoy Natsionalnoy Konferentsii po teploobmenu. V. 4* (*Moscow*)
16. Skripov, V. P., Koverda, V. P. (1982) Spontaneous Crystallization of Supercooled Liquids. "*Nauka*" (*Moscow*)
17. Butorin, G. T., Skripov, V. P. (1972) Crystallization of Supercooled Water. *Kristallografiya* **1**, 379-384
18. Koverda, V. P., Skripov, V. P., Bogdanov, N. M. (1973) Kinetics of Nuclei Formation in Amorphous Films of Water and Organic Liquids. *Doklady akademii nauk SSSR* **212**, 1375-1378
19. Bogdanov, N. M., Koverda, V. P., Skripov, V. P. (1980) Kinetics of Crystallization of Vitrified Layers of Heavy Water, Thiophen and Pseudocumene. *Fizika i Khimiya Stekla* **6**, 395-400
20. Skripov, V. P., Sinitsyn, E. N., Pavlov, P. A., Ermakov, G. V., Muratov, G. N., Bulanov, N. V., Baidakov, V. G. (1988) Thermophysical Properties of Liquids in the Metastable (Superheated) State. *Gordon and Breach Science Publishers* (*New York, London, Paris, Montreux, Tokyo, Melbourne*)
21. Chukanov, V. N., Skripov, V. P. (1971) Specific Volumes of Highly Superheated Water. *Teplofizika vysokikh temperatur* **2**, 739-745.
22. Evstefeev, V. N., Chukanov, V. N., Skripov, V. P., (1977) Specific Volumes of Superheated Water. *Teploenergetica* **9**, 66-67
23. Evstefeev, V. N., Skripov, V. P., Chukanov, V. N. (1979) Experimental Determination of Ultrasound Velocity in Superheated Ordinary and Heavy Water. *Teplofizika vysokikh temperatur* **17**, 299-305
24. Chukanov, V. N. (1971) Dielectric Constant of Superheated Liquid Water. *Teplofizika vysokikh temperatur* **9**, 1071-1073.
25. Vukalovich, M. P., Rivkin, S. L., Aleksandrov, A. A., (1969) Tables of Thermophysical Properties of Water and Steam. *Izdatelstvo standartov* (*Moscow*)
26. Reshetnikov, A. V., Mazheyko, N. A., Skripov, V. P. (2000) Jets of Boiling-Up Liquids. *Prikladnaya mechanika i tekhnicheskaya fizika* **41**, 125-131
27. Reshetnikov, A. V., Isaev, O. A., Skripov, V. P. (1988) Flow-Rate of a Boiling-Up Liquid Flowing out into the Atmosphere. Transition from a Model Substance to Water. *Teplofizika vysokikh temperatur* **26**, 774-777
28. Isaev, O. A., Reshetnikov, A. V., Skripov, V. P. (1988) Study of the Critical Choking of Stationary Nonequilibrium Flows of Boiling-Up Liquid. *Izvestiya AN SSSR. Energetika i transport* **6**, 11-121

29. Reshetnikov, A. V., Skripov, V. P., Koverda, V. P., Skokov, V. N. (2003) Thermodynamic Crisis in Boiling-Up Flows. Detection of Flicker Noise. *Izvestiya Akademii Nauk. Energetika* **1**, 118-125.
30. Reba, I. (1966) Applications of the Coanda effect. *Sci. Amer.* **214**, 84-92
31. Skripov, V. P. (1981) Investigation of Water in Superheated and Supercooled States. *In Teplofizicheskie issledovaniya peregretykh zhidkostey. Ural Scientific Centre of AN SSSR* (*Sverdlovsk*)
32. Rouch, J., Lai, C. C., Chen, C. H. (1977) High frequency sound velocity and sound absorption in supercooled water and thermodynamics singularity at 228 K. *J. Chem. Phys.* **66**, 5031-5034
33. Speedy, R. J., Angell, C. A. (1976) Isothermal compressibility of supercooled water and evidence for a thermodynamics singularity at –45 °C. *J. Chem. Phys.* **65**, 851-858
34. Skripov, V. P., Baidakov, V. G. (1972) Supercooled Liquid – Absence of Spinodal. *Teplofizika visokikh temperatur* **10**, 1226-1230

ESTIMATION OF THE EXPLOSIVE BOILING LIMIT OF METASTABLE LIQUIDS

ATTILA R. IMRE*, GÁBOR HÁZI
KFKI Atomic Energy Research Institute, H-1525 POB 49, Budapest, Hungary(imre@aeki.kfki.hu)
THOMAS KRASKA
Institute for Physical Chemistry, University of Cologne, Luxemburger Str. 116, D-50939, Köln, Germany

Abstract: Condensed matters (liquids, glasses and solids) can be overheated or stretched only up to a limit. Within mean-field approximation, this limit is the so-called spinodal. This is the final limit for overheating, and therefore it is a very important quantity for safety calculations wherever high pressure- high temperature liquids are involved. In temperature-pressure space the spinodal is represented by a curve, starting from the liquid-vapour critical point and decreasing with decreasing temperatures down to the negative pressure region. The determination of the spinodal is a very difficult theoretical and a more-or-less impossible experimental task. By extrapolating chosen quantities, one might get the so-called pseudo-spinodal, a limit close to the real one. Based on a recently developed method, the pseudo-spinodal pressure (for given temperature) of water and helium-3 are determined, using liquid-vapour surface tension, interface thickness and vapour pressure data. The method is already proven to be valid for Lennard-Jones argon (a simple fluid), for carbon-dioxide (a molecular fluid), for helium-4 (a quantum fluid), and the Shan-Chen fluid (a mesoscopic fluid).

Keywords: explosive boiling, overheating, stability limit, metastability, spinodal

Sudden explosive boiling (liquid-vapour phase transition) or sudden condensation (vapour-liquid phase transition) are two important phenomena, wherever high temperature pressurized fluids are involved. Having an overheated and pressurized liquid in a container or in a tube, accidental loss of pressure can initiate very fast boiling, which can cause an explosion-like process (steam explosion or Boiling Liquid Expanding Vapour Explosion). [1,2] In a similar manner, sudden cooling or pressurization of vapour (steam) can cause abrupt condensation; when it is associated with the intrusion of a cold liquid (coolant) it can cause a pressure shock. This is called water hammer in water/steam system, named after the loud metallic bang of liquid filling the space of the former steam phase and hitting the wall of the container.[3] Both processes can cause further damage in the container. Therefore safety calculations require the knowledge of the extent of these processes; mainly the pressure jump associated with them. In this paper we are going to focus only to the liquid-vapour transition (steam explosion) and only mention the inverse process wherever it is necessary.

Sudden liquid-vapour transition starts with the nucleation of the second phase (bubble), therefore the proper calculation requires some nucleation model.[4-8] Nucleation models are widely different and sometimes very inaccurate, therefore we used a different approach. As an ultimate limit of overheating or oversaturation, we used the thermodynamic stability limit,[5] the so-called spinodal. This is a limit, where the compressibility of the initial phase would turn negative, the phase would be unstable, and the system will be forced to form another phase. According to numerous studies,[4-7] the spinodal is always slightly below the homogeneous nucleation limit; the later is the real limit of the overheating of a pure liquid, where bubbles will form due to density fluctuations i.e. the density fluctuation will form minute microbubbles, which can grow to real bubble size and initiate boiling. However, there are indications that the spinodal can be handled as a limit for the homogeneous nucleation limit in case of infinite fast temperature or pressure jumps.[9] We have to distinguish between two kinds of nucleation limits namely the homogeneous and the heterogeneous one. In homogeneous nucleation, the nucleus is generated by the density fluctuation within the pure liquid (as mentioned before), while in heterogeneous nucleation, the nuclei are pre-existing in the form of wall or contamination. In Figure 1 one can see the schematic representation of these stability limits in a liquid. K represents the initial condition. For the sake of simplicity it can be chosen as room temperature and atmospheric pressure. One can initiate phase transition in two different ways, by heating or by de-pressurizing. In case of water, the following sequence can be seen during heating. Reaching point L (100 Celsius, 1 atm) one can reach the saturation curve. From that point on the liquid water is metastable but can remain in liquid form. Reaching point M (heterogeneous nucleation limit) the liquid must boil; the boiling will start on some pre-existing bubbles, formerly hidden in the crevices of the wall or attached to the surface of some floating solid contamination. For water at 1 atm, the heterogeneous nucleation limit can be anywhere between 100 Celsius and approx. 300 Celsius.[5,6] Using very pure liquid, one can reach a higher temperature limit. Recently the experimental limit around 300 Celsius has been reached, which is close to the homogeneous nucleation limit[10] also called improperly as kinetic spinodal. Finally, slightly over the homogeneous nucleation limit, but still below the critical temperature (approx. 374 Celsius for water) one would see the spinodal (O). In similar manner, these limits can be reached by decreasing the pressure. For water at room temperature the saturation curve will be reached at 0.025 bar. By

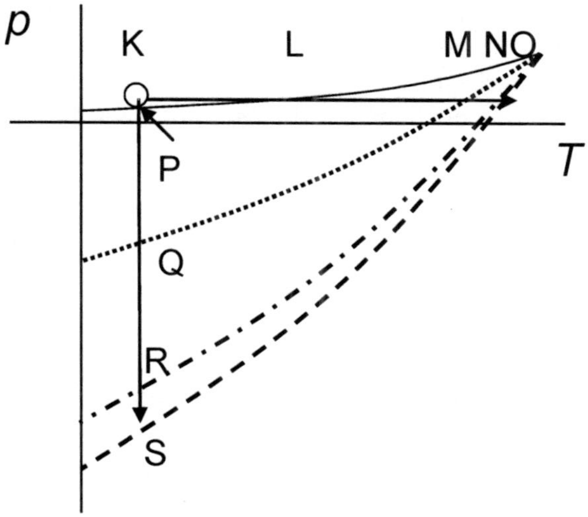

Figure 1. *Schematic diagram of the various stability limits in pT-space, concerning liquid-vapour phase transition. K represents an initial stable point, while L, M, N, O and P, Q, R, S represent the crossing of the vapour pressure curve, the heterogeneous nucleation limit, the homogeneous nucleation limit and the spinodal by heating (LMNO sequence) and by depressurizing (PQRS sequence), respectively. Solid line: saturation curve, dotted line: heterogeneous nucleation limit, dot-dashed line: homogeneous nucleation limit, dashed line: spino-dal. Further details are in the text.*

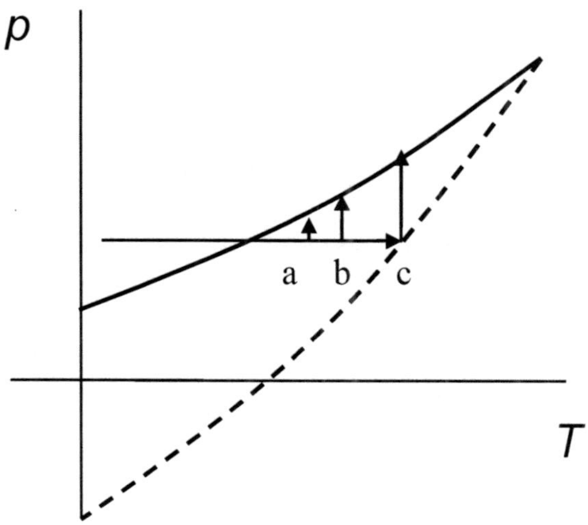

Figure 2. *Schematic diagram showing the extent of the isotherm pressure jump following the vaporisation of a metastable (overheated) liquid with different levels of metastability (a<b<c). The solid curve is the vapour pressure curve, the dashed line is the spinodal, the vertical arrow represents the overheating, and the upright ones represent the nearly isotherm pressure jumps. Further details are in the text.*

stretching the water, one can obtain negative pressure,[5-7,11,12] The experimental limit for water around room temperature is around -1400 bar,[6] this is close to the homogeneous nucleation limit (R); the heterogeneous one (Q) has to be between 0.025 bar and -100 bar, depending on the purity of the sample as well as on the rate of the overheating or the depressurization. Finally somewhere between -2000 and -4000 bar one will see the spinodal (S). The exact location of the spinodal of water is still debated.

Metastable liquid will relax by fast vapourization. The process should be adiabatic, but often can be approximated as an isothermal process.[13,14] After the vapourization, the pressure will jump up to the corresponding vapour pressure. Actually that pressure can be overshoot but here we neglect this effect. In Figure 2 one can see that different levels of overheating (i.e. different levels of metastability) are associated with different amplitude of pressure jump. Point "a" represents small overheating which will relax with a small pressure jump while point "b" marks medium overheating relaxing with a bigger pressure jump. Finally point "c" represents the maximal overheating in the immediate vicinity of the spinodal followed by the highest possible pressure jump. The highest pressure jump can be expected when the vaporisation happens at the immediate vicinity of the spinodal. Therefore knowing the spinodal of a certain liquid, one can estimate as a worst-case scenario resulting in the maximal initial pressure jump without overshot during an accidental overheating or depressurization.

The only problem is that spinodals are not known. Experimental determination is not possible, [5,6] becaus e heterogeneous nucleation always interfere. However one can use the experimental data to extrapolate the spinodal as the vaporisation limit of a very pure liquid with infinite overheating or depressurization level. The related experiments are very difficult; up to know we have only a few data for water and a very few for other liquids. Equation of states (EoS) can describe the behaviour of stable liquids and vapours but may not necessarily reliable for their metastable counterparts.[15] Obviously one can estimate the spinodal from an EoS by calculating the pressures and temperatures where the compressibility turns negative, but depending on the equation of state the results may even be qualitatively incorrect, for example if there are discontinuities in the two phase region. Therefore every method which can give us a fair estimate about the location of the spinodal would be important.

We developed a method when the spinodal can be estimated using the vapour pressure and an interfacial property, namely the maximum of the tangential pressure across the interface:[16]

$$p_{sp} = p_N - c(p_N - p_{T,\min}) \tag{1}$$

where p_{sp}, p_N and $p_{T,min}$ are the spinodal pressure, the normal component of the interfacial pressure (equal to the vapour pressure) and the minimum (i.e. the most negative value) of the tangential element of the pressure across the interface. The factor c=3/2 turns the two-dimensional surface pressure into three-dimensional bulk.

Unfortunately the pressure change across the interface cannot be measured, but we can measure the integral over that pressure change through the interface, which is the surface tension:

$$\gamma = \int_{-\infty}^{+\infty} (p_\mathrm{N} - p_\mathrm{T})dz \tag{2}$$

where z is the direction perpendicular to the interface. In applications of Eq. 2, for example in molecular simulation, it is sufficient to integrate within the interface, i.e. between the z-values where the bulk liquid density and the bulk vapour densities are reached. The surface tension is a measurable quantity and knowing it together with the interfacial thickness, which is also measurable (although with more effort) plus having some approximation about the shape of the $p(z)$ function, one can estimate and $p_{T,min}$. One can derive a new equation from Eq. 1:

$$p_{sp}(T) = p_{vap}(T) - s \cdot \frac{3}{2}\frac{\gamma(T)}{d(T)} \tag{3}$$

where p_{sp}, p_{vap}, γ and d are the spinodal pressure, vapour pressure, liquid-vapour surface tension and the so-called "10-90" liquid-vapour interface thickness at the T temperature, respectively. The factor s is a constant which represents the shape of the tangential pressure profile across the interface; instead of using it as an adjustable parameter, we fixed it as 2 (meaning triangular) or 1 (meaning rectangular) pressure profile.

This method has been tested for several model systems like the Lennard-Jones argon and carbon-dioxide by molecular dynamics simulation, the Shan-Chen fluid by lattice Boltzmann simulation and finally for the experimental system helium-4. [16-18] The method worked well for these systems, therefore here we are trying to apply it here to two other systems, namely for the water in the vicinity of room temperature and for the helium-3 in the whole liquid range.

In Figure 3, the spinodal of helium-3 is shown, calculated with Eq. 3, using triangular pressure profile (dark grey) and rectangular pressure profile (light grey) and compared to the spinodal established by various theoretical methods (see refs in [[6]]). The thickness of the two grey bands represents the error of our method, combined with the scatter of the experimental interface thickness and interfacial tension data used in the calculation. The vapour pressure was taken from,[19]t he interfacial tension from,[20] and the interface thickness from [[21]]. It can be seen that while at low temperature, triangular approximation seems to be better, at high temperature the rectangular gives closer result.

In Figure 4, the spinodal of water is shown around room temperature. Interface thickness and interface tension are taken from the literature,[22,23] and the vapour pressure was calculated by using the IAPWS EoS of Wagner and Pruss.[24] The spinodal represented by the thin dashed line are taken from[8] and based on a molecular dynamic simulation using the TIP5P water potential. The agreement between our results and the TIP5P spinodal is satisfactory, showing

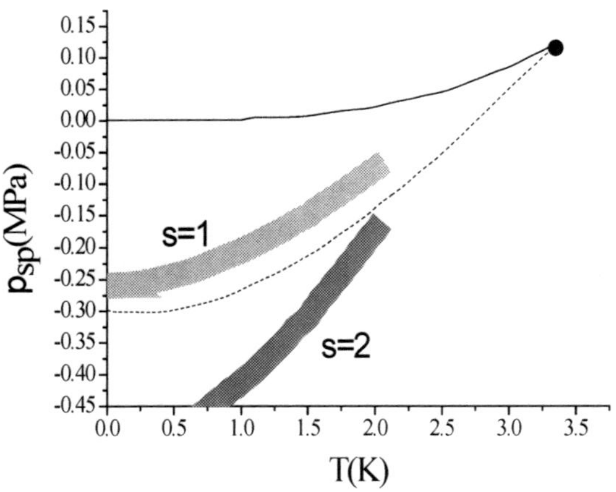

Figure 3. *Spinodal calculated by Eq. 3, using different idealised tangential pressure profiles across the interface, compared to the spinodal given in the literature for helium-3. The solid curve is the vapour pressure curve, the dashed curve is the theoretical spinodal curve, the light grey band represents our results for the triangular pressure profile, the dark grey band represents our results for the rectangular pressure profile.*

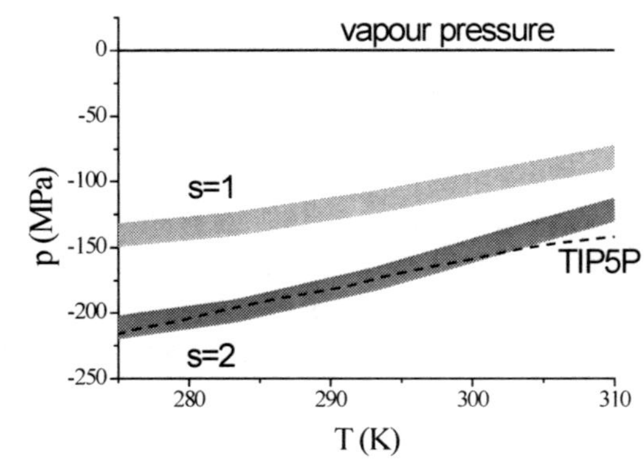

Figure 4. *Spinodal calculated by Eq. 3, using different idealised tangential pressure profiles across the interface, compared to the spinodal given in the literature for water. The solid curve is the vapour pressure curve, the dashed curve is the theoretical spinodal curve, the light grey band represents our results for the triangular pressure profile, the dark grey band represents our results for the rectangular pressure profile.*

the consistency of the method. Lacking high-temperature interfacial data, we were not able to calculate the high-temperature part of the spinodal.

As a conclusion, we can say that the method developed by us is a promising tool to estimate the liquid-vapour stability limit of various liquids. The required data are measurable under stable liquid condition; in this way this is the only method which does not require any measurement under metastable conditions to estimate the spinodal. The most important advantage of this method, that it can be used for any liquid. It is even applicable for liquid mixtures however for the mechanical stability and not for the diffusion stability.

Further studies concerning the stability limit of water at higher pressure and temperature (where this method is expected to be less accurate) with more realistic pressure-profile are in progress. Those data would be particularly important in the safety analysis of power plants (including nuclear ones) where during a so-called LOCA (Loss of Coolant Accident) part of the cooling liquid can reach some degree of metastability due to sudden pressure loss (see e.g. [[25]]).

Acknowledgements

A. R. Imre was supported by Hungarian Research Fund (OTKA) under contract No. K67930, the German Humboldt Foundation, and by the Bolyai Research Grant of the Hungarian Academy of Science. T. Kraska acknowledges support of the DFG by grant Kr1598/24-1.

References

1. Pinhasi, G. A., and Ullman, A., Dayan, A. (2005) Modeling of flashing two-phase flow, *Rev. Chem. Eng.* **21**, 133-264
2. Abbasi, T., and Abbasi, S.A. (2007) Accidental risk of superheated liquids and a framework for predicting the superheat limit, *J. Loss Prevent. Process Ind.* **20**, 165-181
3. Tiselj, I., and Gale, J. (2008) Integration of unsteady friction models in pipe flow simulations, *J. Hydraulic Res.* **46**, 526-535.
4. Baidakov, V.G. (1994) Thermophysical properties of superheated liquids, *Sov. Tech. Rev. B. Therm. Phys.* **5**, 1-88
5. Debenedetti, P.G. (1996) *Metastable Liquids: Concepts and Principles*, Princeton University Press, Princeton, NJ.
6. Imre, A.R., Maris, H.J., and Williams P.R. (Eds.) (2002) *Liquids Under Negative Pressure* (NATO Science Series), Kluwer, Dordrecht
7. Skripov, V.P., and Faizullin, M.Z. (2006) *Crystal-Liquid-Gas Phase Transitions and Thermodynamic Similarity*, Wiley-VCH
8. Herbert, E., Balibar, S., and Caupin, F. (2006) Cavitation pressure in water, *Phys. Rev. E*, **74**, 041603
9. Šponer, J. (1990) The Dependence of Cavitation Threshold on Ultrasonic Frequency. *Czech. J. Phys. B* **40**, 1123-1132
10. Kiselev, S.B., and Ely, J.F. (2001) Curvature effect on the physical boundary of metastable states in liquids, *Physica A* **299**, 357-370
11. Skripov, V.P. (1974) *Metastable Liquids*, Wiley, New York
12. Trevena, D.H. (1987) *Cavitation and Tension in Liquids*, Adam Hilger, Bristol
13. Imre, A., and Van Hook, W.A. (1998) Liquid-liquid equilibria in polymer solutions at negative pressure, *Chem. Soc. Rev.* **27**, 117-123
14. Imre, A., Martinás, K., and Rebelo, L.P.N. (1998) Thermodynamics of Negative Pressures in Liquid, *J. Non-Equilib. Thermodyn.* **23**, 351-375
15. Kraska, T. (2004) Stability limits of pure substances: An investigation based on equations of state, *Ind.&Eng. Chem. Res.* **43**, 6213-6221
16. Imre, A.R., Mayer, G., Házi, G., Rozas, R., and Kraska, T. (2008) Estimation of the liquid-vapor spinodal from interfacial properties obtained from molecular dynamics and lattice Boltzmann simulations, *J. Chem. Phys.* **128**, 114708

17. Imre, A.R., and Kraska, T. (2008) Liquid-vapour spinodal of pure helium-4, *Physica B* **403**, 3663-3666
18. Römer, F., Imre, A.R., and Kraska, T. (2009) The relation of interface properties and bulk phase stability: MD simulations of carbon dioxide, *J. Chem. Phys.*, submitted
19. Huang, Y.H., and Chen G.B. (2006) A practical vapor pressure equation for helium-3 from 0.01 K to the critical point, *Cryogenics* **46**, 833-839.
20. Dyugaev, A.M., and Grigoriev, P.D. (2003) Surface tension of pure liquid helium isotopes, *JETP Letters* **48**, 466-470
21. Barranco, M., Pi, M., Polls, A., and Vinas, X. (1990) The surface tension of liquid He-3 above 200 mK - A density-functional approach, *J. Low Temp. Phys.* **80**, 77-88
22. Rivkin, S.L., and Aleksandrov, A.A. (1980) *Thermal properties of water and steam*, Energia, Moskva
23. Caupin, F. (2005) Liquid-vapor interface, cavitation, and the phase diagram of water, *Phys. Rev. E* **71**, 051605
24. Wagner, W., and Pruss A. (2002) The IAPWS Formulation 1995 for the Thermodynamic Properties of Ordinary Water Substance for General and Scientific Use, *J. Phys. Chem. Ref. Data* **31**, 2, 387-535
25. Todreas, N.E., and Kazimi M.S. (1990) *Nuclear Systems I: Thermal Hydraulic Fundamentals*, Hemisphere, New York

LIFETIME OF SUPERHEATED WATER IN A MICROMETRIC SYNTHETIC FLUID INCLUSION

MOUNA EL MEKKI[1], CLAIRE RAMBOZ[1], LAURENT PERDEREAU[1], KIRILL SHMULOVICH[2], LIONEL MERCURY[1]

[1]*Université d'Orléans-Tours, UMR 6113 CNRS-INSU et Institut des Sciences de la Terre d'Orléans, 1A rue de la Férollerie, 45071 Orléans, France.*
[2]*Institute of Experimental Mineralogy, Russian Academy of Science, 142432 Chernogolovka, Russia*

Abstract: A synthetic pure water fluid inclusion showing a wide temperature range of metastability (T_h - $T_n \approx 50$°C; temperature of homogenization T_h = 144°C and nucleation temperature of T_n = 89°C) was selected to make a kinetic study of the lifetime of an isolated microvolume of superheated water. The occluded liquid was placed in the metastable field by isochoric cooling and the duration of the metastable state was measured repetitively for 7 fixed temperatures above T_n. Statistically, metastability lifetimes for the 7 data sets follow the exponential reliability distribution, i.e., the probability of non nucleation within time *t* equals $e^{-\lambda t}$. This enabled us to calculate the half-life periods of metastability τ for each of the selected temperature, and then to predict τ at any temperature T > T_n for the considered inclusion, according to the equation $\tau(s) = 22.1 \times e^{1.046\times\Delta T}$, ($\Delta$T = T - T_n). Hence we conclude that liquid water in water-filled reservoirs with an average pore size$\approx 10^4$ μm^3 can remain superheated over geological timelengths (10^{13}s), when placed in the metastable field at 24°C above the average nucleation temperature, which often corresponds to high liquid tensions (≈ -50 MPa).

Keywords: microthermometry, experimental kinetics, microvolumes, pure water, natural systems

1. Introduction

Any liquid can exist in three thermodynamic states with regard to the phase diagram: stable, metastable, and unstable. When it is metastable with respect to its vapour, the so-called superheated liquid persists over the more stable vapour owing to the nucleation barrier related to the cost to create the liquid-vapour interface. Practically speaking, a superheated liquid undergoes any P-T conditions located between the saturation and the spinodal curves (Fig. 1). It should be noted that the term "superheating" does not refer to a particular range of temperature, and goes down to temperatures in the melting area. This superheat metastability gives to the liquid a certain "overstability feature" with respect to vapour. Indeed, geologists have long observed that liquid water displays such overstability in certain low and high temperature contexts. For instance, the soil capillary water[1] or the water state in very arid environments like the Mars surface[2,3] are natural examples of low T superheated liquid states whereas certain continental and submarine geysers[4] or the deep crustal rocks[5] can also generate high T superheated solutions.

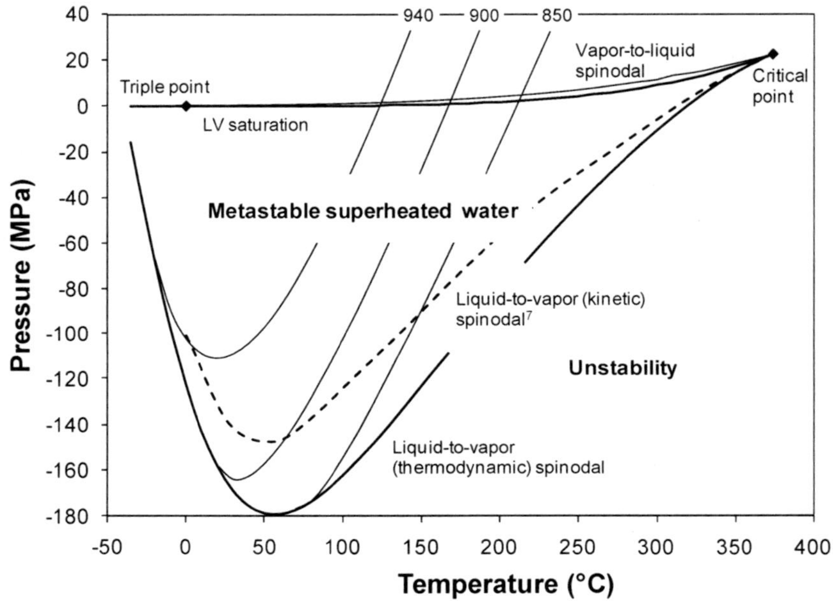

Figure 1. *Pure water phase diagram in (P,T) coordinates calculated from the IAPWS-95 equation of state*[6]*, extrapolated at negative liquid pressures in the superheat domain. The outer lines starting from the critical point are the thermodynamic limits of metastability (spinodal). The dotted line is one of the proposed kinetic metastability limit*[7] *(see text). Three isochoric lines (950, 900 and 850 kg m*$^{-3}$*) are also calculated by extrapolation of the IAPWS-95 equation.*

The shape of the pure water spinodal has long been a matter of debate in the physics community. To date, three competing scenarios are proposed, one with the retracing shape towards positive pressure at low temperatures, the two others with a monotonous decreasing shape. The first model relates to the stability-limit conjecture[8] based on experimental data on supercooled water. It can be demonstrated[9] that the retracing shape corresponds to the intersection of the spinodal curve with the Temperature of Maximal Density (TMD) line. The second proposed scenario derives from molecular simulation[10] and predicts a constant positive slope, hence a spinodal monotonously extending towards negative pressures. That thermodynamically implies that this is the TMD the slope of which changes from negative to positive slope. The experimental data previously mentioned are here interpreted as related to the presence of a second critical point at low temperature and positive pressure[10,11]. The third proposition is the singularity-free hypothesis[12], associating a non-retracing spinodal with thermodynamic divergences. In this model, all the polyamorphic transitions of liquid water at low temperature are just relaxation phenomena and not real phase transitions, referring explicitly in that respect to the percolation model[13].

It is worth noting that most equations of state (EOS) like the van der Waals equation or the IAPWS-95 EOS result in the retracing behaviour when extrapolated in the metastable field (Fig. 1). Obviously, this is not at all an argument in support of the latter conception. However, we want to highlight that our experimental investigations are not concerned for the time being with the low temperature superheating region, so that our calculations are not influenced by this debate. The problem of the extrapolative capability of the IAPWS-95 equation remains but seems to be satisfying enough as already discussed elsewhere[14-16].

2. Scientific context

The chosen experimental technique to investigate the metastability liquid field is to submit to heating-cooling cycles micrometric volumes of fluid trapped in a solid fragment.[14] This method allows measurement of the nucleation temperature (T_n) of every fluid inclusion (IF) in the host crystal, which is an evaluation of the extreme metastability that the given intra-crystalline liquid can undergo before breaking instantaneously (over the experimental time, this occurs within some tenth of seconds). By this method, the extreme tensile strength of one specific liquid (pure water, aqueous solutions...) can be directly recorded. Previous measurements provided evidences that water and in general aqueous solutions can reach very high degrees of metastability (tension reaching hundreds of MPa).[14,17-20] As a matter of fact, every inclusion (geologist dealing with fluid inclusions) is faced to the superheating ability of these micrometric fluid systems, which appears in general as a nuisance as it prevents further measurements of phase equilibria. On the other hand, a superheated liquid experiencing properties in the stretched state (negative pressure domain) displays specific thermodynamic properties. Hence distinctive solvent properties can be expected as compared to bulk water. As an example, in the Red Sea, the abnormal thermal balance of the Atlantis II Deep lower brine[4] has been accounted for by an influx of hypercaloric superheated brines.[4,21]

As noted above, a metastable liquid can be superheated to a very high degree before nucleating a vapour instantaneously. In kinetic terms, this means that the nucleation energetic barrier is of the same order of magnitude as the thermal energy kT of the system. This spontaneous nucleation limit is sometimes called the kinetic spinodal[7] (Fig. 1). When one goes towards PT pairs greater than those of the kinetic spinodal, the nucleation barrier becomes greater than the thermal energy and the lifetime becomes longer. Generally speaking, the Classical Nucleation Theory (CNT) tells us that the nucleation rate of vapour bubble follows an Arrhenius law:[22]

$$J = J_0 \times e^{-\frac{E_0}{kT}} \tag{1}$$

where J is the number of nuclei per unit volume per unit time, J_0 is a kinetic pre-factor, E_b is the nucleation energy barrier related to the energy of interface creation.

The nucleation barrier is easy to formulate considering that the created gas-liquid interface related to bubble nucleation increases the system energy by $2\pi r^2\gamma$ (r is the radius of the spherical bubble, and γ is the liquid-vapour surface tension), while the formation of the most stable phase provides bulk energy ($4/3\pi r^3\Delta P$, with $\Delta P = P_{LIQUID} - P_{VAPOUR}$). According to the CNT, the competition between these two opposite effects results in an energy barrier E_b:[22]

$$E_b = \frac{16\pi}{3\gamma^3\Delta P^2} \tag{2}$$

E_b is reached when the spherical bubble reaches a critical radius $r_C = \frac{2\gamma}{\Delta P}$.

According to (1), P, the probability of nucleation in a volume V and during a given time length t is:

$$P = 1 - e^{-JVt} \tag{3}$$

Nucleation is assumed to occur when P = 0.5. Then, the half-life period τ (median value of duration) required to get at least one vapour nucleus is:

$$\tau = \frac{Ln(2)}{JV} \tag{4}$$

Thus, the time necessary to create a bubble nucleus is inversely proportional to the fluid volume. In other words, if one second is sufficient to create one nucleus in a volume of 1 litre, more than 11 days are required in a volume of 1 mm^3. Statistically, almost 32 000 years are necessary to do so in a volume of 1000 μm^3, a typical size for interstitial pores in most natural porous media. This calculation is made at constant J, namely for a given value of E_b, i.e., at constant metastable "intensity" and heterogeneous nucleation conditions (due to impurities in the liquid or to solid surface singularities). Thus, at constant physico-chemical conditions, nucleation is an event that becomes rarer as the fluid container is smaller. This is why soil scientists recently proposed that capillarity can occur without meeting the Young-Laplace condition, and so capillary water and its special superheating properties can occur in the whole range of microporosity, thus affecting rather large amounts of soil water.[1]

The consequences of superheating in natural systems are twice. In contexts of high thermal anomalies (geyser, phreato-magmatism), the changes in the fluid properties at the nucleation event (relaxation of the superheating features) may have drastic physical consequences: rapid volume changes at the phase transition, potentially accompanied by deep changes in fluid speciation. This implies explosivity and massive solid precipitation (or massive dissolution depending on the role of superheating on the rock-solution equilibria). In dry environments, superheating is a long-lived process controlled by the aridity of the atmosphere (soil capillarity). The rock-water-gas interactions thus involve a superheated liquid component which modifies the chemical features of the resident solutions[1]. Whatever the superheating context, once a natural system becomes metastable, its effective influence will depend on the duration of the metastable state, as both chemical and heat exchanges also require long time to be fully achieved.

3. Sample

The fluid inclusion chosen in this study was selected among the ones previously synthetized to investigate the extreme tensile strength of liquids, depending on composition and density.[14] Our experimental procedure is similar to that adopted in a previous paper.[23] The selected inclusion was placed in a superheated state at a given T > Tn by isochoric cooling, then we waited for the bubble to nucleate. Different intensities of liquid stretching were tested, as the same experiment was performed at 7 different temperatures above Tn. Note that Tn fixes the maximum stretching intensity sustainable by the selected inclusion. According to the CNT and as confirmed by previous experiments,[23] the distribution of metastability lifetimes is expected to display an exponential decrease.

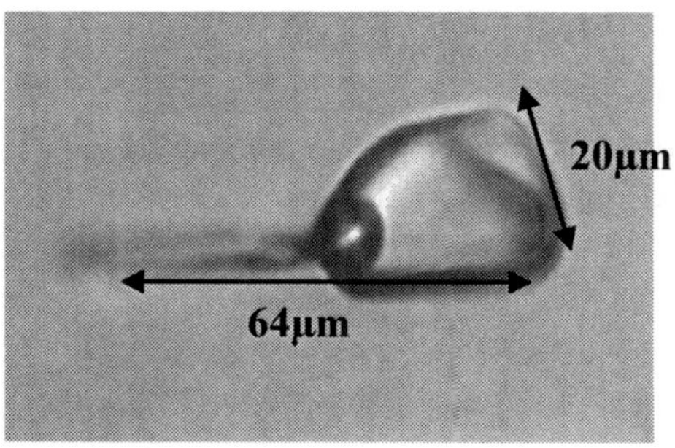

Figure 2. *Microphotograph of the studied pure water synthetic fluid inclusion (x50).*

The chosen fluid inclusion (no 31-7) is located in a 450 μm-thick quartz fragment and is located 77 μm-deep below the crystal surface. It is quite big, 64-μm long and 20 μm-wide (estimated volume ≈ 8600 μm^3) with a long appendix indicating a process of necking down (Fig. 2).

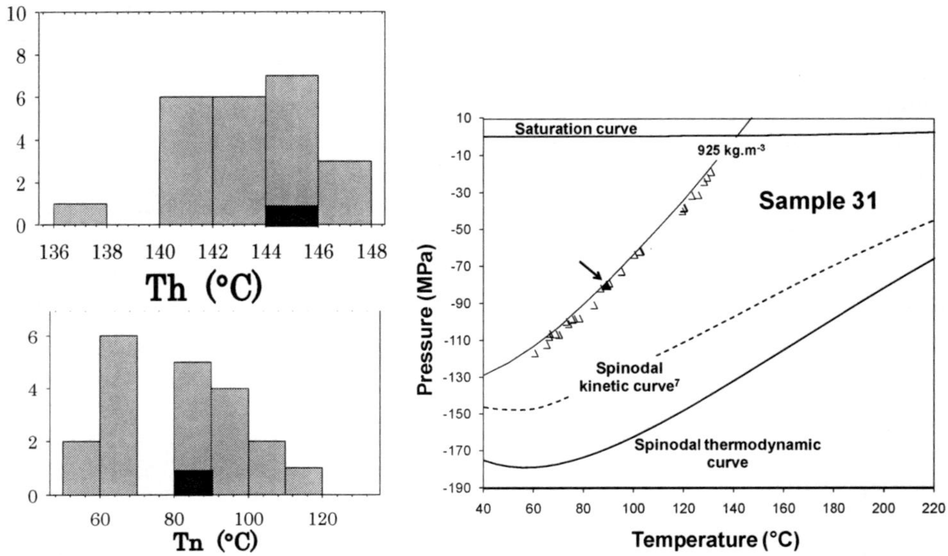

Figure 3. *Properties of sample 31 pure water fluid inclusions. Figs 3a and 3b: Distribution of Th and Tn measurements. Fig. 3c: PT conditions of nucleation of sample 31 inclusions (triangles), with the average corresponding isochore (d = 925 kg m*$^{-3}$*) extrapolated in the metastable field. The PT pure water phase diagram, the isochore and data points are calculated after IAPWS-95 EOS*[6]*. The dark areas and arrow indicate the position of the studied no 31-7 inclusion (see text).*

This inclusion belongs to quartz sample 31, which contains pure water synthetic inclusions with an average density of 925 kgm^{-3}. Note that as quartz is incompressible below 300°C, the PT path followed by the inclusion fluid at changing T is isochoric (constant volume, constant density). The average isochoric PT path of sample 31 inclusions is shown in Figure 3, together with their representative points at T_n. The internal pressure at T_n is calculated from the density-T_n measurements on extrapolating the IAPWS-95 "official" pure water EOS[6]. This equation also allows to derive the thermodynamic fluid properties of pure water at given P-T pairs (see reference 14 for more details).

4. Experimental procedure

Sample 31 quartz fragment was placed on a Linkam heating-cooling stage mounted on a Olympus BHS microscope. Its temperature was allowed to vary (Fig. 4). Phase changes in the inclusion were observed with a x50 LWD objective and were recorded using a Marlin black and white camera (CMOS 2/3" sensor, ≈ 15 pictures/s).

Microthermometry. The key characteristics of the studied no 31-7 fluid inclusion are the homogenization and the nucleation temperatures (T_h and T_n, respectively). T_h is the disappearance temperature of the last drop of vapour in the cavity (at T_h, the saturation conditions are met). T_n ($T_n < T_h$) is the measured temperature when the trapped metastable liquid becomes diphasic (it marks the end of the stretched metastable state). T_h and T_n were measured, in that order, in the course of strictly temperature-controlled heating and cooling cycles (Th: path 1 to 4; T_n: path 4 to 6, Fig. 4).

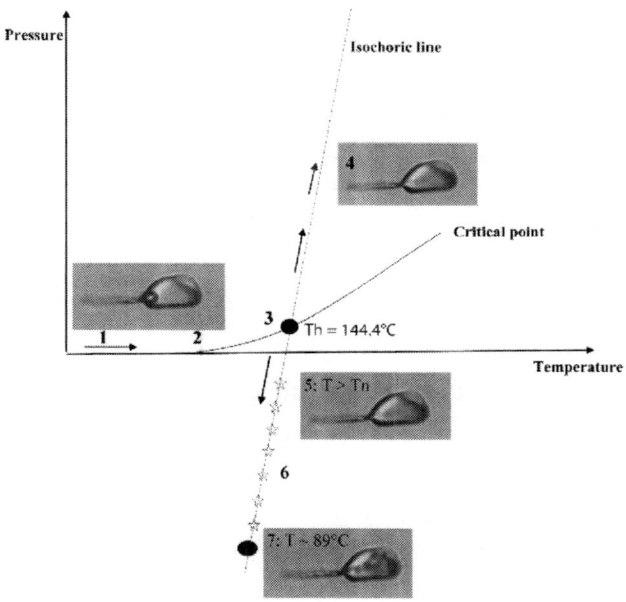

Figure 4. *PT pathways followed by a fluid inclusion heated from ambient conditions (path 1 to 4) then further cooled. Photomicrographs show the successive occluded fluid states observed. The bold curve is the saturation curve and the stars qualitatively represent the seven temperature steps chosen for the kinetic study.*

	Heating cycle (Th measurement)				Cooling cycle (Tn measurement)	
T range (°C)	25-130	130-140	140-150	150-160	160-105	105-80
			Th=144.4			Tn=89
Heating/cooling rate (°C/mn)	30	10	2	10	30	2

Table 1 *Rate-controlled sequences of heating and cooling chosen for T_h and T_n measurements. First, heating along the liquid-vapour curve (diphasic inclusion), then isochoric heating followed by isochoric cooling down (single-phase inclusion).*

Table 1 summarizes the rates of temperature change that were adopted all along the kinetic study. A cooling rate of 2°C/min was chosen to measure T_n as it is offers the best conditions to observe bubble nucleation.

Kinetic measurements consisted in placing inclusion no 31-7 in the metastable field. The procedure was the same as for T_n measurements except that, during cooling, the inclusion was stabilized at 1°, 1.3°, 2°, 3°, 3.5°, 4° and 5°C above T_n, successively (stars Fig. 4). The inclusion was thus kept metastable at 7 fixed temperatures between 90.4° and 94.4°C. For each given temperature, the duration of metastability was measured repetitively (between 5 and 16 metastability lifetime measurements). The beginning of the temperature step was taken as the starting point of the experiment (time 0). Between each set of kinetic measurements at a fixed temperature, we checked that T_h and T_n had not changed significantly.

5. Results

Microthermometry. At the start of the study, T_h and T_n measurements of inclusion no 31-7 were repeated 11 times, following the T procedure summarized in Table 1. Measured T_h and T_n were 144.4° and 88.8°C respectively, with a repeatability of ±0.2°C for T_h and of ± 2.3°C for T_n. Thus the measured range of metastability for inclusion no 31-7 was 54.6° ± 3.3°C (Table 2), corresponding to internal P conditions of – 84 ± 4 MPa.

T(°C)	1	2	3	4	5	6	7	8	9	10	11
Th	144.4	144.4	144.4	144.2	144.2	144.2	145.1	145.1	144.4	144.4	144.4
Tn	89.8	89.8	86.5	86.5	87.0	87.5	87.5	89.8	91.2	87.2	87.2
Th-Tn	54.6	54.6	57.9	57.9	57.24	56.7	56.6	55.3	53.2	57.2	57.2

Table 2 *Repetitive cycles of T_h and T_n measurements.*

We observed more than 50 vapour nucleation events in the IF, which enabled us to identify the main stages of cavitation. The two-phase stable situation was recovered within about 1/3s (5 to 6 images with our camera). In general, nucleation started in the broadest part of the inclusion by a foam, a milky cloud a little more contrasted than the liquid. Then, a burst of tiny bubbles, taking birth in the inclusion appendix, invaded the whole cavity (Microphotograph 7, Fig. 4).

6. Interpretation of the kinetic data sets.

Kinetic results. Figure 6 shows the distribution of kinetic measurements for 7 temperature steps between 90.4° and 94.4°C (duration times in logarithmic scale).

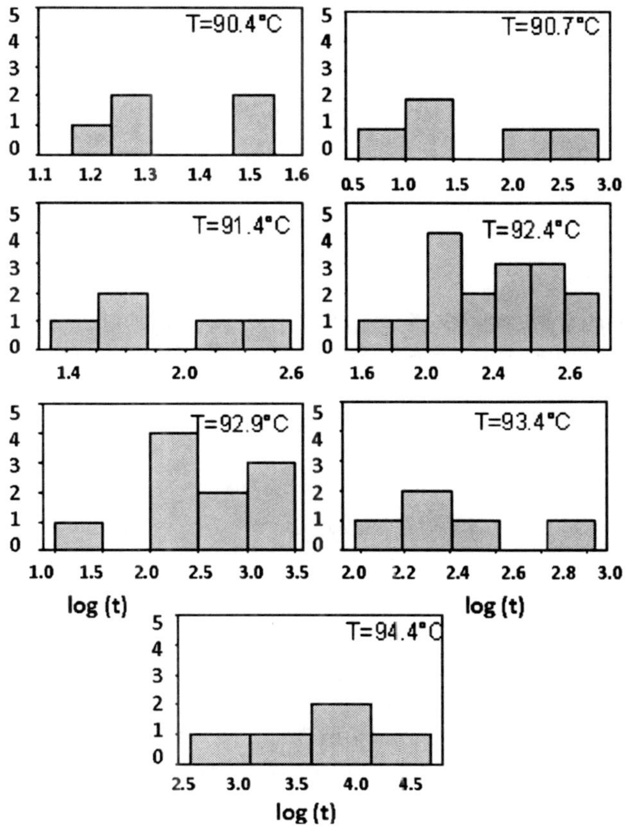

Figure 5. *Distribution of the measured metastability lifetimes (s) of inclusion no 31-7 (logarithmic scale) for 7 temperature steps above T_n.*

Let T_{step} be the temperature above T_n at which inclusion no 31-7 is stabilized in the metastable liquid state. Let t_0 be the time at which the temperature step begins (t_0 is taken as 0). Let t_i be the timelength elapsed between t_0 and the vapour nucleation event ($t > 0$). The variable t is continuous and characterized by a density probability function f(t) such that:

$$\int_0^\infty f(t)dt = 1$$

According to the Classical Nucleation Theory, the repetitive formation of nuclei in a metastable liquid can be considered as a sequence of independent events and the distribution of metastable lifetimes shows an exponential decrease (see also Takahashi *et al.*[23]). This implies that the density probability function f(t) of the nucleation event is:

$$f(t) = \lambda \times e^{-\lambda t} \tag{5}$$

where λ is the exponential decay constant and 1/λ the mean life of the metastable state.

The probability that the vapour bubble nucleates within timelength t is thus:

$$P(E \leq t) = \int_0^t \lambda e^{-\lambda t} dt = 1 - e^{-\lambda t} \quad \text{(Exponential Failure distribution)} \tag{6}$$

The probability of non nucleation of the vapour bubble within timelength t is

$$P(E>t) = e^{-\lambda t} \quad \text{(Exponential Reliability distribution)} \tag{7}$$

Calculation of the decay rate λ and half-life period τ at a fixed temperature step. At each temperature step, we have built the exponential reliability distribution, i.e., the probability of the non nucleation event within timelength *t* (P(E>t). The Ln[P(E>t)] were plotted versus timelength t and the data were fitted by a straight line passing by the origin (Fig. 6; correlation coefficients of the fits ranging between 0.84 and 0.99, Table 3).

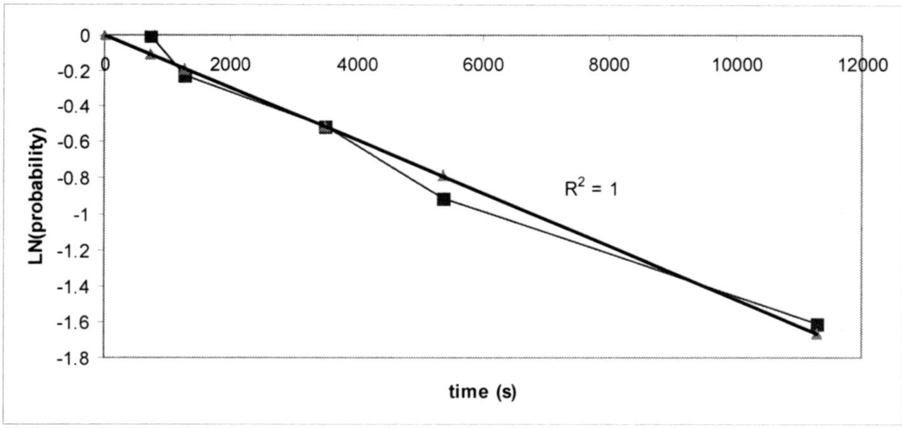

Figure 6. *Observed Fiability law at the temperature step T = 94.4°C.*

Hence we derived the exponential decay constants λ for the 7 T-steps considered (Table 3). The half-life period τ at each T step was then calculated as follows:

$$\tau = \frac{Ln(2)}{\lambda} \tag{8}$$

On Figure 7, the calculated τ are plotted as a function of ΔT, the temperature distance to T_n (i.e. $T-T_n$). The τ values decrease exponentially as a function of ΔT, with a fitted decay constant close to 1.

Temperature (°C)	Exponential Fiability model		
	λ	R^2.	τ
90.4 (5)	0.0169	0.84	40.9
90.7 (5)	0.0046	0.94	149.5
91.4 (5)	0.0021	0.90	322.9
92.4 (16)	0.0024	0.99	286
92.9 (10)	0.00068	0.93	1025
93.4 (10)	0.00059	0.97	1166
94.4 (5)	0.000147	0.99	4702.1

Table 3 *Exponential decay constants λ calculated from the reliability distributions observed at each temperature step, and half-life periods τ related. R is the correlation coefficient of the linear fit of the data (see text). Number between brackets = number of t measurements.*

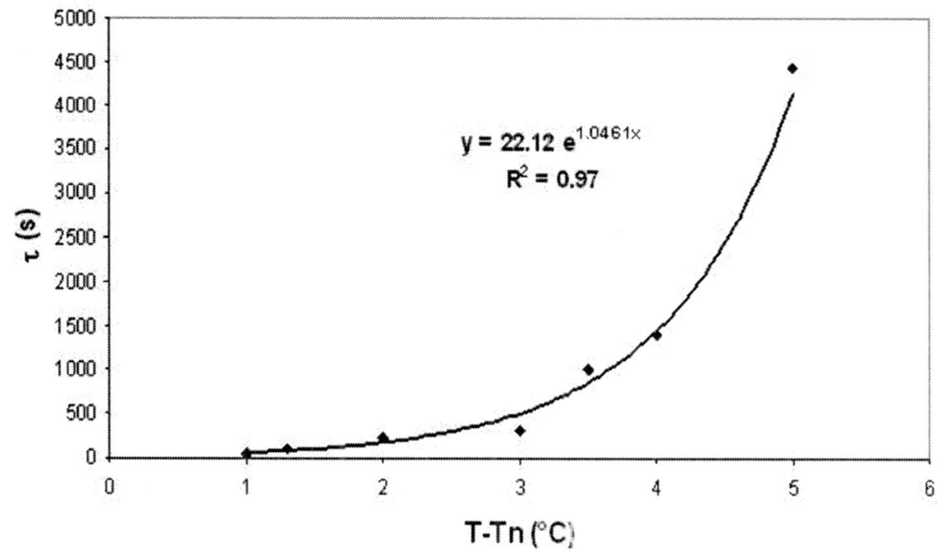

Figure 7. *Inclusion no 31-7: Half-life period of metastability as a function of the intensity of superheating ($T-T_n$). Tn corresponds to the maximum degree of metastability sustainable by the inclusion.*

Due to the fact that the fitted pre-exponential factor is different from 1, we calculate a half-life period at T_n of about 22s instead of 0. On account of the heating rate adopted close to T_n (see Table 1), this indicates that the nucleation event started on average ≈ 0.7°C before the beginning of the temperature step during cooling, our chosen time zero.

Given that the measured variability on T_n is ≈ 2.3°C (Table 2), these results corroborate our choice of placing the starting point of the kinetics experiment at around the beginning of the temperature step, rather than at T_h, as previously proposed[23].

7. Geological implications

Our data show that at a temperature of 24°C above T_n, an occluded liquid with a volume of ≈ 10^4 μm^3, undergoing a tension of ≈ -50 MPa, can sustain such a high superheated state during 10^{13} seconds. A first consequence is that the half-life duration of metastability of such a system, one order of magnitude larger than one million years, is quite relevant to geological timescales. Secondly, it has been recently indicated that the changes in water properties related to superheating significantly influence the rock-water-gas equilibria as soon as the tensile strength of the liquid reaches -20 MPa[1]. Thus, our data prove that the metastability of micrometric fluid volumes is indeed a process of major geochemical importance.

As a conclusion, this paper, together with a companion one, firstly highlights that fluid inclusions are very adapted to the experimental study of superheated solutions at the μm- to mm-scale, both from the metastable intensity and kinetics points of view. In addition, we previously showed that aqueous fluids appear to superheat easily since all the 937 inclusions studied, containing pure water and various aqueous solutions, displayed superheating, some to very high degrees up to -100 MPa. The major point of this paper is to give the first quantitative proof that micro-volumes of highly superheated water can sustain this stretched state for a very long time, infinite at the human scale. The fact that superheating modifies both the thermodynamic and solvent properties of water has already been assessed[15,24-25]. It is here illustrated that such changes can persist over geologically-relevant time-lengths, large enough for superheated fluids to become a possible controlling parameter of the evolution of natural systems.

Acknowledgements

This work has received financial support from the French Agency for Research (Agence Nationale de la Recherche), grant SURCHAUF-JC05-48942 (grant responsible: L. Mercury) and from Russian Fund of Basic Investigations, grant 06-05-64460 (grant responsible: K. Shmulovich). Finally, Jean-François Lenain is greatly acknowledged for his advices and for controlling the statistical treatment of the data.

References

1. Pettenati, M., Mercury, L., and Azaroual, M. (2008) Capillary geochemistry in non-saturated zone of soils. Water content and geochemical signatures, *Applied Geochem.* **23**(12), 3799-3818
2. Meslin, P.Y., Sabroux, J.-C., Berger, L., Pineau, J.-F., and Chassefière, E. (2006) Evidence of 210Po on martian dust at meridiani planum. *J. Geophys. Res. 111*, art. E09012, 14 p
3. Jouglet, D., Poulet, F., Milliken, R. E., Mustard, J. F., Bibring, J. P., Langevin, Y., Gondet B., and Gomez, C. (2007) Hydration state of the Martian surface as seen by Mars Express OMEGA: 1. Analysis of the 3 μm hydration feature, *J. Geophys. Res.* **112**, art. E08S06, 20 p
4. Ramboz, C., and Danis, M. (1990). Superheating in the Red Sea? The heat-mass balance of the Atlantis II Deep revisited, *Earth Planet. Sci. Lett.* **97**, 190-210
5. Shmulovich, K. I., and Graham, C. M. (2004). An experimental study of phase equilibria in the systems H_2O–CO_2–$CaCl_2$ and H_2O–CO_2–NaCl at high pressures and temperatures (500–800°C, 0.5–0.9 GPa): geological and geophysical applications, *Contr. Mineral. Petrol.* **146**, 450-462
6. Wagner, W., and Pruss, A. (2002) The IAPWS Formulation 1995 for the Thermodynamic Properties of Ordinary Water Substance for General and Scientific Use, *J. Phys. Chem. Ref. Data* **31**, 387-535
7. Kiselev, S. B., and Ely, J. F. (2001) Curvature effect on the physical boundary of metastable states in liquids, *Physica A* **299**, 357-370
8. Speedy, R. J. (1982) Stability-limit conjecture. An interpretation of the properties of water, *J. Phys. Chem.* **86**, 982-991
9. Debenedetti, P.G., and D'Antonio, M.C. (1986) On the nature of the tensile instability in metastable liquids and its relationship to density anomalies, *J. Chem. Phys.* **84**(6), 3339-3345
10. Poole, P.H., Sciortino, F., Essmann, U., and Stanley, H.E. (1992) Phase behaviour of metastable water, *Nature* **360**, 324-328
11. Mishima, O., and Stanley, H.E. (1998) The relationship between liquid, supercooled and glassy water, *Nature* **396**, 329-335
12. Sastry, S., Debenedetti, P. G., Sciortino, F., and Stanley, H. E. (1996) Singularity-free interpretation of the thermodynamics of supercooled water, *Phys. Rev. E* **53**, 6144-6154
13. Stanley, H.E., and Teixeira, J. (1980) Interpretation of the unusual behavior of H_2O and D_2O at low temperatures: tests of a percolation model, *J. Chem. Phys.* **73** (7), 3404-3422
14. Shmulovich, K.I., Mercury, L., Thiéry, R., Ramboz, C., and El Mekki, M. (2008) Experimental superheating of water and aqueous solutions. Geochim, *Cosmochim. Acta,* submitted. Shmulovich K.I. (2008) Long-living

superheated aqueous solutions: experiment, thermodynamics, geochemical applications, *this volume*.

15. Mercury, L., Azaroual, M., Zeyen, H., and Tardy, Y. (2003) Thermodynamic properties of solutions in metastable systems under negative or positive pressures, *Geochim. Cosmochim, Acta* **67**, 1769-1785
16. Span, R., and Wagner, W. (1993) On the extrapolation behavior of empirical equation of state, *Int. J. Thermophys.* **18**(6), 1415-1443
17. Roedder, E. (1967) Metastable superheated ice in liquid-water inclusions under high negative pressure, *Science* 155, 1413-1417
18. Green, J. L., Durben, D. J., Wolf, G. H., and Angell, C. A. (1990) Water and solutions at negative pressure: Raman spectroscopic study to -80 Megapascals, *Science* **249**, 649-652
19. Zheng, Q., Durben, D. J., Wolf, G. H., and Angell, C. A. (1991) Liquids at large negative pressures: water at the homogeneous nucleation limit, *Science* **254**, 829-832
20. Alvarenga, A. D., Grimsditch, M., and Bodnar, R. J. (1993) Elastic properties of water under negative pressures, *J. Chem. Phys.* **98**, 11, 8392-8396
21. Ramboz, C., Orphanidis, E., Oudin, E., Thisse, Y., and Rouer, O. (2008), Metastable fluid discharge by the Atlantis Deep submarine geyser: the heat-mass balance of the stratified lower brine revisited in the light of new fluid inclusion data. *This volume.*
22. Debenedetti, P. G. (1996) Metastable liquids. Concepts and principles. *Princeton University Press, Princeton,* 411 p
23. Takahashi, M., Izawa, E., Etou, J., and et Ohtani, T. (2002) Kinetic characteristic of bubble nucleation in superheated water using fluid inclusions, *J. Phys. Soc. Japan* **71**(9), 2174-2177
24. Mercury, L., Pinti, D. L., and Zeyen, H. (2004) The effect of the negative pressure of capillary water on atmospheric noble gas solubility in ground water and palaeotemperature reconstruction, *Earth & Planetary Sci. Lett.* **223**, 147-161
25. Lassin, A., Azaroual, M., and Mercury, L. (2005) Geochemistry of unsaturated soil systems: aqueous speciation and solubility of minerals and gases in capillary solutions, *Geochim. Cosmochim, Acta* **69**, 22, 5187-5201

EXPLOSIVE PROPERTIES OF SUPERHEATED AQUEOUS SOLUTIONS IN VOLCANIC AND HYDROTHERMAL SYSTEMS

RÉGIS THIÉRY[1], SÉBASTIEN LOOCK[1,2], AND LIONEL MERCURY[3]
[1] *Laboratoire Magmas et Volcans, UMR 6524, CNRS/Clermont Université/OPGC, 5, rue Kessler, 63038 Clermont-Ferrand, France.*
[2]*Laboratoire Géoazur, 250 rue Albert Einstein, 06560 Valbonne, France.*
[3]*Institut des Sciences de la Terre d'Orléans, Université d'Orléans, UMR 6113 CNRS-INSU, 1A rue de la Férollerie, 45071 Orléans, France.*

Abstract: Superheated aqueous solutions in volcanic and hydrothermal environments are known to reequilibrate violently through explosive boilings and gas exsolutions. While these phenomena are purely kinetic problems in essence, the explosivity conditions of these demixion processes can be investigated by following a thermodynamic approach based on spinodal curves. In a first part, we recall briefly the concepts of mechanical and diffusion spinodals. Then, we propose to differentiate superspinodal (explosive) transformations from subspinodal (non-explosive) ones. Finally, a quantitative study of spinodal curves is attempted on the binary systems H_2O-CO_2 and H_2O-NaCl with equations of state with solid theoretical basis. It is shown that dissolved gaseous components and electrolytes have an antagonist effect: dissolved volatiles tend to shift the superspinodal region towards lower temperatures, whereas electrolytes tend to extend the metastable field towards higher temperatures. This study may give some clues to understand the explosive destabilization conditions of aqueous solutions in phreatic, phreato-magmatic and hydrothermal eruptions.

Keywords: metastability, equation of state, spinodal, explosivity, aqueous solution, carbon dioxide, sodium chloride, supersaturation, natural systems

1. Introduction

Water is the main natural explosive agent on the Earth. This fact is well demonstrated by all forms of volcanic and hydrothermal explosive manifestations, characterized by a sudden and brutal vaporization of water and other dissolved volatiles from a condensed state, either from aqueous solutions or from supersaturated magmas.[1] This paper is mainly devoted to the first case, i.e. the explosivity of aqueous solutions. Explosions can be defined as violent reactions of systems, which have been perturbed up to transient and unstable states by physico-chemical processes. As such, the traditional approach to such problems is to rely on kinetic theories of bubble nucleations and growths, and this topic has been already the subject of an abundant literature (see references therein[2-3]). We apply here an alternative and complementary method by following a phenomenological thermodynamical point of view. Indeed, an explosive situation is obtained when a boiling transformation perturbs a liquid up to near or through a thermodynamic frontier, i.e., a spinodal, delimiting a thermodynamically forbidden and unstable region of the phase diagram of the system. In a first part, the theoretical grounds of this paradigm are briefly justified, and it will be shown how boilings and gas exsolutions can be differentiated, depending upon the process conditions, either in explosive transformations or non-explosive ones. Then, these concepts will be exploited on two important types of aqueous solutions, which are the H_2O-CO_2 and the H_2O-NaCl systems. This thermodynamic modeling will use equations of state built on solid physical bases, which will allow us to decipher the thermodynamic factors controlling the explosivity of boiling and gas exsolution of aqueous solutions in volcanic and hydrothermal environments.[1]

2. Theoretical concepts of explosivity

The key to an explosive transformation is not the level of mechanical work yielded to the environment, but the rate of mechanical energy release. This latter parameter features the power or yield of the explosion. The higher is this quantity, the stronger are the damages around the explosion focus in terms of fragmentation and other blast effects. In other words, explosive processes are characterized by kinetic rates, which are significantly more elevated by several magnitude orders than in near-equilibrium processes. Therefore, such explosive phenomena can be produced only in strongly disequilibrated systems. Interestingly, the disequilibrium degree of a system can be estimated with the help of the second principle of thermodynamics, which gives us stability criteria that any system must obey.[2,4] The first one is the mechanical stability criterion, which states that any isothermal volume (V) increase of a system must result to a decrease of its internal pressure (P):

$$\left(\frac{\partial P}{\partial v}\right)_T \leq 0. \tag{1}$$

The second one is the diffusion stability criterion, which imposes the net and spontaneous diffusion (i.e. in the absence of any external forces) of species from concentrated regions to less concentrated ones. This criterion is formulated by:

$$\left(\frac{\partial^2 G}{\partial x_i^{\,2}}\right)_{T,P,x_j} \geq 0, \tag{2}$$

where G is the Gibbs free energy and x_i refers to the diffusing species in a fluid mixture. The limiting conditions, i.e. when the above quantities are nil, are of interest, as they characterize highly unstable systems. The locus curves of $(\partial P/\partial v)_T = 0$ and $(\partial^2 G/\partial x^2)_{T,P,x_j} = 0$ can be projected onto any phase diagram and correspond to thermodynamic frontiers separating metastable and unstable domains. The first locus curve refers to the so-called mechanical spinodal curve, whereas the second one is the diffusion spinodal curve.[4,5] Spinodal curves represent the highest disequilibrium degrees, which can be reached by a fluid before its brutal and rapid demixion. Hence, the explosivity of a physical transformation can be assessed, at least qualitatively, by considering the incursion degree of a liquid through its metastable region up to its thermodynamically forbidden domain of instability.

3. Discussion and application to pure water

The paradigm presented in the preceding section can be applied to the case of pure water. Figure 1 shows the stability, metastability and instability fields for water in a pressure-temperature plot, as calculated by the Wagner and Pruss equation of state[6]. Only the mechanical stability criteria is relevant to this one-component system. Limiting stability conditions are encountered along the liquid spinodal curve, noted Sp(L), and along the gas spinodal curve, noted Sp(G). Both spinodal curves meet at the critical point CP with the liquid-gas (LG) saturation curve (also called binodal). The gas spinodal curve indicates the theoretical extreme conditions, which can be attained by a metastable gas (referred to as a supercooled gas). In the other way, the liquid spinodal curve marks the furthest theoretical conditions reachable by a metastable liquid (or superheated liquid) before its explosive demixion into a liquid-gas mixture. Figure 1 depicts also the two main physical processes, which can trigger the boiling of a liquid: these are (1) isobaric boiling and (2) adiabatic decompression (which can be approximated as a quasi-isothermal process for a liquid).

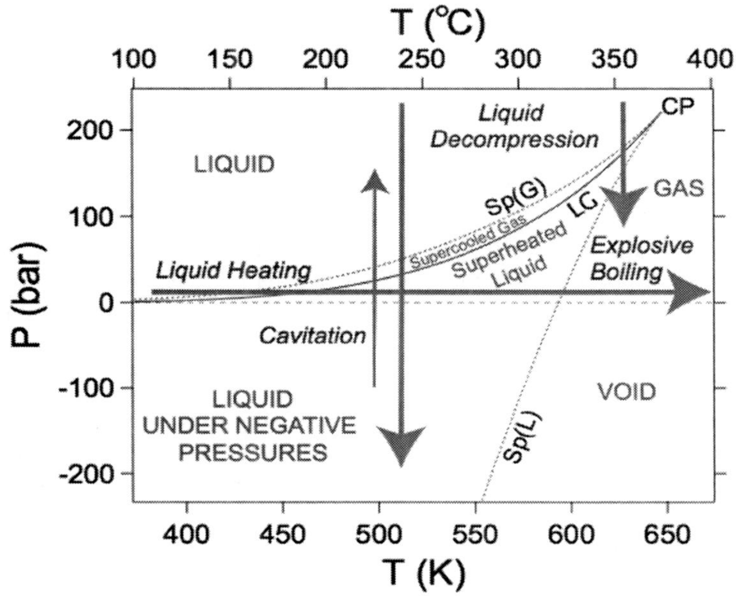

__Figure 1.__ Pressure-temperature diagram illustrating the different perturbation processes of liquid water, and their relations with the stable, metastable and unstable fields of H_2O. Solid line: the saturation curve (LG). Dotted lines: the mechanical liquid spinodal curve Sp(L) and the mechanical gas spinodal curve Sp(G).

In practice, spinodal states of liquid-gas transitions cannot be studied experimentally (at the notable exception of the critical point, which is both a gas and liquid spinodal point). The lifetime of a metastable fluid decreases drastically at the approach of a spinodal curve.[7] Thus, rapid processes, e.g. a very quick heating step or a sudden decompression (Fig. 1), are able to transport a liquid up to spinodal conditions. Energetic barriers of nucleations decrease then to the same magnitude order than molecular fluctuations. Thus, bubble nucleations become active and spontaneous mechanisms, contrasting to the case of weak supersaturation or superheating degrees, where the nucleations of bubbles are known to be a slow process, which must be activated to occur. Kiselev[8], and Kiselev and Ely[9] have calculated precisely the pressure-temperature conditions of this change of nucleation regime for water, introducing the notion of kinetic spinodal. This curve mimics the trend of thermodynamic liquid spinodal curve (but is shifted to lower temperatures in a pressure-temperature plot). Moreover, experimental studies of liquid-liquid demixing in alloys or polymers, as well numerical simulations, have demonstrated that the usual matter separation of nucleation-phase growth is replaced by the faster and more efficient process of spinodal decomposition[2,10] in the instability domain. Hence, the approach of a superheated liquid up to spinodal conditions is synonym for explosive vaporization.

This paradigm has been validated by the analysis of numerous industrial explosions. A first type of explosions is caused by the sudden depressurization of liquids. In the specialized literature, this phenomenon is commonly referred to as a BLEVE[11-16] (acronym for a Boiling Liquid Expansion Vapour Explosion). Another type of explosions is produced by the fortuitous contact of a liquid with a hot body at the origin of FCI (Fuel Coolant Interactions) or MFCI (Molten Fuel Coolant Interactions) explosions.[17-19] In each of these categories (BLEVE and FCI), the explosions are interpreted to result mainly from the destabilization of a fluid at near-spinodal conditions. A schematic illustration is given in Fig. 2 in the case of a sudden liquid decompression. The initial state is a liquid at some temperature *T0* and pressure *P0*, well above the external pressure. The vessel is opened at once, triggering a fast and adiabatic decompression of the liquid. The following depends upon the initial temperature T_0. In the first case (left part of Fig. 2), the depressurization leads only to some bubble nucleations and produces moderate foaming of the liquid surface. In the second case (right part of Fig. 2), relevant to a BLEVE explosion, the opening of the tank is accompanied by a shock wave, and possibly by its failure with emission of projectiles. The boiling proceeds here by active spontaneous bubble nucleations, or conceivably by spinodal decomposition in the case of very high depressurization rates. The thermodynamic interpretation of these two different evolutions is given in the bottom part of Fig. 2, and involves the spinodal temperature T_{sp} at ambient pressure (T_{sp} = 320.45°C = 593.6 K for pure water at one bar, as calculated by the Wagner and Pruss equation of state[1,6]). In the first case, the adiabatic decompression occurs at a temperature below T_{sp}: the depressurization path does not cut the liquid spinodal curve Sp(L) and no explosion occurs. In the second case, the liquid spinodal curve Sp(L) is intersected by the adiabatic depressurization path, as the decompression occurs at a temperature above T_{sp}, triggering a large-scale explosion. Therefore, we suggest to introduce the terms of subspinodal for non-explosive transformations, and superspinodal for the case of explosive ones.

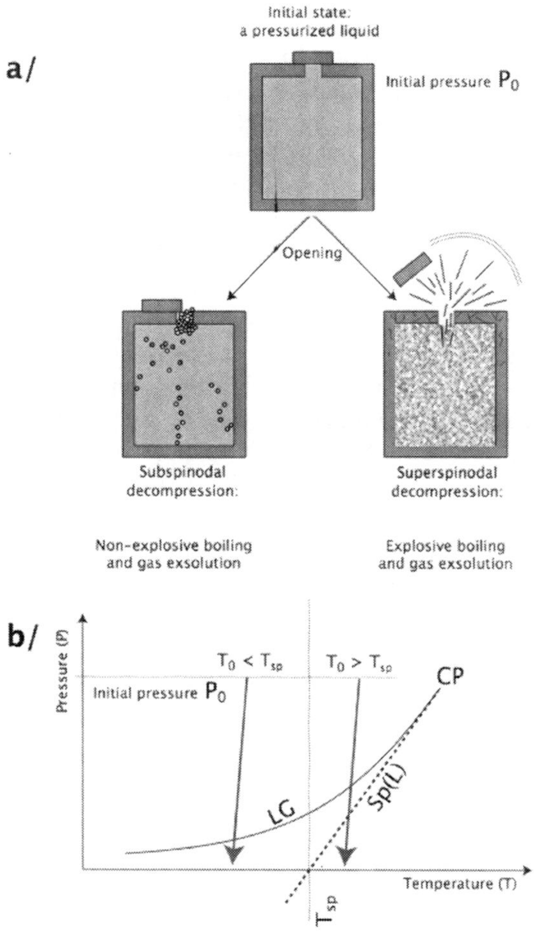

Figure 2. a/ Schematic illustration of a subspinodal (left) and a superspinodal depressurization (right). b/ Thermodynamic interpretation of the transformation explosivity in a pressure-temperature diagram.

Note that this interpretation should not be applied too restrictively. Experience shows that some explosive boilings can already occur at temperatures below T_{sp}.[16] The spinodal temperature T_{sp} is a pure thermodynamical concept, and the temperature T_{hn} of homogeneous nucleation[1,2] (T_{hn} = 304°C, 577 K at one bar for pure water), which is a kinetic parameter, could be more appropriate. Moreover, depending upon the circumstances, a decompression under sub-spinodal conditions does not always trigger boiling, and the solution becomes then supersaturated. In the case of a transient decompression in a confined system, cavitation (Fig. 1) can take place.[1]

Nevertheless, thermodynamics provides us with a simple concept which can help us to analyze the possible evolution, explosive or not, of a boiling or gas exsolution process. However, while the liquid spinodal curve of water is presumably well known, at least in its high-temperature part,[5] the topology of spinodal curves of aqueous solutions is poorly known. The purpose of the next two sections is to fill in this gap for CO_2 and NaCl aqueous solutions.

4. The H_2O-CO_2 system

The representation of spinodals is a highly demanding task for an equation of state, as calculations are done beyond their fitting range with experimental data. As a consequence, this requires a model with good extrapolation capabilities. The corollary is that we must restrict ourselves to equations of state with a good physical basis, and which do not rely on illfounded empirical correlations. Moreover, the H_2O-CO_2 system involves rather complex molecular interactions, which are not easy to describe rigorously[20,21]: indeed, H_2O is a strong dipolar molecule, which associates to neighboring water molecules through hydrogen bounds, whereas CO_2 is a quadrupole. A first approximation is to use van der Waals like equations of state (the so-called cubic equations of state), but which incorporate into their attractive *a* parameter the effects of hydrogen bounds, dipole-dipole and dipole-quadrupole interactions. A preliminary selection leads us to choose the Peng-Robinson-Stryjek-Vera (PRSV) equation of state,[22-24] which gives good results for mixtures of polar and nonpolar components.[25] A quadratic mixing rule with a zero binary interaction parameter between H_2O and CO_2 has been retained to describe the mixing properties of water and carbon dioxide. Therefore, results given here have to be considered as semi-quantitative. Nevertheless, they should give a reasonable idea of the topology of spinodal curves in water-gases systems. All calculations (thermodynamic properties, binodals, spinodals, critical curves) have been made with the help of the LOTHER library[20,21,26] for fluid phase equilibria calculations.

Figure 3A gives solubility curves and spinodals calculated by the PRSV equation of state at 323 K, 50°C in a pressure-mole fraction of CO_2 diagram. For comparison, the solubility curve L(G) of CO_2 in water, calculated by the more accurate model of Duan and Sun,[27] is also drawn and shows that the PRSV equation of state underestimates the CO_2 solubility in water. The mechanical spinodal curves for the liquid, noted mSp(L), and for the gas, noted mSp(G) are plotted too with diffusion spinodal curves Sp(L) and Sp(G). The relations between spinodal curves can be observed more clearly on a molar volume-mole fraction of CO_2 (Fig. 3B). Mechanical spinodal curves mSp(L) and mSp(G) meet at a pseudo-critical point (pCC).[5,28] The diagram shows also that the mechanical instability field (and the pCC) is included in the diffusion instability domain. This result can be generalized and has been demonstrated by Imre and Kraska[5]. Thus, for mixtures, the relevant stability criterion is not the mechanical one, but the diffusion one.

The projections of the L(G) and Sp(L) isotherms in a P-x_{CO2} diagam (Fig. 3A) are almost vertical and parallel. As a consequence, the depressurization of a CO_2-supersaturated solution cannot perturb the fluid up to near-spinodal conditions: gas exsolution will always proceed only by moderate bubble nucleations and any decompression process will be subspinodal.

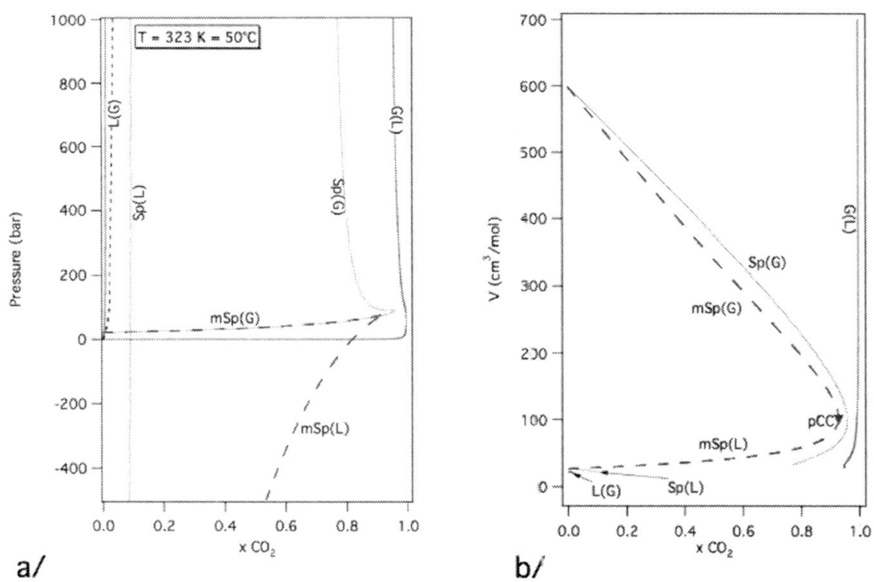

Figure 3. a/ Pressure - CO_2 mole fraction diagram showing the boundaries of stable, metastable and unstable fields, as calculated by the PRSV equation of state in the H_2O-CO_2 system at 323 K, 50°C. Solid lines: the solubility curve L(G) of CO_2 in liquid water, the solubility curve G(L) of H_2O in gaseous CO_2. Dotted lines: the diffusion liquid Sp(L) and gas Sp(G) spinodal curves. Long dashed curves: the mechanical liquid mSp(L) and gas mSp(G) spinodal curves. Short dashed curve: solubility curve L(G) calculated by the Duan and Sun model[27]. b/ Molar volume-CO_2 mole fraction diagram illustrating the relations between the diffusion and the mechanical metastable fields. Triangle marker: the pseudo-critical point (pCP) at 50°C.

Superspinodal depressurizations are possible in the H_2O-CO_2 system at much higher temperatures. An example is given in Fig. 4, where binodal and spinodal curves at 623 K, 350°C are plotted in a P-x_{CO2} diagram. The four curves L(G), G(L), Sp(L) and Sp(G) join at one critical point CP of the H_2O-CO_2 critical curve. Moreover, L(G) and Sp(L) curves are not spaced out. For example, a CO_2 aqueous solution with $x_{CO2} = 0.04$ at 350°C is saturated at 220 bar, but is already in a spinodal state at 195 bar. Therefore, any brutal decompression of a CO_2-saturated solution should lead to a large scale destabilization at this temperature.

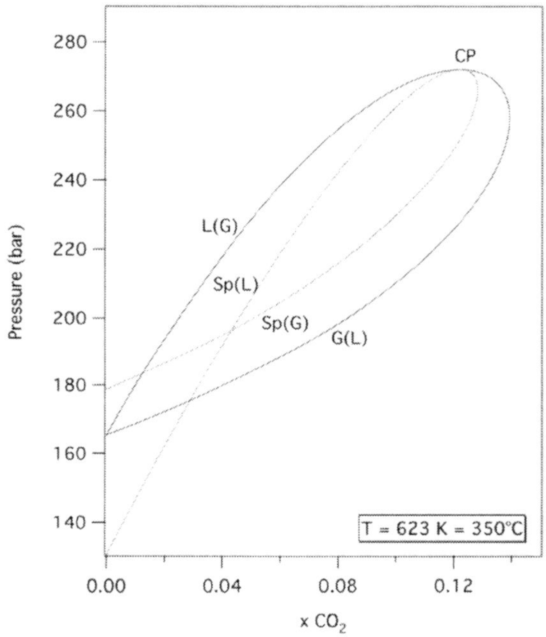

Figure 4. *Pressure-CO_2 mole fraction diagram showing the extent of stable, metastable and unstable fields in the H_2O-CO_2 system at 623 K, 350°C.*

Figure 5 depicts the liquid spinodal curves Sp(L) in a pressure-temperature diagram for fixed CO_2 compositions. The region of negative pressures, which is of interest for describing the capillary properties of CO_2 aqueous solutions,[7] has been also included. Interestingly, it can be noted that spinodal Sp(L) isopleths present a pressure-temperature trend, which looks similar to the liquid spinodal curve of pure water.[1,2,6,7] At low temperatures, the Sp(L) isopleths are decreasing steeply before to reach a pressure minimum. Then at subcritical temperatures, isopleths are less spaced and sloped, and they finish to meet the H_2O-CO_2 critical curve. The temperature appears as a determining parameter in the explosivity control of CO_2 aqueous solutions. Like for water, the easiest way to generate an explosive vaporization is a sudden depressurization in the superspinodal domain, where spinodal curves have a gentle slope in a P-T diagram (Fig. 5). This superspinodal field can be estimated theoretically from the PRSV equation of state for temperature T above 425 K (150°C), whose value is to be compared with the spinodal temperature T_{sp} of pure water at 1 bar at 320.45°C.[1,6] Depressurizations are expected to be subspinodal below this temperature threshold, and superspinodal above. Therefore, the presence of dissolved volatiles in aqueous solutions reduces strongly their metastability field towards lower temperatures and accentuate their explosivity potential with respect to pure water. In the subspinodal region (T<150°C), a very fast and important CO_2 dissolution is necessary to shift an aqueous solution from saturation conditions to spinodal ones: such a process seems to be unlikely.

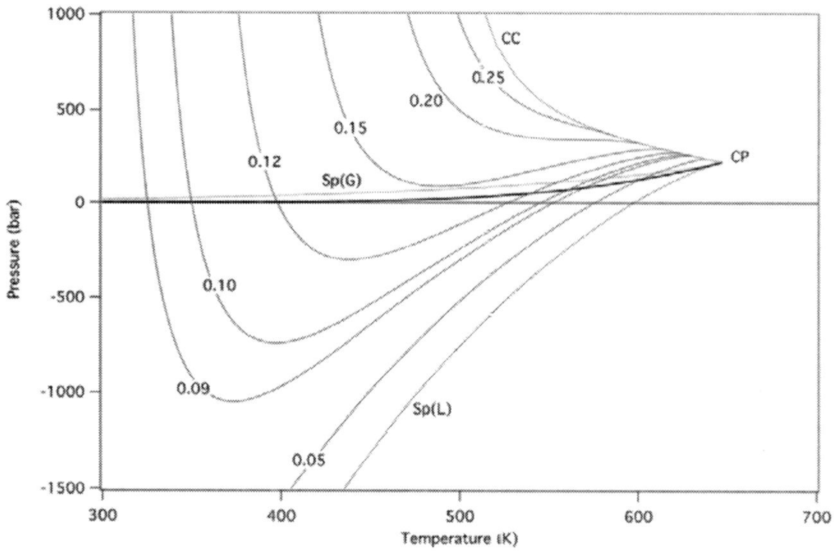

Figure 5. *The liquid spinodal curves in a pressure-temperature diagram for the H_2O-CO_2 system, as calculated by the PRSV equation of state. Numbers refer to the mole fraction xCO_2 of dissolved CO_2 in the aqueous solution.*

5. The H_2O-NaCl system

The same approach can be applied to investigate the explosivity conditions of the H_2O-NaCl system. We have selected the Anderko-Pitzer (AP) equation of state,[29] which is based on realistic physical hypotheses. It describes H_2O-NaCl by means of statistical thermodynamic models[30,31] developed for dipolar hard spheres. This assumption is reasonable at high temperatures, where NaCl is known to form dipolar ion pairs. However, for this reason, this equation of state is only applicable above 573 K, 300°C.

Figure 6 displays a first P-x_{NaCl} diagram depicting binodal and spinodal isotherms at 623 K, 350°C. The equation of state reproduces well the tabulated data by Bischoff[32] for the solubility curves L(G) and G(L). The pressures of the diffusion spinodal curves Sp(L) and Sp(G) decrease with increasing x_{NaCl} mole fractions (note also that the spinodal curves run through the stability field of halite at pressures below 100 bars). As a consequence, the H_2O-NaCl system presents a dual behaviour concerning its stability during a rapid depressurization. For x_{NaCl} between 0 and 0.04, a decompression path up to atmospheric pressures will intersect the liquid spinodal curve Sp(L): this will trigger an explosive exsolution of H_2O from the brine, featuring a superspinodal process. Differently, for higher x_{NaCl}, the spinodal curve Sp(L) runs to negative pressures, and any brutal depressurization will generate only non-explosive boiling of the concentrated brines, characterizing a subspinodal transformation.

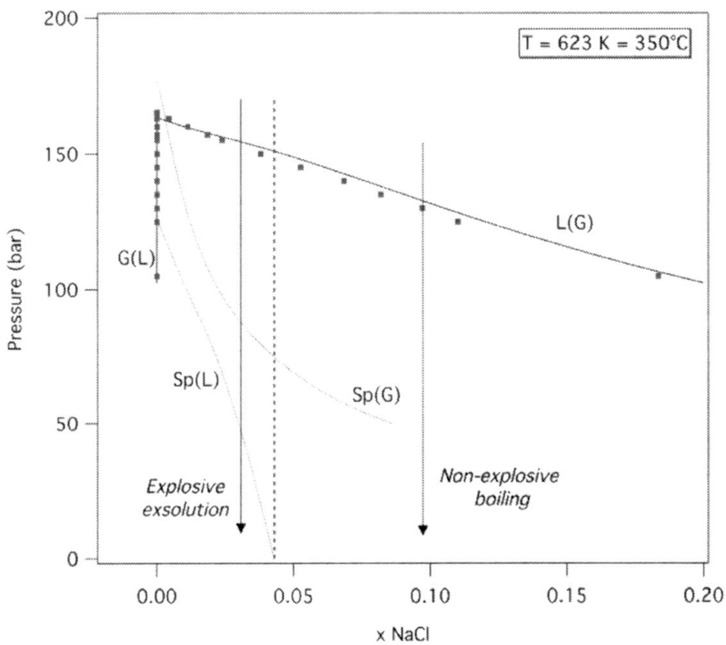

Figure 6. *The stable, metastable and unstable fields in a pressure-NaCl mole fraction diagram for the H_2O-NaCl system at 623 K, 350°C, as calculated by the Anderko and Pitzer equation of state*[29]*. Solid lines: the solubility curves L(G) and G(L). Dotted lines: the spinodals Sp(L) and Sp(G). Squared markers: experimental data compiled by Bischoff*[29]*.*

Two other portrays of the metastability fields of the H_2O-NaCl system are given in the P-x_{NaCl} diagrams of Fig. 7A (at 380°C) and Fig. 7B (at 500°C), i.e. at temperatures above the critical point of H_2O. Now, both spinodal Sp(L) and Sp(G) isotherms join at a critical point (CP). Another intersection point can be observed at lower pressures, but beware, this is only an artefact generated by the projection of spinodal isotherms onto P-x_{NaCl} planes. Again, H_2O-NaCl brines present the same contrasting behaviour, when they are submitted to a sudden depressurization. Below some x_{NaCl} threshold (e.g. for $x_{NaCl} < 0.08$ at 380°C and $x_{NaCl} < 0.19$ at 500°C), fast decompressions will result to superspinodal vaporizations. Thus, subspinodal boilings concern only rather concentrated brines at these elevated temperatures. Note also that the metastability field of supercooled vapours (i.e. between the G(L) and Sp(G) curves) extends over a non negligible range of pressures and NaCl compositions.

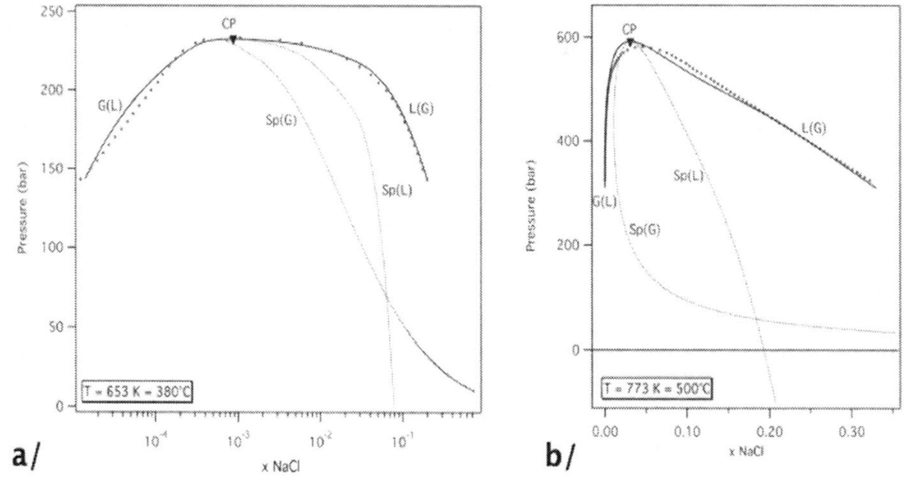

Figure 7. *Pressure-NaCl mole fractions diagrams of the H_2O-NaCl diagram, as calculated by the Anderko and Pitzer equation of state*[29]*. a/ At 653 K, 380°C. b/ At 773 K, 500°C.*

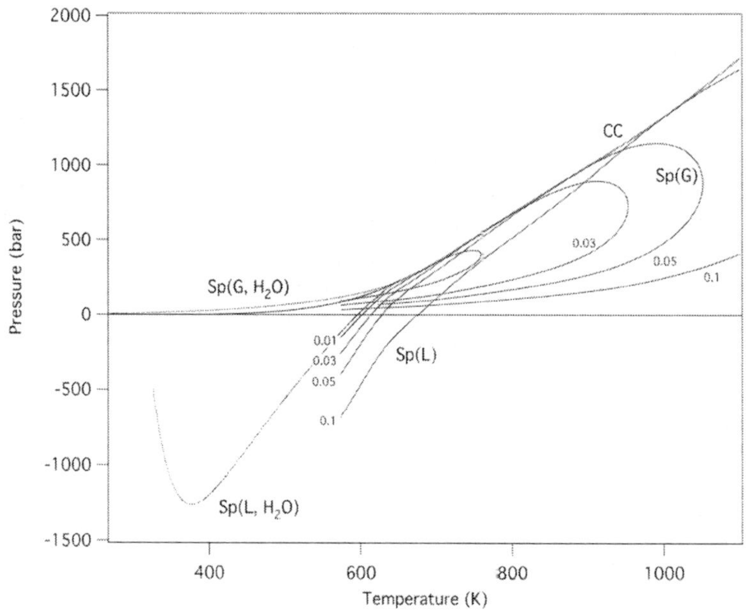

Figure 8. *Stability fields, calculated by the Anderko and Pitzer equation of state*[29] *of the H_2O-NaCl system in a pressure-temperature diagram. Solid lines: the saturation curve (LG) of pure water and the spinodal isopleths of H_2O-NaCl fluids (numbers refer to the mole fractions of NaCl). Dotted lines: the liquid spinodal curve Sp(L, H_2O) and gas spinodal curve Sp(G, H_2O) of pure water.*

Figure 8 illustrates the metastability fields of the H_2O-NaCl system in a *P*-*T* diagram by means of a series of spinodal isopleths calculated respectively for x_{NaCl} = 0.01, 0.03, 0.05 and 0.1. Each isopleth depicts a loop, which is tangential to one point of the H_2O-NaCl critical curve. At this critical point, the nature of the isopleth changes from a liquid spinodal Sp(L) to a gas spinodal Sp(G). The diagram shows clearly that the addition of NaCl translates progressively the liquid spinodal curve Sp(L) to higher temperatures: e.g., the spinodal temperature at 1 bar increases from 320.45°C at $x_{NaCl} = 0$ to 403°C at $x_{NaCl} = 0.1$ (i.e. 26.5 NaCl wt %, but in the stability field of halite). Hence, high concentrations of electrolytes favour the metastability of aqueous solutions. This conclusion is in agreement with experimental data of synthetized fluid inclusions of Shmulovich et al.[7] or with the observed behaviour of natural hydrothermal systems involving the circulations of brines[33].

6. Geological implications

This preliminary theoretical work can give some insights on the functioning of volcanic and hydrothermal systems. We shall mention here three application examples of eruptive phenomena, which are still poorly understood and need further investigations.

The first one is related to the Lake Nyos disaster (Cameroon) in 1986, which was produced by an explosive and massive release of CO_2 from a crater lake and killed about 1700 people.[34,35] To explain this phenomenon, the preferred hypothesis amongst researchers is the model of a limnic eruption,[36] which results from the explosive CO_2 exsolution of dormant supersaturated waters[37] from the lake hypolimnion (CO_2-rich and dense lower layer) at a depth of 210 m.[35] The gas exsolution is supposed to release the pressure of the the water column, and thus to exert a positive feedback on further CO_2 demixing, which would be able to create an overturn of lake waters and to sustain a steady gas flow.[37] However, the triggering mechanism is unknown (sinking of cold rain water, landslide, volcanic CO_2 influx, etc ...). Moreover, such a model requires massive bubble nucleations and growths from the lake bottom. Fig. 3A suggests that a depressurization at the mild temperatures, which are typically found at the bottom (around 23 and 26°C),[38] might not be sufficient to generate such large-scale bubblings. An additional constraint would be the occurrence of higher temperatures (at least, above 150°C, see Fig. 5). Such conditions could favour increasingly the intense CO_2 exsolution and can easily by produced by injections of hydrothermal hot waters.

The second example is linked to phreato-magmatic eruptions, that appear to be predominant over basements constituted by finely porous formations like shales and siltstones[39]. This observation suggests that porosity may play an important role in the explosive behaviour of these boiling systems. The boiling in porous media can be described by:[40]

$$\mu_i^{\text{liquid}}\left(T, P_{\text{liquid}}\right) = \mu_i^{\text{gas}}\left(T, P_{\text{gas}}\right) \quad (3)$$

where μ_i^k refers to the chemical potential of component i ($i = H_2O$, CO_2 or NaCl) in the phase k. The capillary pressure $\Delta P = P_{\text{liquid}} - P_{gas}$ can be appreciated by a capillary equation, like the Laplace relation, which takes into account interfacial effects between the liquid wetting phase and the gas bubble. Fig. 9a displays the boiling curves calculated for different bubble radii with the Wagner and Pruss equation of state[6], combined with the empirical relation of the same authors describing the temperature dependence of the liquid-gas interfacial tension. In this *P-T* plot, only the pressure-temperature properties of the liquid are drawn (the gas pressure is almost confounded with the saturation pressure of bulk water). Rigorously speaking, other effects, like solid-liquid-gas interactions or curvature effects should be allowed for, but this will not change the conclusions below. The main result of this diagram is that capillary forces shift liquid water into the metastable field of superheated waters. As a consequence, boiling in finely porous media is more prone to explosive reactions in the case of an external destabilization. An illustration of this aspect can be given in Fig. 9b, which displays a peperitic intrusion, i.e. a mixture of clasts and poorly consolidated wet sediments, into a lava flow (Pardines, 30 km south of Clermont-Ferrand, France). The injection of this peperitic dyke was permitted by the lower density of the peperite and the overpressure generated by the confinement of this boiling vapour-liquid-solid system. Remarkable facts are (1) the brecciated and altered aspect of the host basalt with radial and concentric joints separating rounded clasts, attesting pervasive percolation of steams, and (2) the subhorizontal apophyses, which developped at several levels from the peperites.

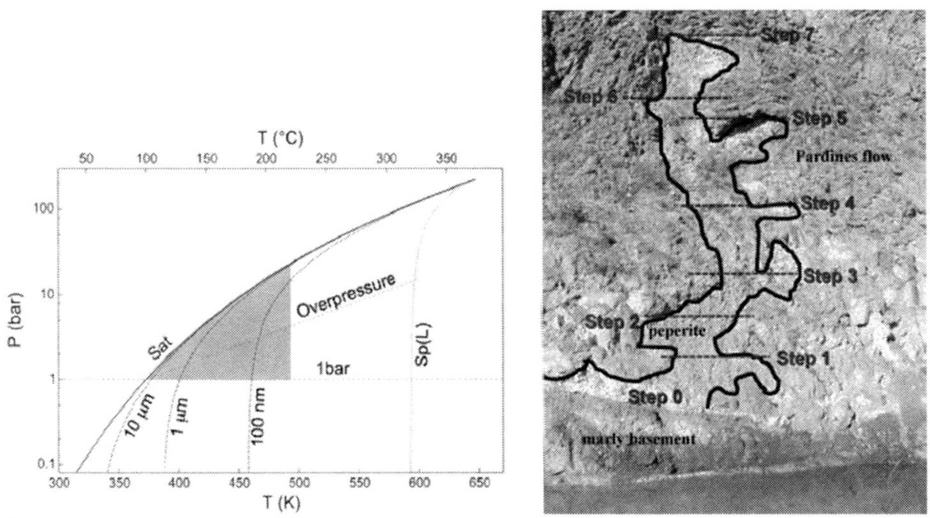

a) **b)**

***Figure 9**. a) A P-T diagram showing the boiling curves of adsorbed water in porous media (numbers refer to the radius of the pore). The dotted curve is the liquid spinodal curve Sp(L). b) The peperitic dyke in the lava flow of Pardines (the boundaries are outlined by the thick curve). The horizontal dashed lines indicate the positions of the successive growth pulses of the dyke.*

These apophyses may be interpreted as successive steps of progression accelerations of the dyke. Thus, the intrusion was not a continuous phenomenon, but instead an alternance of (1) dormant, but recharging phase, where the peperite receives additional heat from the lava flow and (2) explosive boiling episode, providing the net impulse required to push the viscous basaltic flow. The fine-grained peperitic matrix is rather mesoporous (pore diameters between 100 nm and 10 μm) and microporous (pore diameters below 100 nm). Thus, according to Fig. 9a (shaded area), capillary effects are sufficient to generate capillary pressures between 0 and 10 bar (this latter value is commonly found in argillaceous and silty soils featured by low water contents). Note that the porosity of these natural materials is rather heterogeneous, defining a multitude of microsystems with different ebullition temperatures. Hence, destabilization of larger pores by an external perturbation (shock, lava displacement, ...) may trigger a positive feedback on the boiling of smaller pores and be at the origin of the chaotic behaviour of this small hydrothermal system.

The last application example is related to hydrothermal systems, sustained by a magmatic chamber, either in the oceanic lithosphere (at an accretion ridge[41]) or in a continental crust (e.g. a porphyry intrusion[42]). Fig. 10a displays the main parts of these systems, involving an upper brittle region overlying the ductile field. Both domains are featured by fundamental differences, which are summarized in Table 1. Of interest is the transition zone, which is characterized by a strong pressure gradient. As a consequence, a liquid, which has been exsolved by a magma, can follow here a depressurization path, cutting the Sp(L) liquid spinodal curve (see the *P-T* diagram of Fig. 10b). Thus, it can boil explosively in a way, which is similar to BLEVE accidents in the industry[11-18]. Another case is the flushing out of pockets of aqueous solutions under pression (e.g. the "water sills" of Fyfe et al.[43]), which can lead also to superspinodal decompressions. Note that such explosive phenomena are not systematic, and depend on the initial salinity and the temperature of the fluid, as depicted in Fig. 10b. Nevertheless, this mechanism of explosive boiling play probably an important role in the brecciation of hydrothermal reservoirs.

Properties	The ductile region	The brittle region
Pressure regime	Lithostatic	Hydrostatic
Permeability	Low	High
Temperature range	≥ 400°C	100-360°C
Pressure range	≥ 300 bar	1-300 bar
Fluid connectivity	No	Yes
Main mechanism of heat transport	Conduction	Fluid advection

Table 1 Main differences between the brittle zone and the ductile region of large hydrothermal systems. Pressure and temperature ranges are only indicative values.

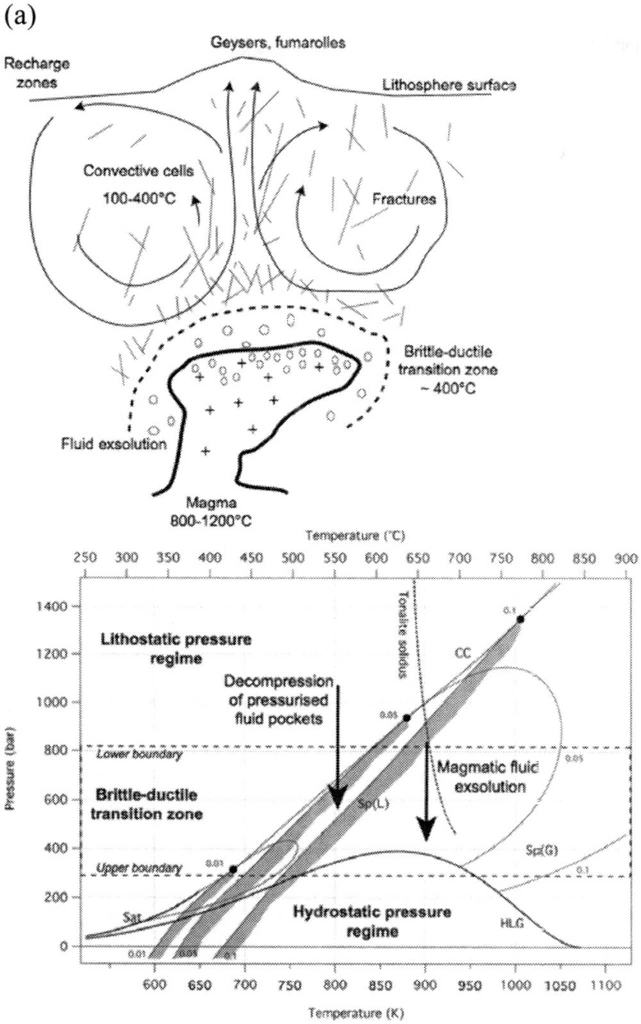

Figure 10. *a) Conceptual sketch of hydrothermal systems. b) P-T diagram illustrating potentially explosive processes for the H_2O-NaCl in an hydrothermal environment (see text). Thick solid lines: the saturation curve for pure water (Sat) and the three-phase halite-liquid-vapour curve (HLG). Dotted line: the critical curve (CC). Thin solid lines: the spinodal curves for x_{NaCl} = 0.01 (3.2 wt % NaCl), 0.05 (14.6 wt % NaCl) and 0.1 (26.5 wt % NaCl) with their corresponding critical points (filled circles). The gray zones along liquid spinodal curves Sp(L) indicate the onset of the instability field of superheated NaCl aqueous solutions. The boundaries of the brittle-*

7. Conclusions

Spinodals represent an appealing concept, which allows to link the kinetics of physical transformations to the thermodynamic properties of the system. Explosive vaporizations can be identified as processes, which perturb a liquid up to near-spinodal singularities. This simple criteria, which is commonly applied by safety engineering in the industrial field, has been generalized here to the case of aqueous solutions. This modeling study shows the antagonist effects of dissolved volatiles and electrolytes: gaseous species tend to shift the explosivity conditions to lower temperatures, whereas dissolved salts tend to displace spinodal conditions to higher temperatures. As a result, this work can give useful indications to constrain the modeling of hydrothermal, phreatic and phreato-magmatic eruptions.

Acknowledgements

This work has been financially supported by the Agence Nationale de la Recherche (ANR) for the project SURCHAUF-JC05-48942.

References

1. Thiéry, R. and Mercury, L. (2009) Explosive properties of water in volcanic and hydrothermal systems, *J. Geophys. Res.* (accepted).
2. Debenedetti P. G. (1996) *Metastable liquids. Concepts and principles*, Princeton University Press, Princeton, NJ, 411 p.
3. Lasaga, A. (1998) *Kinetic theory in the Earth Sciences*, University Press, Princeton, NJ, 811 p.
4. Rowlinson, J., and Swinton, F. (1982) *Liquid and Liquid Mixtures*, Butterworth Scientific, 3rd edition.
5. Imre, A., and Kraska, T. (2005) Stability limits in binary fluid mixtures, *J. Chem. Phys.* **122**, 1–8.
6. Wagner, W., and Pruss, A. (2002) The IAPWS formulation 1995 for the thermodynamic properties of ordinary water substance for general and scientific use. *J. Phys. Chem. Ref. Data* **31**, 387–535.
7. Shmulovich, K., Mercury, L., Thiéry, R., Ramboz, C., and El Mekki, M. (2009) Superheating ability of water and aqueous solutions. Experiments and geochemical consequences, *Geochimica et Cosmochimica Acta* (accepted).
8. Kiselev, S. (1999) Kinetic boundary of metastable states in superheated and stretched liquids, *Physica A* **269**, 252–268.
9. Kiselev, S., and Ely, J. (2001) Curvature effect on the physical boundary of metastable states in liquids, *Physica A* **299**, 357–370.
10. Debenedetti, P. (2000) *Phase separation by nucleation and by spinodal decomposition: fundamentals*, In: Kiran, E. et al. (eds), Supercritical Fluids, pp 123–166. Kluwer Academic Publishers, The Netherlands.
11. Abbasi, T., and Abbasi, S. (2007) The boiling liquid expanding vapour explosion (BLEVE): Mechanism, consequence assessment, management, *Journal of Hazardous Materials* **141**, 480–519.
12. Casal, J., and Salla, J. (2006) Using liquid superheating for a quick estimation of overpressure in BLEVEs and similar explosions, *Journal of Hazardous Materials* **A137**, 1321–1327.
13. Planas-Cuchi, E., Salla, J., and Casal, J. (2004) Calculating overpressure from BLEVE explosions, *Journal of Loss Prevention in the Process Industries* **17**, 431–436.
14. Pinhasi, G., Ullmann, A., and Dayan, A. (2007) 1D plane numerical model for boiling liquid vapor explosion (BLEVE), *International Journal of Heat and Mass Transfer* **50**, 4780–4795.

15. Salla, J., Demichela, M., and Casal, J. (2006) BLEVE: a new approach to the superheat limit temperature, *Journal of Loss Prevention in the Process Industries* **19**, 690–700.
16. Reid., R. C. (1979) Possible mechanism for pressurized-liquid tank explosions or BLEVE's, *Science* **203**, 1263–1265.
17. Reid, R. C. (1976) Superheated liquids, *Am. Scientist* **64**, 146–156.
18. Reid, R. C. (1983) Rapid phase transitions from liquid to vapour, *Advances in Chemical Engineering* **12**, 105–208.
19. Corradini, M. L., Kim, B. J., and Oh, M. D. (1988) Vapor explosions in light water reactors: A review of theory and modelling, *Progress in Nuclear Energy* **22**(1), 1–117.
20. Perfetti, E., Thiéry, R., and Dubessy, J. (2008) Equation of state taking into account dipolar interactions and association by hydrogen bonding. I- Application to pure water and hydrogen sulphide, Chem. Geol. **251**, 58–66.
21. Perfetti, E., Thiéry, R., and Dubessy, J. (2008) Equation of state taking into account dipolar interactions and association by hydrogen bonding: II- Modelling liquid-vapour equilibria in the H_2O-H_2S, H_2O-CH_4 and H_2O-CO_2 systems, *Chem. Geol.* **251**, 50–57 (2008).
22. Stryjek, R., and Vera, J. (1986) An improved Peng-Robinson equation of state with new mixing rules for strongly non ideal mixtures, *Can. J. Chem. Eng.* **64**, 334–340.
23. Stryjek, R., and Vera, J. (1986) PRSV2: a cubic equation of state for accurate vapour-liquid equilibrium calculations, *Can. J. Chem. Eng.* **64**, 820–826.
24. Stryjek, R., and Vera, J. (1986) Vapour-liquid equilibria of hydrochloric acid and solutions with the PRSV equation of state. Fluid Phase Equilibria **25**, 279–290.
25. Duan, Z., and Hu, J. (2004) A new cubic equation of state and its applications to the modeling of vapor-liquid equilibria and volumetric properties of natural fluids, *Geochimica et Cosmochimica Acta* **14**, 2997–3009.
26. Thiéry, R. (1996) A new object-oriented library for calculating high-order multivariable derivatives and thermodynamic properties of fluids with equations of state, *Computers & Geosciences* **22**(7), 801–815.
27. Duan, Z., and Sun, R. (2003) An improved model calculating CO_2 solubility in pure water and aqueous NaCl solutions from 273 to 533 K and from 0 to 2000 bar, *Chem. Geol.* **193**, 257–271.
28. Asselineau, L., Bogdanic, G., and Vidal, J. (1979) A versatile algorithm for calculating vapour-liquid equilibria, *Fluid Phase Equilibria* **3**, 273–290.
29. Anderko, A., and Pitzer, K. (1993) Equation-of-state representation of phase equilibria and volumetric properties of the system NaCl-H_2O above 573 K, *Geochimica et Cosmochimica Acta* **57**, 1657–1680.

30. Boublik, T. (1970) Hard sphere equation of state, *J. Chem. Phys.* **53**, 471–472.
31. Stell, G., Rasaiah, J., and Narang, H. (1972) Thermodynamic pertubation theory for simple polar fluids. J. Mol. Phys. **23**, 393–406.
32. Bischoff, J. (1991). Densities of liquids and vapors in boiling NaCl-H_2O solutions: A PVTX summary from 300 to 500°C, *Am. J. Sci.* **291**, 369–381.
33. Orphanidis, E. (1995) Conditions physico-chimiques de précipitation de la barytine épigénétique dans le bassin sud-ouest de la fosse Atlantis II (Mer Rouge): données des inclusions fluides et approche expérimentale. Implications pour le dépôt des métaux de base et métaux précieux. *Thèse Université d'Orléans*, 180 p.
34. Schenker, F., and Dietrich, V. J. (1986) The Lake Nyos gas catastrophe (Cameroon): a magmatological interpretation, Schweiz. *Mineral. Petrogr. Mitt.* **66**, 343–384.
35. Evans, W. C. (1996) Lake Nyos: knowledge of the fount and the cause of disaster, *Nature* **379**(6560), 21–22.
36. Zhang, Y. (1996) Dynamics of CO_2-driven lake eruptions, *Nature* **379**(6560), 57–59.
37. Rice, A. (2000) Rollover in volcanic crater lakes: a possible cause for Lake Nyos type disasters. *J. Volcan. Geotherm, Res.* **97**, 233–239.
38. Kantha, L. H., and Freeth, S. J. (1996) A numerical simulation of the evolution of temperature and CO_2 stratification in Lake Nyos since the 1986 disaster, *J. Geophys. Res.* **101**(B4), 8187–8203.
39. Grunewald, U., Zimanowski, B., Büttner, R., Philipps, L. F., Heide, K., and Büchel, G. (2007) MFCI experiments on the influence of NaCl-saturated water on phreato-magmatic explosions, *J. Volc. Geotherm. Res.* **159**, 126–137.
40. Shapiro, A., and Stenby, E. (2001) Thermodynamics of the multicomponent vapor-liquid equilibrium under capillary pressure difference, *Fluid Phase Equilibria* **178**, 17–32.
41. Nehlig, P. (1993) Interactions between magma chambers and hydrothermal systems: oceanic and ophiolitic constraints, *J. Geophys. Res.* **98**(B11), 19621–19633.
42. Driesner, T., and Geiger, S. (2007) Numerical simulation of multiphase fluid flow in hydrothermal systems, *Reviews in Mineralogy & Geochemistry* **65**, pp. 187–215.
43. Fyfe, W. S., Price, N. J., and Thompson, A. B. (1978) *Fluids in the Earth's crust. Developments in Geochemistry* **1**, 383 pp., Elsevier Scientific, Amsterdam.

VAPOUR NUCLEATION IN METASTABLE WATER AND SOLUTIONS BY SYNTHETIC FLUID INCLUSION METHOD

[1]KIRIL SHMULOVIC, AND [2]LIONEL MERCURY
[1]*Institute of Experimental mineralogy RAS, 142432 Chernogolovka, Russia*
[2]*Institut des Sciences de la Terre d'Orléans, UMR 6113 CNRS/Université d'Orléans, 1A rue de la Férollerie, 45071 Orléans Cedex, France.*

Abstract: Experimental data for temperatures of homogenization (Th, L+V→L) and vapour phase nucleation (Tn, L→L+V) presented after recalculation to P-T parameters with equation of state for water (Wagner, Pruss, 2002) or Duan's equations for salt solutions. Samples were prepared by method of synthetic fluid inclusions in quartz with densities > 0.8 g/cm^3. The spontaneous nucleation begin at pressure –20 MPa and in the same sample some inclusions keep homogeneous state up to – 150 MPa. Increasing of quartz solubility in trapped liquids lead to decreasing of quantity of high temperature boiling inclusions.

Keywords: metastability, liquid, water, aqueous solutions, negative pressure, vapour nucleation

1. Introduction

Metastable state of liquid water takes place between curves of the Liquid-Vapour (L+V) equilibrium and a spinodale. Spinodals are curves where $(\delta V/\delta P)_T=0$ and the position of the curve on P-T plot depends on the equation of state. It has been studied from 18th century. The early history is published on site (http://www.kfki.hu/~pressure/history.html). Modern situation in the metastable state of liquids regularly have been discussed on specialized conferences, for instance in NATO conference in Budapest, 2002 and in Yekaterinburg, Russia, 2007. Here we present data for the spontaneous nucleation and disappearing of vapour phase at cooling and heating. Kinetic measurements and size effects are subject of other topic (El Mekki et. al., this volume).

Last review related to the nucleation of vapour phase in water (Herbert et. al., 2006) at metastable state and original these authors' data have shown:

1. The vapour nucleation takes place in water under negative pressures (tension) from –20 to –30 MPa while the minimum on the spinodale curve is arranges near -200 MPa. Skripov (1972, tabl. 15) observed similar situations for liquid water, mercury and chloroform, where the cavitation pressure is 5-7 times smaller than predicted by the theory of homogeneous nucleation for the spinodale pressure minimum. For benzol, vinegar acid, aniline and CCl_4 the same ratio is 1.5-2 only.
2. The methods for determining the nucleation pressure were different but all of them are characterised by one common property - the samples has one volume and then the nucleation anywhere in the sample means equalizing pressure to L-V equilibrium with speed of sound. The measurements were done using Berthelot tubes, shock waves, acoustic cavitation with semi- spherical source, U-type tubes in centrifuges, etc.
3. The exception is created by fluid inclusions in minerals, quartz mainly. These inclusions are liquid trapped by growing quartz or at sealing cracks in the monocrystal seeds at high P and T. At ambient conditions the inclusions have typical size 10–500 μm with bulk density 0.8-0.95 g/cm^3 and content liquid and small vapour bubble. Vapour nucleation was studied by Austin Angell group (Green et. al., 1990, Zheng et. al., 1991, 2002) in synthetic fluid inclusions. In spite of large splitting of experimental points, these authors found very negative pressures, in one inclusion the nucleation was observed at –140 MPa.

The difference between -20 MPa and -30 MPa ("big" volume) and –140 MPa (fluid inclusions) is principal for geochemical application as calculations of shifts for constant of equilibrium for 23 reactions (dissociation, dissolution, hydration, ion exchange etc., Zilbelbrandt, 1999) demonstrate large effect of negative pressures. If nucleation take place at -20 to -30 MPa, the ratio K_p/K_0 (where K_p and K_0 are constants in metastable and stable states) is near 0.8-0.9, but if the negative pressure could be -100 to -150 MPa the ration will be 0.1-0.3. It could be dramatically change rate and sequence of important geochemical reactions.

Firstly the negative pressure appearance in fluid inclusions was found by Roedder (1967), who observed melting in pure water inclusion in quartz at + 6.5°C and from slope of L-S equilibrium estimate pressure in the inclusion as – 80 MPa.

2. Experimental

All fluid inclusions were synthesized using an internal heating apparatus ("gas bomb") at Edinburgh University. The runs were done with quartz matrix at 750 MPa and temperatures 530-700°C with duration from 8 to 13 days. Main part of the runs used Bodnar-Sterner method (Bodnar, Sterner, 1985) and this approach was better for measurements than overgrowing

(Shmulovich, Graham, 2004). Some runs were done with quartz seed without cracks, it overgrow by new quartz at thermal gradient 5°C per seed. In this case we have had many inclusions in one focus distance of microscope, but sometime the quantity was too much for separate phase transition in one inclusion from other. Pt-capsules were charged by quartz (2x2x12 mm, with long dimension is parallel to quartz "c" axis), water or solution and amorphic SiO_2. Capsules placed to Ta or Mo holder, inside the holder temperature was measured and controlled by two Pt-Pt-(13% Rh) thermocouples. Pressure was measured by manganine gauges, calibrated via mercury melting point and up to 200 MPa checked by Bourdon gauge. The thermocouples did not calibrate specially, from producer certificates the error could be estimated as +/- 1°C.

Since the gas apparatus has relatively large thermal gradient and constant leakage each run contented 3 or 4 capsule, one with pure water as standard and others – with solutions. The results were compared inside the group of runs mainly.

Comparison of densities in pure water fluid inclusions with calculated by equation of state (Wagner, Pruss, 2002) demonstrates systematic deviation in the nominal run parameters. Pressure could be smaller or/and temperature higher than marked in Table 1. Usually a pressure in the bomb was putted as 750 MPa, but day's leakage lead to loss of 15-17 MPa, which were compensated by intensifier. So, real pressure at the runs was between 730 and 750 MPa. The thermocouples measured the argon temperature in the capsule holder, but capsules were lying on wall of holder and real T in capsule could be higher from conductive heat exchange. All together the systematic error recalculated to temperature was estimate as 20-30°C, and real T was on the value higher than marked.

The runs quenched by switching out power supply and temperature fall down to 100°C during 1 minute and in the same time the pressure decreased by intensifier to 300 MPa. After quenching the bomb cooled down to room temperature, pressure decreased to atmospheric and bomb opened. Capsules checked for hermetic, opened and quarts seed cleaned by water, dried, and fixed by Canadian balsam inside glass tubes. The tubes were cut on 5-7 discs with 0.6-0.7 mm thickness and each ones was two side polished (0.5 mm thickness after).

The samples cleaned by acetone in ultrasonic bath 3-4 times and measured by similar stages produced by "Linkam" (UK). In the stage at measurements of runs from N. 4 to 36 the silver cup cover sample to minimize thermal gradient (stage of Edinburgh University) the samples from 37 to 42 have measured in IEM RAS (Russia) and ISTO (Orleans, France). Last stage was calibrated and the data corrected.

At the marked parameters all fluid inclusions has densities more than 0.8 g/ccm and homogenized at heating into liquid phase ($L+V \rightarrow L$). The main part of measurements of homogenization (Th) and vapour nucleation (Tn) temperatures were done at speed of heating/cooling as 2°C/min. Special test to reproducibility the values was done for sample from run N. 23, where one relatively large inclusions (visible size 100x8 μm) was measured 7 times with different rate of heating/cooling from 1 to 5°C/ min. The values of Th are variable from 200.7 to 200.9°C, and Tn changed from 146.7 to 146.9°C and the difference was 54 +/- 0.1°C. So, thermal gradient inside our samples with 0.5 mm thickness (after polishing) and rate of heating/cooling do not influent on results seriously.

3. Results

Usually we select 2-3 polished samples from each run and tried to measure 20-30 inclusions for each run. Some samples content only 10-25 measurable inclusions, some - more than 50. All together Th and Tn were measured for more than 1000 inclusions. The inclusions near top surface are better for observation as near Th gas bubble is too small. In some inclusions Th cannot be measured as bubble going to dark part, formed shadow from faces of the inclusion. Temperature Tn determinate much easy as bubble appears as large one, sometime typical boiling with many bubbles observed. Film with intensive retrograde boiling inclusion at overcooling on 51°C below equilibrium (Th= 221°C) could be seen on (http://www.iem.ac.ru/staff/kiril). It is visual presentation of tension in metastable area before spontaneous nucleation and equilibration.

4. Pure water

The statistic of measurements in the histogram form have been published in the volume devoted to V. Skripov memory (Shmulovich et. al., 2008). Distributions of Th values are Gaussian-type usually, with standard deviation 2-3°. The normal distribution means that average values could be taken as characteristic of liquid density and deviation is result of gradients in run holders and error of measurements of the sample on Linkam stage.

The scattering of Tn and, accordingly, Th-Tn values, especially for runs with 530-550°C, do not corresponds to normal distribution. The difference Th-Tn used instead Tn to minimize influence of contact surface of samples with silver plate and inter laboratory shift. Distribution of Th and Th-Tn demonstrate:

1. Inclusions with pure water have equal distribution of Th-Tn in wide interval of temperatures. It is mean that these distributions reflect some physical phenomena but not random errors. Reasons for that require special discussion.
2. Increasing of run temperatures (decreasing density of inclusions) lead to decreasing of intervals Th-Tn. In the samples, synthesized at 530-550°C, some inclusions has Th-Tn near 60-70°C, but in the runs at 650-700°C

(runs 18 and 34) this values decrease almost in 3 times. It is expectable effect as the value must be zero at critical point.

3. Minimal values Th-Tn for different samples are between 12 and 20°C. It is mean, that nucleation of vapour phase never observed before overcooling inclusion on 10°C (or smaller) below L-V equilibrium. At average slops of water isochors as 1.5-1.6 MPa per 1°C, it corresponds to nucleation pressures from –20 to –30 MPa. Last values are very similar to cavitation pressures, estimated by other methods (see review Herbert et. al., 2006). By other words, in some fluid inclusions formation of vapour phase begin at the same pressure (rate of tension) as in capillaries, optic or other cells, but in other inclusions with same water density the cavitation take place at much higher tension (larger values of Th-Tn).

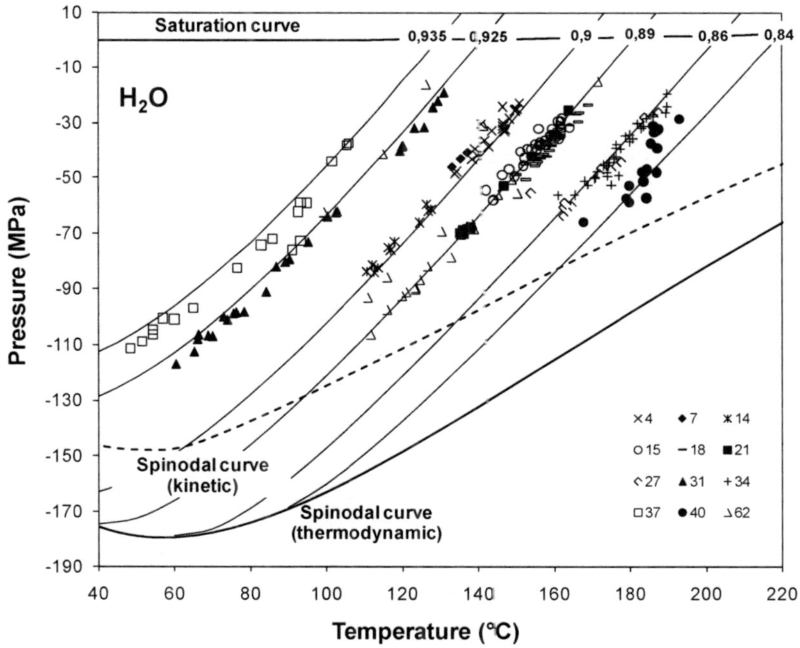

Figure 1. *Pressure of spontaneous nucleation in syntetic fluid inclusions with pure water. Legend content numbers of runs. Densities of isochores marked on top.*

Calculations of pressure related to the spontaneous nucleation for each inclusions were done with equation of state (EoS, Wagner, Pruss, 2002), recommended IAPWS in 1995 and presented on fig. 1. This many parametric EoS is very precise reproduce of experimental data and good for interpolation, but as many EoS of similar type could produce large errors at extrapolation. These calculations were done for area where this EoS was not calibrated.

Indirect support for application the EoS in the metastable state could be finding in the review of Baidakov (this volume, fig. 6), where isotherms don't change the slope at crossing of saturation curve (L+V). All points for each sample are located along a single isochoric curve, splitting in the horizontal direction what corresponds the variation in Th. The interval of nucleation pressures along isochors corresponds to Th-Th, recalculated by EoS in pressure values. This pressure interval is increasing with increasing densities of water and decreasing to zero at critical point.

Minimal pressure of nucleation was found in samples from run N. 31 (550°C) and correspond –117 MPa. In the sample from N. 37 (530°C) minimal pressure determinate as –105 MPa. It is some higher than was found before (-140 MPa, Zhang et. al., 1991), but authors noted very bad reproducibility of this measurements and some cooling procedure the inclusion do not nucleate vapour at all. The maximal pressure of nucleation was estimated as –22 MPa, at higher pressure the vapour phase did not formed.

The large temperature interval of vapour nucleation (and, consequently, pressure interval) could be result of 3 main effects:

1. Heterogeneous nucleation. Vapour phase formation could provoke by singularities in the inclusions – microcrystals, sharp ledges, etc. The heterogeneous nucleation takes place at tension smaller or much smaller than limiting tension for homogeneous nucleation. It seems that decreasing of centers for heterogeneous nucleation must lead to shift Th-Tn to larger values or, as minimum, decrease quantity of inclusions with small Th-Tn.
2. Gas impurities. Limited gas solubility can provoke formation gas phase (nucleation), contenting not a vapour but this gas mainly. This effect will decrease Th-Tn as well. In our runs this effect is not important as concentrations of dissolved air and hydrogen jr oxygen are negligible.
3. Non isochoric behavior of the inclusions. It is mean that volume of inclusions can change at tension and the point sprays on fig. 1 must go out from isochors. For large (50*50*1 μm) water inclusion the changing of density was confirm by Brilluen spectroscopy (Alvarega et. al., 1993). This effect will work in opposite direction to compare with first two and will lead to increasing Th-Tn.

These effects can work together or separately.

Runs with 0.1 and 0.5 m NaOH solutions checked influence of heterogeneous nucleation on Th-Tn. The matrix quartz is well soluble in alkali solutions and, consequently the solutions must decrease singularities on inner surface of the inclusions. The diagram in P-T coordinates for inclusions with alkali solutions presented on fig. 2. The runs in legend on fig. 2 were done at the same conditions as nearest number on fig. 1. Comparing with fig. 1 show that good fluid solvent for matrix crystal dramatically decrease the quantity of inclusions with relatively high Tn (or small Th-Tn). Practically all points of P-T nucleation are grouping near –100 MPa at density 0.92 g/cm^3. This result was reproduced in the similar runs at different densities and later in the 3 months runs by autoclave method at low P and T (200 MPa, 300°C). It is mean that nucleation at small tension (small Th-Tn on fig.1) is result of singularities on inner surface of inclusions (surface quality) rather than properties of liquid inside.

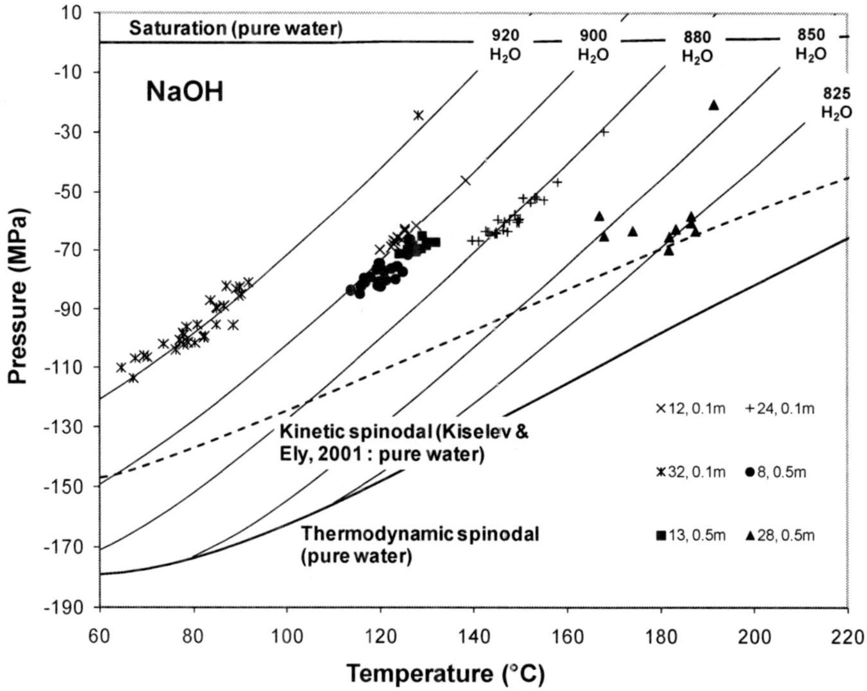

Figure 2. Pressures of spontaneous vapour phase nucleation in alkali solutions. Legend content numbers of runs and NaOH concentrations.

4. Salt solutions

We studied 1 and 5 m solutions of NaCl, CsCl and $CaCl_2$ and 1.3 m solutions $NaClO_4$, Na_2SO_4, Na_2WO_4 and Na_2MoO_4. The supersaturated at room temperatures solutions of Na_2SO_4, Na_2WO_4 and Na_2MoO_4 were trapped in inclusions also. Fig. 3 and 4 demonstrate the data for NaCl and $CaCl_2$ solutions. The calculation of homogenization density and pressure was performed using the software available on the website of the Duan Research Group (http://www.geochem-model.org/fluidinc/h2o_nacl/calc.php), based on the most recent papers about the subject. For each sample, an isochoric point was calculated at positive pressure with the same temperature interval than the Th-Tn difference (but with the opposite sign, of course). The isochor thus defined is extrapolated down to the nucleation temperature. This procedure is analogous but somewhat more empirical than the direct extrapolation of isochors calculated from a complex equation of state, as done for pure water.

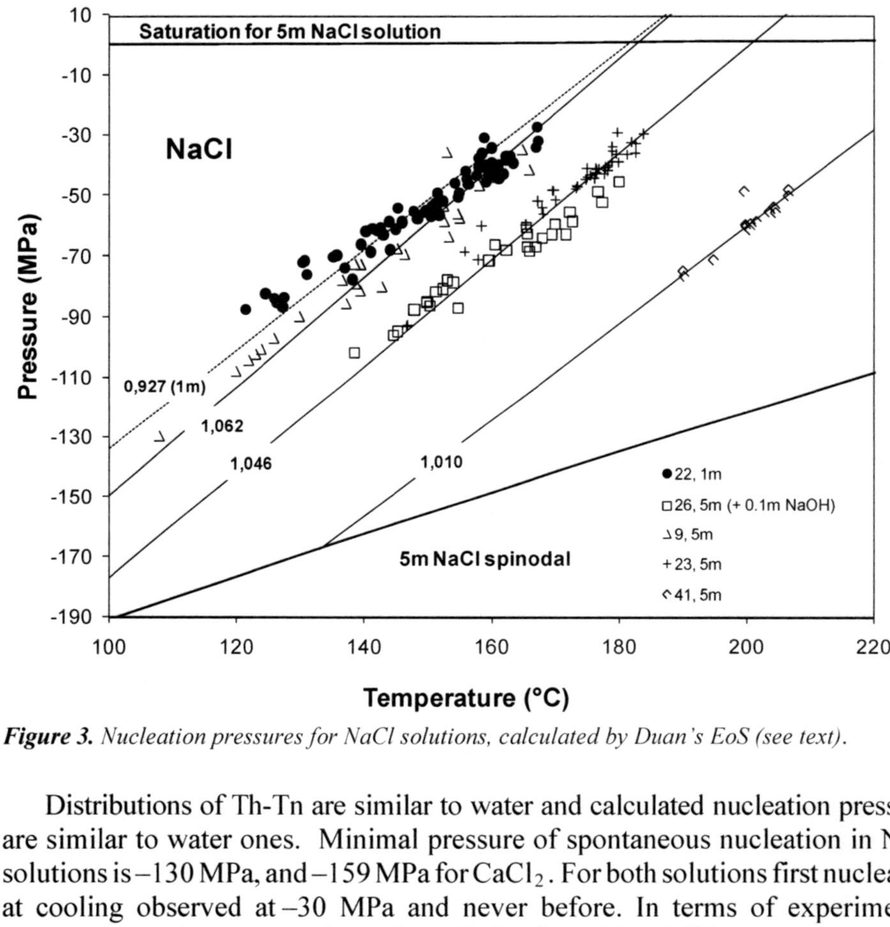

Figure 3. *Nucleation pressures for NaCl solutions, calculated by Duan's EoS (see text).*

Distributions of Th-Tn are similar to water and calculated nucleation pressures are similar to water ones. Minimal pressure of spontaneous nucleation in NaCl solutions is –130 MPa, and –159 MPa for $CaCl_2$. For both solutions first nucleation at cooling observed at –30 MPa and never before. In terms of experimental measurements it correspond to values Th-Tn from 20 to 90°C.

We cannot calculate nucleation pressure for CsCl solutions as EoS are absent up to now, but because measured values of Th-Tn sometime was more 130°C visualization have sense. Some inclusions with 5 m CsCl solution (run N. 35) has Th = 135°C and did not nucleate vapour phase at cooling up to 0°C. Vapour bubble was appeared only after 2 hours keeping at the temperature.

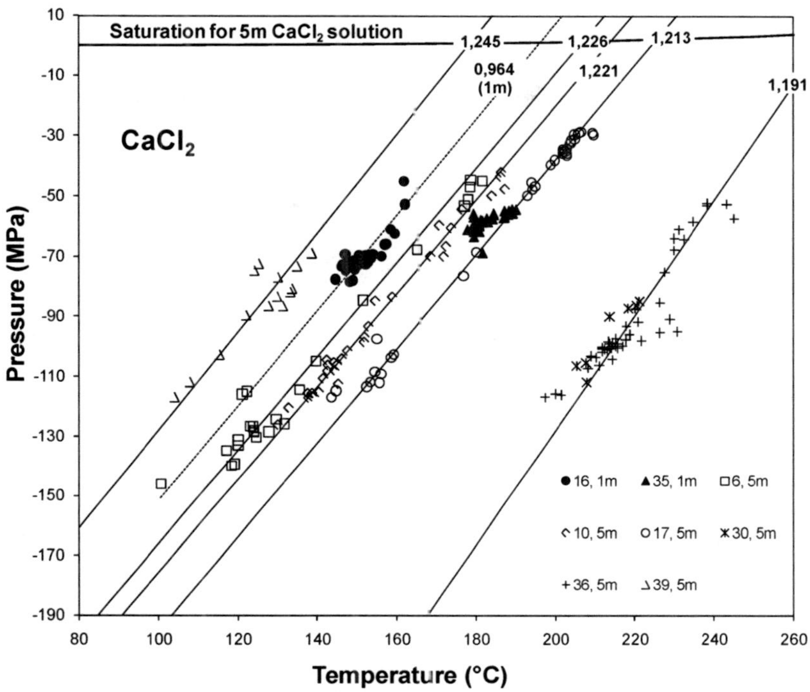

Figure 4. *Nucleation pressures for $CaCl_2$ solutions. See text for scheme of calculation.*

As seen on fig. 5 the very rough estimations of nucleation pressure shifted to more negative than on fig. 1–4. The nucleation in CsCl solution never observed above −40 MPa and this process extended up to −200–250 MPa. Main part of the experimental points are grouping around −100 MPa as on fig. 2 for alkali solutions. It could be result of similar quartz solubility in these liquids: SiO_2 solubility in CsCl solutions is higher than in any other chlorides solutions including even salt-in effect in wide P-T area (Shmulovich et. al., 2006).

The results for aqueous solutions on base of 0.1 m NaOH with 1.3 m $NaClO_4$, Na_2SO_4, Na_2WO_4 and Na_2MoO_4 did not produce principally new information. Compound $NaClO_4$ partly destroyed in the runs at 300°C. Samples in the runs with Na_2MoO_4 usually content very small and bad quality inclusions. The values of Th and Tn for 1.3 m solutions Na_2SO_4 or Na_2WO_4 are similar to the values of pure 0.1 m NaOH solution in parallel runs at the same conditions.

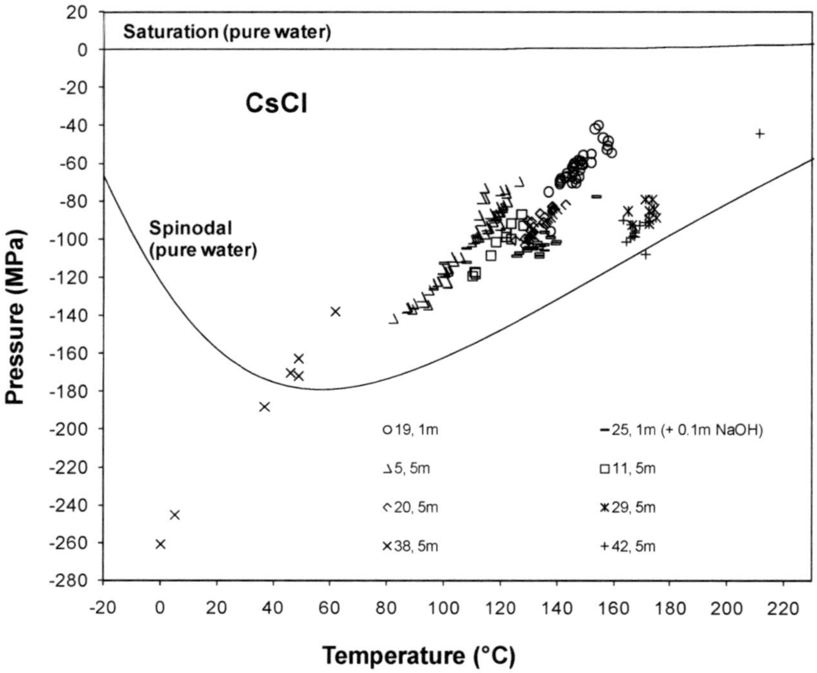

Figure 5. *Vapour nucleation pressures for inclusions with CsCl solutions, calculated by EoS for pure water. Water spinodale included for comparing.*

The supersaturated at room temperatures solutions of Na_2SO_4, Na_2WO_4 and Na_2MoO_4 were trapped in inclusions also. The inclusions with Na_2WO_4 and Na_2MoO_4 solutions did not content solid phase at room temperature, i.e. double metastability could be observed. Thermometric measurement show very small values Th-Tn, corresponding –20 MPa or around for spontaneous nucleation of a vapour phase. The inclusions with supersaturated Na_2SO_4 content at room temperature 2 and 3 phase combinations (L+V or L+V+S) and data of thermometry are contradictory. In the run by quartz overgrowing the Th-Tn was large enough, up to 65°C, but in the parallel run with sealing of cracks the values was near 15+/- 5°C.

Antracene was trapped into inclusions as well to measure pressure by shift of Raman spectra with pressure. It attempt was unsuccessful as antracene is very hydrophobic and the inclusions did not going deep to metastable area.

6. Conclusions

1. Large tension in water and aqueous solutions at typical sizes of the system 3-300 micrometers is real physical phenomena but not artifact.

2. The spontaneous vapour phase nucleation in one inclusion is accidental process which could start at –20 MPa and take place at any tension up to ~ –150 MPa.
3. Better quality of inner surface of the inclusions (increasing quartz solubility in the liquid) lead to shift of nucleation to larger tension.
4. Salt solutions (NaCl, $CaCl_2$) with concentration up to 5 m do not change principally metastable area to compare with pure water except CsCl solutions where SiO_2 solubility larger than in water.

Acknowledgements

This work has received financial support from Russian Fund of Basic Investigations, grant 06-05-64460 and from the French Agency for Research (Agence Nationale de la Recherche), grant SURCHAUF-JC05-48942. The authors warmly thank Bruce Yardley for help at start of the projects and Colin Graham, who welcome KS at the School of Geosciences in Edinburgh to carry out the hydrothermal synthesis. I will be happy to explain any "dark" place of the article by e-mail: kiril@iem.ac.ru

References

1. Skripov, V. P. (1972) *Metastable liquid* (*Moscow, Nauka*), in Russian.
2. Alvarenga, A. D., Grimsditch, M., and Bodnar, R. J. (1993) Elastic properties of water under negative pressures, *J. Chem. Phys.* **98**, 8392-8396
3. Baidakov, V. G. (2009) Experimental investigation of superheated and supercooled water (*this volume*)
4. Bodnar, R. J., and Sterner, S. M. (1987) Synthetic fluid inclusions, in Ulmer G., Barns H. L. (eds.) "Hydrothermal experimental techniques" (*J. Wiley and Sons, NY*), 423-457
5. Green, J. L., Durben, D. J., Wolf, G. H., and Angel, C. A. (1990) Water and solutions at negative pressure: Raman spectroscopy study to –80 MPa, *Science* **249**, 649-652
6. Herbert, E., Balibar, S., and Caupin, F. (2006) Cavitation pressure in water, *Phys. Rev. E*, **74**, 041603
7. Kiselev, S. B., and Ely, J. F. (2001) Curvature effect on the physical boundary of metastable states in liquids, *Physica A* **299**, 357-370
8. Roedder, E. (1967) Metastable superheated ice in liquid-water inclusions under high negative pressure, *Science* **155**, 1413
9. Shmulovich, K. I., Mercury, L., Ramboz, C., and El Mekki, M. (2008) Metastable water in synthetic fluid inclusions in quartz, in book "Metastable state and phase transitions" (*memory of P.Scripov*), Yekaterinburg, v. 8, p. 210-219 (in Russian)

10. Shmulovich, K. I., Yardley, B.W.D., and Graham, C.M. (2006) Solubility of quartz in crustal fluids: experiments and general equations for salt solutions and H_2O–CO_2 mixtures at 400–800°C and 0.1–0.9 GPa, *Geofluids* **6**, 154-167
11. Wagner, W., and Pruss, A. (2002) The IAPWS formulation 1995 for the thermodynamic properties of ordinary, substance for general and scientific use, *J. Phys. Chem. Ref. Data,* v. 31, n. 2, p. 387-535
12. Zheng, Q., Durben, D. J., Wolf, G. H., and Angel, C. A. (1991) Liquid at large negative pressures: water at the homogeneous nucleation limit, *Science* **254**, 829-832
13. Zheng, Q., Green, J., Kieffer, J., Poole, P. H., Shao, J., Wolf, G. H., and Angel, A. (2002) Limiting tensions for liquid and glasses from laboratory and MD studies. In "Liquids under negative pressure", eds. Imre A., Maris, M., Williams, P. R. *NATO Sci. Ser. II, Math. Phys. and Chem.* **84**, p. 33-46
14. Zilberbrand, M. (1999) On equilibrium constants for aqueous geochemical reactions in water unsaturated soils and sediments, *Aquatic Geochem.* **5**, 195-206

METHOD OF CONTROLLED PULSE HEATING: APPLICATIONS FOR COMPLEX FLUIDS AND POLYMERS

PAVEL V. SKRIPOV
Institute of Thermal Physics, Ural Branch of Russian Academy of Sciences, ul. Amundsena 106, Ekaterinburg, Russian Federation

Abstract: We are developing the method of controlled pulse heating of a thin wire probe to the investigation of heat exchange and phase stability for complex liquids (mixtures and thermally unstable fluids) under conditions of powerful heat release. Some applications complementing the regular measurements of spontaneous boiling-up temperature under definite volume-time values and heat-flux density through a liquid before and after the boiling-up have been discussed. The role of the traditional Odessa Thermophysical Meetings for generation of "metastable" ideas has been emphasized.

Keywords: pulse heating, heating function, superheated states, mixtures, oils, polymers

1. Introduction

As an introduction to the main part of the paper, it is pertinent to mention this year as the anniversary of a few significant events in thermal physics. Let us note the centenary of the first helium liquefaction by Heike Kamerlingh Onnes at Leiden. In his plenary report at the 18th European conference on thermophysical properties Dr. Arno Laesecke called this event a breakthrough in research on thermophysical properties of substances and, among other problems, he related the studies into metastable states of substances to "uncharted territories" in thermophysics.[1] In this connection our workshop seems to be well-timed.

In line with the workshop topic let us mention the 150th birthday anniversary of Alexandr Ivanovich Nadezhdin (1858-1886) who was a prominent researcher in the second half of the 19th century. Nadezhdin was born in the family of the military physician in the village Verkhopenie in the Kursk province (Russia) and went to a grammar school in Kiev. Being a student at St. Vladimir Kiev University, he already worked at the well-known Kiev physical laboratory headed by M. P. Avenarius and added one of the outstanding pages to the lab's history.[2] Let us note a very elegant method for determination of the critical temperature of a substance in nontransparent tube developed by Nadezhdin. Based on the concept that the densities of a liquid and the vapor equalize at the critical point, he designed an instrument and called it "a differential densimeter".[3] Incidentally, this instrument was not replicated by anyone else. By that time he obtained more than one-fourth of all the data on critical parameters including the first measurement of the critical temperature of water. With the method and the instrument for determination of critical parameters in hand, Nadezhdin was the first to closely test the principle of corresponding states and, actually, became one of the founders of the thermodynamic similarity theory. It is indicative that Nadezhdin's thesis, which was published in "Proceedings of Kiev University" in 1885-1886, was entitled "Etudes of Comparative Physics". The second and third parts of the thesis were translated into German by Avenarius and published after the author's death.[4]

2. Background

To start with, let us consider a particular case of metastable states – a liquid superheated with respect to the liquid-vapor equilibrium temperature. For simplicity let us take a pure liquid at positive pressures, see Fig. 1. The region of superheated states is limited from below by the binodal $T_s(p)$ and from above by the experimental line of attainable superheat, or, in other words, the line of spontaneous boiling-up $T^*(p; t_{\rm exp})$ of the liquid. An understandable limitation is imposed on the volume of superheated sample V and the time period $t_{\rm exp}$ of experiment. Naturally, the experimental time should be shorter than the life time $t_{\rm l}$ of the metastable state.

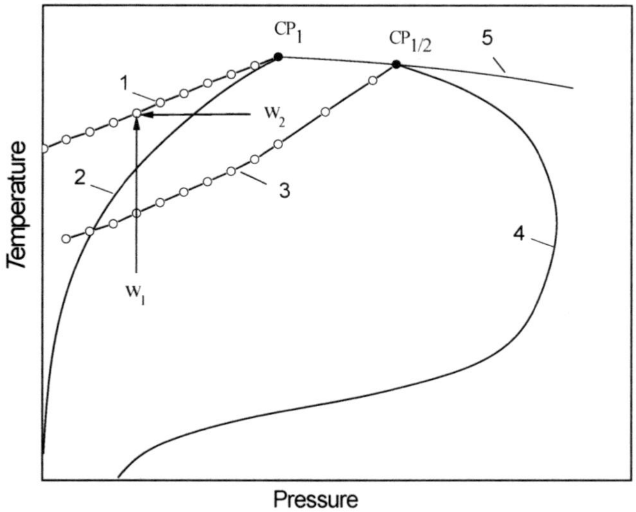

Figure 1. *Liquid-vapor phase diagram for pure liquid (1, 2) and binary mixture (3, 4): the lines of attainable superheat (1, 3), binodals $T_s(p;\ c = const)$ (2, 4) and liquid-gas critical curve (5). Symbols "CP" indicate the critical point, "w" – the way of superheating of pure liquid.*

The phase transition is accompanied by the change in substance density up to an order of magnitude. This provides a basis for detection of the boiling-up moment by a certain experimental feature. For a sufficiently small value of $V \cdot t_{exp}$ product one can achieve the attainable superheat values predicted by the homogeneous nucleation theory.[5]

The superheated states are reached in experimental practice by various ways. Two basic ways (w_1 and w_2) are shown in the Fig. 1. Originally, the constant-pressure heating process was realized with the method of a droplet rising in a host liquid. For the process of isothermal decrease in a pressure a small bubble chamber has been designed. These methods, developed in the complete form in the Ural thermophysical school,[5-8] have proved to be helpful. They provided receiving the data collection related to the spontaneous boiling-up temperature versus pressure, or, in more exact terms, to the mean life time of the superheated liquids with respect to the definite p-V-T parameters. Moreover, the second method has provided a basis for performing measurements of substance properties in superheated states. Indeed, step by step, the corresponding methods have been developed and the essential results for properties of pure liquids in superheated states have been obtained.[5-8]

However, in nature, engineering and private life we deal with solutions. Addition of the second component considerably complicates the problem. In this connection, attention may be paid, firstly, to the transfer of the liquid-vapor critical curve and, hence, the binodals T_s (p; c = *const*, c – the concentration in liquid phase) to the region of elevated pressures, see Fig. 1, and secondly to changes in the thermodynamic compatibility of components with temperature and pressure. These factors lead to considerable extent of the two-phase equilibrium region with respect to that of pure liquid [9,10] and, consequently, to the principal increase in the requirements on the experimental methods and devices used to study this region.

Another point relevant to our context deals with thermal instability of a wide class of liquids. For example, most if not all of polymeric liquids are thermally unstable ones. The line of its attainable superheat for these liquids exceeds the onset temperature of thermal decomposition of molecules. So, the liquid-vapor phase transition ceases to be point-like with respect to temperature and proves to be dependent on the heating time, or more exactly, on the heating trajectory in time-temperature plane.[11,12] This determines the difference of the phenomenon of spontaneous nucleation in complex fluids, compared to that of simple ones.

As the size of the metastable region and the number of relaxation times by all the signs increase, the use of quasi-static methods becomes unacceptable. It is probably for this reason in part that research into not fully stable states of complex systems (mixtures and thermally unstable fluids[9-12]) progresses very slowly. To solve the problem, we are developing the method of controlled pulse heating of a thin wire probe – resistance thermometer.

3. Method of controlled pulse heating

Initially, the method was designed for elucidation of spontaneous boiling-up kinetics on the background of artificial vapor formation centers.[6,7] The current version of the method[13] gives a possibility to achieve the controlled superheat (with respect to liquid-vapor or liquid-liquid equilibrium temperature as well as to the onset temperature of thermal decomposition of molecules) taking into account a number of thermal properties of the sample, to select the characteristic boiling-up signal on the monotonic heating curve (see Fig. 2), to determine the substance temperature and the heat-flux density from the probe through the substance at any moments we are interesting. The essential features of the method are as follows:

the relatively small value of heated volume being in accordance with the wire diameter (10 μm and 20 μm);
the relatively small value of heating time being from 1 μs to 10 ms;
the probe's combination of a heater and resistance thermometer functions makes it a convenient to use digital means for the pulse shape creation and for recording the response to a particular pulse. Details of the use of the method for a study of superheated states of water/[poly(propylene glycol) - PPG-425] mixture[14] – a typical system with the lower critical solution temperature – can be found in the first poster.

Finally, the method allows one to select the entrance trajectory (in time-temperature or power-temperature variables) into the region of the substance superheated states being in accordance with a chosen model of heat exchange and with a set of relaxation times for the system. In the general case the heating function can be arbitrary. The most significant particular cases are as follows:

the constant power mode, see Fig. 2a; the temperature plateau mode, see Fig. 2b; cooling of the impact-heated probe mode.

The first technique makes up the most convenient case for modeling the dependence of thermophysical properties on temperature from the data of pulse experiment. Moreover, it gives grounds for the direct comparison of thermal resistance for different samples of fluids. The second one is defined as a combination of a short "heating" pulse and a more longer "thermostating" pulse which balances out the heat flux into the sample in the course of experiment. It provides essential definiteness in comparison of the heat transfer parameters and the mean life-time of superheated systems up to their boiling-up. The comparison is based on the analysis of changes in time of the power that must be released in the probe for maintenance of its temperature at a chosen level. The boiling-up signal is resolved just on the power curve. With the third technique we attempted to reduce the heating time in order to increase as much as possible the depth of reproducible penetration into the region of superheated states of a substance while maintaining its structure.

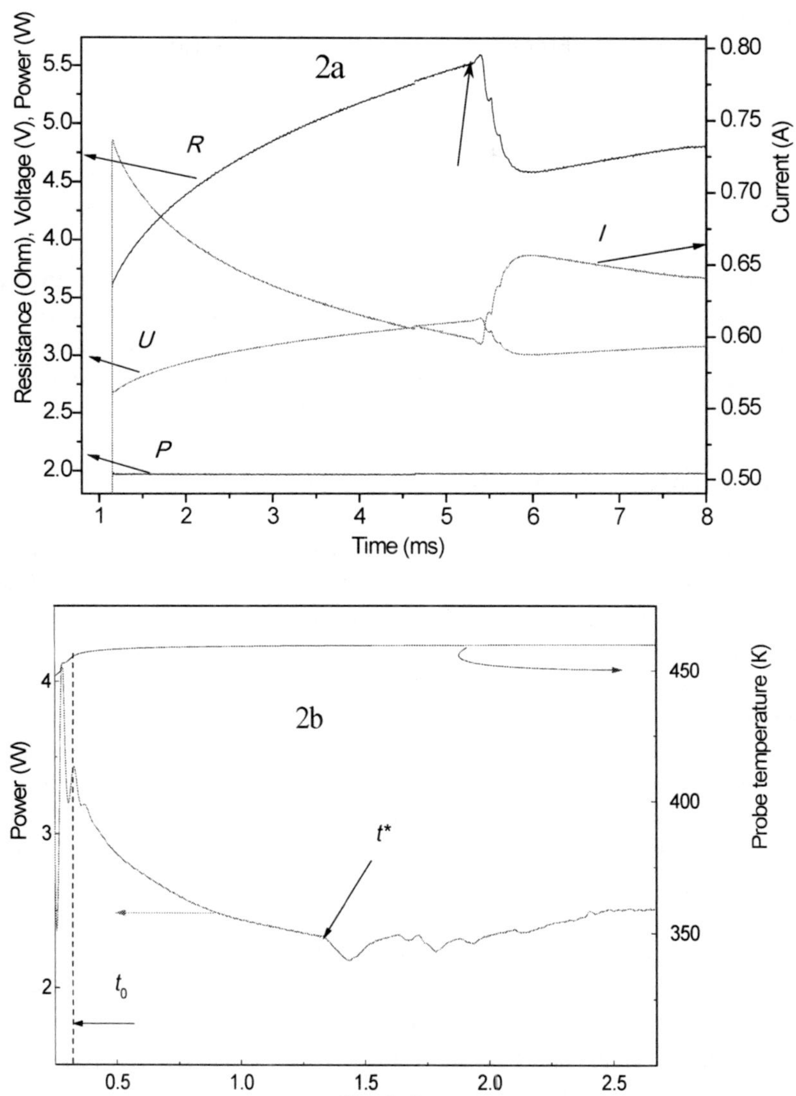

Figure 2. *The method of controlled pulse heating of a thin wire probe: characteristic heating curves in the constant power mode* $P(t) \approx const$ *(2a) and the temperature plateau one* $T_{pl} = T(t > t_0) \approx const$ *(2b). Here* t_0 *is the time period required for transition to the regime. Here and further, arrows show the moment of spontaneous boiling-up (*$t = t^*$*) for the liquids.*

Due to the short length of a probe temperature rise (about 1 μs) the measurement stage is shifted to the "cooling tail" followed by the shock heating pulse.[13] The cooling process is recorded due to relatively low, so-called monitoring current across the probe. The shock pulse may be superimposed on a basic heating function (for example, on the temperature plateau one) at a selected instant of time.

4. Applications

Firstly, the applications include estimation of the liquid-vapor critical point coordinates.[15,16] The absence of reliable methods for prediction of critical parameters of thermally unstable liquids is the main barrier to thermodynamic simulation as applied to these liquids. The essence of our approach is as follows. By definition, the boiling line $T^*(p)$ terminates at the critical point. Indeed, the amplitude of the boiling-up signal shows the monotonic decrease with pressure, and on achieving a certain value of pressure p_{c*}, it is no longer resolved, see Fig. 3. This pressure value is taken as an approximation for the critical pressure. The corresponding value of temperature $T^*(p = p_{c*})$ is taken as an approximation for the critical temperature of the system, see Fig. 4. In the course of moving along the boiling line of a pure liquid, the proximity of the approach to the critical point is determined by the resolving power of the device, an appropriate choice of heating rates set and opportunities of software for the useful signal selection and summation as well. But such a procedure becomes less reliable in passing to binary systems (not to mention multi-component systems[11,17]) due to the temperature dependence of the degree of compatibility for components and specific shape of binodals for mixtures.

Secondly, the applications include the comparison of subsecond thermal stability of polymeric liquids, which do not boil without their decomposition. Let us return to the temperature plateau mode. It allows the creation of nearly isothermal conditions and the determination of the mean life-time of a substance $t_l(T_{\mathrm{pl}})$ before its decomposition (marked by boiling-up signal) at a given probe temperature T_{pl}. Fig. 5 clears up the procedure of the mean life-time of a superheated liquid determination. Here the values of T_{pl} serve as a parameter. In the course of measurements on oils we have revealed that the experimental values of life-time are controlled mainly by the content of volatile impurities rather than basic properties of the selected oil. Thus, we have received an indirect method of rapid analysis for the concentration of volatile impurities independent on their nature in commercial oils, see Fig. 6. The method is based on the existence of an unambiguous dependence of the life-time of a superheated substance on the volatile impurities content at a given temperature T_{pl} value. This dependence has proved to be the steepest in the region of negligibly small, at the level of traces, contents of volatile impurities. The experimental fact of extremely high sensitivity of the characteristic life-time of superheated oil to the presence of soluble low-boiling component has been applied in the device for monitoring of an actual state of an oil, see Fig. 7.

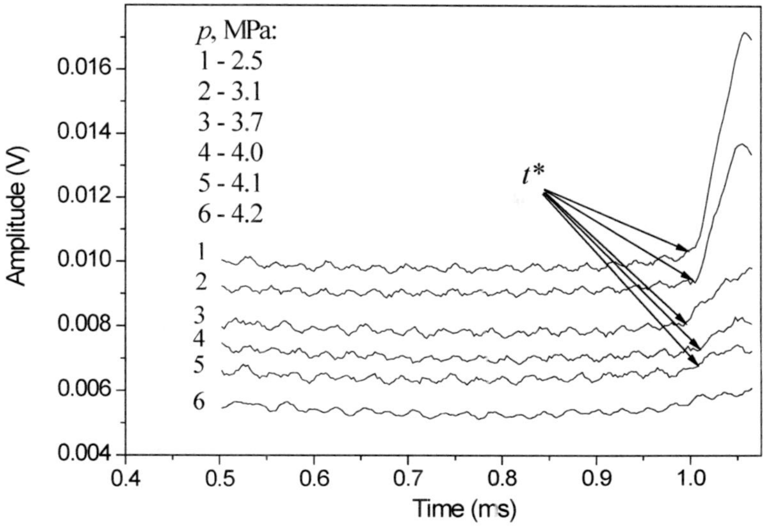

Figure 3. *The amplitude of on-line boiling-up signals separated from the background of smooth heating for PPG-425/water mixture at different pressures and given heating time.*

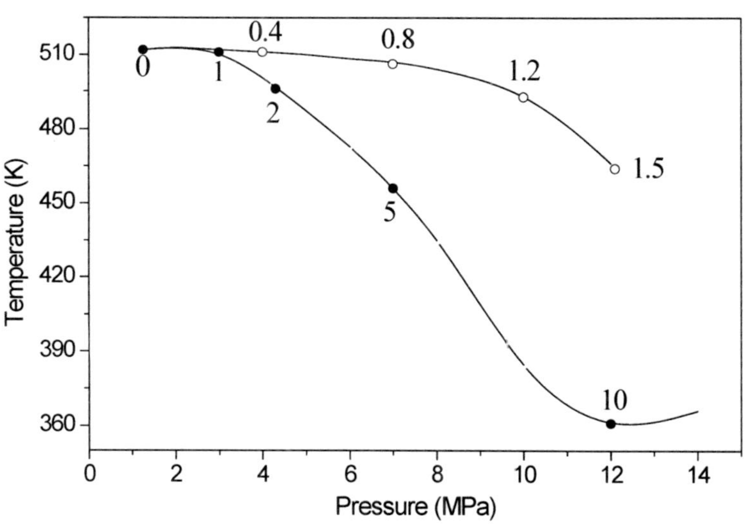

Figure 4. *The approximation for the liquid-vapor critical curve for PPG-425/water and PPG-425/CO_2 mixtures. The indicated numbers are the CO_2-saturation pressure values in MPa (open circles) and water weight fraction (filled circles).*

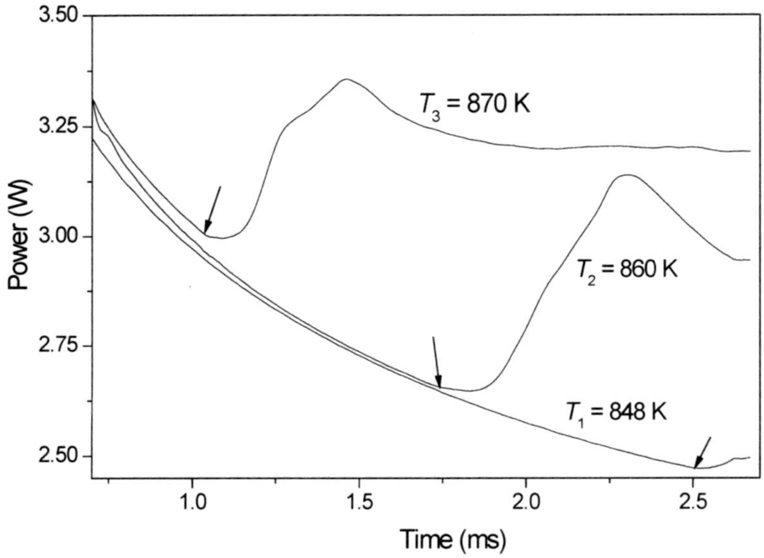

Figure 5. *Heat power called for the probe thermostabilization in PPG-2000 vs. time.*

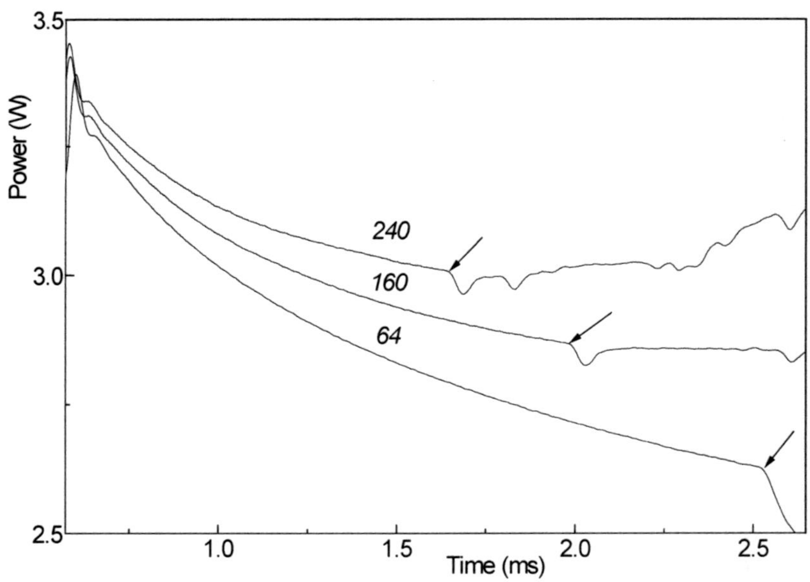

Figure 6. *Heat power called for the probe thermostabilization in oil "Bitzer" vs. time at a given probe temperature T_{pl} = 760 K. Water content (in gram of water in ton of oil) serves a parameter.*

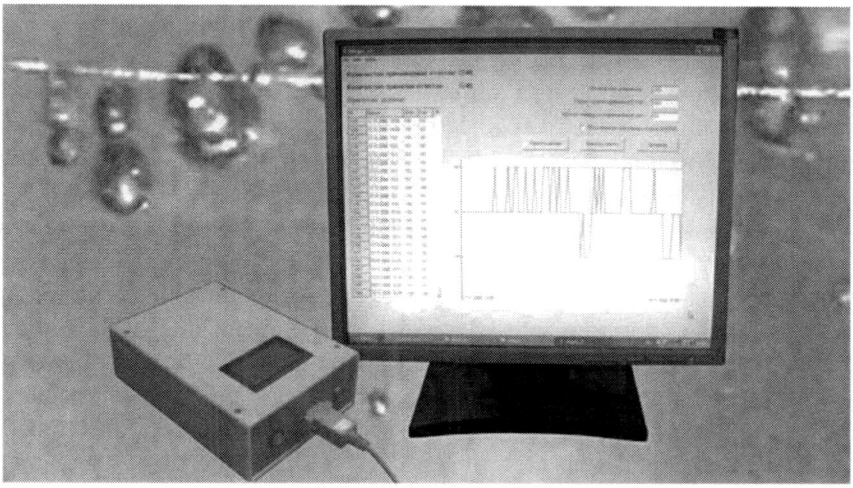

Figure 7. The device for local monitoring of volatile impurities in technological oils.[18]

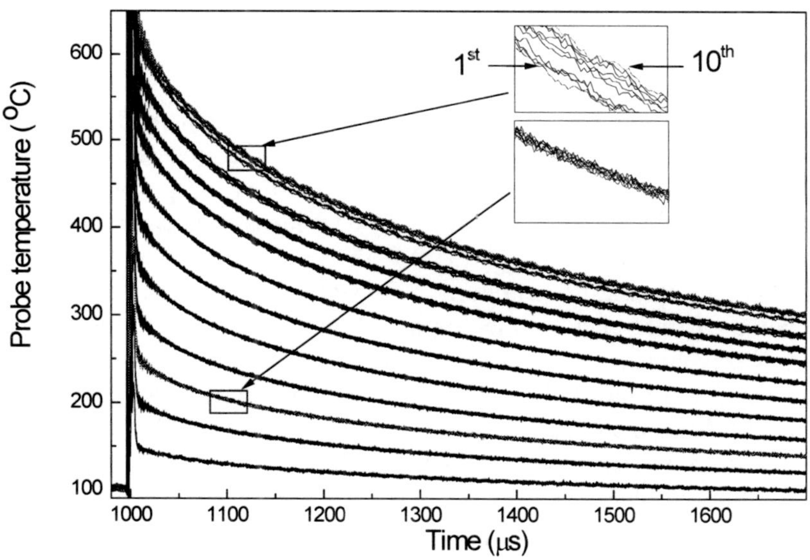

Figure 8. *The probe temperature vs. time in experiments with 10 pulses per series for each value of power* P_{sh}. *Arrows in the top inset indicate curves which correspond to the first and tenth pulses.*

As for the third mode which is based on combining the thermal impact and the monitoring heating functions with characteristic pulse lengths of the order of 1 μs and 1 ms, respectively. We have developed this approach for the comparison of thermal resistance and short-time thermal stability of polymers and binders under conditions of shock heating. Thermal contact between the probe and polymer is achieved in the following way. The probe together with current supply construction was immersed in the cell with liquid monomer (for example, methyl metacrylate) or reacting fluid and implanted into the bulk of sample during polymerization. In the course of experiment the power of shock heating pulse P_{sh} is increased in step by step manner, see Fig. 8.

Measurements were based on the comparison of cooling curves related to the reproducible pulses with a given power value. In experiments on glassy polymers we used 10 sequential pulses per series usually. The recorded equivalent of the cooling rate changes is the mean integral temperature corresponding to the chosen time interval on the response curves. The cooling tail run may be controlled by the monitoring function parameters choice. At a certain step of the probe temperature increment the response curves were no longer coincided, see insets of the Fig. 8. Systematic increase in the thermal resistance of a substance in a series (which is taken as a sign of the substance thermal decomposition) became evident. The characteristic data on the thermal decomposition onset for polymers have been presented elsewhere.[19]

5. Conclusions

In conclusion I would like to emphasize the contribution made by the host city of the workshop to the development of the research area under consideration. Odessa is, as judged from the history of problem, the most favorable place for generation of thermophysical ideas. It is enough to recollect the chain of the well-known in the USSR thermophysical meetings (so-called Schools) held in Odessa. Let us return to the starting point of the chain, namely, to the open-air Conference on the Applied Thermodynamics Problems chaired by Ya. Z. Kazavchinskiy in September 1962.

A working moment of the Conference is shown on Fig. 9. Professor V.K. Semenchenko from the Moscow State University is surrounded by five young researchers from Ekaterinburg (former Sverdlovsk). All of them managed to solve essential problems within the next few years. Doctors V.P. Skripov (1927-2006) and P.E. Suetin (1927-2003) have developed their own directions in thermal physics. Following the workshop topic, let us note that V.P. Skripov has been investigating metastable states for 45 years[20,21] and has proved to be the founder of the Ural School of thermophysics. The upper row on the Fig. 9 presents the "first wave" of his disciples. Yu.D. Kolpakov (1929-2006) has developed the method of light scattering in substances above and below the critical temperature as an instrument for revealing phase states with a reduced stability.[6] E.N. Dubrovina (1937) has investigated the boiling crisis.

Figure 9. *The lower row: V.P. Skripov, V.K. Semenchenko and P.E. Suetin; the upper row: Yu.D. Kolpakov, G.V. Ermakov and E.N. Dubrovina.*

Now she is the editor of the annual volume "Metastable States and phase Transitions",[21] possibly, the only edition devoted exclusively to the phenomenon of metastability. Finally, G.V. Ermakov (1938), the coauthor of the report "Investigation of the attainable superheat of liquids in a wide region of pressure" at that Conference, has developed the method of a rising droplet to study the nucleation kinetics under pressure and a number of methods to determine the thermodynamic properties of liquids in superheated states.[8] To my opinion, his attitude to science in a broad sense corresponds to "the spirit of Heike Kamerlingh Onnes".[1]

Acknowledgements

This work was supported by the Russian Foundation for Basic Research, project no. 06-08-01324-a), grant of the President of RF (NSh-2999.2008.8) and the Joint (for Ural and Siberian Branches of Russian Academy of Sciences) Project or Basic Research.

References

1. Laesecke, A. (2008) 100 Years Liquefaction of Helium – A Breakthrough in Thermophysical Properties Resarch and Its Contemporary Significance, *in 18th European Conference on Thermophysical Properties. Book of Abstracts*, p. 249, University of Pau, France.
2. Kipnis, A. Ya. (1990) From the history of molecular physics: A.I. Nadezhdin 1858-1886, *in Studies on history of physics and mechanics. 1990*, Nauka, Moscow, 5-36 (in Russian).
3. Khvol'son, O. D. (1923) *Course in Physics*, **5th** Ed., V. **3**, Berlin, 648-649 (in Russian).
4. Nadejdin, A. I. (1887) Ueber die Ausdehnung der Flüssigkeiten und den Uebergang der Körper aus dem flüssigen in der gasförmigen Zustand. *Exner's Rep. Phys*. **23**, 617-649, 685-718. Ueber die Spannkraft der gesättigten Dämpfe. *Ibid*. 759-790.
5. Skripov, V. P. (1992) Metastable States, *J. Non-Equilib. Thermodyn*., **17**, 193-236.
6. Skripov, V. P. (1974) *Metastable Liquids*, Halsted Press, John Wiley & Sons, Inc., (New York).
7. Skripov, V. P., Sinitsyn, E. N., Pavlov, P. A., Ermakov, G. V., Muratov, G. N., Bulanov, N. V., and Baidakov, V. G. (1988) *Thermophysical Properties of Liquids in the Metastable (Superheated) State* (Gordon and Breach Science Publishers, London).
8. Ermakov, G. V. (2002) Thermodynamic Properties and Boiling-Up Kinetics of Superheated Liquids, UrO RAN, (Ekaterinburg,), *in Russian*.
9. Skripov, P. V. and Puchinskis, S. E. (1996) Spontaneous Boiling-Up as a Specific Relaxation Process in Polymer-Solvent Systems, *J. Appl. Polym. Sci.* **59**, 1659-1665.
10. Skripov, P. V., Puchinskis, S. E., Starostin, A. A., and Volosnikov, D. V. (2004) New Approaches to the Investigation of the Metastable and Reacting Fluids, *in S. J. Rzoska and V. Zhelezny (eds.), Nonlinear Dielectric Phenomena in Complex Liquids NATO Sci. Series II,* vol. **157**, 191-200 (Kluwer, Brussel).
11. Pavlov, P. A., and Skripov, P. V. (1999) Bubble Nucleation in Polymeric Liquids under Shock Processes, *Int. J. Thermophys*., **20** (6), 1779-1790.
12. Puchinskis, S. E., and Skripov, P. V. (2001) The Attainable Superheat: From Simple to Polymeric Liquids, *Int. J. Thermophys.* **22** (6), 1755-1768.
13. Skripov, P. V., Smotritskiy, A. A., Starostin, A. A., and Shishkin, A. V. (2007) A Method of Controlled Pulse Heating: Applications, *J. Eng. Thermophys.,* **16** (3), 155-163.
14. Skripov, P. V., Smotritskiy, A. A., and Volosnikov, D. V. (2008) Thermophysical properties of doubly metastable fluid. Experiment and modeling,

in 18th European Conference on Thermophysical Properties. Book of Abstracts, P. 152-153, University of Pau, France.

15. Nikitin, E. D., Pavlov, P. A., and Skripov, P. V. (1993) Measurement of the critical properties of thermally unstable substances and mixtures by the pulse-heating method, *J. Chem. Thermodyn.* **25**, 869-880.
16. Nikitin, E. D., and Popov, A. P (2007) Using the Phenomenon of Liquid Superheat to Measure Critical Properties of Substances, *J. Eng. Thermophys.* **16** (3), 200-204.
17. Skripov, P. V., Starostin, A. A., Volosnikov, D. V., and Zhelezny, V. P. (2003) Comparison of thermophysical properties for oil/refrigerant mixtures by use of pulse heating method, *Int. J. Refrig.* **26** (8), 721-728.
18. Shangin, V. V., Il'inykh, S. A., Puchinskis, S. E., Skripov, P. V., and Starostin, A. A. (2008) Method of heat pulse testing for technological liquids monitoring, *Izvestiya vuzov. Gorniy Zhurnal*, No. 8, *in Russian.*
19. Volosnikov, D. V., Efremov, V. P., Skripov, P. V., Starostin, A. A., and Shishkin, A. V. (2006) An Experimental Investigation of Heat Transfer in Thermally Unstable Polymer Systems, *High Temperature* **44** (3), 463-470.
20. Nakoryakov, V. E., and Baidakov, V. G. (2007) To the Reader, *J. Eng. Thermophys.* **16** (3), 107-108.
21. From Editors (2008) *in E.N. Dubrovina (ed.), Metastable States and Phase Transitions,* **9**, 4-11, Ekaterinburg, UrO RAN, *in Russian.*

PART V

OTHER METASTABLE SYSTEMS

COLLECTIVE SELF-DIFFUSION IN SIMPLE LIQUIDS UNDER PRESSURE

[1]NIKOLAY. P. MALOMUZH, [2]KONSTANTIN S. SHAKUN, [3]VITALIY YU. BARDIK
[1]*Odesa National University, Dvoryans'ka Str., 2, Odesa 65026, Ukraine*
[2]*Odesa National Maritime Academy, Didrikhson Str., 8, Odesa 65023, Ukraine*
[3]*Taras Shevchenko Kyiv National University, Acad. Glushkov Prosp., 2, Kyiv 03127, Ukraine*

Abstract: The behavior of the self diffusion coefficient in simple liquids under pressure is discussed. It is taken into account that the self-diffusion coefficient is the sum of the collective and one-particle contributions. From our reasons it follows that the collective contribution monotonously increases with pressure. The comparison with the computer simulation data for the full self-diffusion coefficient of argon shows that the relative value of the collective part increases from 0.2 for the pressure of saturated vapor up to 0.76 and larger for pressure 10 GPa.

Keywords: self-diffusion coefficient, thermal hydrodynamic fluctuations, dynamic viscosity, kinematic viscosity, Maxwellian relaxation time

In the general case, the thermal motion in liquids represents a combination of shifts of molecules with respect to their nearest surrounding and the collective drift in the field of thermal hydrodynamic fluctuations[1-3]. It is clear that an increase of the pressure is accompanied by the growth of the liquid density and, as a result, by essential increase of the relative role of the collective contribution to the self-diffusion coefficient. Indeed, due to the geometric restrictions, the relative motion of molecules is reduced to oscillations in the cell formed by the nearest neighbors. At the same time, an increase in the density influences the vortical modes of the thermal motion of molecules to a much smaller extent (Fig. 1). Since the collective transport in liquid is related just to vortical (transversal) hydrodynamic modes[1-3] (see Fig. 2), one can conclude that the role of the collective drift in the self-diffusion increases as the pressure grows.

S. Rzoska et al. (eds.), *Metastable Systems under Pressure*,

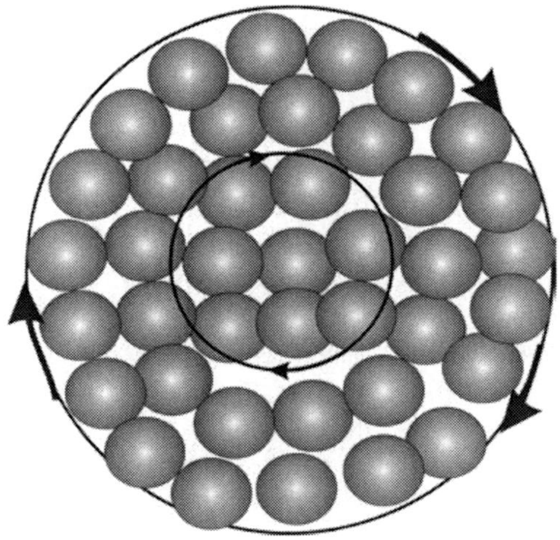

Figure 1. *Il lustration of the relative displacements of molecules in the extremely dense liquid. They are only possible due to thermal vortical excitations.*

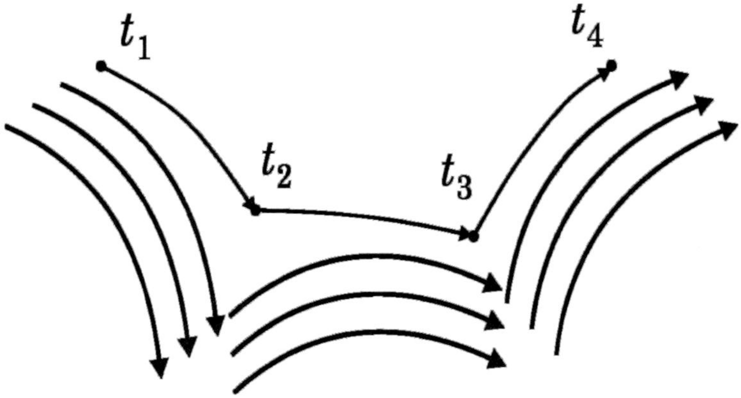

Figure 2. *Illustration of the collective drift* ($\vec{r}(t_1) \rightarrow \vec{r}(t_2) \rightarrow \vec{r}(t_3) \rightarrow \vec{r}(t_4)$) *of a molecule caused by fluctuation vortices in hydrodynamic velocity field in liquids.*

It is worth noting that an increase of the degree of "collectivization" for the molecular motion in associated liquids, leads to the considerable change of the self-diffusion and the shear viscosity. That is why it is appropriate to consider, first of all, the influence of the pressure on the processes of self-diffusion in simple liquids, in particular, in liquid argon.

In correspondence with refs.[1-3], the self-diffusion coefficient of molecules in liquids can be presented in the form

$$D = D_c + D_r, \tag{1}$$

where D_c and D_r stand for the collective and one-particle contributions, respectively. In accordance with ref.[1,2] the collective part D_c of the self-diffusion coefficient is identified with the self-diffusion coefficient D_L of a Lagrange particle with the suitable radius r_*:

$$D_c = D_L\big|_{r_L = r_*}, \tag{2}$$

where

$$D_L = \frac{1}{3}\int_0^\infty \varphi^{(L)}(t)dt, \tag{3}$$

and $\varphi^{(L)}(t) = <\vec{V}_L(t)\vec{V}_L(0)>$ is the velocity autocorrelation function of a Lagrange particle with the radius r_L. It was shown in ref.[2,3] that

$$D_L = \frac{k_B T}{5\pi\eta r_L}$$

and

$$r_* = 2\sqrt{\nu\tau_M},$$

where η and ν are the dynamic and kinematic shear viscosities correspondingly, and τ_M is the Maxwell relaxation time for the transversal modes of liquids. As the result the collective part of the self-diffusion coefficient for molecules is put to be equal:

$$D_c = \frac{k_B T}{10\pi\eta\sqrt{\nu\tau_M}}, \tag{4}$$

The temperature dependence of the Maxwell relaxation time is approximated by the expression[3,4]

$$\tau_M = \tau_M^{(0)}\left(\frac{\nu(T)}{\nu_0}\right)^{2/3}, \tag{5}$$

where $\nu_0 = 0.00134$ cm^2/s [5] and $\tau_M^{(0)} = 2.22\cdot 10^{-13}$ s are the values of the corresponding parameters at a temperature of 100 K.[6]

The behavior of the self-diffusion and viscosity coefficients of liquid argon during the increase of the external pressure in the interval (1,3-52) GPa was an object of the molecular-dynamic investigation performed in ref. [[7]]. The molecular motion was simulated as a motion of spheres characterized by the Buckingham intermolecular potential

$$U(r)=\varepsilon\left(Ae^{-\alpha\frac{r}{r_0}}-B\left(\frac{r_0}{r}\right)^6\right),\tag{6}$$

where $A=\frac{6e^{\alpha}}{\alpha-6}$, $B=\frac{\alpha}{\alpha-6}$. The parameters of the potential have the following values: $\frac{\varepsilon}{k_B}=122\ K$, r_0=3,85 Å α =13,2. The density of the investigated system changes within the limits from $\approx 0.6\,\mathrm{g/cm^3}$ at T = 298 K and P = 1,3GPa up to 4,05 g/cm^3 at T = 3000K, P = 52GPa. The computer simulation[7] are presented in Fig. 3 and Fig. 4.

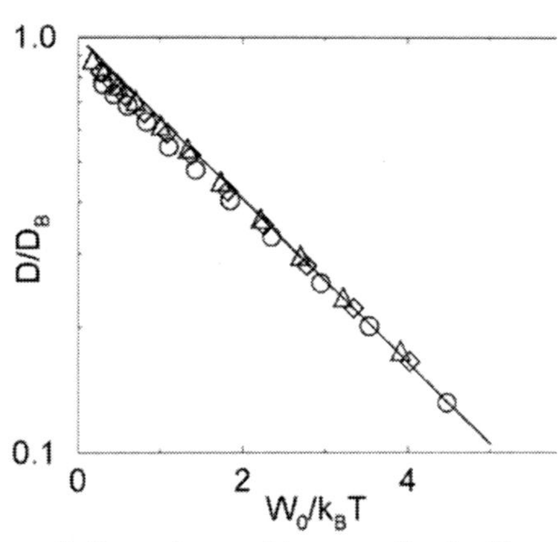

Figure 3. *Dependence of the normalized self-diffusion coefficient of argon on the dimensionless combination* W_0/k_BT (o – T = *298 K,* ◊ – T = *1000 K, and* Δ – T = *3000 K) according to ref. [[7]].*

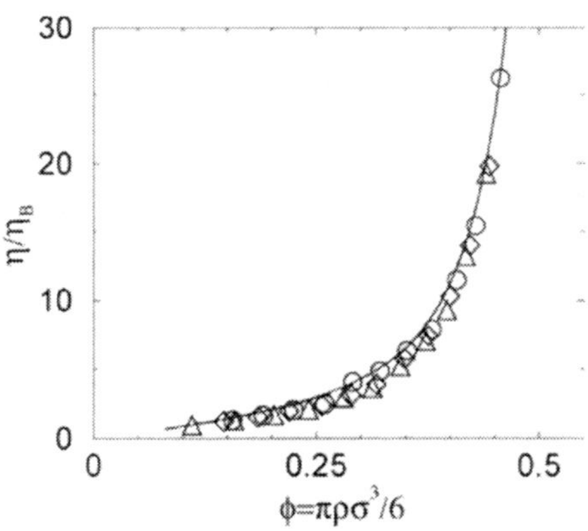

Figure 4. *Dependence of the normalized viscosity coefficient of argon on the packing factor* ϕ *(o – T = 298 K,* ◊ – *T = 1000 K, and Δ – T = 3000 K) according to ref. [[7]].*

The quantities D_B and η_B denote the Boltzmann coefficients of self-diffusion and viscosity determined by the expressions[8,94]

$$D_B=1.019\,\frac{3}{8n\sigma^2}\left(\frac{k_BT}{\pi m}\right)^{1/2},\tag{7}$$

$$\eta_B = 1.016 \frac{5}{16\sigma^2}\left(\frac{mk_BT}{\pi}\right)^{1/2}. \tag{8}$$

The dimensionless combination W_0 / k_BT used for the description of the dependence of the kinetic coefficients on the temperature and density is directly connected with the Carnahan-Starling equation [10]

$$\frac{W_0}{k_BT} = \frac{pv_0}{k_BT} = \frac{\phi(1+\phi+\phi^2-\phi^3)}{(1-\phi)^3}, \tag{9}$$

where, $v_0 = \pi\sigma^3/6$ and $\phi = \frac{\pi\rho\sigma^3}{6}$ denotes the packing factor (n is the numerical density of particles), σ is the hard sphere diameter of an argon. It is necessary to note that the temperature dependencies of all quantities studied in ref. [11,12] are presented as functions of the dimensionless density φ and so they should be invariant about the density values and hard sphere diameter, for which the combination $n\sigma^3$ remains to be constant.

In connection with this fact we should take into account that for natural argon the effective diameter σ is a function of temperature and density or pressure.[11,12] The careful analysis of this problem carried out in ref. [11,12] shows that modeling of the repulsive potential for argon using the inverse power function $U_r(r) \sim \left(\frac{\sigma}{r}\right)^m$ leads to the different values of σ and the steepness parameter m at high and low pressure and temperatures. So, for T = 298 K, P = 1,3 GPa and ρ = 1,95 g/cm^3, the effective diameter σ = 3,0965 Å and m =12. At the same time for T = 1000 K, P = 9,3GPa and ρ = 2,75 g/cm^3 – σ = 2,7348 Å and $m \approx (23 \div 24)$. From here it follows that the diameter of hard spheres, substituted to the Carnahan-Starling equation and Enskog formulas for the self-diffusion and shear viscosity coefficients, should be taken as a function of temperature and density.

We can reestablish values of σ at different temperatures and densities by fitting the computer data [7] for the self-diffusion coefficient by the equation

$$\frac{D}{D_B} = e^{-\xi\phi\frac{1+\phi+\phi^2-\phi^3}{(1-\phi)^3}}, \quad \xi = 0.45, \tag{10}$$

and entropy per molecule with the help of expression [10]

$$\frac{s_e}{k_B} = \phi\frac{4-3\phi}{(1-\phi)^2}. \tag{11}$$

Here it is necessary to note, that the exponential dependence in (10) immediately corresponds to the linear dependence of D/D_B on W_0/k_BT in Fig. 3. The respective values of the effective diameter σ at different densities and temperatures are presented in the Table. These results are in quite good agreement with those obtained in refs. [11,12].

Table 1 Main Thermodynamic and Kinetic Characteristics of Liquid Argon in Accordance with Ref. [7]

$D\cdot10^5$, см²/с	$\eta\cdot10^6$, г/(см с)	ρ, г/см³	σ, A	$D\cdot10^5$, см²/с	$\eta\cdot10^6$, г/(см с)	ρ, г/см³	σ, A
–	–	–	–	24.36	2255	1,789	2,816
10,72	1967	1,42	3,151	16.51	2775	2,02	2,79
8,17	2363	1,54	3,145	12.72	3230	2,23	2,778
5,94	2895	1,69	3,124	9.25	3801	2,4	2,762
4,39	3494	1,807	3,107	6.88	4387	2,58	2,747
3,14	4313	1,95	3,096	5.07	5018	2,75	2,745
P_{max}= 1,3GPa, T=298K, σ = 3,0965 A				P_{max}= 9,3GPa, T= 1000K, σ = 2,7346 A			

Moreover, for comparison the values of the effective diameter σ*, obtained from the expression for the dynamic shear viscosity for a system of soft spheres, are also presented in Table. More definitely, the formula

$$\frac{\eta}{\eta_0} = \frac{\left(1+\frac{2}{5}\varphi_* S_1\right)\left(1+\frac{2}{5}\varphi_* S_2\right)}{S_0} + \frac{48}{25\pi}\varphi_*^2 S_3, \tag{12}$$

for the normalized shear viscosity was used. It was obtained in ref. [13] with the help of the Enskog equation for soft spheres with power repulsive potential of type, described above. Here $\eta_0 = \frac{5\sqrt{m_{mol}k_BT}}{16\sqrt{\pi}\sigma_0^2}$, σ_0 is the hard sphere diameter, $\phi_* = \frac{2\pi}{3}n\sigma^3$ and the coefficients S_i can be calculated by the formula

$$S_0 = \frac{\sqrt{\pi}}{24}s_2^{(5)}, \quad S_1 = \frac{5}{6}s_3^{(2)} - \frac{1}{5}s_3^{(4)}, \quad S_2 = \frac{2}{15}s_3^{(4)},$$

where

$$s_k^{(q)} = \frac{8\kappa^{-k/2m}}{\sqrt{\pi}}\int_0^\infty e^{-x^2} g\left(\varphi_*\kappa^{-3/2m}x^{-3/m}\right)x^{2+q\frac{k}{m}}dx, \quad \kappa = \frac{2k_BT}{\varepsilon},$$

and $g(u)$ is the equilibrium binary correlation function. The model expression for $g(u)$ was taken from ref. [14]. The some divergence between σ and σ^* seems to be natural because of many approximations made to get (10)-(12).

The direct calculation of the collective contribution D_c/D_s to the self-diffusion coefficient is complicated by the inadequate temperature dependence of the shear viscosity in ref. [3]. Indeed, it is easy to verify that the ratio η/η_B for the model argon increases with temperature on isochors. From the physical viewpoint, this result is inadequate. It is worth noting that for $\phi \leq 0.4$ the values of η from ref. [7] and those determined on the basis of the Enskog theory[9] for hard spheres diameter of which coincides with the effective diameter

$$\frac{\eta}{\eta B} = 4\phi\left(\frac{(1-\phi)^3}{4\phi^2 - 2\phi} + \frac{4}{5} + 1.5228\,\frac{2\phi^2 - \phi}{(1-\phi)^3}\right), \tag{12}$$

of molecules, are practically coinciding. For $\varphi > 0.4$ the essential divergence between them is observed. It is not difficult to verify that the shear viscosity given by (12) has quite satisfactory temperature dependencies on isochors. Note, that the value $\phi = 0.4$ corresponds to $\rho \approx 1.8$ g/cm^3, i.e. it can be related to high pressures. In connection with this we suppose that the temperature and density dependencies of the shear viscosity can be approximated by the Enskog formula (12) up to $\phi = 0.47$. The corresponding curves are presented in Fig. 5.

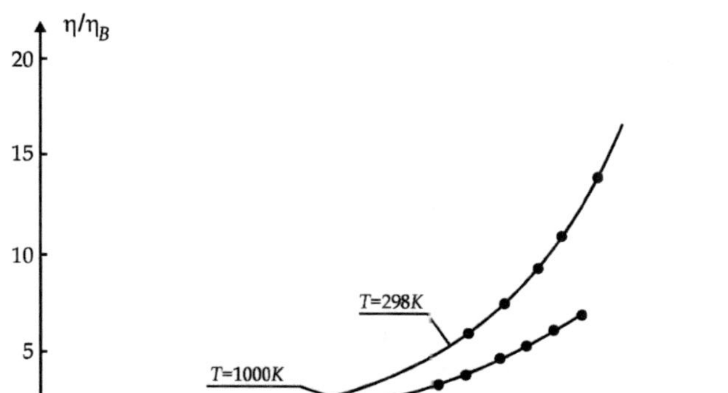

Figure 5. *The normalized shear viscosity for argon[7] vs. the dimensionless density ϕ according to (12). The circles denote the points used for the calculation of D_c by the formula (2).*

The applicability of the Enskog theory for high pressures is explained by the vortical character of the thermal motion of molecules. For molecular motions presented in Fig. 1 the relative motion of two neighboring molecules is only essential. In this case all molecules being on some sphere (circle) interact with their neighbors on the next spheres (circles) identically. So, the conditions for the applicability of two-particle approximation arise.

The comparative behavior of the self-diffusion coefficient from ref. [7], as well as its collective part calculated according to (4) and (5) with the shear viscosity from Fig. 5 is presented in Fig. 6. As we see, the relative value D_c / D_s of the collective contribution to the self-diffusion coefficient monotonically increases with density. At ρ =2,75 g/cm^3 and the temperature T = 1000 K it reaches 76%. At fixed density the ratio D_c / D_s also increases with temperature. This fact has the natural explanation: the intensity of vortical motions of molecules increases with temperature.

Thus, our initial prediction about the increase of the collective drift of molecules at high pressures is confirmed by quantitative estimates.

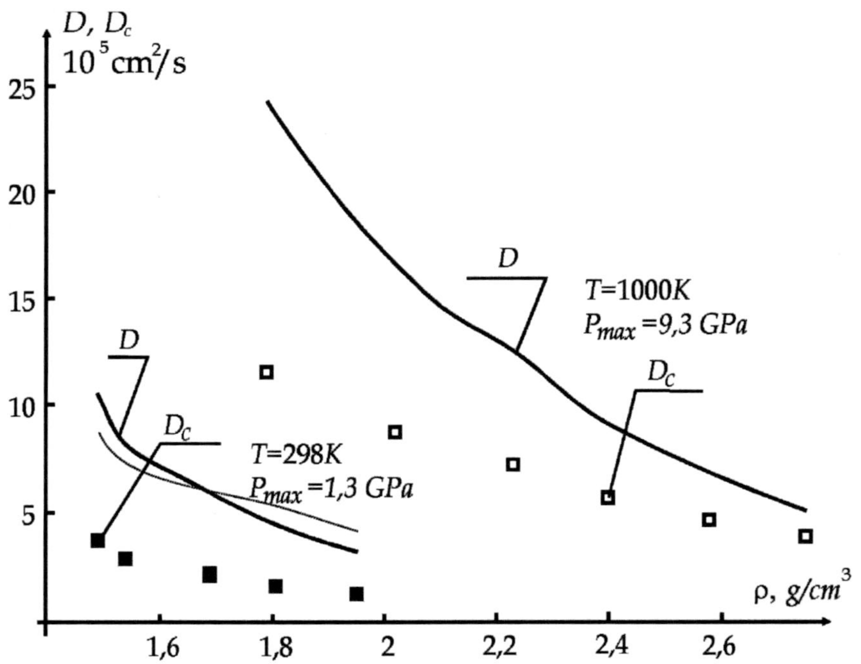

Figure 6. Self-diffusion coefficient of argon as a function of density. Solid lines correspond to experiments[7], slim line - to modified Enskog theory. Squares - results of calculations according to (2).

The simplification of the relative motions of molecules at high pressure should also lead to the simple interconnection between the shear viscosity and the self-diffusion coefficient. From the dimensionality reasons it follows:

$$D \sim \frac{k_B T}{\eta r_m(n,T)}, \tag{13}$$

r_m is the effective radius of a molecule. We expect that at high pressures $r_m(n,T) \approx const$, so

$$R(T_1,T_2) \approx 1, \tag{14}$$

where $R(T_1,T_2) = \frac{D(T_1)}{D(T_2)} \cdot \frac{T_2}{T_1} \frac{\eta(T_1)}{\eta(T_2)}$. Estimating $R(T_1,T_2)$ with the help of the shear viscosity from Fig .5 and the self-diffusion coefficients from Fig. 3 for different densities we obtain the values

$$R(T_1 = 298K, T_2 = 1000K)\big|_{\rho=1.8g/cm^3} = 1.05\ ,$$

$$R(T_1 = 298\,K, T_2 = 1000\,K)\big|_{\rho=1.9g/cm^3} = 1.03\ ,$$

which are in quite good agreement with (14).

References

1. Fisher, I. Z. (1971) *Zh. Eksp. Teor. Fiz.* **61**, 1647
2. Lokotosh, T. V., and Malomuzh, N. P. (2000) *Physica A* **286**, 474
3. Bulavin, L. A., Lokotosh, T. V., and Malomuzh, N. P. (2008) *J. Mol. Liq.* **137**, 1
4. Bulavin, L. A., Malomuzh, N. P., and Pankratov, K. N. (2006) *Zh. Strukt. Khim.* **47**, 54
5. *CRS handbook of chemistry and physics: a ready-reference book of chemical and physical data* (1996) 67th ed Ed.-in-chief R.C.West (Boca Raton: CRS Press, p. 894)
6. Dexter, A. R., and Matheson, A. J. (1971) *J. Chem. Phys.* **54**, 203
7. Bastea S. (2004) *Cond. Met.* **1**, 1153
8. Resibois, P., and de Leener, M. (1977) Classical Kinetic Theory of Fluids (Wiley, New York)
9. Chapman, S., and Cowling, T. G. (1970) *The Mathematical Theory of Non-Uniform Gases* (Cambridge Univ. Press, Cambridge)
10. Hansen, J. P., and McDonald, I. R. (1986) Theory of Simple Liquids (Academic Press, London)
11. Bulavin, L. A., Lokotosh, T. V., Malomuzh, N. P., and Shakun, K. S. (2004) *UJP*, **49**, 556

12. Bardic, V. Yu., Malomuzh, N. P., and Sysoev, V. M. (2005) *JML*, **120**, 27
13. Kurochkin, V. I. (2002) *Journ. of Techn. Phys.* **11**, 72
14. Ferziger, J. H., and Kaper, H. G. (1972) Mathematical theory of transport processes in gases, (Amsterdam-L., Nort-Holland P.C.)

THERMAL CONDUCTIVITY OF METASTABLE STATES OF SIMPLE ALCOHOLS

A.I. KRIVCHIKOV[a]*, O.A. KOROLYUK[a], I.V. SHARAPOVA[a], O.O. ROMANTSOVA[a], F.J. BERMEJO[b], C. CABRILLO[b], I. BUSTINDUY[b], AND M.A. GONZÁLEZ[c]

[a]*B. Verkin Institute for Low Temperature Physics and Engineering of NAS Ukraine, Kharkov, Ukraine*

[b]*Instituto de Estructura de la Materia, C.S.I.C., and Dept. Electricidad y Electrónica-Unidad Asociada CSIC, Facultad de Ciencia y Tecnología Universidad del País Vasco/EHU, E- 48080 Bilbao, Spain*

[c]*Institute Laue Langevin, 6 Rue Jules Horowitz, F-38042-Grenoble Cedex 9, France*

Abstract: The thermal conductivity $\kappa(T)$ of glassy and supercooled liquid methanol, ethanol and of 1-propanol has been measured under equilibrium vapor pressure in temperature interval from 2 K to 160 K by the steady-state method. The metastable orientationally disordered crystal of ethyl alcohol is found to exhibit a temperature dependence of $\kappa(T)$ that is remarkably close to that of the fully amorphous solid, especially at low temperatures. In the case of propyl alcohol, our results emphasize the role played by internal molecular degrees of freedom as sources of strong resonant phonon scattering. For all samples here explored, the glass-like behavior of $\kappa(T)$ is described at the phenomenological level using the model of soft potentials. The thermal transport is then understood in terms of a competition between phonon-assisted and diffusive transport effects. The thermal conductivity κ is thus a sum of two contributions: $\kappa = \kappa_I + \kappa_{II}$, where κ_I is the acoustic phonon component dependent on the translational and orientational ordering of molecules, κ_{II} – is the phonon diffusion component corresponding to a non – acoustic phonon heat transfer in accordance with the Cahill – Pohl model.

Keywords: thermal conductivity, glassy state, supercooled liquid, phonon scattering

The primary monohydric (aliphatic) alcohols can be formed by substituting alkyl groups within the hydrocarbon chain by OH moieties able to form hydrogen bonds with a nearby molecule. The general formula of such alcohols is $H(CH_2)_nOH$, where n is the number of carbon atoms in the

S. Rzoska et al. (eds.), *Metastable Systems under Pressure*,

hydrocarbon chain and the OH substitution can take place either at the chain ends or at some atom at mid-chain positions. Unlike water ice, which is an associated substance with strong tetragonally – directed cooperative H-bonds, monohydric alcohols have less-stronger H-bonds and their chain structure leads to chain-like structures within both liquid and solid states. The strength of the H-bond interactions upon different physical properties decreases as the number n of the carbon atoms increases. In common with other H-bonded systems, the structure and dynamics of simple alcohols is characterized by an interplay between directional- (i.e. electrostatic and H-bond interactions) and dispersion interactions, which force the molecules to adopt linear chain structures, whereas the available evidence tells that inter-chain interactions are governed by Van-der-Waals forces, the strength of which increases with the size of the molecule.

Glass-formation in these materials is known to depend on the balance of forces referred to above. In fact, the glass-forming ability of the primary alcohols increases with the molecular length. An example for this is the difficulty of formation in methanol by rapid cooling if compared to ethyl and propyl alcohols. As a matter of fact, pure methyl alcohol can be easily prepared as a glass by means of vapor deposition onto a cold substrate below the glass-transition temperature, T_g= 103.4 K.[1,2] The poor glass-forming ability of this substance is determined by the structure of its liquid and crystalline phases, which consist of zigzag chains of alternating H-bonded molecules. The glass phase of methanol can also be obtained by addition of a small quantity of water (~ 6.5 mol. % H_2O) [3] which leads to a calorimetric glass-transition taking place over a wide temperature interval T_g= 100 K -120 K.[4-6]

On the other hand, methanol has the shortest and most mobile molecule, which makes it a suitable object for modeling the properties of alcohols having more complex structures.[7-12] U nder equilibrium vapor pressure methanol crystallizes at T_m=175.37 K into an orientationally disordered high-temperature state (β-phase). Variation of temperature and pressure[10-13] unveils a rich polymorphism resulting from H-bond interactions which are stronger than their dispersive counterparts. In turn, solid ethanol also shows a complex phase diagram under equilibrium vapor pressure. It can either form a single stable thermodynamic–equilibrium phase (T_m=159 K), which is an orientationally – ordered monoclinic crystal; or can exhibit in three metastable long lasting phases – a positional, fully disordered glass, an orientationally - disordered crystal with a static disorder (orientational glass) and a crystal with a dynamic orientational disorder. The fully amorphous (glass) state in ethanol (as well as in the propanol isomers) can be formed easily by fast supercooling the liquid below the glass transition temperature T_g = 97 K. Slower cooling leads to the orientationally disordered states which makes ethanol a remarkable material offering us the possibility of comparing the thermal conductivities of three phases – crystal, positional and orientational glasses within the same temperature interval.[14,15]

In stark contrast with the case of the lower alcohols, the propanol isomers exhibit far less polymorphism.[16] Under equilibrium vapor pressure, the materials can be prepared within their ordered crystals and glass (amorphous) states only. The glass transition temperature of 1-propanol (1-Pr) is T_g =98 K and its melting temperature is T_m= 148 K.

The present survey jointly reports on data on the thermal conductivities $\kappa(T)$ of methanol, ethanol and 1-propyl alcohol in the glass and supercooled liquid states. The data were measured under equilibrium vapor pressure in the temperature interval from 2 K - 160 K.

The thermal conductivity of different materials was measured under equilibrium vapor pressure with a setup[17] that uses the steady-state potentiometric method. The different phases were prepared within the container using different cooling – heating cycles for the same sample and taking into account the thermal history.[6,14-16,18] In short, the glasses were prepared by very fast cooling (above 50 K min^{-1}) of the room-temperature liquids through their glass transition regions to the boiling temperature of liquid N_2. Since the glass transition temperature T_g of methanol, ethanol and propanol is higher than the boiling point of nitrogen, the glass samples were prepared by immersing the container with the sample, directly into liquid nitrogen. The $\kappa(T)$ of the glasses of ethanol, propanol and methanol, the latter with a water impurity of 6.6 mol. % H_2O, were measured in these experiments. The measurements were performed with gradually decreasing temperature. After reaching the lowest temperature, the measurement was continued with increasing temperatures. Above T_g the glass samples transform into a supercooled liquid. Further increases in temperature and above $T \approx 121$ K leads the supercooled liquid of methanol to spontaneous crystallization and the thermal conductivity of the sample increased sharply.[6]

The crystalline bcc phase of ethanol with dynamic orientational disorder was prepared at 125 K starting from the supercooled liquid by slow cooling. Further cooling caused a transition to the orientational glass state at T_g =92 K . The transition triggered the release of heat. The cooling was then continued down to liquid helium temperatures. Heating the sample above 116 K produced a rather fast transformation into an equilibrium completely ordered phase. The good reproducibility of the $\kappa(T)$ results for solid phases[14,15] proves that the phase transformations were full and completed.

The supercooled liquids of ethanol and propanol were investigated starting either by cooling the normal liquid or melting the structural glass, while supercooled liquid methanol was prepared by melting the structural glass.

The temperature dependences $\kappa(T)$ of the three alcohol samples,[6,14,15,18,16] within glass, supercooled liquid and normal liquid are shown in Fig. 1. The normal-liquid data are taken from ref.[19]. The temperature interval includes the glass-transition (T_g) and the melting (T_m) temperatures of the alcohols.

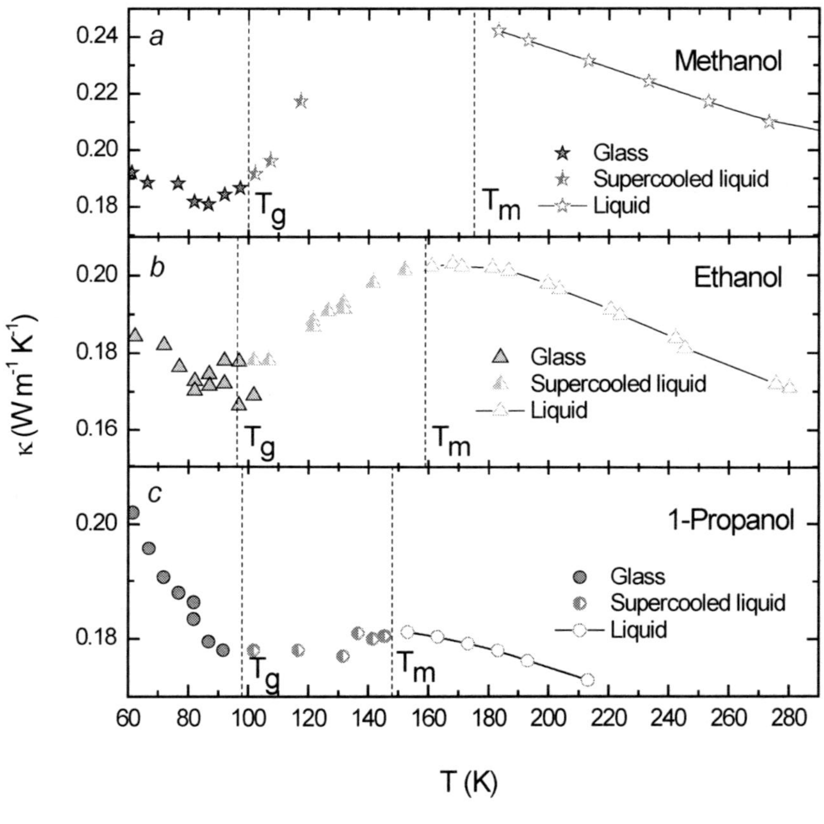

Figure 1. *Thermal conductivity of elementary alcohols. Fig. 1a. Methanol with 6.6.% H_2O: ξ - glass, - supercooled liquid,[6] ψ - liquid.[19] Fig.1b. Ethanol: 7 - glass, Λ - supercooled liquid,[14,15,18] 8 - liquid.[19] Fig.1c. 1-propanol: , - glass, 6 - supercooled liquid,[16] – - liquid.[19]*

Examination of Fig. 1 shows that the behavior of κ(T) displays remarkably similar features for the tree substances: it increases in the normal liquid with decreasing temperature and has a distinct maximum near T_m. Further lowering temperature leads to a decrease of κ(T) in the supercooled liquid region and passes through a broad minimum which roughly matches the region of the glass - supercooled liquid transformation. Such a minimum in κ(T) is thus a truly anomalous feature exhibited by these materials. Once within the glassy states, $\kappa(T)$ increases with decreasing temperature. The thermal conductivity of the water – methanol solution was measured at gradually increasing temperature on transformation from the glass state to a supercooled liquid. The fig.1a does not show $\kappa(T)$ – data for supercooled methanol at $T = 120$ K ÷ T_m because at $T = 121$ K it transforms into the crystal phase, which is evident in the $\kappa(T)$ – data: at this temperature, that is the $\kappa(T)$ – value coincides exactly with results taken from crystal grown from the liquid at $T \approx T_{m.}$[6] It is interesting to point out here that for the supercooled liquids, the growth of $\kappa(T)$ with increasing temperature is stronger for methanol, moderate for ethanol and weak for propanol.

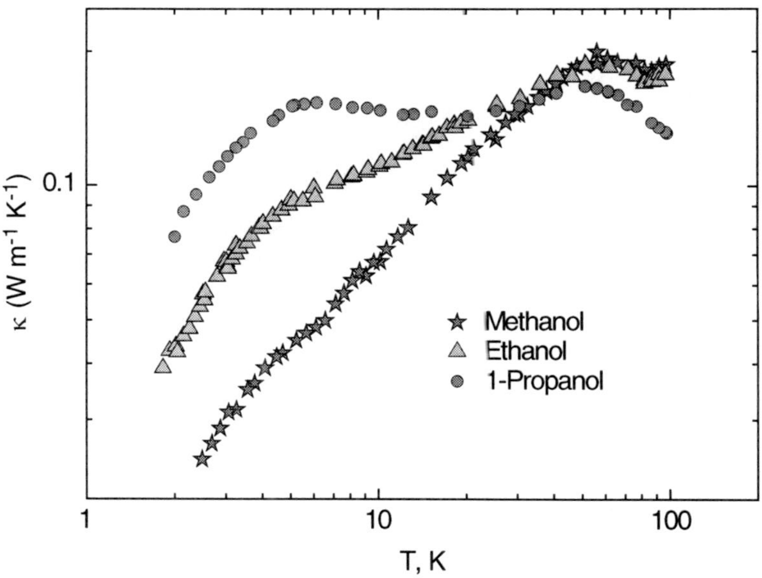

Figure 2. *Thermal conductivity of glass – state alcohols: ξ - methanol,*[6] *▲ –ethanol,*[14,15] *, - 1-propanol.*[16]

The differences in the heat transfer processes for the glass, supercooled liquid and normal liquid phases of the three materials are attributed to the competition between the phonon transport and diffusive heat-transfer effects that are governed by dynamical processes taking place within the GHz-range.

The thermal conductivity of the three samples within their glass states from T = 2 K up to T_g is shown in Fig. 2. The data shown there exhibits a temperature dependence characteristic of most amorphous solids.[20] The thermal conductivity increases with temperature with a maximum rate below 4 K. A smeared «plateau» then follows within 5–10 K, which becomes more marked with increasing alcohol-chain length and is followed by a further increase in conductivity which becomes smoother as the chain-length increases. Such an increase lasts up to $T = 50$ K, where a smeared maximum appears. On further heating, a smeared minimum is observed at $T \approx 80$ K.

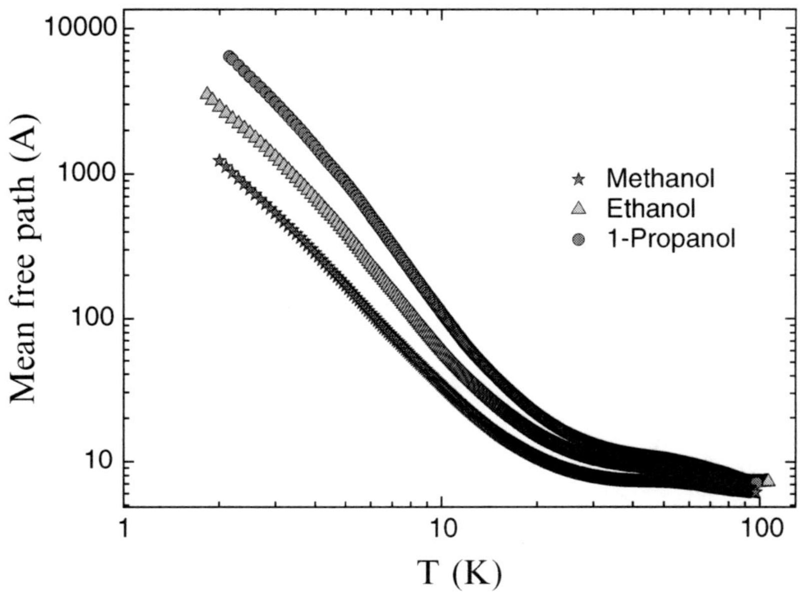

Figure 3. *Phonon mean free path in methanol, ethanol and 1-propanol.*[16]

Most data pertaining to the thermal conductivity of glasses show an increase with temperature in the region above the plateau. In all the alcohols investigated, the thermal conductivity starts to increase with temperature again after passing through a smeared minimum.

At low temperatures the behavior of $\kappa(T)$ for the three alcohols displays a systematic trend. Data for 1-propanol are much higher than those for ethanol and in turn these are higher and less steep than those for methanol. In what follows we interpret such a finding as an evidence of the role played by internal molecular degrees of freedom as sources of strong resonant phonon scattering.

The thermal conductivity is described as a sum of two contributions $\kappa(T) = \kappa_I(T) + \kappa_{II}(T)$ arising from propagating acoustic phonons and from localized short-wavelength vibrational modes, or phonons with the mean free path equal to the phonon half-wavelength, respectively. The temperature dependence of $\kappa_I(T)$ for all glasses was first analyzed on phenomenological grounds using the soft-potentials-model (SPM),[21-23] which portrays phonon scattering as mainly caused by low-energy excitations of a strongly anharmonic ensemble of particles.

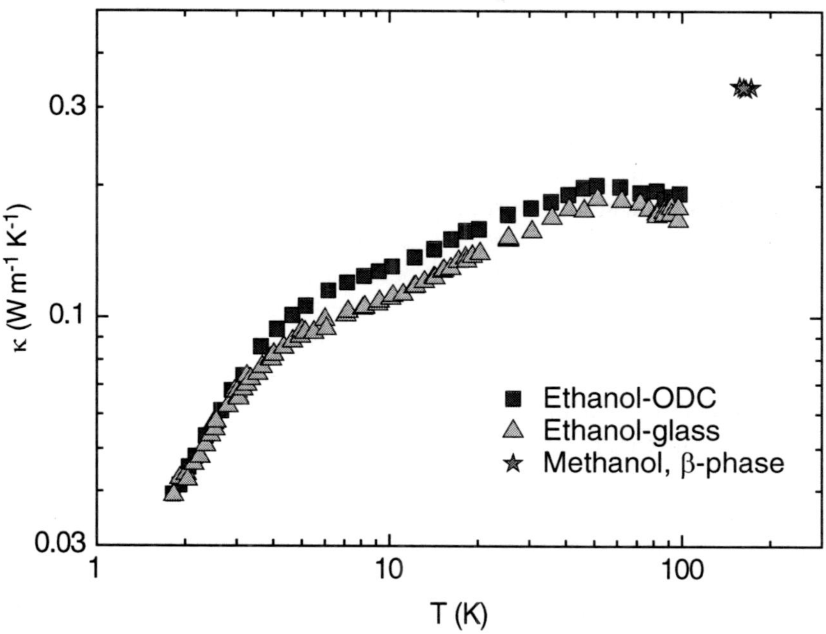

Figure 4. *Thermal conductivity of ethanol in the states of orientational and structural glasses.*[14,15] *The thermal conductivity of β–phase methanol*[6] *is shown for comparison.*

There, the scattering rate of acoustic phonons in a disordered system is given by the sum of three terms describing scattering by the tunnel states, classical relaxors, and soft quasi-harmonic vibrations. The simplest description of $\kappa_{II}(T)$ is provided by the phenomenological Cahill-Pohl model.[24,25] The $\kappa_{II}(T)$ contribution increases with growing T and becomes dominant in the temperature region corresponding to energies of the boson peak.

From the measured values of $\kappa(T)$ as well as from the phonon specific heat, estimates for the temperature dependence of the phonon mean free paths l have been derived and the results are shown in Fig. 3. Data for glassy 1-Pr show larger values for the mean free path than those for the other two glasses (methanol and ethanol). On the other hand, the temperature dependence of l shows three well differentiated regions, varying as T^{-2} – below 3 K, a T^{-3} - regime within 3-10 K, and, finally, a high temperature region for $T > 40$ K where it goes as $T^{-1/2}$. The crossover points (i.e., the intercepts between double logarithmic lines) are about 4 and 20 K, respectively.

The temperature dependence $l(T)$ of the set of monohydric alcohols here studied exhibits a change from a ballistic regime of phonon propagation to a diffusive one. The transition temperature is within the range 20-40 K. At present there is still a discussion if acoustic excitations can travel within the glass beyond the spectral feature localized at low frequencies, known as a boson peak and typically appearing at some hundreds of GHz. Its origin is not clear yet, but it is related to the hump in the heat capacity curve plotted as C_p/T^3. The energy of the broad boson peaks observed in the inelastic neutron scattering spectra of the materials investigated here is about 30 K [14,26,27], which comes rather close to the characteristic temperature where changeover from ballistic to diffusive heat transport takes place.

We can thus conclude that the thermal conductivity of monohydric alcohol in the structural-glass state is dependent on the number of carbon atoms in its molecule, i.e. the ratio of the number of hydroxyl groups to that of carbon atoms (the hydrogen-bond density). The observation shows that $\kappa(T)$ grows most intensively in the temperature region where scattering of propagating acoustic phonons by molecular degrees of freedom is dominant.

Within the three alcohols here investigated, only ethanol can be prepared within the orientational glass state, that is a cubic bcc crystal where the lattice is formed by the molecular-centers-of-mass but retaining a static orientational disorder. Such a phase although metastable has a relatively long live, enabling its detailed study along a wide range of temperatures. The $\kappa(T)$ - curve of the orientational glass of ethanol is similar to that of its structural glass (see Fig. 4) but the thermal conductivity magnitude is somewhat higher in the orientational glass. As in the case of the structural glass, the thermal conductivity increases with temperature and there is a smeared plateau at T = 5 ÷ 10 K. A further rise of temperature leads to an increase of the thermal conductivity; it further passes through a smeared maximum at $T \approx 51$ K, then through a smeared minimum (anomaly of thermal conductivity) at $T \approx 86$ K and finally starts to grow slightly. For comparison, Fig. 4 illustrates the dependence $\kappa(T)$ for the orientationally – disordered phase (β–phase) of methanol.[6] In contrast to the orientational glass, the β–phase features a dynamic orientational disorder of molecules. $\kappa(T)$ in the β–phase is basically temperature – independent and appreciably larger than the thermal conductivity of the ethanol and methanol glasses. The results obtained indicate that the molecular orientational disordered, which is the main source of acoustic phonon scattering is a dominant factor of the heat transfer in monohydric alcohols.

To conclude, the investigation of the thermal conductivity of monohydric alcohols within the interval 2 K-160 K has revealed a number of new features in its temperature dependence which are related to the rotational degrees of freedom. The measurement of κ(*T*) is here proven to be a sensitive tool for investigating the considerable distinctions in the space-time correlations between a stable liquid and a metastable supercooled one. Within the structural-glass state κ(*T*) is shown to be dependent on the number of carbon atoms in its molecule, i.e. the ratio of the number of hydroxyl groups to that of carbon atoms (the hydrogen-bond density). The observation shows that κ(*T*) grows most intensively in the temperature region where scattering of propagating acoustic phonons by molecular degrees of freedom is dominant. As a final remark, the present data show that the molecular orientational disorder constitutes the main source of acoustic phonon scattering and thus the limiting factor for the heat transfer in these materials.

Acknowledgements

The authors are sincerely grateful to Prof. V.G. Manzhelii for helpful discussions and interest in this study. The investigations are made on the competition terms for joint projects of NAS of Ukraine and Russian Foundation for Fundamental Research (Agreement N 9-2008, Subject: "Collective processes in metastable molecular solids").

References

1. Sugisaki, M., Suga, H., and Seki, S. (1968) *Bull. Chem. Soc. Japan*, **41**, 2586
2. Susan M. Dounce, Julia Mundy, and Hai-Lung Dai (2007) *J. Chem. Phys.* **126**, 191111
3. Bermejo, F. J., Martin, D., Martínez, J. L., Batallan, F., García-Hernández, M., and Mompean, F. J. (1990) *Phys. Lett.* A **150**, 201
4. Bermejo, F. J., García Hernández, M., Martínez, J. L., Criado, A., and Howells, W. S. (1992) *J. Chem. Phys.* **96**, 7696
5. Bermejo, F. J., Alonso, J., Criado, A., Mompean, F. J., Martinez, J. L., Garcia-Hernandez M., and Chahid, A. (1992) *Phys. Rev.* B, **46**, 6173.
6. Korolyuk, O. A., Krivchikov, A. I., Sharapova, I.V., and Romantsova, O.O., (2009) to be published in *Low Temp. Phys.*
7. Steytler, D. C., Dore, J. C., and Montague, D. C. (1985) *J. Non Cryst. Solids* **74**, 303
8. Doba, T., Ingold, K. U., Reddoch, A. H., Siebrand, W., and Wildman, T.A. (1987) *J. Chem. Phys.* **86**, 6622
9. Brown, J. M., Slutsky, L. J., Nelson, K. A., and Cheng, L.-T. (1988) *Science* **241**, 4861-4865
10. Torrie, B. H., Weng, S.-X., and Powell, B. M. (1989) *Molecular Physics*, **67**, 575

11. Lucas, S., Ferry, D., Demirdjian, B., and Suzanne, J. (2005) *Journal of Physical Chemistry* B, **109**, 18103
12. Gromnitskaya, E. L., Stal'gorova, O. V.,Yagafarov, O. F., Brazhkin, V. V., Lyapin A. G., and Popova, S. V. (2004) *JETP Letters* **80**, 597
13. Torrie, B. H., Binbrek, O. S., Strauss, M., and Swainson, I .P. (2002) *J. Solid State Chem.*, **166**, 415
14. Bermejo, F. J., Fernandez-Perea, R., Cabrillo, C., Krivchikov, A. I., Yushchenko, A. N., Korolyuk, O. A., Manzhelii, V. G., Gonzalez, M. A. and Jimenez-Ruiz, M. (2007) *Low Temp. Phys.*, **33**, 790
15. Krivchikov, A. I., Yushchenko, A. N., Manzhelii, V. G., Korolyuk, O. A., Bermejo, F. J., Fernandez-Perea, R., Cabrillo, C., and Gonzalez, M. A. (2006) *Phys. Rev.* B **74**, 060201
16. Krivchikov, A. I., Yushchenko, A. N., Korolyuk, O. A., Bermejo, F. J., Fernandez-Perea, R., Bustinduy, I., and Gonzalez, M. A. (2008) *Phys. Rev.* B 77, 024202
17. Krivchikov, A. I., Manzhelii, V. G., Korolyuk, O. A., Gorodilov, B. Ya., and Romantsova, O. O. (2005) *Phys. Chem. Chem. Phys.* **7**, 728; Krivchikov, A. I., Gorodilov, B. Ya., and Korolyuk, O. A. (2005) *Instrum. Exp. Tech.* **48**, 417
18. Krivchikov, A. I., Yushchenko, A. N., Korolyuk, O. A., Bermejo, F. J., Cabrillo, C., and González, M.A. (2007) *Phys. Rev. B* **75**, 214204.
19. Vargaftic N. B. et al. (eds.) (1994) *Handbook of thermal conductivity of liquids and gases, CRC Press,* [in Russian], *(Moscow).*
20. Pohl, R. O., Liu, X., and Thompson, E. (2002) *Rev. Mod. Phys.* **74**, 991.
21. Buchenau, U., Galperin, Yu. M., Gurevich, V. L., Parshin, D. A., Ramos, M. A., and Schober, H. R., (1992) *Phys. Rev.* B **46**, 2798; Parshin, A. (1993) *Phys. Scr.* **49A** , 180
22. Ramos, M. A., and Buchenau, U. (1997) *Phys. Rev.* B **55**, 5749.
23. Bermejo, F. J., Cabrillo, C., Gonzalez, M. A., and Saboungi, M. L., (2005) *J. Low Temp. Phys.* **139**, 567
24. Cahill, D. G., and Pohl, R. O., (1988) *Ann. Rev. Phys. Chem.* **39**, 1, 93
25. Cahill, D. G., Watson, S. K., and Pohl, R. O. (1992) *Phys. Rev. B* **46**, 6131
26. Osamu, Y., Kouji, H., Takasuke, M., and Kiyoshi, T. (2000) *J. Phys.: Condens. Matter* **12**, 5143
27. Surovtsev, N. V., Adichtchev, S. V., Rössler, E., and Ramos, M. A. (2004) *J. Phys.: Condens. Matter* **16**, 3, 223

TRANSFORMATION OF THE STRONGLY HYDROGEN BONDED SYSTEM INTO VAN DER WAALS ONE REFLECTED IN MOLECULAR DYNAMICS

K. KAMIŃSKI[1], E. KAMIŃSKA[1], K. GRZYBOWSKA[1], P. WŁODARCZYK[1], S. PAWLUS[1], M. PALUCH[1], J. ZIOŁO[1], S. J. RZOSKA[1], J. PILCH[2], A. KASPRZYCKA[3] AND W. SZEJA[3]

[1] *Institute of Physics, Silesian University, ul. Uniwersytecka 4, 40-007 Katowice, Poland;* [2]*Academy of Physical Education, Dept. Biological Sci., Raciborska 1, 40-074 Katowice, Poland;* [3]*Silesian University of Technology, Department of Chemistry, Div. Org. Chem., Biochem. and Biotechnology, ul. Krzywoustego 4, 44-100 Gliwice, Poland*

Abstract: Dielectric relaxation studies on disaccharides lactose and octa-O-acetyl-lactose are reported. The latter is a hydrogen bonded system while the former is a van der Waals glass former. The transformation between them was arranged by substituting hydrogen atoms in lactose by acetyl groups. Hereby the influence of differences in bounding on dynamics of both systems is discussed. We showed that the faster secondary relaxation (labeled γ) in octa-O-acetyl-lactose has much lower amplitude than that of lactose. The relaxation time and activation energy remain unchanged in comparison to the γ- relaxation of lactose. We did not observe the slow secondary relaxation (labeled β), clearly visible in lactose, in its acethyl derivative. Detailed analysis of the dielectric spectra measured for octa-O-acetyl-lactose in its glassy state (not standard change in the shape of the γ- peak with lowering temperature) enabled us to provide probable explanation of our finding. No credible comparative analysis of the α- relaxation process of the lactose and octa-O-acetyl-lactose are presented, because loss spectra of the former carbohydrate were affected by the huge contribution of the dc conductivity. Notwithstanding, one can expect that octa-O-acetyl-lactose has lower glass transition temperature and steepness index than lactose.

Keywords: lactose, octa-O-acetyl-lactose, molecular "glassy" dynamics

1. Introduction

Saccharides are a huge group of compounds, playing a key role in many bio-chemical reactions of living organisms.[1,2] They control recognition process of the cell, store energy and have positive effect on functioning of the human body. This made carbohydrates very important group of materials. In literature one can find a lot of reports devoted to the examination of the physico-chemical properties of the saccharides. Notwithstanding, new results of measurements revealed previously unknown features of this group of compound and enable us to propose some explanations of commonly known phenomena in carbohydrates. Saccharides are excellent glass-forming liquids. After melting they can be easily supercooled. Therefore, they can be examined by the broad band dielectric spectroscopy. Molecular dynamics of sugars has been extensively investigated by many authors, so far.[3-15] One can conclude from these studies that two secondary relaxation modes can be detected in the glassy state of the mono- and di-saccharides. The faster one, labeled by us as a γ- relaxation process, is clearly visible in dielectric loss spectra, both for mono- and disaccharides, whereas the slower one, β- relaxation, seems to be visible only in two monosaccharides sorbose and galactose[12], all disaccharides[9,13-17] and polysaccharides[18-22]. In four types of monosugars, studied by us, glucose, fructose, ribose and 2-deoxy D- ribose the β-mode appears in dielectric loss spectra as an excess wing in the high frequency flank of the structural α-relaxation peak[11,12].

Concerning the γ- relaxation process in saccharides, there are different points of view on a molecular mechanisms responsible for this relaxation. Some authors associated γ-process with an intermolecular origin[9,16,23], while the other claim that γ-process is of an intramolecular character[8,24]. Another opinion has been expressed by Faivre *et al.*, who claimed that the γ- relaxation has more complicated nature and both intra- and intermolecular motions contribute to this process[25]. It should be mentioned that despite the use of sophisticated methods, such as NMR[26,27] or molecular dynamics simulations[28], it was not possible to propose an explanation of the molecular mechanism that governs these secondary relaxations. New experimental evidences about origin of this process were provided in our previous investigations. We carried out high pressure dielectric measurements on fructose and leucrose.[12,14] Results of this experiment showed that a maximum of γ- peak shifts only slightly (about 0.3 decade) towards lower frequencies with increasing pressure from 0.1 MPa to 500 MPa. Thus it was concluded that the considered mode is insensitive to pressure and then its nature is intramolecular. Further confirmation of our interpretation came from the latest measurements performed in monosaccharide fructose[29]. We studied changes in molecular dynamics of this carbohydrate in the supercooled liquid as an effect of the chemical transformation of the β- pyranose form of D- fructose (this form is present in crystals of this sugar) to the α- pyranose, α, β- furanose and to the non-cyclic fructose. In chemistry this reaction is known as a mutarotation.

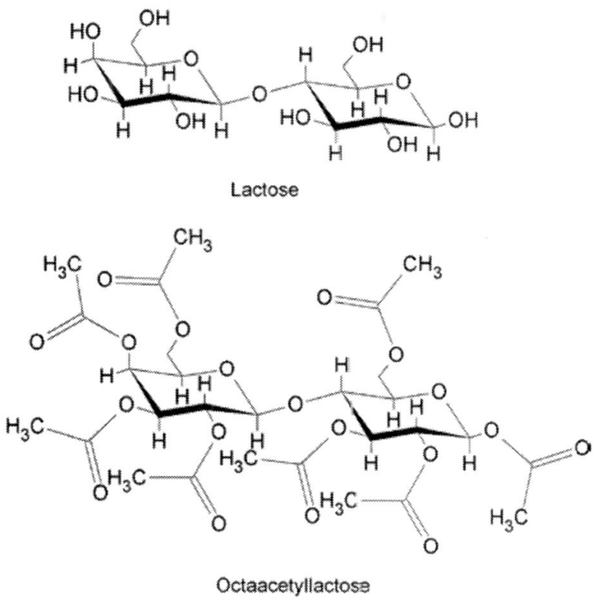

Figure 1. *Chemical structures of the lactose and octa-O-acetyl-lactose.*

It was further shown that the structural α- relaxation peak shifts to the lower frequencies as the reaction proceeds, while the frequency of maximum secondary relaxation loss remains constant, and only the significant increase in amplitude of this relaxation process was observed. Such a behavior of γ-relaxation confirms that in fact the secondary process takes its source from intramolecular motions occurring within the monosaccharide unit.

It is worth noting that a secondary relaxation of intermolecular character should move towards lower frequencies as the α- relaxation does. Recent studies[30-34] indicates that it is a natural implication of a relationship between the secondary relaxation of intermolecular nature and the α- relaxation process. This issues can be expressed within the framework of the Coupling Model proposed by K. Ngai[35-38], by using the primitive relaxation time:

$$\tau_0 = (t_c)^n (\tau_\alpha)^{1-n} \tag{1}$$

where t_c= 2ps (small glass-forming liquids). A fair agreement between the considered secondary relaxation time and τ_0 made it possible to classify the secondary relaxation as intermolecular in origin. Thus, one can state that γ-relaxation is of intramolecular origin. However, the question about molecular mechanism responsible for this process is still an open issue.

In disaccharides one can also observe a slower secondary β-relaxation. Recently, it was shown that twisting rotation of the monosugar units around glycosidic bond is responsible for occurrence of this relaxation[13]. This supposition was confirmed by theoretical conformational analysis. Additionally, it was proven that the activation energy provides direct information about structural rigidity of the examined disaccharides.

In this paper we focused on investigation of the molecular dynamics of two structurally similar systems with completely different type of interactions, such as lactose and its derivative – octa-O-acetyl-lactose. The former is a highly hydrogen bonded sugar, whereas the latter is a normal van der Waals material. It is a commonly known fact that the presence of the hydrogen bonds in the sample results in a dramatic change of chemical and physical properties. The influence of hydrogen bonding is strongly demonstrated in relaxation dynamics in the supercooled and glassy states. It is well documented for the series of polyalcohols[39-42] and sugars[15]. It is observed that with increasing number of the hydroxyl group (it implies greater ability to formation of the hydrogen bonds) the glass transition temperature and steepness index increase significantly, separation between secondary relaxation and main α- process becomes greater. Moreover with increasing number of hydroxyl group the activation energies of the secondary relaxation increase. The target of this paper is also a comparison of dynamical properties of the lactose and octa-O-acetyl-lactose, enabling to examine the influence in type of interactions (hydrogen bonding vs. van der Waals) on dynamics. This can provide new facts important for explaining the molecular mechanism responsible for γ- relaxation.

2. Experiment

Lactose (98% of purity) was supplied from Fluka, octa-O-acetyl-lactose was synthesized for the purpose of this paper[[i]]. The chemical structures of both investigated saccharides are presented in Fig 1. In order to avoid caramellization of lactose (this disaccharide caramelizes relatively easy) the sample was heated very quickly to its melting point and when whole sample was molten it was supercooled very quickly. Isobaric dielectric measurements at ambient pressure were carried out using a Novo-Control GMBH Alpha dielectric spectrometer (10^{-2}-10^{7} Hz). The samples were placed between two stainless steel flat electrodes of the capacitor with gap 0.1 mm. The temperature was controlled by the Novo-Control Quattro system, with use a nitrogen-gas cryostat. Temperature stability of the samples was better than 0.1 K

3. Results and Discussion

Loss dielectric spectra of both investigated saccharides were arranged into three panels, as shown in Fig. 2. In the panel (a) data measured above glass transition temperature T_g for lactose (filled symbols) and octa-O-acetyl-lactose (open symbols) are displayed. In panels (b) and (c) spectra measured below T_g for both carbohydrates are collected. It can be seen that in loss dielectric spectra above the glass transition temperature for lactose there is no α- relaxation peak in the experimental window. This is due to the huge contribution of the dc conductivity to measured spectra. An explanation of this phenomenon can be deduced from studies by the Crofton and Pethrick[44]. They showed that the dc-conductivity contribution for sugars is an effect of the proton migration amongst hydroxyl groups. This is consistent with the experimental findings that with increasing hydroxyl groups in the series of D-ribose, glucose, lactose, and any oligosaccharide or polysaccharide a significant increase in dc conductivity is observed[11,12,15]. In fact, in ours studies well separated structural α-relaxation peaks appeared only for D-ribose and glucose, whereas for other carbohydrates the observation of the α-process was impossible. Probably a growth in the number of hydroxyl groups, implying formation of the stronger hydrogen bonded network, favors and makes more efficient the migration of protons. Another confirmation of connection between proton hopping in the hydrogen bonded network of saccharides and the enormous dc conductivity comes from the comparison of the loss dielectric spectra of lactose and its acetyl derivative. In the latter system all hydrogen atoms were substituted by the acid groups and consequently hydrogen bonds were destroyed. There were no paths for the migration of protons in octa-O-acetyl-lactose. It can be seen in Fig. 2(a) that in octa-O-acetyl-lactose the dc contribution is significantly lower than in the case of lactose. Therefore, it is possible to monitor the α- relaxation peak in the wide range of temperatures for octa-O-acetyl-lactose. The maximum of this process is clearly visible even in the vicinity of the glass transition temperature.

Another very interesting behavior is observed when comparing loss spectra of both studied herein carbohydrates in the range below T_g (see Fig. 3). There are two, well separated, secondary relaxation processes in lactose. In octa-O-acetyl-lactose only one is detected. A maximum of the secondary relaxation of octa-O-acetyl-lactose is almost the same as that of the γ- relaxation of lactose at the chosen temperature. This can suggest that the considered secondary modes in lactose and its acetyl derivative may be of the same origin. Thus, in the further part of this paper we will label the secondary relaxation as a γ- processes (just as in the case of lactose).

The transition from the hydrogen bonded system into the van der Waals one is also reflected in the dynamics of the γ- relaxation. At the first glance, it is visible that amplitudes of the considered relaxation are much lower in octa-O-acetyl-lactose than in lactose. Moreover, from fitting the γ- peaks of both saccharides to the Havriliak–Negami function one can note that the dielectric strength of this process changes with temperature in the opposite ways. For lactose a decrease of $\Delta\varepsilon_\gamma$ with lowering temperature occurs, while in the case of octa-O-acetyl-lactose the dielectric strength of the γ mode slightly increases.

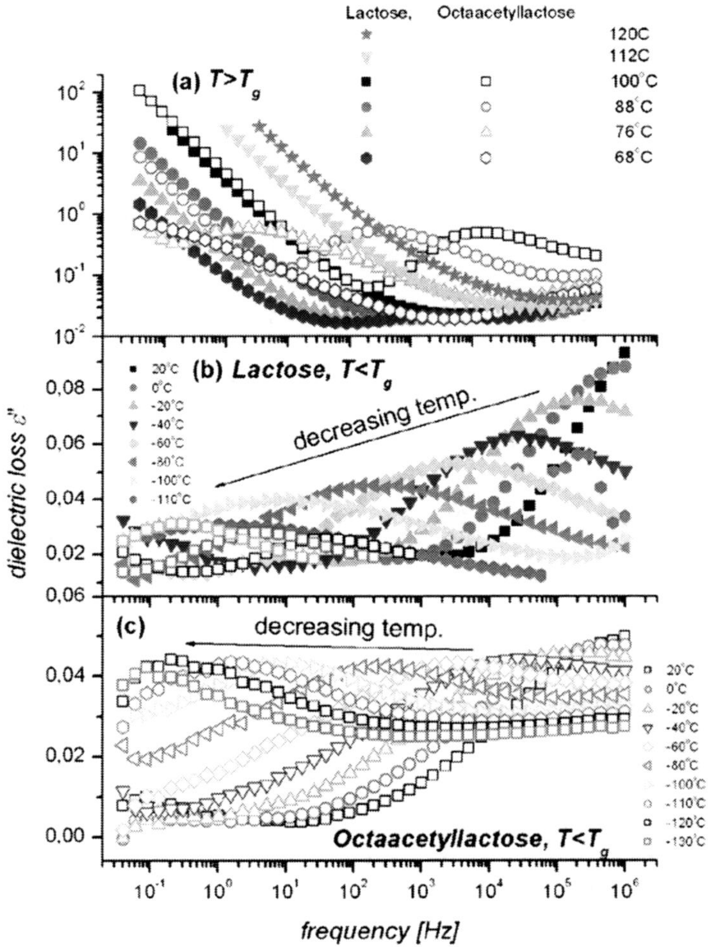

__Figure 2.__ (a) Dielectric loss spectra measured above the glass transition temperature for lactose (filled symbols) and octa-O-acetyl-lactose (open symbols). Panel (b) and (c) represent dielectric loss spectra obtained below T_g for lactose and for octa-O-acetyl-lactose, respectively.

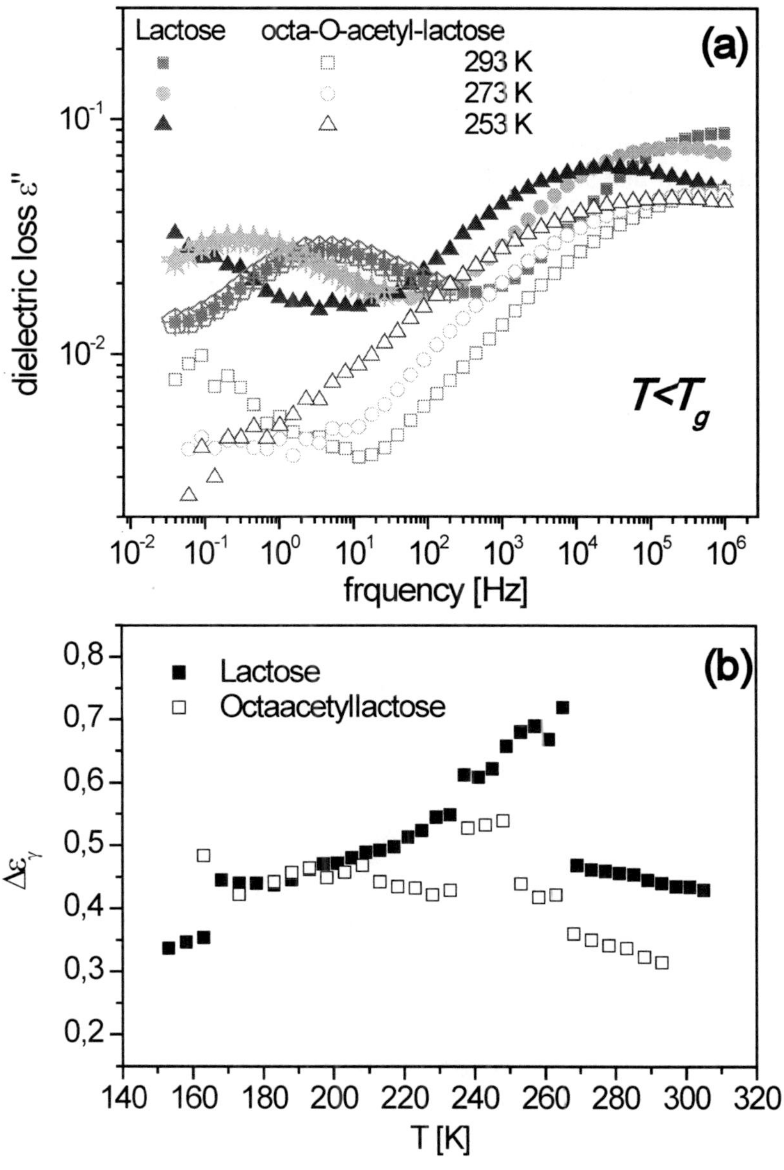

Figure 3. *(a) Comparison of the dielectric loss spectra measured for lactose (filled symbols) and octa-O-acetyl-lactose (open symbols) obtained below their* T_g *at three indicated temperatures. (b) Temperature dependence of dielectric strengths of the γ- relaxation processes of lactose and its acethyl derivative.*

The shape of γ- relaxation peak of both carbohydrates is asymmetric. In the introduction it was stated that earlier studies of the authors indicate that this relaxation is linked to internal degrees of freedom (motions within the sugar ring). However, basing on literature data[45-48] one could expect that such a kind of motions will manifest in loss dielectric spectra as a symmetric peak. From the other hand it is noted that in the whole sugar family the γ- peak cannot be described by the Cole-Cole function because it is clearly asymmetric. However, it has to be pointed out that saccharides are strongly hydrogen bonded systems. Thus one can try to suppose that probably such a kind of complex interactions may modify the shape of the γ- peak and make it asymmetric in loss dielectric spectra.

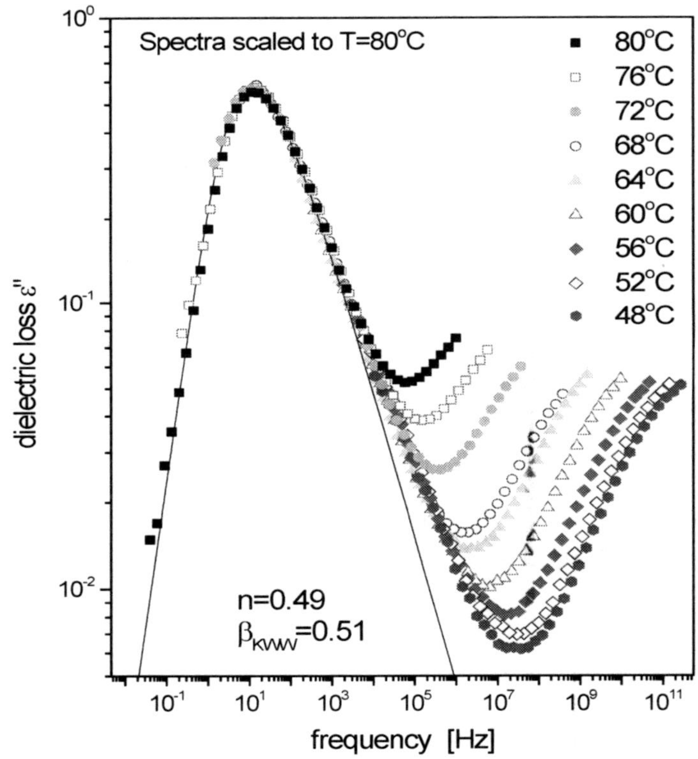

Figure 4. *Superimposed representative loss spectra of the octa-O-acetyl-lactose measured at nine indicated temperatures are presented. The dielectric loss spectra were superposed to that obtained at T=353 K. The solid curve is a Kohlrausch-Williams-Watts function with $n=(1-\beta)=0.49$*

On the other hand the secondary relaxation in octa-O-acetyl-lactose is *also asymmetric. We attempt to provide a credible explanation what is the pro*bable reason of asymmetric shape of the γ- relaxation in acetyl-lactose in the further part of this paper. An interesting observation one can make after analyzing the loss dielectric spectra of the considered saccharides. For octa-O-acetyl-lactose there is no sign of presence of any equivalent relaxation process to the β-relaxation present in lactose.

The molecular origin of β mode in disaccharides was extensively investigated by us in previous papers. We have shown that the rotation of monosugar rings around the glycosidic bond is the real molecular mechanism of this relaxation[13]. Our supposition was supported by the theoretical conformational computation performed on sucrose (it belongs to the disaccharides). For the lactose one can also find results of theoretical modeling of the conformational changes occurring via the glycosidic bond[49]. Therefore it is possible to compare data derived from dielectric investigations to that presented in Ref. [[48]].

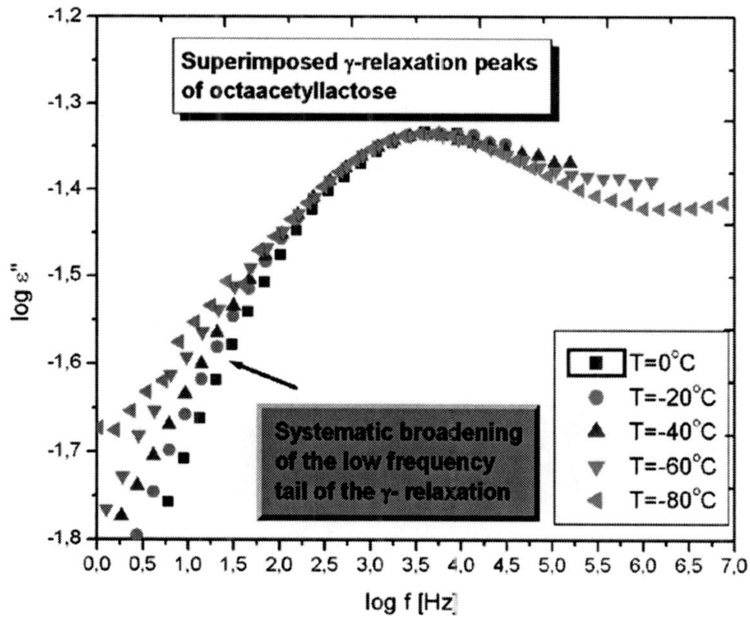

Figure 5. Superimposed representative dielectric loss spectra measured at five indicated temperatures much below the glass transition temperature of octa-O-acetyl-lactose. The loss spectra were superposed to that obtained at T= 233 K.

Recently we reported[15] that activation energy obtained for β- relaxation in lactose is equal to E_a=76 kJ/mol. The authors calculated theoretically the energy barrier which has to be overcome to change a torsion angle between two monosaccharides (twisting motions around glycosidic bond) as E_a=56 kJ/mol.[48] One can see that both reported values differ. However, this discrepancy between experimentally and theoretically estimated activation energies can be easily explained. Theoretical computations were performed only for an isolated molecule, i.e. intermolecular interactions and hydrogen bonds were not considered during calculations.

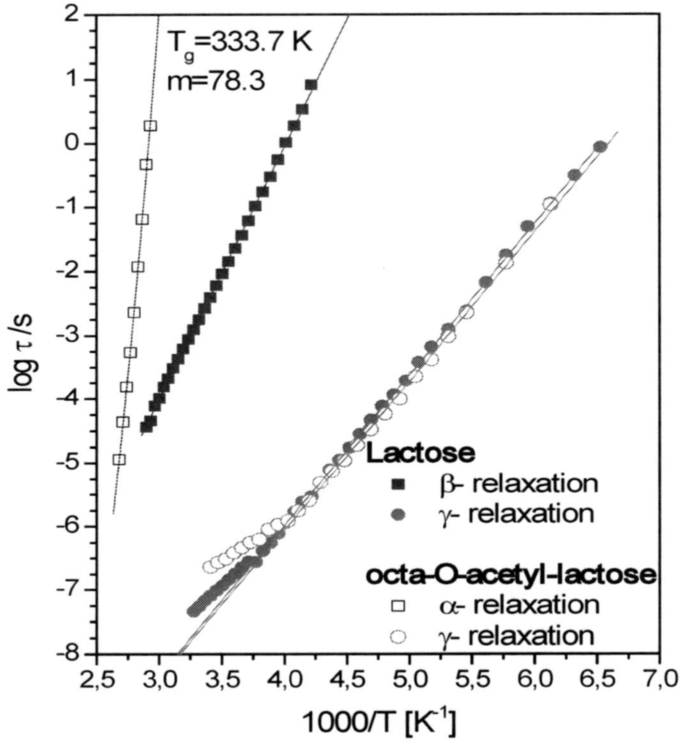

Figure 6. *Temperature dependences of the structural (α) and the secondary γ-relaxation times for octa-O-acetyl-lactose, as well as the temperature dependence of secondary β- and γ-relaxation times for lactose. The solid lines were obtained from fitting experimental data for the secondary relaxations to the Arrhenius law whereas the dashed line represents a fit the temperature dependence of α-relaxation times to the VFTH equation.*

Nevertheless, if the β-relaxation were of intramolecular origin (the process governed only by its own energy barrier) one could expect much better agreement between experimental and theoretical data. Moreover, the influence of intermolecular interactions on the β-relaxation is supported by our results of high pressure measurements, which show that the slower secondary relaxation of disaccharides is sensitive to pressure.[14]

Therefore, one can claim that interactions coming from surrounding molecules affect the twisting motions of monosugars around glycosidic bond. In this way one may be explained the difference in reported values of the theoretically calculated energy barrier for rotation of the monosugars around glycisdic bond and the β- process activation energy obtained from dielectric measurements.

Now let us try to account for the lack of the slower secondary relaxation in octa-O-acetyl-lactose. In order to do this, two opposite scenarios have to be taken into account. In the first one we consider a situation in which the acetyl groups introduced into the sugar rings cause an increase in rigidity of the disaccharide (greater steric hindrances in octa-O-acetyl-lactose than those in lactose). Hence the activation energy of the slower secondary relaxation increases significantly. Consequently, the β- process moves towards the structural α-relaxation peak. In the second scenario we suppose that substituting all hydrogen atoms by the acetyl groups results in an increase in flexibility of the disaccharide. To explain this supposition one can refer to the theoretical modeling calculation made on lactose. It was shown that in lactose there is one effective internal hydrogen bond between second and third carbon atoms of glucose and galactose units respectively which makes the structure of lactose more rigid.[48] Thus the elimination of the hydrogen bonds from the systems should results in the greater mobility of the monosugar units around glycosidic linkage in octa-O-acetyl-lactose than in the case of lactose.

If the first scenario were real, the slower secondary relaxation should express its presence as an excess wing on high frequency side of the α- relaxation peak. To check this we superimposed dielectric loss spectra of octa-O-acetyl-lactose measured above and below T_g to that obtained at T=353 K. Next we fitted a master curve constructed in this way to the Kohlrausch-Williams-Watts function

$$\phi(t) = \exp[-(t/\tau_{\alpha})^{1-n}] \tag{2}$$

with n=0.49 Such a procedure is very useful to check whether an excess wing, regarded as a hidden slower secondary relaxation, is present in the studied system. In the case of the acetyl derivative of lactose one can observe a slight deviation of the experimental data from the KWW fit at the frequencies of 4 decades higher than the maximum of the α-peak. This may suggest the presence of the hidden secondary relaxation under the structural relaxation peak.

However, it is difficult to interpret this in such a way. One can see that the coupling parameter *n* estimated for the α- relaxation peak of octa-O-acetyl-lactose is significant (it means that the distribution of α-relaxation times is quite broad). It implies that the separation between maxima of dielectric loss of the expected secondary relaxation and main structural relaxation peaks should be significant. Consequently, the β- mode should be clearly visible in the dielectric loss spectra. However, in the case of octa-O-acetyl-lactose the γ- relaxation peak is only visible. Therefore, we can state that the first scenario is not true. On the basis of dielectric data we can show that the substitution of hydrogen atoms by the acetyl groups does not result in increasing of the structural rigidity of the disaccharide.

Now let us consider the second scenario. Theoretical studies on lactose showed that there is one effective internal hydrogen bond in lactose. Substituting hydrogen atoms by the acethyl groups results in elimination of all hydrogen bonds present in lactose (including this internal one connecting glucose and galactose unit). Consequently, it should result in a greater mobility of the monomeric units around the glycosidic linkage in octa-O-acetyl-lactose than in the case of lactose. Recently we reported that the activation energy barrier E_a, which has to be overcome to make possible the twisting motion of the monosaccharide units via glycosidic bond, is equal to76 kJ/mol for lactose.[15]

Thus, taking into account the fact that the internal hydrogen bond does not exist in octa-O-acetyl-lactose, one can expect that the activation energy of the secondary relaxation related to the twisting motion of the glucose and galactose units via glycosidic bond should be at least 20 kJ/mol (energy needed to make possible breaking of the internal hydrogen bond) lower than that of lactose.

However, it has to be pointed out that it is only an approximate calculation, because we did not consider intermolecular hydrogen bonds which surely also make the structure of lactose more rigid. In such a case the activation energy of the slower secondary relaxation should be comparable to that for the γ-relaxation (E_a=44 kJ/mol). This may imply that the slower secondary relaxation seen in lactose may be undetectable in the case of acethyl derivative of this disaccharide, because maxima of both secondary relaxations can be too close to each other. In fact, the inspection of the dielectric loss spectra obtained for octa-O-acetyl-lactose below its glass transition temperature (see Fig. 2) showed that there is only one secondary relaxation peak. However, a detailed analysis of the γ- loss peak revealed that probably two secondary processes contribute to it.

The first surprising observation which confirms our supposition is that the γ-loss peak is still asymmetric in octa-O-acetyl-lactose just like in the case of lactose. However, the latter one is highly hydrogen bonded system, so we can try to speculate that these particular interactions may modify the shape of the γ-peak and make it asymmetric. In the former system the situation is far clearer. Octa-O-acetyl-lactose is van der Waals liquid, and it is commonly observed for such systems that the secondary relaxation of intramolecular nature has a symmetric shape in loss dielectric spectra.[44-47] Thus, according to our recent findings the γ- loss peak of octa-O-acetyl-lactose should be symmetric. Such an unusual shape of γ- relaxation peak may be caused by the presence of–two secondary relaxation modes which are strongly overlapped. The first secondary process is related to the motions of the monosaccharides around the glycosidic linkage (β- relaxation of lactose) whereas the second one is of the same origin as the faster secondary relaxation of lactose. Another observation indicating the presence of the two secondary relaxations exhibited in the γ- loss peak is an asymmetric broadening of this peak with lowering temperature. It can be seen in Fig. 5 that the slope of high frequency side of the γ- peak does not change while its low frequency tail slightly broadens . This unusual finding enabled us to conclude that the second scenario is the real one. Consequently, we can state that the substitution of all hydrogen atoms in lactose by acetyl groups resulted in the enhanced flexibility of the monosugar units around the glycosidic linkage. This, in turn, caused that the β- relaxation loss peak moved toward higher frequencies and overlapping with the dominant γ- loss peak.

The relaxation map of lactose and octa-O-acetyl-lactose can be seen in Fig. 6. The structural relaxation times is not presented for lactose because we have not been able to extract such information from our dielectric data. As it has been pointed out earlier in this paper there is a huge contribution of dc conductivity to the loss spectra of this disaccharide, which covers the structural relaxation peak. Moreover, we performed measurements only at temperatures little bit higher than the glass transition temperature reported in literature for this carbohydrate (see Table 1), because lactose is a sugar easy to caramelize. Thus, carrying out measurements at higher temperatures is very risky due to the non-controlled change occurring in the investigated sample.

In order to determine the structural relaxation times for the octa-O-acetyl-lactose we analyzed dielectric loss spectra of this carbohydrate with use of the Havriliak- Negami function. The temperature dependence of $\log_{10}\tau_\alpha$ was fitted to the Vogel- Fulcher- Tammann (VFT) function

$$\tau_\alpha = \tau_\infty \exp\left(\frac{D_T T_0}{T - T_0}\right) \tag{3}$$

From the VFT fit one can calculate the glass transition temperature $T_g = 334$ K (T_g is defined as a temperature at which $\tau_\alpha = 100$s) and the steepness index m=78 estimated from the following equation

$$m = d\log_{10}\tau_\alpha / d\,(T_g/T)\big|_{(Tg/T)=1} \tag{4}$$

Although we are not able to get such information for lactose from dielectric measurements we can compare the glass transition temperature for octa-O-acetyl-lactose to the data reported by other authors for lactose. In literature one can find a lot of DSC studies on lactose. The glass transition temperatures obtained from these investigations are within the range T_g= 374-387 K.[50-53]

Thus it is clearly visible that they are about 40-50 K higher than the value of T_g evaluated on the basis of dielectric measurements for acetyl derivative of lactose. It is surely related to the fact that molecules of the reduced disaccharide form strong hydrogen bonds and mainly these interactions are responsible for so significant difference in the glass transition temperatures of the both investigated herein carbohydrates. Influence of hydrogen bonds on structural dynamics should be also reflected in fragility indexes. Unfortunately, it is impossible to compare directly values of *m* estimated for lactose and acetyl-lactose, because no one reported *m* for the lactose up to now. However, we can try to compare the steepness index calculated for octa-O-acetyl-lactose to those obtained for other disaccharides[Błąd! Nie zdefiniowano zakładki.]. In our recent paper we estimated *m* for different disaccharides. The values of fragility were within the range m = 100-140.[15] One can see that steepness indexes calculated for reduced disaccharides are significantly greater than that obtained for octa-O-acetyl-lactose This experimental outcome is consistent with the commonly noted rule that with increasing number of hydrogen bonds the glass transition temperature and the fragility index increase (for instance see Ref. [39]).

Parameters		Lactose	Octaacetyllactose
α - relaxation VFT relation	$log_{10}\ \tau_{VFT}$ [s]	-	-25.70
	$D_T\ T_0$ [K]	-	3270
	T_0 [K]	-	215
β- relaxation Arrhenius relation	$log_{10}\ \tau_\infty$ [s]	15.90	-
	E_a [kJ/mol]	76	-
γ- relaxation Arrhenius relation	$log\ \tau_\infty$ [s]	-15.02	-15.40
	E_a [kJ/mol]	44	45
T_g[K]		374 K[49], 385 K[50]387 K[51], 387[52]	334
m		-	78

Table 1 *Values of fitted parameters for the evolution of relaxation processes*

The temperature dependences of relaxation times of the β and γ processes of lactose and the secondary mode of octa-O-acetyl-lactose are presented in Fig. 6. In order to determine relaxation times of β- and γ- modes of lactose and octa-O-acetyl-lactose the Cole-Cole and Havriliak–Negami functions were used respectively. Activation energies of all secondary relaxations were estimated from the Arrhenius fits

$$\tau_\beta = \tau_0 \exp\left({E_a}/{k_B T} \right) \tag{5}$$

of the temperature dependences of the secondary relaxation times. All fitted parameters are collected in Table 1.

It is very surprising that the activation energies and relaxation times of the γ-processes for lactose and its acetyl derivative are the same. One can expect that the elimination of the hydrogen bonds in the latter system should manifest more prominently in the dynamics of its secondary mode. Especially that the structural similarities of both investigated herein carbohydrates indicate that the considered relaxation processes are of the same molecular origin. Moreover, it is reasonable to suppose that in order to enable any motion, even intramolecular, (as it is in the case of γ- relaxation) one should expect that at least one hydrogen bond has to be broken in so strongly hydrogen bonded system as it is for lactose. A natural implication of this assumption is that the γ-relaxation in octa-O-acetyl-lactose (van der Waals liquid) should have a lower activation energy. It should be also added that Noel *et al.*[10] and Ermolina *et al.*[54] observed a great influence of water on dynamics of the γ- relaxation process.

They showed that the amplitude of this mode increases significantly with moisture content. Moreover, they found out that activation energy of this relaxation increases. Thus the result of their investigations is a clear indication that hydrogen bonds may be connected in some way to the observed relaxation process. On the other hand our experimental data revealed something opposite. However, on the basis of only dielectric data we are not able to find a reasonable explanation of this experimental observation.

4. Conclusions

In this paper we investigated two structurally similar systems lactose and octa-O-acetyl-lactose. However, there are completely different types of interactions in both saccharides. The first one is a hydrogen bonded system while the second one is a typical van der Waals liquid. The change in type of molecular interactions has a great influence on dynamics of the structural relaxation process. The glass transition temperatures of both carbohydrates differ significantly about 40-50 K. The higher T_g is noted for lactose as expected. The transition from the hydrogen bonded system to the van der Waals one is also reflected in the steepness index which is much lower for acetyl derivative of lactose than for any other disaccharide.

The substitution of all hydrogen atoms by the acetyl groups resulted also in the change of mobility of the glycosidic linkage. On the basis of our dielectric data and theoretical conformational calculations we were able to demonstrate that the rotational energy of the monosaccharide units via the oxygen bridge is greater in acetyl-lactose than in the case of lactose. Finally we found out that dynamics of the γ- relaxation in both examined herein systems is almost the same (the same relaxation times at chosen temperature as well as the activation energies). The only one difference between considered modes in lactose and its acethyl derivative is the amplitude of the γ- loss peak. In the former one the dielectric strength of the γ- relaxation peak is significantly greater than in the case of the latter. Our findings are very surprising, because we expected that in so strongly hydrogen bonded disaccharide (lactose) the potential barriers for all motions even intramolecular ones are surely governed by the energy needed to break at least one hydrogen bond. Consequently, it was reasonable to suppose that the activation energy and relaxation times of the γ- peak should be different in octa-O-acetyl-lactose and lactose. Especially that an enormous dielectric strength of the γ- peak in lactose indicated that motions of a greater group (probably motions of the part of the monosaccharide ring) can be a probable molecular mechanism of this relaxation. However, at this point we are not able to find an explanation of so little change of dynamics of the γ- relaxation process in lactose in comparison to its acetyl derivative.

Acknowledgements

The research was supported by the Foundation for Polish Science within the framework of the TEAM programme from the European Union Structural Funds (Poland) within the Innovative Economy Operational Programme framework. K.K. acknowledges financial assistance from FNP (2008).

References

1. Gabius, H. J. (2000) *Naturwissenschaften*, **87**, 108
2. Lindhorst, T. L. (2000) *Chem. Z.* **34**, 38
3. Moran, G. R., Jeffrey, K. R., Thomas, J. M., and Stevens, J. R. (2000) *Carbohydr. Res.* **328**, 573
4. Gangasharan, S., and Murthy, S. S. N. (1993) *J. Chem. Phys.* **99**, 9865
5. Chan, R. K., Pathmanathan, K., and Johari, G. P. (1986) *J. Phys. Chem.* **90**, 6358
6. Tombari, E., Cardelli, C., Salvetti, G., and Johari, G. P. (2001) *J. Mol. Struct.* **559**, 45
7. Gangasharan, S., and Murthy, S. S. N. (1995) *J. Phys. Chem.* **99**, 12349
8. Ermolina, I., Polygalov, E., Bland, C., and Smith, G. (2007) *J. Non-Cryst. Solids* **353**, 4485–4491
9. Gusseme, A. D., Carpentier, L., Willart, J. F., and Descamps, M. (1986) *J. Phys. Chem.* **107**, 10879
10. Noel, T. R., Parker, R., and Ring, S. G. (2000) *Carbohydr. Res.* **329**, 28
11. Kaminski, K., Kaminska, E., Wlodarczyk, P., Paluch, M., Ziolo, J., and Ngai, K. L. (2008) *J. Phys.: Condens. Matter*, **20**, 335104

12. Kaminski, K., Kaminska, E., Paluch, M., Ziolo, J., and Ngai, K. L. (2006) *J. Phys. Chem. B* **110**, 25045
13. Kaminski, K., Kaminska, E., Hensel-Bielowka, S., Chelmecka, E., Paluch, M. Ziolo, J., Wlodarczyk, P., and Ngai, K. L. (2008) *J. Phys. Chem. B* **112**, 7662
14. Kaminski, K., Kaminska, E., Hensel-Bielowka, S., Pawlus, S., Paluch, M., and Ziolo, J. (2008) *J. Chem. Phys.* **129**, 084501
15. Kaminski, K., Kaminska, E., Wlodarczyk, P., Pawlus, S., Kimla, D., Kasprzycka, A., Paluch, M., Ziolo, J., Szeja, W., and Ngai, K. L. (2008) *J. Phys. Chem. B* **112**, 12816–12823
16. Carpentier, L., and Descamps, M. (2003) *J. Phys. Chem. B* **107**, 271
17. Noel, T. R., Parker, R., and Ring, S. G. (2000) *Carbohydr. Res.* **329**, 839
18. Kubát, J., and Pattyranie, C. (1967) *Nature* **215**, 390
19. Kimura, M., and Nakano, J. (1976) *J. Polym. Sci., Polym. Lett.* **14**, 741
20. Montes, H., and Cavaille, J. Y. (1999) *Polymer*, **40**, 2649
21. Meissner, D., Einfeldt. J., and Kwasniewski, A. (2000) *J. Non-Cryst. Solids* **275**, 199
22. Starkweather, H. W. (1991) *Polymer* **32**, 2443
23. Gangasharan, G., and Murthy, S. S. N. (1995) *J. Phys. Chem.* **99**, 12349
24. Noel, T. R., Parker R., and Ring, S. G. (1996) *Carb. Res.* **282**, 193
25. Faivre, A., Niquet, G., Maglione, M., Fornazero, J., Jal, J. F., and David, L. (1999) *Eur. Phys. J. B* **10**, 277
26. van Dusschoten, D., Tracht, U., Heuer, A., and Spiess, H. W. (1999) *J. Phys. Chem. A* **103**, 8359
27. Moran, G. R., Jeffrey, K. R., Thomas, J. M., and Stevens, J. R. (2000) *Carbohydrate Res.* **328**, 573–584
28. Molinero, V., and Goddard, W. A. (2005) *Phys. Rev. Lett.* **95**, 045701
29. Wlodarczyk, P., Kaminski, K., Paluch, M., and Ziolo, J. (2009) submitted to *J. Phys. Chem. B*
30. Böhmer, R., Diezemann, G., Geil, B., Hinze G., Nowaczyk A., and Winterlich, M. (2006) *Phys. Rev. Lett.* **97**, 135710
31. Ngai, K. L. (1998) *J. Chem. Phys.* **109**, 6982
32. Ngai, K. L. (2003) *J. Phys.: Condens. Matter* **15**, S1107.
33. Mierzwa, M., Pawlus, S., Paluch, M., Kaminska, E., and Ngai, K. L. (2008) *J. Chem. Phys.* **128**, 044512
34. Kessairi, K., Capaccioli, S., Prevosto, D., Lucchesi, M., Sharifi, S., and Rolla, P. A. (2008) *J. Phys. Chem. B* **112**, 4470
35. Ngai, K. L. (2003) *J. Phys.: Condens. Matter* **15**, S1107
36. Ngai, K. L., and Paluch, M. (2004) *J. Chem. Phys.* **120**, 857
37. Ngai, K. L. (2006) *J. Non-Cryst. Solids* **351**, 2635
38. Ngai, K. L. (1998) *J. Chem. Phys.* **109**, 6982

39. Döß, A., Paluch, M., Sillescu, H., and Hinze, G. (2002) *Phys. Rev. Lett.* **88**, 095701
40. Grzybowska, K., Pawlus, S., Mierzwa, M., Paluch, M., and Ngai, K. L. (2006) *J. Chem. Phys.* **125**, 144507
41. Paluch, M., Grzybowska, K., and Grzybowski, A. (2007) *J. Phys.: Condens. Matter* **19**, 205117; Grzybowski, A., Grzybowska, K., Zioło, J., and Paluch, M. (2006) *Phys. Rev. E* **74**, 041503
42. Grzybowski, A., Grzybowska, K., Zioło, J., and Paluch, M. (2007) *Phys. Rev. E* **76**, 013502
43. Wolform, M. L., and Thompson, A. (1963) in *Methods in Carbohydrate Chemistry*, Whistler, R.L., Wolfrom, M.L., BeMiller, J.N. eds., (*Academic Press Inc., New York and London, Vol. II, p. 212*)
44. Crofton, D. J., and Pethrick, R. A. (1981) *Polymer* **22**, 1048
45. Kaminska, E., Kaminski, K., Hensel-Bielowka, S., Paluch, M., and Ngai, (2006) *J. Non-Cryst. Solids* **352** 4672–4678
46. Kaminska, E., Kaminski, K., Paluch, M., and Ngai, K. L. (2006) *J. Chem. Phys.* **124**, 164511
47. Paluch, M., Pawlus, S., Hensel-Bielowka, S., Kaminski, K., and Psurek, T., Rzoska, S. J., Ziolo, J., and Rolad, C. M. (2005) *Phys Rev B* **72**, 224205
48. Sekula, M., Pawlus, S., Hensel-Bielowka, S., Ziolo, J., Paluch, M., and Rolad, C. M. (2004*) J. Phys. Chem. B* **108**, 4997
49. Clarissa, O., da Silva, C., and Nascimento, M. A. C. (2004) *Carbohydrate Res.* **339**, 113–122
50. Roos, Y. (1993) *Carbohydr. Res.* **238**, 39
51. Miller, D. P., and de Pablo, J. J. (2000) *J. Phys. Chem B.* **104**, 8876
52. Dudognon, E., Willart, J. F., Caron, V., Capet, F., Larsson, T., and Descamps, M. (2006) *Solid State Comm.* **138**, 68
53. Taylor, L. S., and Zografi, G. D. (1998) *J. Pharm. Sci.* **87**, 1615
54. Ermolina, I. et al. (2009) *manuscript in preparation*

EFFECTS OF PRESSURE ON STABILITY OF BIOMOLECULES IN SOLUTIONS STUDIED BY NEUTRON SCATTERING

MARIE-CLAIRE- BELLISSENT-FUNEL[1],
MARIE-SOUSAI APPAVOU[1,2] AND GABRIEL GIBRAT[1]
[1]Laboratoire Léon Brillouin (CEA-CNRS) CEA Saclay 91191 Gif-sur-Yvette-Cedex, France
[2]Forschungszentrum Jülich GmbH, IFF-JCNS, Lichtenbergerstrasse 1., 85747 Garching, Germany.

Abstract: Studies of the pressure dependence on protein structure and dynamics contribute not only to the basic knowledge of biological molecules but have also a considerable relevance in full technology, like in food sterilization and pharmacy. Conformational changes induced by pressure as well as the effects on the protein stability have been mostly studied by optical techniques (optical absorption, fluorescence, phosphorescence), and by NMR. Most optical techniques used so far give information related to the local nature of the used probe (fluorescent or phosphorescent tryptophan). Small angle neutron scattering and quasi-elastic neutron scattering provide essential complementary information to the optical data, giving quantitative data on change of conformation of soluble globular proteins such as bovine pancreatic trypsin inhibitor (BPTI) and on the mobility of protons belonging to the protein surface residues.

Keywords: protein folding, globular proteins, pressure, small angle neutron scattering, quasi-elastic neutron scattering, optical techniques, NMR

1. Introduction

The protein folding is the focus of many researchers since the fundamental work of Anfinsen[1]. How can a freshly synthesized chain adopt its native functional conformation in a short time frame? A random conformation search of polypeptide would take more than the lifetime of the universe for an average sized protein (Levinthal's paradox).[2] It is generally accepted that folding is a directed process rather than a stochastical one. The funnel model allows to accounting for the directed folding.[3,4] In this model the Gibbs free energy of the protein and the surrounding water molecules as a function of the

S. Rzoska et al. (eds.), *Metastable Systems under Pressure*,

conformational coordinates can be represented by a rugged, funnel-shaped surface that is valid only under native conditions (temperature, pressure, pH,).

The figure 1 represents the Gibbs free energy landscape of protein. One can notice the ruggedness of the surface: the molecule can be trapped in several energetic minima. In a few deeper minima, the protein can be trapped for considerable time. These minima are assigned to the intermediate states observed experimentally and recognized recently to be at the origin of Neurodegenerative diseases, such as Alzheimer's and Parkinson's diseases. The common feature of these illnesses is the formation of aggregates that are organized in a fibril[5].

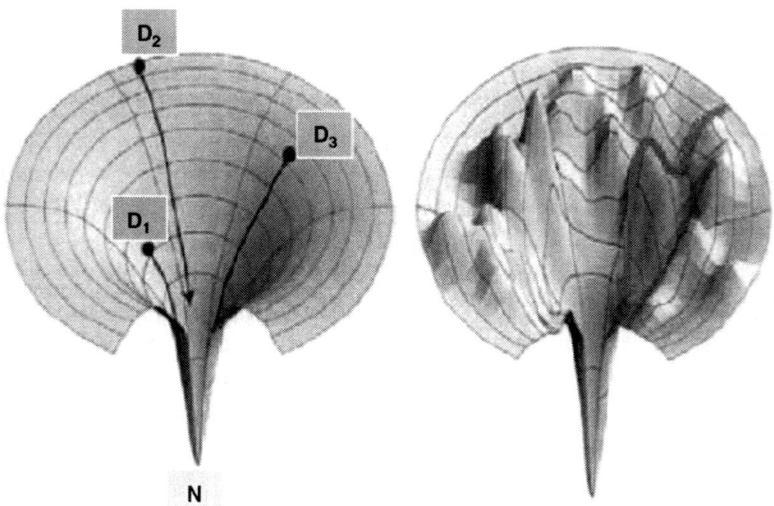

Figure 1. *The funnel shape Gibbs-free energy landscape of protein. The global minimum corresponds to the native state. Left: There are no local minima. Two-state folding approximation Right: There are many local minima. Presence of intermediate states, the red line corresponds to a folding pathway giving rise to a metastable intermediate state*[3,4].

The paper is organized as follows: After describing the folded state of the globular protein, the various ways of unfolding protein will be presented and discussed. We will focus our attention into the effects of pressure upon protein solutions. The results of experiments using various techniques (FTIR, fluorescence, small angle neutron scattering) to probe the structure and the dynamics of bovine pancreatic trypsin inhibitor (BPTI) protein solutions will be given. In fact, our aim purpose is to study the relationship between the structure and the dynamics of a globular protein such as BPTI in the native state and in the unfolded states where the protein is not functional.

2. The globular protein: Folded state

The interactions between the amino acids and the solvent (electrostatic, hydrophilic, hydrophobic, S-S) determine the globular conformation. We can give some naive picture of the folded state in terms of a liquid-hydrocarbon model where the hydrophobic core stabilizes globular proteins. The hydrophilic (polar and charged) amino acids are exposed to the solvent and the hydrophobic (polar) amino acids are less exposed to the solvent and buried in the interior of the protein.

This model was established by Chen and Schoenborn[6] when identifying water molecules and ions in a crystal of CO myoglobin protein by combining neutrons and x-ray crystallography. The surface structure exhibits 85 water molecules and 5 ions while the access path of the CO to the heme is devoid of bound water. Thus there is a subtle interplay between the hydrophilic and hydrophobic phenomena that control the stability of the folded state.

3. How to unfold a protein?

As stated above there are several ways to get an unfolded protein state.

3.1. TEMPERATURE AND CHEMICAL EFFECTS

The effect of temperature is to disrupt the interactions. The water penetrates as a result of protein unfolding.

The effect of chemicals is more complex and depends on the denaturant (guanidinium chloride, urea, acids). In this case, we have to deal with an additional component to the protein. For example, acids break salt bridges of protein while urea contributes to break H-bonds and β-mercaptoethanol and dithiotreitol reduce S-S bonds. (For a review see reference7)

3.2. PRESSURE EFFECTS

The effects of pressure on organic systems are known since the 19th century from qualitative experiments. The effects of high hydrostatic pressure treatment on the inactivation of microorganisms were reported 100 years ago by Hite[8]. On 1899, Hite uses pressure for milk preservation. On 1914, Bridgman[9] notices that egg white looks “cooked” after pressure treatment. Though it is not intuitive, proteins also unfold with pressure.

First let us recall that at ambient conditions, in case of globular soluble proteins, the globule is relatively loosely packed with cavities and a hydrophobic core. By applying pressure, it results by adding more water into the same volume that such some “efficient” packing becomes necessary. Water penetrates the protein interior. One comes to the following conclusions:

i) The variation of volume of the protein ΔV is negative because water molecules go into the protein while the hydrophobic groups do not come into the water.
ii) The protein unfolds as a result of water penetration.
iii) The protein becomes more solvated by water molecules.

3.3. WHY TO USE PRESSURE?

There are several reasons to measure the effect of pressure on a wide variety of thermodynamic systems. Perhaps the most important argument is that one can separate the effects of volume and thermal energy changes, which appear simultaneously in temperature experiments[10]. Moreover, high pressure can induce unfolding of protein in a different way from thermal denaturation. The pressure studies have considerably increased in the last decades[11,12,13].

4. The protein stability diagram

The first systematic observations of the behavior of proteins with respect to temperature and pressure came from the kinetic work of Suzuki[14] and the thermodynamic work of Hawley[15]. The elliptic phase diagram is shown in Fig. 2.

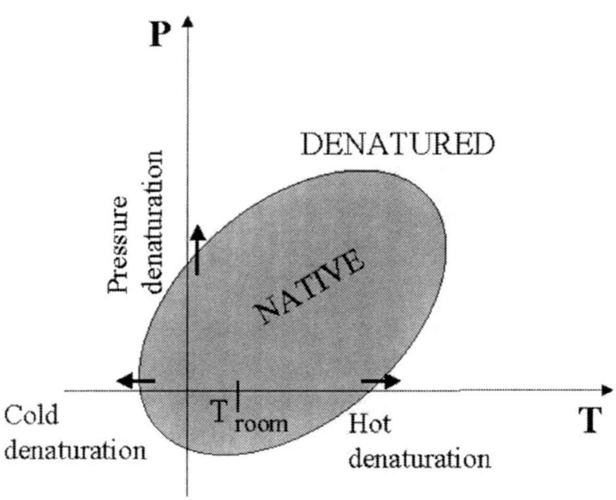

Figure 2. *Schematic representation of the elliptic phase diagram characteristic of proteins. In this Pressure-temperature diagram, the ellipse defines pressure and temperature range inside which the protein is stable and not denatured (Suzuki[14], Hawley[15]).*

The shape originates from a Taylor expansion of the free energy difference between the unfolded and the native state of the protein as a function of temperature and pressure, and the second order terms make a significant contribution. Thus, the volume change is not only pressure but also temperature dependent:

$$\Delta V(P,T) = \Delta V^0 + \Delta\alpha(T - T_0) - \Delta\beta(P - P_0) \qquad (1)$$

The term ΔV^0 refers to reference conditions, $\Delta\beta$ is the compressibility factor difference and $\Delta\alpha$ the difference of the thermal expansion between the denatured and native state.[16] The physics behind the elliptic phase diagram can be related to the degree of correlation between the enthalpy and the volume change of the unfolding transition and is discussed by Leisch[17]. It is now well established that all proteins are characterized by an ellipsoidal P, T phase diagram[18]. For myoglobin, it has been observed with infra-red spectroscopy strong similarities between the cold and pressure-denatured state and a quite different conformation for the heat-denatured state.[19]

We have studied the effect of the temperature and pressure on the conformations and dynamics of a soluble, globular BPTI protein.

5. Results

We present results from small angle neutron scattering performed on NEAT spectrometer at Helmholtz Zentrum Berlin (HZB) using the Small Angle Neutron Scattering (SANS) configuration and from quasi-elastic neutron scattering (QENS) on IN5 spectrometer at ILL (Grenoble), on a 85 mg/ml concentrated solution of BPTI, as a function of applied pressure between 1 bar and 6000 bar and after pressure release.

A detailed description of the neutron scattering theory can be found in references 20 and 21.

5.1. INFLUENCE OF PRESSURE ON THE CONFORMATIONS OF A BPTI PROTEIN

The evolution of the radius of gyration and of the shape of the protein under pressures up to 6000 bar has been studied by small angle neutron scattering. Small angle neutron scattering principles can be found in references 22 and 23. When increasing pressure from atmospheric pressure up to 6000 bar, the pressure effects on the global structure of BPTI result on a reduction of the radius of gyration from 13.4 Å down to 12.0 Å. Between 5000 and 6000 bar some transition already detected by FTIR[24] is observed. The pressure effect is not reversible because the initial value of the radius of gyration is not recovered after pressure release. By extending the range of wave-vectors to high q, we have observed a change of the form factor (shape) of the BPTI under pressure. At atmospheric pressure BPTI exhibits an ellipsoidal form factor that is characteristic of the native state. When the pressure is increased from atmospheric pressure up to 6000 bar, the protein keeps its ellipsoidal shape. The parameters of the ellipsoid vary and the transition detected between 5000 and 6000 bar in the form factor of BPTI is in agreement with the FTIR results. After pressure release, the form factor of BPTI is characteristic of an ellipsoid of revolution with a semi-axis a, slightly elongated with respect to that of the native one, indicating that the pressure-induced structural changes on the protein are not reversible[25,26].

The evolution of the specific volume as a function of pressure is plotted in figure 3. We observe that this evolution is linear. At ambient pressure the specific volume is equal to 0.726 cm^3/g and very close to that found in the literature[27] (0.730 cm^3/g). The diminution of the specific volume between 1 bar and 6000 bar, gives a value of the compressibility[25,26] of $4.8.10^{-2}$ $kbar^{-1}$. This value is close to the value of $4.5.10^{-2}$ $kbar^{-1}$ found by Kitchen and co-workers[28].

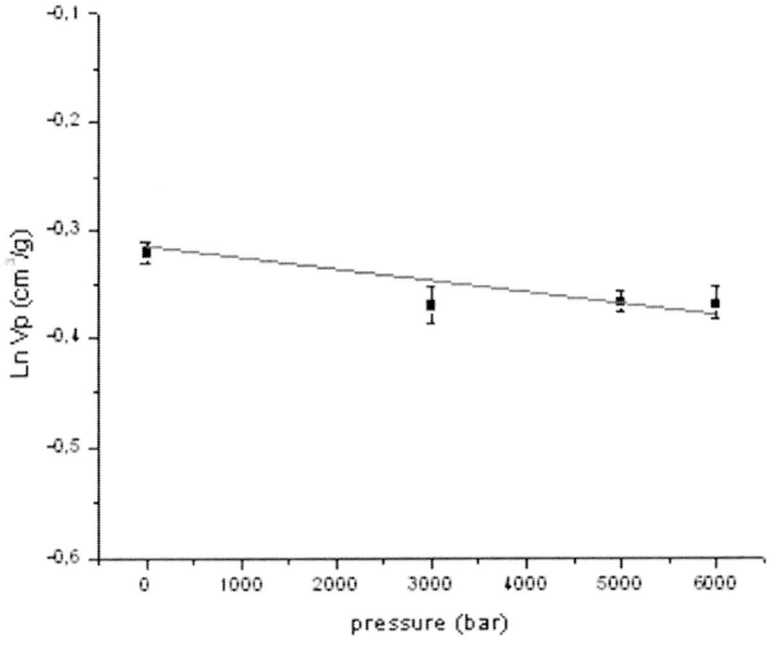

Figure 3. *Evolution of the specific volume of BPTI v_P as a function of applied pressure.*[25,26]

5.2. INFLUENCE OF PRESSURE ON THE DYNAMICS OF A BPTI PROTEIN

Quasi-elastic neutron scattering principles can be found in reference 29. The figures 4 and 5 present for q=1 $Å^{-1}$ the dynamic structure factor S(q,ω) and the intermediate scattering function I(q,t) of BPTI when increasing pressures at three different values up to 6000 bar and when releasing pressure at 1 bar.

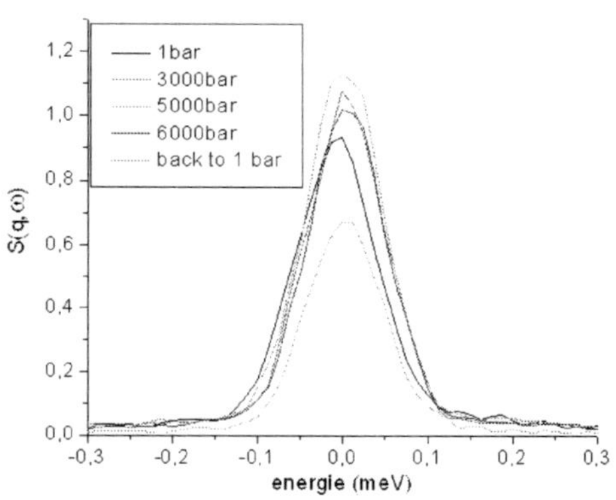

Figure 4. *S(Q,ω) spectra at room T for q=1 $Å^{-1}$.*

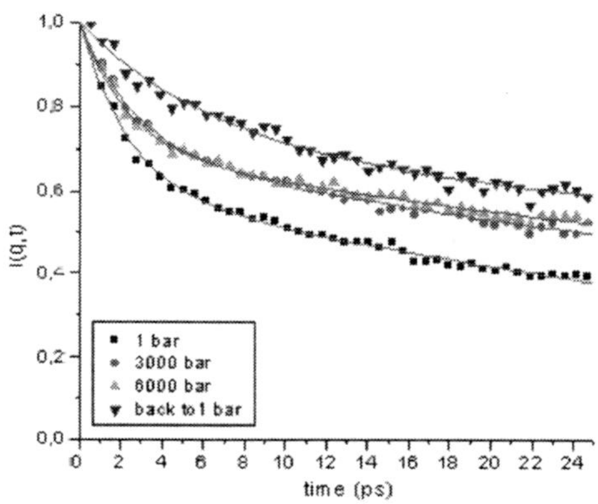

Figure 5. *I(q,t) of BPTI in solution at room T for q=1 $Å^{-1}$. Solid lines fit using eq. 2.*

I(q,t) is the Fourier transformation of S(q,ω) and can be expressed by eq. 2.

$$S(q,t) = A_0 * \exp\left(-\frac{t}{\tau_1}\right) + (1 - A_0) * \left(-t * \frac{\tau_1 \tau_2}{\tau_1 + \tau_2}\right) \quad (2)$$

$\tau_1(q)$ and $\tau_2(q)$ are the two relaxation times describing respectively the global diffusion and the internal motions. $A_0(q)$ is the pseudo Elastic Incoherent Structure Factor (EISF), $A_0(q) = p + (1 - p)\left[\frac{3 j_1(qa)}{qa}\right]^2$ (3)

where j_1 is the first order spherical Bessel function and p is the proportion of immobile protons, at the resolution of the spectrometer.[26]

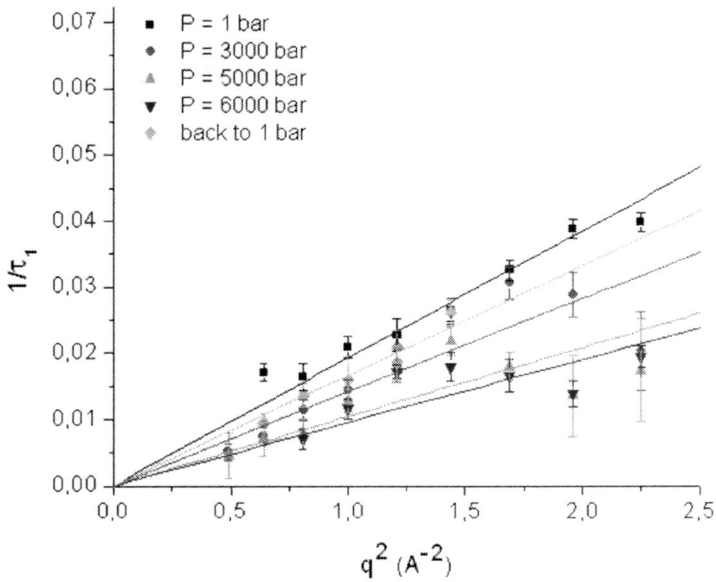

Figure 6. *Inverse of Relaxation time $1/\tau_1$ as a function of P. From fitting of I(q,t), eq. 2.*

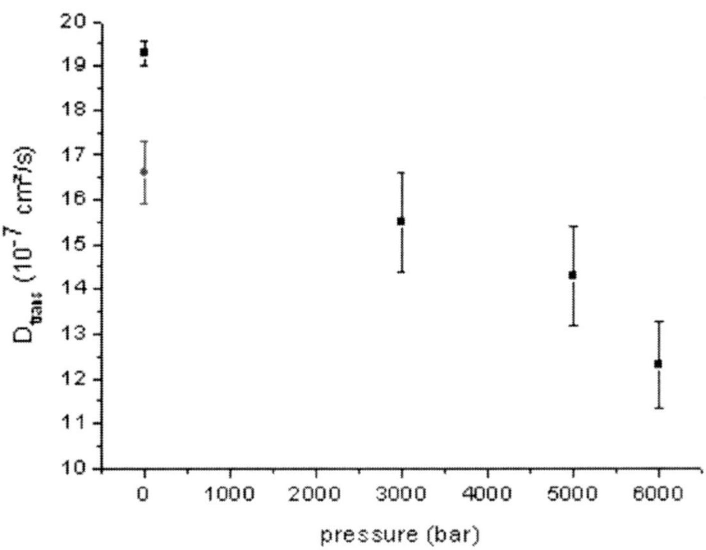

Figure 7. *D_{trans} of BPTI in solution as a function of P. Full black squares -, D_{trans} under P increase, red one - back to 1 bar.*

The fitting of I(q,t) spectra with the previous model give us A_0, τ_1 and τ_2 as a function of pressure. Figure 6 represents the evolution of $1/\tau_1$ as a function of

q^2. It is characterized by a linear behavior the slope of which gives the global diffusion coefficient D_{trans} of BPTI in solution. Figure 7 displays the evolution of D_{trans} as a function of pressure. As a result, at high pressure, global diffusion motions of protein are slowed down. At atmospheric pressure, the obtained value is in agreement with that given by dynamic light scattering[30]. As for structural changes, the effect of pressure on global motions is not reversible because the value of D is not recovered when the pressure is released.

Figure 8 shows the evolution of the relaxation time τ_2 of internal motions as a function of pressure.

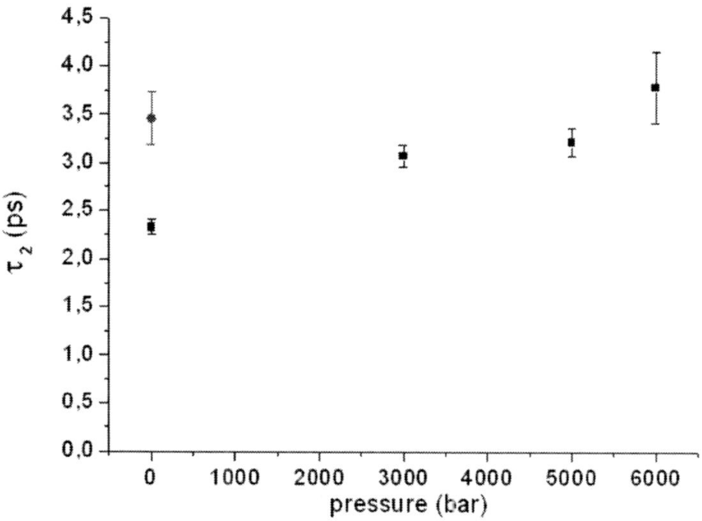

Figure 8. *Relaxation time τ_2 as a function of P. From fitting of I(q,t), eq. 2.*

We also observe a slowing down of the relaxation time τ_2 related to internal motions. The relaxation time τ_2 increases from 2.3 ps at ambient pressure to 3.6 ps at 6000 bar. This increase of τ_2 is accompanied by a slight decrease of the proportion of mobile protons (from 43% at ambient pressure to 39% at 6000 bar) and of the radius of the sphere in which motions occurs (from 3.6 Å at ambient pressure to 3.1 Å at 6000 bar). At atmospheric pressure, QENS experiments on other proteins have shown that the motions observed, at similar wave vector and energy transfer ranges, concern protons belonging to the side chains at the surface of the protein[31,32,33]. Thus the observed modifications of internal dynamics can only concern lateral chains. Moreover, molecular dynamics simulations have shown that backbone of BPTI is not affected by pressures up to 5000 bar[34]. A possible explanation about the pressure effects on internal protein motions is that the effect of pressure is to increase the density of first hydration layer at the protein surface[35]. This water density increase induces highest constraints on lateral chain motions from residues at the surface of the protein leading to a slowing down of motions.

6. Conclusion

We have investigated the influence of the pressure on the structure and the dynamics of a model protein, such as a globular soluble BPTI protein. Structural investigation by small angle neutron scattering on a BPTI solution at 85 mg/ml allowed us to follow the protein conformational changes. Between atmospheric pressure and 6000 bar, the protein keeps its ellipsoidal shape with some significant variation of ellipsoid parameters, in particular at 6000 bar. Concerning quasi-elastic neutron scattering experiment, we observe a slowing down of global and internal motions of BPTI with pressure. Effects of pressure on both structure and dynamics of BPTI seem to be not reversible. Some further experiments using small angle and quasi-elastic neutron scattering up to 10 kbar are needed to investigate more precisely the protein denaturation of BPTI.

Moreover, the experimental difficulties inherent to scattering techniques (neutrons, X-rays) can explain the relative rareness of the pressure denaturation studies of protein solutions[25,26,36,37].

References

1. Anfisen, C. B. (1973) Principles that govern the folding of protein chains, *Science* **181**, 223-230
2. Karplus, M. (1997) The Levinthal paradox: yesterday and to day, *Fold. Des.* **2**, S69-S75
3. Frauenfelder, H., Sligar, S.G., and Wolynes P.G. (1991) The energy landscapes and motions of proteins, *Science* **254**, 1598-1603
4. Dill, K. A., and Chan, H. S. (1997) From Levinthal to pathways to funnels, *Nat. Struct. Biol.* **4**, 10-9
5. Dobson, C. M. (2001) The structural basis of protein folding ans its links with human disease, *Philos. Trans. R. Soc. Lond.* **B356**, 133-145
6. Chen, X., and Schoenborn, B. P. (1990) Hydration in protein crystals. A neutron diffraction analysis of carbonmonoxymyoglobin, *Acta Crystallographica* **B46**, 195-208
7. Yon, J. M. (2004) Aggregation, protein. In : Meyers RA, ed. Encyclopedia of molecular cell biology and molecular medicine, (New York: John Wiley) 23-52
8. Hite, B. H. (1899) The effect of pressure in the preservation of milk, *Bulletin of West Virginia University Agricultural Experimental Station* **58**, 15-35
9. Bridgman, P. W. (1914) The coagulation of albumen by pressure, *J. Biol. Chem.* **19**, 511-512

10. Weber, G., Drickamer, H. G. (1983) The effect of high pressure upon proteins and other biomolecules, *Q. Rev. Biophys.* **16**, 89-112
11. Heremans, K. (Ed.) (1997) High Pressure Research in Bioscience and Biotechnology, (Leuven University Press, Louvain)
12. Taniguchi, Y., Stanley, H. E., and Ludwig, H. (Eds.) (2001) *Biological Systems under Extreme Conditions: Structure and Function*, (Springer, Heidelberg)
13. Heremans, K. K. (2004) Biology under extreme conditions, *High Press. Res.* **24** (1), 57–66
14. Suzuki, K. (1960) Studies on the kinetics of protein denaturation under high pressure, *Review of Physical Chemistry of Japan* **29**, 91-98
15. Hawley S. A. (1971) Reversible pressure-temperature denaturation of chymotrypsinogen, *Biochemistry* **10**, 2436-2442,
16. Smeller, L. (2002) Pressure-temperature phase diagram of biomolecules, *Biochimica et Biophysica Acta* **1595**, 11-29
17. Lesch, H., Hecht, C., and Friedrich, J. (2004) Protein phase diagrams: the physics behind their elliptic shape, *Journal of Chemical Physics* **121**, 12671-12675
18. Winter, R., and Dzwolak, W., (2004) Temperature-pressure configurational landscape of lipid bilayers and proteins, *Cell. Mol. Biol.* **50**, 397-417
19. Meersman, F., Smeller, L., and Heremans, K. (2002) Comparative Fourier transform infrared spectroscopy of cold-, pressure-, and heat-induced unfolding and aggregation of myoglobin, *Biophysical Journal* **82**, 2635-2644
20. Squires, G. L. (1978) *Introduction to the theory of Thermal Neutron Scattering* (Cambridge, University Press)
21. Lovesey, S. M. (1984) Theory of neutron scattering from condensed matter (Oxford University Press Eds)
22. Jacrot, B. (1976) The study of biological structures by neutron scattering from solution, *Rep. Prog. Phys.* **39**, 911-935
23. Zaccaï, G., and Jacrot, B. (1983) Small Angle Neutron Scattering, *Ann. Rev. Biophys. Bioeng.* **12**, 139-157
24. Takeda, N., Nakano, K., Kato, M., and Taniguchi, Y. (1998) Pressure-induced structural rearrangements of bovine pancreatic trypsin inhibitor studied by FTIR spectroscopy, *Biospectroscopy* **4**, 209-216
25. Appavou, M.S., Gibrat, G., Bellissent-Funel, M.-C., Plazanet, M., Pieper, J., Buschteiner, A., and Annighofer, B. (2005) Influence of a medium pressure on structure and dynamics of a BPTI protein, *J. Phys. Condens. Matter.* **17**, S3093-S3099

26. Appavou, M. S., Gibrat, G., and Bellissent-Funel, M.-C. (2006) Influence of pressure on structure and dynamics of bovine pancreatic trypsin inhibitor (BPTI): Small angle and quasi-elastic neutron scattering studies, *Biochimica et Biophysica Acta* **1764**, 414-423
27. Gallagher, W. H., and Woodward, C. K. (1989) The concentration dependence of the diffusion coefficient for bovine pancreatic trypsin inhibitor: a dynamic light scattering study of a small protein, *Biopolymers*. **28**, 2001-2024
28. Kitchen, D. B., Reed, L. H., and Levy, R. M. (1992) Molecular dynamics simulation of solvated protein at high pressure, *Biochemistry* **31**, 10083-10093
29. Bée, M. (1988) *Quasi-elastic neutron scattering, principles and applications in solid state chemistry, biology and materials science* (Adam Hilger, Bristol, Philadelphia 0
30. Gallagher, W. H., and Woodward, C. K. (1989) The concentration dependence of the diffusion coefficient for bovine pancreatic trypsin inhibitor: a dynamic light scattering study of a small protein, *Biopolymers* **28**, 2001-2024
31. Zanotti, J.-M., Bellissent-Funel, M.-C., and Parello, J. (1999) Hydration-coupled dynamics in proteins studied by neutron scattering and NMR. The case of the typical EF-hand calcium-binding parvalbumin, *Biophys. J.* **76**, 2390-2411
32. Perez, J., Zanotti, J. M., and Durand, D. (1999) Evolution of the internal dynamics of two globular Proteins from Dry Powder to solution. *Biophysical Journal* **77**, 454-469
33. Gibrat, G., Blouquit, Y., Craescu, C.T., and Bellissent-Funel, M-C. (2008) Biophysical studies of thermal denaturation of calmodulin protein: Dynamics, *Biophysical Journal* **95**, 5247-5256
34. Brunne, R. M., and Van Gunsteren, W. F. (1993) Dynamical properties of bovine pancreatic trypsin inhibitor from a molecular dynamics simulation at 5000 atm, *FEBS Letters* **323**, 215-217
35. Mentré, P., and Hui Bon Hoa, G. (2000) Effects of High Hydrostatic Pressures on Living Cells: A Consequence of the Properties of Macromolecules and Macromolecules-associated Water, *Intern. Rev. Cyt.* **201**, 1-84
36. Doster, W., and Gebhardt, R. (2003) High pressure-unfolding of myoglobin studied by dynamic neutron scattering, *Chem. Phys.* **292**, 383-387
37. Loupiac, C., Bonetti, M., Pin, S., and Calmettes, P. (2006) β-lactoglobulin under high pressure studied by small-angle neutron scattering, *Biochimica et Biophysica Acta* **1764**, 211-23

GENERALIZED GIBBS' THERMODYNAMICS AND NUCLEATION - GROWTH PHENOMENA

JÜRN W. P. SCHMELZER,
Institut für Physik, Universität Rostock, 18051 Rostock, Germany, e-mail: juern-w.schmelzer@uni-rostock.de

Abstract: Nucleation processes – the formation of aggregates of critical sizes allowing their further deterministic growth - are one of the basic mechanisms first-order phase transitions (like condensation and boiling, segregation in solutions or crystallization and melting) may proceed. In order to describe theoretically the kinetics of nucleation-growth processes, the so-called work of cluster formation – the change of the thermodynamic potential due to the formation of a cluster of a given size and composition - has to be known. This quantity is conventionally determined in the framework of Gibbs' thermodynamic theory of heterogeneous systems employing certain additional assumptions. However, Gibbs restricted his analysis exclusively to "equilibrium states of heterogeneous substances" and, already by this limitation, does not supply us with a fully satisfactory solution of mentioned problem. Generalizing Gibbs' method, recently a thermodynamic description of non-equilibrium states consisting of clusters of arbitrary sizes and composition in the otherwise homogeneous ambient phase was developed by us. This approach leads not only to a sound foundation of the thermodynamic aspects of the theoretical description of cluster growth processes but also to a variety of principally new insights into the course of nucleation-growth or spinodal decomposition processes, in general. In particular, it leads to a different set of thermodynamic equilibrium conditions for the determination of the properties of the critical clusters as compared with Gibbs' classical treatment. Some of the most important consequences of this new approach in application both to nucleation and growth-dissolution processes are analyzed in the present contribution.

Keywords: crystallization, crystal growth, nucleation, thermodynamics, nano-crystals

1. Introduction

In his fundamental papers [1], published first in the period 1875 – 78, J. W. Gibbs extended classical thermodynamics to the description of heterogeneous systems consisting of several macroscopic phases in thermodynamic equilibrium. As an additional application, he analyzed thermodynamic aspects of nucleation phenomena and the dependence of the properties of critical clusters – aggregates being in unstable equilibrium with the ambient phase - on supersaturation. Regardless of the existing impressive advances of computer simulation techniques and density functional computations (cf. e.g. [2, 3]), the method developed by Gibbs is predominantly employed till now in the theoretical interpretation of experimental data on nucleation-growth phenomena (cf. e.g. [4,5]). However, as evident from the title of his work [1] ("On the equilibrium of heterogeneous substances"), Gibbs' directed his analysis exclusively to equilibrium states of heterogeneous systems. It follows immediately as a first consequence that Gibbs' thermodynamics in its original from is not applicable – without developing more or less founded additional assumptions - to the description of growth and dissolution processes of clusters of super- and sub-critical sizes. The description of the properties of clusters of such arbitrary sizes in the ambient phase is not covered by Gibbs' classical method, restricting its applicability to the specification of the properties of critical clusters (or clusters in stable equilibrium with the ambient phase).

2. Generalized gibbs approach: basic ideas and application to nucleation

The thermodynamic state parameters – size and composition - of a critical cluster are determined in Gibbs classical approach via a subset of the well-known thermodynamic equilibrium conditions (equality of temperature and chemical potentials of the different components) identical in Gibbs' classical treatment to those obtained for the description of phase equilibriums of macroscopic systems. Employing these dependencies it turns out that – following Gibbs' classical approach – the bulk properties of the critical clusters are widely the same as the respective properties of the newly evolving macroscopic phase (at least, as far as the formation of condensed phases is considered). Once this is the case, one can assume then that also the properties of sub- and supercritical clusters deviate only slightly from the respective parameters of the newly evolving macroscopic phases. This assumption is commonly employed in the description of growth and dissolution processes [6]. However, above mentioned result of Gibbs' theory concerning the properties of critical clusters is in contradiction to results of molecular dynamics and density functional computations of the respective parameters as demonstrated first by Cahn and Hilliard [7]. In such approaches it can be shown that the properties of critical clusters deviate, in general, significantly from the properties of the newly evolving macroscopic phases, in particular, for large supersaturations. This way, the question arises what the origin of such discrepancies is and how they can be removed eventually.

Restricting the analysis to equilibrium states, Gibbs considers exclusively variations of the state of heterogeneous systems proceeding via sequences of equilibrium states. For such quasi-stationary reversible changes of the states of a heterogeneous system, Gibbs' adsorption equation is valid [1]. This equation describes in the framework of Gibbs' theory the dependence of the surface tension on the state parameters of the system under consideration. It leads to the consequence that the surface tension has to depend (in the simplest case) on $(k+1)$ independent thermodynamic parameters, where k is the number of components in the system. This result implies that – according to Gibbs' original approach – the surface tension depends on the state parameters of one of the coexisting phases merely. This limitation is not restrictive for equilibrium states and quasi-stationary processes in between them. For such cases, the properties of one of the phases are uniquely determined via the equilibrium conditions by the properties of the alternative coexisting phase. However, in the search for the critical clusters (or for the saddle point of the appropriate thermodynamic potential) we have to compare not different equilibrium states but different non-equilibrium states of the heterogeneous system under consideration. For the considered different non-equilibrium states, the surface tension has to depend, in general, on the state parameters of both coexisting phases, i.e., on $2(k+1)$ independent parameters. Gibbs' adsorption equation does not allow one, in principle, to account for such dependence. Since Gibbs' adsorption equation is a consequence of Gibbs' fundamental equation for the thermodynamic parameters describing the contributions of the interface to the thermodynamic functions, latter relation has to be changed in order to allow us to develop a thermodynamic description of thermodynamic non-equilibrium states of the considered type and to allow the surface tension to depend on the sets of state parameters of both ambient and cluster phases. Such generalization of Gibbs' description to non-equilibrium states was performed recently (cf. Ref. 9 and 10 for an overview). This generalized Gibbs' approach employs Gibbs' method of dividing surfaces as well. However, Gibbs fundamental equation for the superficial or surface quantities is generalized allowing one to introduce into the description the essential dependence of the surface tension on the state parameters of both coexisting phases. In this theory, first the thermodynamic potentials for the respective non-equilibrium states are formulated. After this task is performed, the general equilibrium conditions are derived. They have the following form (in application to the surface of tension).

$$(T_\alpha\text{-}T_\beta)s_\alpha + (p_\beta\text{-}p_\alpha) + \sigma\frac{dA}{dV_\alpha} + \sum_{j=1}^{k}\rho_{j\alpha}(\mu_{j\alpha} - \mu_{j\beta}) = 0, \tag{1}$$

$$(\mu_{j\beta}\text{-}\mu_{j\alpha}) = \frac{3}{R}\left(\frac{\partial\sigma}{\partial\rho_{j\alpha}}\right)_{\{\rho_{j\beta}\},T_\beta}, \qquad (T_\beta\text{-}T_\alpha) = \frac{3}{R}\left(\frac{\partial\sigma}{\partial s_\alpha}\right)_{\{\rho_{j\beta}\},T_\beta}.$$

Here T is the temperature, p is the pressure, σ is the surface tension, A is the surface area, V is the volume, s is the entropy density, ρ_i are the particle densities, and μ_i the chemical potentials of the different components, R is the radius of the critical cluster referred to the surface of tension, the index α specfies the parameters of the cluster while β refers to the ambient phase. The equilibirum conditions coincide with Gibbs' expressions for phase coexistence at planar interfaces ($R \to \infty$) or when, as required in Gibbs' classical approach, the surface tension is considered as a function of only one of the sets of intensive variables of the coexisting phases, either of those of the ambient or of those of the cluster phase. In such limiting cases, Gibbs' equilibrium conditions

$$(p_\beta\text{-}p_\alpha) + \sigma\frac{dA}{dV_\alpha} = 0, \qquad \mu_{j\alpha} = \mu_{j\beta}, \qquad T_\alpha = T_\beta. \tag{2}$$

are obtained as special cases from Eq. (1).

$$W_c = \frac{1}{3}\sigma A\ . \tag{3}$$

It has the form as given via Eq. (3) provided in both approaches the surface of tension is chosen as the dividing surface. However, since the parameters of the critical clusters are determined in a different way in both the classical and generalized Gibbs' approaches, Eq. (3) leads, consequently, also to different results for the work of formation of clusters of critical sizes and other characteristics of the nucleation process. Some results are illustrated on Fig.1. In particular, it is evident from Fig.1 that the critical cluster properties change significantly with supersaturation (here expressed via the density of the liquid (Fig.1a). Moreover, while – according to the classical theory – the thermodynamic driving force of nucleation increases monotonically with increasing supersaturation (curve 1 in Fig.1b), the generalized Gibbs approach – accounting for changes of the properties of the critical clusters – shows a non-monotonic behavior: the thermodynamic driving force of nucleation increases first and after reaching a maximum it tends to zero (curve 2 in Fig.1b) in the approach of the spinodal curve (where the properties of the critical clusters approach the properties of the ambient phase). Different size parameters are shown on Fig.1c. The expression for the work of critical cluster formation W_c remains in the generalized Gibbs' approach the same as in the classical Gibbs' approach.

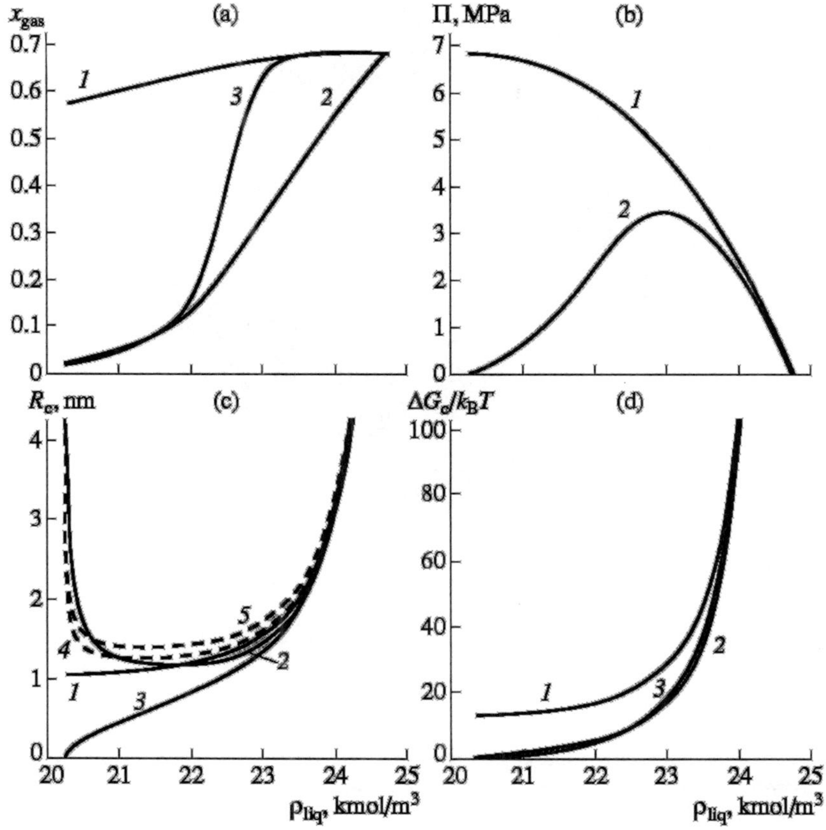

Figure 1. *Composition of the critical cluster (bubble of critical size), x_{gas}, thermodynamic driving force of critical bubble formation, □, radius of the critical bubble, R_c, and work of critical bubble formation, □G_c, computed for the case of boiling in binary liquid-gas solutions in dependence on supersturation here expressed via the density of the liquid ρ_{liq} (for the details see Ref.21). By the number (1), the results are shown computed via the classical Gibbs approach employing the capillarity approximation, number (2) refers to computations via the generalized Gibbs approach and number (3) to computations via the van der Waals square gradient density functional method.*

It is evident that the question of determination of the size of the critical cluster is a rather non-trivial problem, it depends qualitatively on the definition of the size, i.e., for which of the dividing surfaces the parameter has to be computed and which assumptions are employed in its determination. Employing the classical Gibbs method and the capillarity approximation, we arrive at curve 1 in Fig.1c.

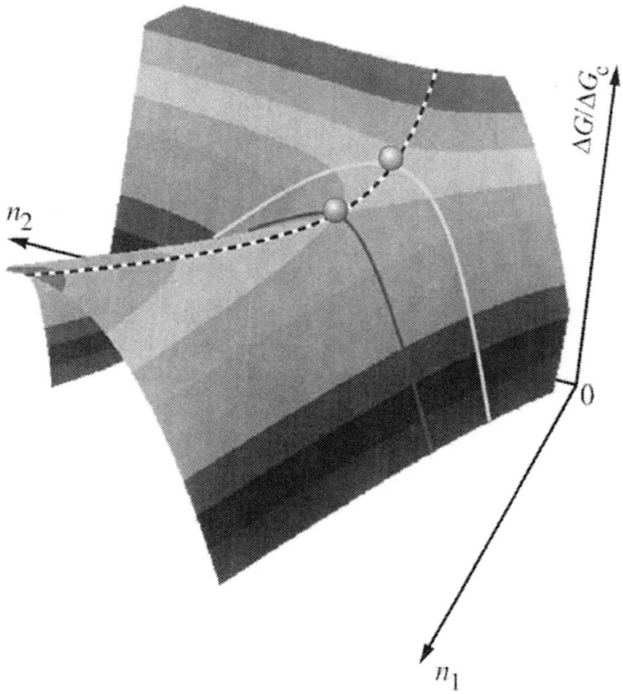

Figure 2. *Hyper-surface of the thermodynamic potential difference between the heterogeneous state consisting of a cluster in the otherwise homogeneous ambient phase and the homogeneous initial state (here demonstrated for the case of a binary system). The curve via the saddle corresponds to the generalized Gibbs approach while the classical Gibbs method corresponds to ridge crossing.* n_1 *and* n_2 *are here the number of particles in the cluster.*

If instead for its determination the work of critical cluster formation is used computed via van der Waals density functional methods, we arrive at curve 3, i.e., the critical cluster size has to tend to zero. In contrast, the radius of the critical cluster referred to the surface of tension in the generalized Gibbs approach tends to infinity (curve 2 in Fig.1c) similarly to estimates of the size of the critical cluster in van der Waals square gradient density functional computations (curves 4 and 5 in Fig.1c). Finally and most importantly, the work of critical cluster formation computed via the classical Gibbs approach employing the capillarity approximation remains finite at the spinodal curve while both van der Waals and generalized Gibbs approaches lead to widely similar results and a vanishing of the work of critical cluster formation in approaching the spinodal curve (as it has to be the case according to the meaning meaning of the spinodal curve defining the boundary between thermodynamically metastable and unstable homogeneous initial states).

A detailed analysis shows[10] further that the classical Gibbs approach employing the capillarity approximation i.e. assuming that the surface tension is equal to the respective value for a planar coexistence of both phases at planar interfaces overestimates the work of critical cluster formation as compared with the generalized Gibbs approach not only for the particular application discussed here but generally. Independent on the number of components in the system and the application discussed, the work of critical cluster formation – computed via the classical Gibbs approach – is larger as compared with the results of the generalized Gibbs method. In more detail, the situation is illustrated in Fig. 2. In Fig.2, the hyper-surface of the thermodynamic potential difference between the heterogeneous system – cluster in the otherwise homogeneous ambient phase - and the homogeneous initial state is shown computed via the generalized Gibbs method allowing one to account for a dependence of the state parameter of the cluster phase on supersaturation and for a dependence of the surface tension on the state parameters of both ambient and newly evolving phases. The line via the saddle illustrates the most probable path of evolution of the system to the new state. In comparison, the respective path is also given if the classical Gibbs approach is employed for its determination. As it turns out this path of evolution does not correspond to a path via the saddle point but to ridge crossing.

Assuming that the kinetic pre-factor, J_0, in the expression for the steady-state nucleation rate J

$$J=J_0\exp\left(-\frac{W_c}{k_B T}\right). \tag{4}$$

does not depend significantly on the chosen path, it follows as a direct consequence that the classical Gibbs approach to the determination of the steady-state nucleation rate leads to too low values of the nucleation rate and related quantities like the limit of accessible supersaturations (i.e., the value of the supersaturation at which nucleation proceeds at a measurable at normal experimental time scales rate). Consequently, experimentalists interpreting their data on nucleation in terms of the classical theory should be aware that the process proceeds in reality as a rule with higher nucleation rates and/or earlier as predicted by classical theory.

3. Application to cluster growth processes

Above performed analysis leads to the consequence that clusters of critical sizes have properties which are widely different, in general, from the pro-properties of the newly evolving macroscopic phases. By this reason, also the properties of sub- and supercritical clusters have to depend, in general, both on supersaturation and cluster size. In order to develop an appropriate description of the course of the phase transitions, one has to develop thus as a next step a method to establish the dependence of composition of clusters of arbitrary sizes on mentioned parameters. In order to establish the most probable trajectory of evolution of the clusters in the space of thermodynamic state variables, we proposed recently that the preferred most probable path of evolution of the clusters is determined by the deterministic equations of cluster growth and dissolution starting with initial states slightly above and below the critical cluster size[11].

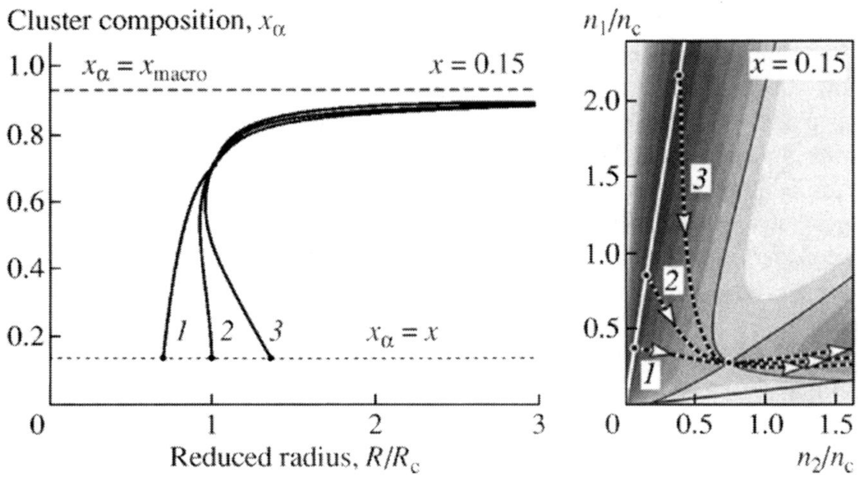

Figure 3. *Cluster composition x_α in dependence on cluster size R/R_c (in reduced units) and trajectory of the most probable path of the clusters in their evolution to the new phase.*

In its simplest tentative version [12], the behavior of the system resembles then the motion of a mass in a viscous fluid (i.e., with a velocity proportional to the force acting on the mass) in some force field determined by the shape of the thermodynamic potential surface. These methods allow one a straightforward determination of the most probable path of evolution independent on the particular kind of phase transformation considered. An example is shown in Fig. 3. The change of the composition of the clusters in dependence on their sizes leads to a size-dependence of almost all thermodynamic (in particular, the driving force of cluster growth and surface tension) and kinetic parameters (diffusion coefficients and growth rates) determining the course of the phase transition (for the details, see again Ref. 12). Some first experimental analyses confirming these theoretical predictions are given in Refs.12-15. Taking into account such size dependence, it can be easily explained why thermodynamic and kinetic parameters obtained from nucleation experiments may not be appropriate for the description of growth or dissolution and vice versa (cf. [11,12]). Following the thermodynamic analysis in the framework of the generalized Gibbs' approach, as analyzed here, and the method of determination of the most probable trajectory of evolution employed we come to the conclusion that the kinetics of nucleation and growth in solutions does not proceed according to the classical picture but exhibits features typical for spinodal decomposition. Moreover, essential features of the process of spinodal decomposition and the phase transformation kinetics in the vicinity of the classical spinodal curve can be interpreted in term of the generalized Gibbs approach as well. [16]

4. The temperature of the critical clusters

According to Eqs.(1), the generalized Gibbs approach leads to the conclusion that, not as an exception (if the system follows a trajectory not passing the saddle but a ridge point in the classical description) but as the general rule, the temperature in the critical cluster has to be different from the temperature of the ambient phase[5,9,10] while according to classical theory the temperature has to be the same both in the critical cluster and the ambient phase.

The possibility of a direct proof of the question which of the mentioned predictions reflects the situation more correctly was opened by recent molecular dynamics simulations of argon condensation performed by Wedekind et al.[17, 18]. The difference of the average temperature of the critical droplets $\Delta T = T_{\alpha} - T_{\beta}$, according to their simulations, is shown in Fig. 4 (upper curves in the Fig.4, taken from Refs. 17-18). It seems to be obvious that the (average) temperature of the critical clusters is larger as compared with the temperature of the surrounding vapor.

The authors interpret their results based on the work of Feder et al. [19]. According to Feder et al.[19], which follow in their analysis the classical Gibbs method, the temperature difference $\Delta T = T_{\alpha} - T_{\beta}$ between the droplet and the vapor can be written in the form

$$\frac{(T_\alpha - T_\beta)}{T_\beta} = \frac{q}{b^2 + q^2}\left(-\frac{\partial \Delta G(n)}{\partial n}\right), \tag{5}$$

where q and b^2 are positive parameters. Wedekind et al. note then that, according to Feder et al.[19], "clusters with sizes smaller than the critical size are predicted to be colder than the bath, and clusters bigger than the critical size will be warmer, *just as we have found it in the simulations*...". While the first part of this statement is a direct consequence of Eq. (5), the second part seems to contradict the molecular dynamics results obtained by the authors and reproduced here in Fig. 4. In contrast to the classical approach, the generalized Gibbs approach predicts a difference of the temperatures of the critical drops in comparison with the surrounding vapour (cf. Eqs. (1)). In this way, the results of the mentioned molecular dynamics simulations can be interpreted as giving another example supporting the generalized Gibbs method. In order to obtain a more quantitative comparison of the predictions of the molecular dynamics simulations with the results of an analysis performed within the generalized Gibbs method, in latter one the dependence of the surface tension on the entropy density of the cluster phase has to be known. In order to establish the required kind of dependence, results for the dependence of the surface tension $\square_{LV}$ of liquids in equilibrium with the vapour may be employed in a form as reported by Skripov and Faizullin.[20]

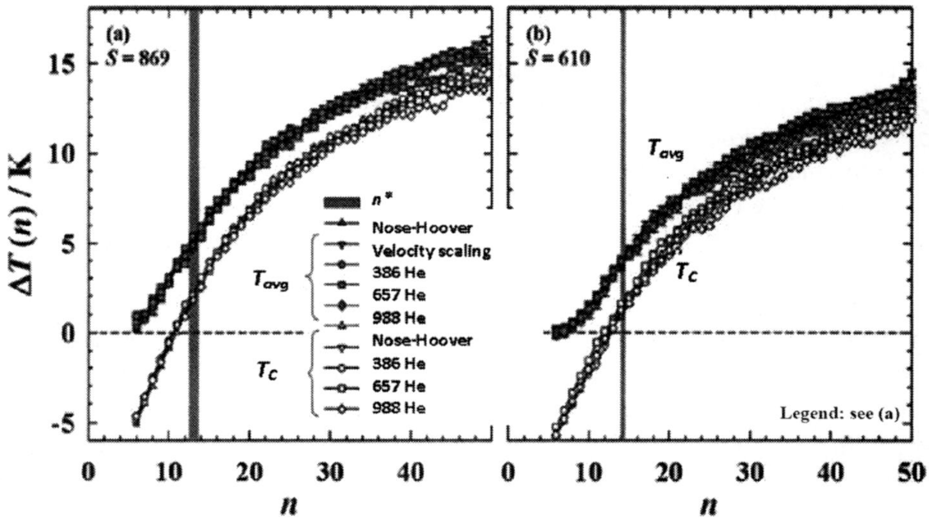

Figure 4. *Deviations $\Delta T = T_{\square} - T_{\beta}$ of the average cluster temperatures (T_{avg} upper curves) curves) and the so-called by the authors local equilibrium cluster temperatures (T_c lower curves) from the bath temperatures in dependence on the droplet size n for two realizations of the molecular dynamics simulations of Wedekind et al.*[17, 18]*. The vertical lines specify the location of the critical cluster sizes.*

Skripov and Faizullin showed for 25 different liquids that the surface tension can be described nearly perfectly via a dependence of the form (see Fig. 5):

$$\sigma_{LV} \cong A\left(\frac{\Delta H_{LV}}{V_L}\right)^{\mathrm{m}} \quad \text{with} \quad \mathrm{m}=2.15 \,. \tag{6}$$

Here ΔH_{LV} is the change of the enthalpy of the system when the volume V_L of the liquid is transferred into the gas.

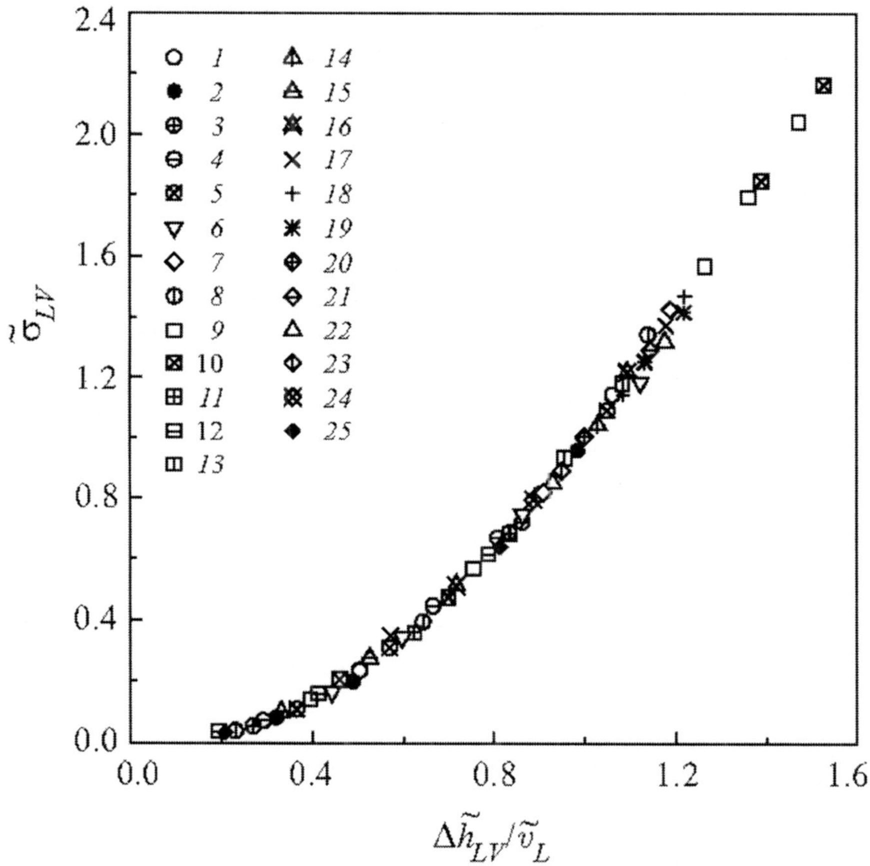

Figure 5. Dependence of the reduced liquid-vapour surface tension on the reduced volume of the liquid obtained for 25 different substances by Skripov and Faizullin[20]. The data are well-approximated by Eq.(6).

The parameter A can be expressed here via the respective value of the surface tension in an appropriately chosen reference (taken by the authors to refer to the temperature $(T/T_c)=0.6$), where T_c is the critical temperature) state via

$$A = \sigma_{LV}\left(\frac{T}{T_c} = 0.6\right)\left\{\frac{V_{LV}\left(\frac{T}{T_c} = 0.6\right)}{\Delta H_{LV}\left(\frac{T}{T_c} = 0.6\right)}\right\}^m . \tag{7}$$

In order to compute the temperature difference between the argon critical droplet and the vapour, we have to express ΔH_{LV} via the entropy density of the liquid phase. The change of the enthalpy of a given volume V_L of the liquid at an equilibrium coexistence of liquid and vapour phases can be expressed generally as

$$\Delta \mathrm{H}_{\mathrm{LV}} = (\mathrm{T_V} S_V + \mu_{\mathrm{V}} n_V) - (\mathrm{T_L} S_L + \mu_{\mathrm{L}} n_L). \tag{8}$$

Taking into account that, in such reversible transformation of the liquid into the gas, the relation $n_V = n_L$ holds (the liquid is completely transformed into the gas) and pressure, p, temperature, T, and chemical potential, □, are the same in both phases, we get straightforwardly

$$\Delta \mathrm{H}_{\mathrm{LV}} = \mathrm{T}(S_V - S_L) = T(s_V V_V - s_L V_L), \tag{9}$$

or

$$\frac{\Delta \mathrm{H}_{\mathrm{LV}}}{V_L} = \mathrm{T}\left(s_V \frac{V_V}{V_L} - s_L\right). \tag{10}$$

Taking into account that $s_{□}$ in Eq. (1) is in the considered here application the entropy density of the liquid ($s_{□} = s_L$), we obtain after some straightforward transformations

$$\frac{(\mathrm{T}_\alpha - \mathrm{T}_\beta)}{\mathrm{T}_\beta} = -\frac{3}{\mathrm{T}_\beta \mathrm{R}}\left(\frac{\partial \sigma}{\partial s_\alpha}\right)_{\{\rho_{i\beta}\},\, \mathrm{T}_\beta} = \frac{3\mathrm{mA}}{\mathrm{R}}\left(\frac{\Delta \mathrm{H}_{\mathrm{LV}}}{\mathrm{V_L}}\right)^{\mathrm{m}-1} \geq 0 . \tag{11}$$

It turns out that the temperature of the critical drops has to be, according to the generalized Gibbs approach, higher as compared with the temperature of the surrounding vapour phase, a result, which seems to us to be in full agreement with the molecular dynamics simulations of Wedekind et al. [17,18].

Similarly, for the case of boiling (bubble formation, $s_{□} = s_V$), we would get get instead

$$\frac{(\mathrm{T}_\alpha - \mathrm{T}_\beta)}{\mathrm{T}_\beta} = -\frac{3}{\mathrm{T}_\beta \mathrm{R}}\left(\frac{\partial \sigma}{\partial s_\alpha}\right)_{\{\rho_{i\beta}\},\, \mathrm{T}_\beta} = -\frac{3\mathrm{mA}}{\mathrm{R}}\left(\frac{\mathrm{V_V}}{\mathrm{V_L}}\right)\left(\frac{\Delta \mathrm{H}_{\mathrm{LV}}}{\mathrm{V_L}}\right)^{\mathrm{m}-1} \leq 0 . \tag{12}$$

Here we have to expect a lower value of the temperature of the critical gas bubble as compared to the surrounding it liquid. In addition, the effect is – as compared to condensation at otherwise equivalent conditions – amplified by a factor (V_V/V_L). A more detailed study of these and related problems is in progress.

5. Discussion and conclusions

The generalized Gibbs approach is able to lead to a variety of further new insights into the course of first-order phase transformations. In addition to mentioned results, it allows one, for example, a new interpretation of the problem of existence or non-existence of metastable phases in crystallization of different glass-forming melts and the evolution of bimodal cluster size distributions for intermediate stages of segregation processes (cf. Ref. 11), a new approach to the description of spinodal decomposition. This way, we believe that the further development of the generalized Gibbs' approach in application to the description of phase formation processes may serve in future – combining the simplicity of the classical Gibbs' approach with the accuracy of density functional approaches and computer simulation methods - as a quite powerful new and generally applicable tool in order to resolve problems in the comparison of experimental results and theoretical predictions which have not found a satisfactory solution so far.

Acknowledgement

The financial support from the Deutsche Forschungs-gemeinschaft (DFG) is gemeinschaft (DFG) is gratefully acknowledged.

References

1. Gibbs, J. W. (1928) The Collected Works, vol. 1, *Thermodynamics* (Longmans & Green, New York).
2. Hale, B. N. (2004) Computer Simulations, Nucleation Predictions, and Scaling. In: *Nucleation and Atmospheric Aerosols 2004: 16th International Conference, Conference Proceedings* (Eds. M. Kasahara, M. Kulmala) 3-14
3. Obeidat, A., Li., J.-S. & Wilemski, G. (2004) Binary Nucleation of a Non-ideal System from Classical and Density Functional Theories. In: *Nucleation and Atmospheric Aerosols 2004: 16th International Conference, Conference Proceedings* (Eds. M. Kasahara, M. Kulmala) 81-84
4. Wölk, J., Strey, R. & Wyslouzil, B. E. (2004) Homogeneous Nucleation Rates of Water: State of the Art. In: *Nucleation and Atmospheric Aerosols 2004: 16th International Conference, Conference Proceedings* (Eds. M. Kasahara, M. Kulmala) 101-114

5. Schmelzer, J. W. P., Röpke, G. & Priezzhev, V. B. (Eds.) (2008) Nucleation Theory and Applications. (Joint Institute for Nuclear Research Publishing House, Dubna, Russia, 1999, 2002, 2005, 2008); copies of the proceedings can be ordered via the author (Email: juern@theor.jinr.ru or via juern-w.schmelzer@uni-rostock.de).
6. Kelton, K. F. (1991) *Solid State Physics* **45**, 75-177
7. Cahn, J. W. & Hilliard, J. E. (1958) *J. Chem. Phys.* **28**, 258-267
8. Cahn, J. W. & Hilliard (1959) **31**, 688-699
9. Schmelzer, J. W. P., Boltachev, G. & Baidakov, V. G.: Is Gibbs Theory of Heterogeneous Systems Really Perfect? In: *Nucleation Theory and Applications*, J. W. P. Schmelzer (Ed.) (2004) (WILEY-VCH, Berlin-Weinheim)
10. Schmelzer, J. W. P., Boltachev, G. Sh., & Baidakov, V. G. (2006) *J. Chem. Phys.* **124**, 194503
11. Schmelzer, J. W. P., Abyzov, A. S. & Möller, J. (2004) *J. Chem. Phys.* **121**, 6900-6917
12. Schmelzer, J. W. P., Gokhman, A. R. & Fokin, V. M. (2004) *J. Colloid Interface Science* **272**, 109-133
13. Zanotto, E. D. & Fokin, V. M. (2003) *Phil. Trans. Royal Society London* **A 361**, 591-613
14. Fokin, V. M., Yuritsyn, N. S., & Zanotto, E. D. (2004) Nucleation and Crystallization Kinetics in Silicate Glasses: Theory and Experiment. In: *Nucleation Theory and Applications*, J. W. P. Schmelzer (Ed.) (WILEY-VCH, Berlin-Weinheim,)
15. Fokin, V. M., Yuritsyn, N. S., Zanotto, E. D., & Schmelzer, J. W. P. (2006) *J. Non-Crystalline Solids* **352**, 2681-2714
16. Abyzov, A. S. & Schmelzer, J. W. P. (2007) *J. Chem. Phys.* **127**, 114504
17. Wedekind, J., Reguera, D., & Strey, R. (2007) *J. Chem. Phys.* **127**, 064501
18. Wedekind, J., Reguera, D., & Strey, R. (2007) The temperature of nucleating droplets. In: C. O'Dowd, P. E. Wagner (Eds.), *Nucleation and Atmospheric Aerosols.* (Springer, Berlin)
19. Feder, J., Russell, K. C., Lothe, J., & Pound, G. M. (1966) *Advances in Physics* **15**, 111-178
20. Skripov, V. P. & Faizullin, M. Z. (2006) *Crystal-Liquid-Gas Phase Transitions and Thermodynamic Similarity* (WILEY-VCH, Berlin - Weinheim)
21. Schmelzer, J. W. P., Baidakov, V. G., & Boltachev, G. Sh. (2003) *J. Chem. Phys.* **119**, 6166-6183

SELF-ASSEMBLING OF THE METASTABLE GLOBULAR DEFECTS IN SUPERHEATED FLUORITE-LIKE CRYSTALS

L. N. YAKUB[1)], E. S. YAKUB*[2)]
[1)] Thermophysics Dept., Odessa State Academy of Refrigeration, 1/3 Dvoryanskaya St., Odessa, 65082, Ukraine
[2)] Cybernetics Dept., Odessa State Economic University, 8 Preobrazhenskaya St., Odessa, 65082, Ukraine
E-mail: yakub@oseu.odessa.ua

Abstract: Molecular dynamics simulation of relaxation processes in metastable stoichiometric crystals with fluorite structure based on a partly-ionic model after a thermal shock was carried out. Simulation was pointed on the study to the evolution of the defects concentration in (a short overheating period, which started at the temperature well above the melting temperature) followed by a fast quenching. A short period of overheating was initiated to completely destroy the less stable anionic sublattice. To avoid melting of the cationic sublattice this period was limited to a few picoseconds. During the whole simulation, we monitored the number and type of defects forming in the cell. This way of the metastability simulation requires a very large MD cell. After the thermal shock and a sudden (or step by step) temperature drop, the defects interaction under conditions when the thermal motion is restricted, leads to self-assembling of the long-living metastable globular clusters, containing 12 or 13 interstitial anions and eight lattice vacancies, forming a cuboctahedral structures. These clusters are alike those known in hyper-stoichiometric fluorite crystals, but have been never observed at stoichiometric conditions yet.

Keywords: superheated crystals, molecular dynamics, defects, relaxation, cuboctahedral clusters

1. Introduction

In this paper we are considering the metastable states of some of the chemical compounds with formula RX_2 which crystallize in a lattice similar to that of calcium fluoride (CaF_2) known as fluorite (*Hayes*). In such structure crystallizes pure stoichiometric UO_2 as well as its non-stoichiometric counterparts known as hyperstoichiometric UO_{2+x} and hypostoichiometric

S. Rzoska et al. (eds.), *Metastable Systems under Pressure*,

UO_{2-x}, which are our main points of interest due to their technological importance. In a compound RX_2 each ion of species R is surrounded by eight equivalent nearest-neighbour ions of species X forming the corners of a cube of which R is the centre. Each ion of species X is surrounded by a tetrahedron of four equivalent R ions. More fundamentally the structure has a FCC (face-centred-cubic) translational group. If the structure is interpreted in terms of a primitive cube of side a, it comprises three inter-penetrating FCC lattices. The first is a lattice of species R, the X species are located on two further lattices with similar translational vectors forming simple cubic (SC) sub-lattice. The site of the R ion has O_h symmetry and the site of the X ion has T_d symmetry. Fourth FCC structure, complementing the first FCC lattice to SC structure, is a virtual sub-lattice build by a set of empty interstitial sites. Each interstitial site again has O_h symmetry, being at the centre of a cube of eight X ions.

Presence of these interstices provides to the fluorite structure extremely specific features. In UO_2 particularly, it allows for placement of some radioactive decay products, these sites are responsible for existence of hyperstoichiometric UO_{2+x} phase, where the extra oxygen ions fill the empty interstitial sites in the fluorite lattice *etc*. First case is extremely important in radiation damaged UO_2. Second one is crucial in oxidation of pure UO_2 in atmospheric conditions. Diffusion of atmospheric oxygen into the bulk of crystal brings excess oxygens into empty interstices. These become filled more or less randomly only at low *x*, at higher concentration of extra anions they form different types of clusters, including so-called 2:2:2 Willis dimers (*Willis*), tetra- and pentameric defects clusters of cuboctahedral symmetry (*Allen and Tempest*). Last defects appear due to interaction of extra anions with intrinsic crystal FP defects (anion Frenkel pairs, i.e. anion vacancies and anion interstitials).

Formation and interaction of unstable FP defects, the dominant intrinsic defects in UO_2, alkaline earth fluorides and related compounds generally plays an important role in many diffusion mechanisms

Mechanisms of formation of these defects in pure and damaged UO_2 are different. In pure UO_2, oxygen Frenkel pairs (OFP) defects appear spontaneously as thermally induced excitations of anionic crystalline sublattice, in damaged dioxide they normally produced by fast fission products.

In this work, we studied formation and decay of such defects by molecular dynamics (MD) simulation technique under conditions of “thermal shock”, i.e. very short and intensive local heating of crystal, when the temperature rises well above the equilibrium thermodynamic melting temperature. In such metastable conditions the oxygen sub-lattice melts much faster then the more stable FCC structure build from heavy uranium ions.

This short "thermal shock", followed by fast quenching to relatively low temperature, like thermal spike, originated by a fast fission fragment, produces a lot of OFP defects. Below, after description of potential model applied, we present and discuss results of the above computer simulation experiment.

2. Potential model

Interionic interaction energy consists of Coulombic and short-range terms:

$$U_N = U_N^{(C)} + U_N^{(S)}.$$

Coulombic interaction energy $U_N^{(C)}$ was evaluated using recent modification of the Ewald scheme (*Yakub and Ronchi, 2003*) based on angular averaging of Ewald sums over all orientations of the reciprocal lattice eliminates the periodicity artifacts imposed by conventional Ewald scheme and provides much faster computation of electrostatic energy in computer simulations of disordered condensed systems.

This approach results in simple expression for the total Coulomb energy for N point charges Z_i (i=1,...N) placed in the main cubic cell of the size L :

$$U_N^{(C)} = \sum_{i=1}^{N} \frac{3Q_i^2}{16\pi\varepsilon_0 r_m} + \sum_{1\le i<j\le N} \phi_{ij}^{(C)} \tag{1}$$

where

$$\phi_{ij}^{(C)} = \begin{cases} \dfrac{Q_i Q_j}{4\pi\varepsilon_0 r_{ij}} \left\{ 1 + \dfrac{1}{2}\left(\dfrac{r_{ij}}{r_m}\right)\left[\left(\dfrac{r_{ij}}{r_m}\right)^2 - 3\right] \right\} & r_{ij} < r_m \\ 0 & r_{ij} \ge r_m \end{cases} \tag{2}$$

are effective potentials of electrostatic interaction and r_m is radius of the volume-equivalent sphere of the main cell. This method is more accurate and much faster than the conventional Ewald scheme.

Table 1. Potential parameters of semi-ionic model for stoichiometric UO_2, the ionicity parameter ζ = 0.5552 is common for all species.

Ion pairs	**A$_{ij}$**, eV	**b$_{ij}$**, 10^{-10} m	**C$_{ij}$**, **eV**·10^{-60} m	**D$_{ij}$**, **eV**	β_{ij}	r^*_{ij}, 10^{-10} m
O^{2-} - O^{2-}	883.12	0.3422	3.996	0	-	-
U^{4+} - U^{4+}	187.03	0.3422	0	0	-	-
U^{4+} - O^{2-}	432.18	0.3422	0	0.5055	1.864	2.378

It allows for extensive simulations of ionic systems, containing up to few thousand of particles in the main cell, even on PC. New MD simulation software, implementing the above-mentioned approach, was recently developed (*Yakub and Ronchi, 2005*) and tested on simple model systems like one-component plasma and primitive ionic model of electrolyte.

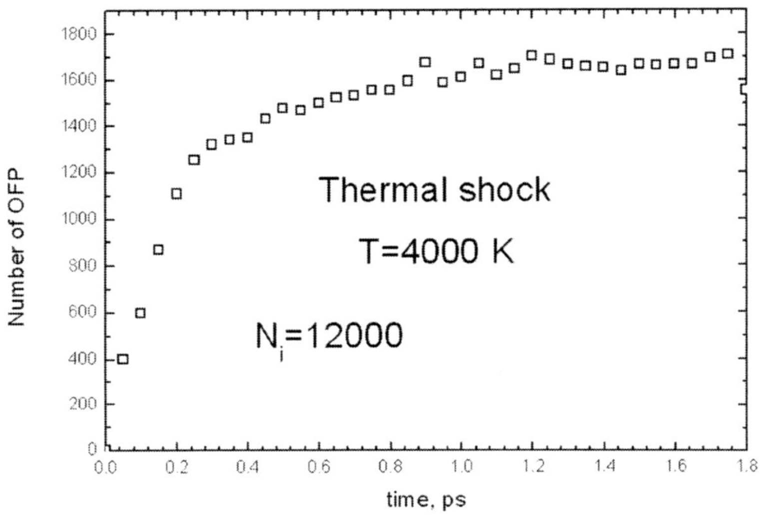

Figure 1. *Number of OFP defects as a function of simulation time during thermal shock period.*

The total charges of ions q_α are non-formal: $Q_\alpha = Z_\alpha^{eff} |e|$, and the effective charges $Z_\alpha^{eff} = \zeta Z_\alpha$ are proportional to the respective formal charges (Z_α =+4 for U^{4+} and Z_α =-2 for O^{2-}). The ionicity ζ is considered as one of the free parameters of the model.

The short-range contribution $U_N^{(S)}$ is pair additive:

$$U_N^{(S)} = \sum_{1 \le i < j \le N} \Phi_{\alpha\beta}^{(S)}\left(r_{ij}\right). \tag{3}$$

The pair short-range interionic interaction potential is written in the form:

$$\Phi_{\alpha\beta}^{(S)}\left(r_{ij}\right) = \Phi_O\left(r_{ij}\right) + \Phi_B\left(r_{ij}\right) + \Phi_D\left(r_{ij}\right). \tag{4}$$

The potential energy $\Phi_D\left(r_{ij}\right)$ of van der Waals attraction (dispersion forces) is represented by a simple inverse-power function:

$$\Phi_D\left(r_{ij}\right) = -C_{\alpha\beta}\big/ r_{ij}^6, \tag{5}$$

and $\Phi_D\left(r_{ij}\right)$ is attributed to O^{2-}- O^{2-} interaction only.

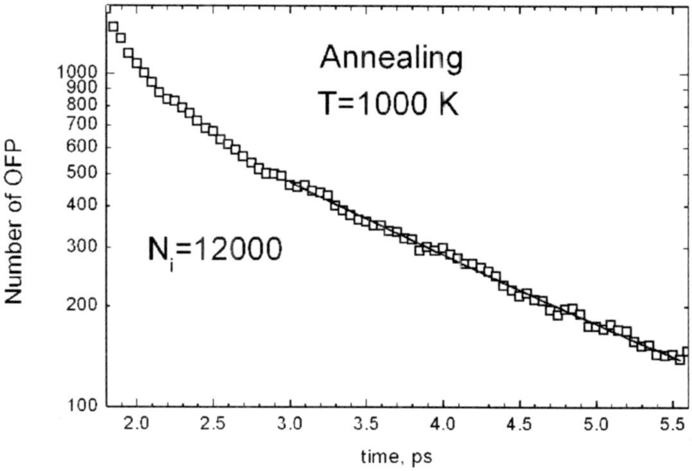

Figure 2. *Number of OFP as function of simulation time during annealing period at T=1000 K.*

Short-range overlap (exchange) repulsion of closed electronic shells is characterized by an exponential term:

$$\Phi_O\left(r_{ij}\right) = A_{\alpha\beta}\exp\left(-\frac{r_{ij}}{b_{\alpha\beta}}\right). \tag{6}$$

The energy of covalent bonding is attributed only to U^{4+}- O^{2-} interactions and and represented by the Morse function:

$$\Phi_B\left(r_{ij}\right) = D_{\alpha\beta}\left[\exp\left(-2\beta_{\alpha\beta}\left(r_{ij}-r^*_{\alpha\beta}\right)\right)-2\exp\left(-\beta_{\alpha\beta}\left(r_{ij}-r^*_{\alpha\beta}\right)\right)\right]. \tag{7}$$

Potential parameters of this partly ionic model for UO_2 (*Yakub, Ronchi and Staicu, 2007*) are presented in Table 1.

3. MD simulation

The simulation method applied, which might be termed "thermal shock and quenching", starts from a short ~ 2 ps (1 ps = 10^{-12} s) MD simulation run at a very high initial temperature (T=4000K). Thermal shock temperature applied is about 1000 K higher than the melting temperature and about 1.5 times higher then the Bredig transition temperature (where oxygen sub-lattice melts under equilibrium conditions). To avoid strong influence of periodic boundaries, we applied a big simulation cell, containing 12000 ions.

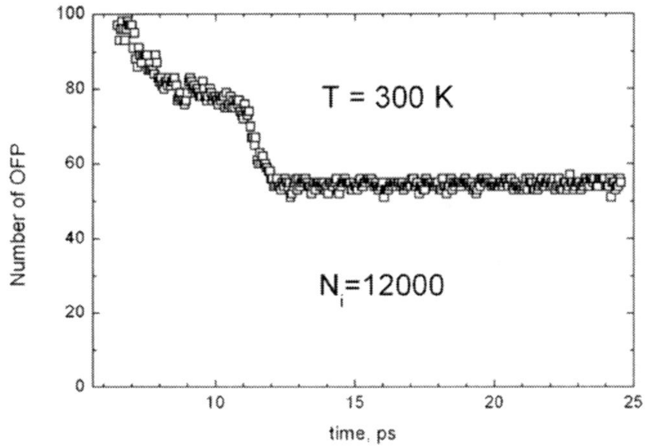

Figure 3. *Number of OFP as a function of simulation time during second annealing at T=300 K.*

Thermal shock produces an essential number of OFP defects in the oxygen sublattice. The concentration of defects grows rapidly during first picoseconds of simulation and than tends to stabilize. In the metastable state obtained about 40% of interstitial sites are populated by displaced anions (see Fig. 1). Kinetics of evolution of defects may be effectively studied by an instant dropping of temperature (quenching) of overheated solid up to temperature about 1000 K. At this temperature another short MD run (about ~3-5 ps) was performed to study the relaxation of defect concentration. As it follows from Fig. 2, the defect concentration decreases almost exponentially with a relaxation time of ~ 2 ps.

During a long MD run at this temperature the initial fluorite structure may be completely recovered. To study geometry and energetics of different defects and defect clusters, which may appear spontaneously during thermal spike, originated by a fast fission fragment, we interrupted the annealing stage in our computer experiment and dropped the temperature again, this time up to T=300 K. At room temperature, after 10-15 ps the concentration of defects stabilizes at a rather low level (see Fig. 3). Only few percents of interstices remain populated by anions, but their geometry is far from random. Fig. 4 presents a typical configuration produced after a few picoseconds of thermal shock at T=4000 K, annealing at T=1000 K and instant quenching to T=300 K. Along with many single interstitials and vacancies quite randomly distributed, appear few charged cuboctahedral clusters, containing from 12 or 13 interstitials and eight vacancies, centred on interstices. The only difference between these CO(12) and CO(13) clusters is presence of an anion in the central interstitial site.

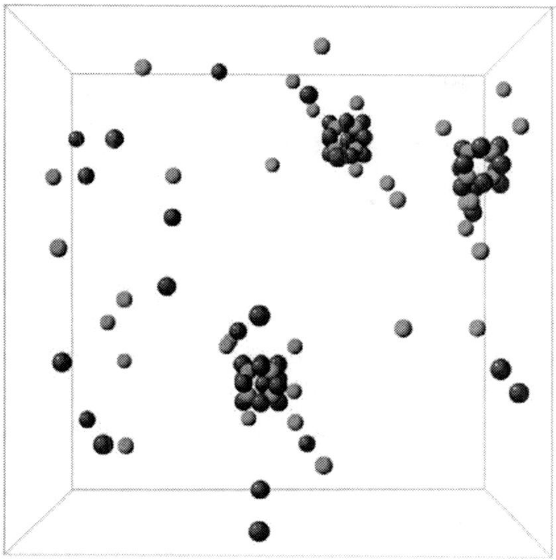

Figure 4. *Defects clusterings in UO_2 after thermal shock, quenching and annealing periods. Dark spheres symbolize oxygen ions in interstitial positions, light grey spheres represent vacancies in oxygen sub-lattice. Two interstitial-centred CO(13) and one CO(12) self-assembled globular clusters are clearly seen. Uranium and oxygen ions in their lattice positions are not shown for the sake of clarity.*

Energy of formation for different types of globular clusters was investigated under static and dynamic conditions. Analyzing results of several simulations, we found that CO(13) cluster appear more frequently than CO(12). The total energy of a globular cluster is rather high (from 10 to 30 eV). However, it is noticeably less than the sum of energies of separated OFP defects. The value of formation energy calculated per one defect is usually 3-5 times less than the OFP formation energy (~5 eV).

4. Disscussion

The interaction of lattice defects in fluorite-like solids under conditions when the thermal motion is restricted, leads to self-assembling of the long-living metastable globular clusters. The same effect was observed in hyper-stoichiometric and radiation-damaged UO_{2+x} (*Yakub, 2008*). At elevated temperatures, metastable globular clusters may survive for hundreds of picoseconds.

Two types of such globular clusters were discovered in this computer experiment. The first one CO(12) has a no interstitial anion in the middle, the second one CO(13) is interstitial-centred. Both types of clusters have non-zero negative net charge and this feature at least partly determines their stability.

Additional static energy calculations (*Yakub, 2008*) reveal that rather high (about 10 eV) total formation energy of globular clusters is noticeably less than the total energy of separated defects. Binding energy calculated per one defect is about 1 eV, i.e. noticeably less than the OFP formation energy (~5 eV).

In radiation damaged UO_2, the equilibrium form of the cluster depends on the presence of uranium vacancies and interstitials. For instance, two uranium defects in close proximity may initiate formation of a chain-like clusters build from OFPs. Generally, the shape of globular defect clusters formed around uranium interstitial ion is more open. Sandwich-like clusters build of four oxygen vacancies, six oxygen, and single uranium interstitial have been found during simulation of radiation damaged UO_2. Note that in such type of simulations use of a big simulation cells, containing at least ten thousand of ions is crucial. In a smaller cell, the influence of periodic boundaries restricts formation of globular and long chain-like clusters. Molecular dynamics simulations based on partly ionic models like proposed recently for uranium dioxide (*Yakub, Ronchi and Staicu, 2007*) may provide a promising way to model formation of such metastable structures and study their relaxation processes in radiation-damaged fluorite-like solids.

Acknowledgment

Authors acknowledge the cooperation of Claudio Ronchi and Dragos Staicu from Joint Research Centre of European Commission, Institute for Transuranium Elements, Karlsruhe, Germany.

References

1. Allen G.C., and Tempest P. A. (1986) *Proc. Roy Soc. A* **406**, 325
2. Hayes W.(ed.) (1974) "*Crystals with the fluorite structure*", Clarendon, 448 P
3. Willis B.T.M. (1978) *Acta Cryst*. **A34**, 38
4. Yakub, E., and Ronchi, C., 2003, *J. Chem. Phys.* **119**, 11556
5. Yakub, E., and Ronchi, C., 2005, *J. Low Temp. Phys*, **139**, 633
6. Yakub, E., Ronchi, C., and Staicu, D. (2007) *J. Chem. Phys.* **127**, 094508
7. Yakub, E. (2008) *Computer simulation of non-stoichiometric ionic crystals*, 7th Conference on Cryocrystals and Quantum Crystals, July 31st - Aug 5th, Wroclaw, Poland

STUDY OF METASTABLE STATES OF THE PRECIPITATES IN REACTOR STEELS UNDER NEUTRON IRRADIATION

[1]A. GOKHMAN [*], [2]F. BERGNER
[1]*South Ukrainian Pedagogical University, Staroportofrankovskaya 26, 65020, Odessa, Ukraine*
[2]*Forschungszentrum Dresden-Rossendorf, P.O.Box 510119, 01314 Dresden, Germany*

Abstract: The Lifshitz - Slezov theory is applied to study the metastable states of the matrix damage clusters, MD_s, and the copper enriched clusters, CEC_s, in neutron irradiated steels. It was found that under irradiation conditions the CEC_s are at the Ostwald stage for a neutron fluence of about 0.0002 dpa. The time dependence of number density, MD_N, is determined by summarizing all differential equations of the master equation for MD_s with neglecting of dimmers concentration in comparison with concentration of the single vacancies and subtraction of the number CEC_s that replace the MD_s, namely vacancy clusters, due to the diffusivity of copper and other impurity atoms to them. For binary Fe-0.3wt%Cu under neutron irradiation with dose 0.026, 0.051, 0.10 and 0.19 dpa the volume content of the precipitates from the SANS experiment is found to be about 0.229, 0.280, 0.237 and 0.300 vol% respectively. The volume fraction of CEC_s in these samples is 0.195 vol% and the calculated volume fraction of MD_s is 0.034, 0.085, 0.042 and 0.105 vol% for doses 0.026, 0.051, 0.10 and 0.19 dpa respectively.

Keywords: metastable states, neutron irradiation, clusters, Ostwald' stage

1. Introduction

Thermal annealing of embrittled reactor steels is well-known option for lifetime extension of light water reactors[1]. In fact, large-scale annealing of the core belt region for the purpose of life extension was successfully applied in a number cases including the first – generation VVER440-type reactor pressure vessel steels. In the last three decades, major scientific efforts have been focused on the estimation of the temperature-time regime for optimum recovery of mechanical properties [2,3], a methodology for verification of annealing effect using non-destructive techniques [2,3], small-specimen tests [3,4] and the

[*] Corresponding author. Tel.:+38-048-7-32-51-07; fax: +38-048-7-32-51-03; e-mail: gokhman@paco.net (A. Gokhman)

phenomenological description based on mechanical testing. On the other hand the lack of the thermodynamic consideration of this problem is observed. Irradiation effect on the reactor steels is characterized by the increase of the temperature of the ductile-brittle transition from (-20°C) up to the reactor surveillance temperature. The embrittled behaviour of the reactor steels is determined by the evolution of their nanodefects that are formed due neutron irradiation. It is well-known approaches to estimate the increase in yield stress and subsequently the shift of the temperature of the ductile - brittle transition from the distribution of the irradiated induced defects[5-10]. Two types of features have primarily been taken into account: copper enriched clusters, CEC_s, and matrix damage clusters, MD_s, of not identified nature[11-16]. The first feature is radiation-enhanced and appears in steels with mild and high content of copper and already after lower fluencies whereas the second type arises from irradiation-induced effects. The existence and evolution of these defects is unexpected from the thermodynamic point of view[17] and they can be considered as metastable states with long lifetime. Both features exhibit the same or a very similar size distribution with a clearly pronounced peak near radius of approximately 1nm. Related to the small-angle neutron scattering, SANS, effects, there is only a difference in the ratio between the total and the nuclear scattering section (the so-called A-ratio)[18]. So it is not possible to distinguish distribution of CEC_s and MD_s experimentally only and to estimate the corresponding impacts into the mechanical properties. The complimentary SANS experiment and rate theory[19-22] simulation is found sufficient to solve this problem but the approach[19-22] needs in an integration of the stiff system of about 10000 ordinary differential equations. Moreover it results in the more detailed description of the microstructure that is necessary to predict the mechanical behaviour of irradiated steels. The objective of the present paper is to suggest the approach to determine the volume content and mean size of CEC_s and MD_s from the SANS data and analytical results of the theory of metastable states.

2. Analytical description of the CEC_s and MD_s kinetics in irradiated reactor steels

2.1. THE NUCLEATION AND DETERMINISTIC STAGE

Summarizing of all differential equations of the master equation[22], related to CEC_s, results in Eq. (1) for number density of CEC_s, CEC_N:

$$\frac{\partial CEC_N}{\partial t} = \beta_1 \cdot C_{Cu1} - \alpha_2 \cdot C_{Cu2}, \tag{1}$$

where β_1 is the attachment rate for single copper atom; C_{Cu1} is the concentration of free copper monomers in the alloy subtracted by copper threshold; α_2 is the emission rate of single copper atoms from Cu clusters of two copper atoms and C_{Cu2} is the number density of these clusters.

In the assumption of the diffusivity controlled kinetics regime and applicability of detailed balanced principle the Eq. (1) transforms to the Eq. (2):

$$\frac{\partial CEC_N}{\partial t} = \sqrt[3]{3\cdot\pi^2}\,\frac{4\cdot D_{Cu}\cdot C^2_{Cu_1}(t)}{a_{0Cu}^{\;2}}\cdot\left[1-\frac{C_{Cu2}(t)}{C^2_{Cu_1}(t)}\cdot\exp\left(-\frac{E_{bCu}(2)}{k_B\cdot T}\right)\right] \qquad (2)$$

where k_B is the Bolzmann' constant, T is the temperature in K; the copper diffusivity, D_{Cu}, is calculated as the product of the thermal copper diffusivity on the ratio between steady state and thermal concentrations of the single vacancies; binding energy, E_{bCu} (2), in eV is determined on Eq. (3) in accordance with[23]:

$$E_{bCu}(2) = 0.294 - 0.368\cdot(2^{\frac{2}{3}}-1) \qquad (3)$$

Functions CEC_N(t) and C_{Cu1}(t) are connected by Eq. (4)[10]:

$$\frac{\partial C_{Cu_1}(t)}{\partial t} = -n^{mean}_{Cu}(t)\cdot\frac{\partial CEC_N}{\partial t}, \qquad (4)$$

where the mean number of copper atoms in one CEC, n_{Cu}^{mean}, is givenby:

$$n^{mean}_{Cu} = \chi_1\cdot C_{Cu_1} + \chi_2 \quad \text{and} \quad \chi_1 = D_{Cu}\cdot(\omega/\phi) \qquad (5)$$

Here χ_1, χ_2 and ω are the adjusting parameters, ϕ is the neutron flux.

In order to integrate the differential equations (2)-(5) the additional relation is need because number density of the dimmers is determined from number density of monomers in case of steady state only. In general case, the time dependence of the number densities of all clusters can be found as solving of master equation of cluster dynamics. Now we have interest to the short description and we are looking for the additional approximation relation. The fluence dependence of the volume fraction of CEC_s is estimated by saturation (*tanh*) form[24] that is roughly equivalent to a Johnson_Mehl-type function:

$$f(\phi t) = 1-\exp[-k(\varphi t)^n], \qquad (6)$$

where (ϕt) is the neutron fluence, k and n are the adjust parameters.

The exponential dependence of the fluence is used also in our approach on the basis of the discussion of the time dependence of the right side of Eq. (2).

At the nucleation stage the right side of Eq. (2) is positive but it decreases with time up to zero, when deterministic stage starts and equal zero during this stage. In order to account this behavior we replace the expression in quadratic brackets of Eq. (2) on the exp(-ξϕt), where ξ is the adjust parameter in dpa^{-1} units, for study of the nucleation stage. Integration of Eqs. (2)-(6) results in:

$$\frac{\chi_1}{\chi_2^2}\cdot\ln\frac{C_{Cu_1}(0)\cdot\left(\chi_2+\chi_1\cdot C_{Cu_1}(0)\right)}{C_{Cu_1}(t)\cdot\left(\chi_2+\chi_1\cdot C_{Cu_1}(t)\right)}-\frac{\left(C_{Cu_1}^{-1}(0)-C_{Cu_1}^{-1}(t)\right)}{\chi_2}=-\xi\phi\cdot\exp\left(-\frac{t}{\xi\phi}\right) \quad (7)$$

At the deterministic stage Eq. (7) transforms to Eq. (8):

$$\frac{\chi_1}{\chi_2}\cdot\ln\frac{C_{Cu_1}(0)\cdot\left(\chi_2+\chi_1\cdot C_{Cu_1}(0)\right)}{C_{Cu_1}(t)\cdot\left(\chi_2+\chi_1\cdot C_{Cu_1}(t)\right)}=C_{Cu_1}^{-1}(0)-C_{Cu_1}^{-1}(t) \quad (8)$$

From Eq. (8) the time dependence $(t)C_{Cu1}$ is found easy and then the time dependence CEC_N(t) can be calculated in accordance with Eq. 4. CEC_s replace the MD_s, namely vacancy clusters under neutron irradiation due to the flux of copper and other impurity atoms to them. This can be described by[10]:

$$MD_N(t)=\lambda\cdot(\phi t)-CEC_N(t), \quad (9)$$

where λ is the adjusting parameter

Equations (2), (8) are (9) are the formal solving of the paper subject but they need in the set of adjusting parameters: λ1, λ1, ω and λ.

2.2. THE OSTWALD'S STAGE

In accordance with Lifshitz-Slezov theory[25] the time dependence of the number density of CEC_N at the Ostwald' stage can be presented by Eq. (10):

$$CEC_N(t)=\frac{0.99}{4}\cdot\frac{C_{Cu_1}(0)-C_{Cu}^{th}}{C_{Cu}^{th}}\cdot\frac{k_BT}{D_{Cu}\cdot\gamma}\cdot\frac{1}{t}, \quad (10)$$

where $C_{Cu1}(0)$ is the initial concentration of free copper atoms in the alloy; C_{Cu}^{th} is the copper threshold estimated as the copper solubility limit or free copper monomers thermal equilibrium concentration; γ is the surface tension for CEC_s that depends on the CEC_s composition. From the continuity demand for the function of CEC_N(t) at the moment of time, t_0, when the Ostwald' regime starts, the Eq. (10) can be transformed into Eq. (11):

$$CEC_N(t)=CEC_N(t_0)\cdot\frac{t_0}{t}, \tag{11}$$

where the value of t_0 and $CEC_N(t_0)$ are determined by Eqs. (12), (13):

$$t_0=\frac{0.99}{4}\cdot\frac{C_{Cu_1}(0)-C_{Cu}^{th}}{C_{Cu}^{th}}\cdot\frac{n_{Cu}^{mean}(t_0)}{C_{Cu_1}(0)-C_{Cu_1}(t_0)}\cdot\frac{k_BT}{D_{Cu}\gamma} \tag{12}$$

$$CEC_N(t_0)=\frac{C_{Cu_1}(0)-C_{Cu_1}(t_0)}{n_{Cu}^{mean}(t_0)} \tag{13}$$

Due neutron irradiation the copper diffusivity increases as in Eq. (14):

$$D_{Cu}^{irr}=D_{Cu}^{th}\cdot\frac{C_{v1}}{C_{veq}} \tag{14}$$

where the thermal single vacancy equilibrium, C_{veq}, is about 10^{-14} atom^{-1} for $T=$ 543K; the copper thermal diffusivity D_{Cu}^{th} = $2.11\cdot10^{-26}$ m^2/s for T=543K according[26] and γ is about 0.3 J/m^2 for pure copper[20] and depends on the CEC_s content for the case of *Ni* and others additional impurities; the actual single vacancy concentration, C_{v1}, is estimated by it's steady state value, C_{vst},:

$$C_{vst}=\frac{G_v}{D_v\cdot S_v^T}, \tag{15}$$

where the diffusivity of vacancy, D_v, is about 10^{-17}m^2/s for T=543K, sink strength of dislocation net for vacancies, S_v^T, is estimated as the dislocation density, ρ_d, about 10^{14} m^{-2} and the production rate of single vacancies after cascade stage, G_v, is determined in accordance with Eq. (16):

$$G_v=\lambda\cdot\phi \tag{16}$$

G_v is about $0.649\cdot10^{-9}\,s^{-1}$ in study[22] and C_{vsat} is about 10^{-7} atom^{-1} respectively on Eq. (15). The upper limit of value t_0 is about 15 hours for conditions[22] in accordance with relations (11)-(16) and known from SANS upper limit of n_{Cu}^{mean} is 1489 for the maximal fluence. The corresponding neutron fluence (ϕt) value is about 0.0002 dpa. Hence CEC_s in reactor steels investigated[22] achieve the Ostwald' stage and the volume of them is determined from the copper content of these steels. The volume content of MD_s is calculated then by subtraction of CEC_s volume from the SANS data. Determination of the irradiation induced change in mechanical properties on[27] with using of data on the volume content of CEC_s and MD_s is found in a good agreement with experiment for reactor steels[22]. The data for REVE material Fe-0.3wt%Cu, T=573K, ρ_d is about 0.910 ^{14}m^{-2}, rate dose is 139x10^{-9} dpa/s (ref. 28) are considered too. From SANS it was found the total volume content: 0.229, 0.280, 0.237 and 0.300 for irradiation doses 0.026 dpa, 0.051dpa, 0.10 dpa and 0.19 dpa respectively, n_{Cu}^{mean} is about 4000. Value of t_0 is found about 2464 seconds from the Eqs. (10)-(16) that corresponds to 0.0003 dpa. Hence *CEC*$_s$ are at Ostwald' stage in all Fe-0.3wt%Cu samples[28]. The same conclusion is obtained in the tentative rate theory simulations[28]. The calculated volume fraction of *MD*$_s$ is 0.034, 0.085, 0.042 and 0.105 vol% for doses 0.026, 0.051, 0.10 and 0.19 dpa respectively.

3. Discussion

The obtained estimation of the irradiation time t_0, when Ostwald' stage starts, is in agreement with Eq. (17)[29]:

$$t_0 = \frac{C_{veq}}{C_{vst}} \cdot t_0^{'} \tag{17}$$

Here $t_0^{'}$ is the ageing time, when *CEC*$_s$ achieve the saturation in the absence of irradiation. For Magnox steels this time in years is determined from Eq. (18):

$$\log t_0^{'} = 1.1 \cdot \frac{10^4}{T} - 18.17 \tag{18}$$

For T is about 543K $t_0^{'}$ is about 10 years on Eq. (17) and for irradiation conditions[28] t_0 is about 1 hour on Eqs. (17), (18).

Calculation of the so-called fluence factor of the copper impact into the transition temperature shift[27] confirms that for the value fluence about 0.0002 dpa saturation of this factor is observed already. The attempt to separate out the embitterment data into the two components of *CEC*$_s$ and *MD*$_s$ was carried out[8,9]. It was found[9] that the copper component remains unchanged already due irradiation 0.0005 dpa at 190-200°C of weld surveillance material. Similar analysis for PWR submerged-arc weld material following accelerated irradiation at temperatures in the range 255-315°C shows that copper contribution remains unchanged after irradiation with dose of 0.01 dpa.

4. Conclusions

For reactor steels the determination of the change in mechanical properties due neutron irradiation as well as recovering annealing procedure can be done on the basis of SANS experiment and some analytical results of theory of metastable states related to the volume content of new phase. The following procedure is suggested to find the volume content of the *CEC*$_s$ and *MD*$_s$:

- To estimate the irradiation time t_0, when Ostwald' stage starts, on Eqs. (12)-(16).
- If SANS measurements are carried out for samples that have been irradiated during the time great than t_0, to find the volume fraction of CEC_s as copper content in at% subtracted by copper threshold in at%.
- The volume fraction of MD_s is determined then as the volume fraction from SANS subtracted by the volume fraction of CEC_s.

Under irradiation conditions[22,28] the CEC_s in reactor steels and iron alloy Fe-0.3wt%Cu are at the Ostwald stage already for neutron fluence about 0.0002 dpa, 0.0003 dpa respectively and suggested procedure can be applied to study of the mechanical properties change in these steels due neutron irradiation.

References

1. Planman, T., and Pelli R., Törrönen, K. (1995) Irradiation Embrittlement Mitigation, *AMES Report*, No **1**, EUR 16072 EN (European Commision)
2. Popp, K., Brauer, G., and Leonhardt, W.-D. *et al.*, (1989) in Radiation Embrittlement of Nuclear Reactor Pressure Vessel Steels: *An International Review, ASTM STP 1011*, ed. By L.E. Steele (American Society fro Testing and Materials, Philadelphia)
3. Pelli, R., and Törrönen, K. (1995) State of the Art Review on Thermal Annealing, *AMES Report*, No. 2, EUR 16278 EN (European Commision,).
4. Kryukov, P., Platonov, Y., and Shtrombakh, *et al.* (1996) *Nucl. Engng Design* **160**, 59
5. Week, R. W., Pati, S.R., Asby, M.F., and Barrand, P. (1990) *Acta Met.* **17**, 1403
6. Russel, K. C., and Brown, L. M. (1972) *Acta Met.* **20**, 969
7. Office of Nuclear Regulatory Research, U.S. Nuclear Regulatory Commission, Washington: NRC (1988) "Improved Embrittlement Correlations for Reactor Pressure Vessel Steels", p. 569, Washington DC.
8. Williams, T., Burch, P., English, C., and Ray, P. (1988) *Proceedings of the Third International Symposium on Environment Degradation of Materials in Nuclear Power Systems- Water Reactors*, Traverse City, 121, The Metallurgical Society,
9. Jones, R., and Williams, T. (1996) The dependence of radiation hardening and embrittlement on irradiation temperature. In: Gelles DS, Nanstead RK, Kumar AS., Little EA, editors, *Effects of Radiation on Materials: 17th International Symposium*, vol. **1270**, ASTM,
10. Hiranuma, N., Soneda, N., Dohi, K., Ishino, N., Dohi, N., and Ohata, H. (2004) Mechanistic Modeling of Transition Temperature Shift of Japanese RPV Materials, 30th MPA-Seminar in conjuction with the *9th German-Japanese Seminar, Stuttgart*, October 6 and 7, , p. 3.1-3.19.

11. Odette, G. R. (1983) *Scr. Metall.* **17**, 1183
12. Odette, G. R. (1995) *MRS Symposium Proceedings*, Pittsburgh, **373**, 173
13. Odette, G. R., Lucas, Radiat. G. E. (1989) *Eff. Def. Solids* **144**, 189
14. Odette, G. R. (1998) in: M. Davies (Ed.), Neutron Irradiation Effects in Reactor Pressure Vessel Steels and Weldments, IAEA-JWG-LMNPP-98/3, Vienna, 438
15. Buswell, J. T., Phytian, W. J., and Mc Elroy, R. J. et al., (1995) J. Nucl. Mater. **225**, 196
16. Fisher, S. B., and Buswell, J. T. (1987) *Int. J. Pressure Vessel Piping* **91**, 27
17. Gokhman, A., Böhmert, J., and Ulbricht, A. (2004) *J. Nucl. Mater.* **334**, 195
18. Böhmert, J., Gokhman, A., Grosse, M., Ulbricht, u., and Nachweis, A. (2003) Interpretation und Bewertung bestrahlungsbedingter Gefügeänderungen in WWER–Reaktordruckbehälterstälen, Forschubngszentrum Rossendorf, Rossendorf, *Wissenchaftlich-technische, Bericht, FZR-381, June.*
19. Duparc, H., Moingeon, C., Smetaninsky-de-Grande, N., and Barbu, A. (2002) *J. Nucl. Mater.* **302**, 143
20. Christien, F., Barbu, A. (2004) *J. Nucl. Mater.* **90**, **324**
21. Gokhman, F., Bergner, A., and Ulbricht, U. (2005) Iron matrix effects on cluster evolution in neutron irradiated reactor steels, *9th Research Workshop Nucleation Theory and Applications*, Joint Institute for Nuclear Research, 25 June – 3 July, Dubna, in Schmelzer, J. W., Roepke P., Priezzhev, V. B. (eds.) (2006) Nucleation Theory and Applications, *JINR Publ.*, 408-419
22. Gokhman, F., Bergner, A., and Ulbricht, U., Birkenheuer, *Defect and Diffusion Forum* **277**, 75 (2008).
23. Takahashi A., Soneda, N., Ishino, S., and Yagawa, G. (2003) *Phys. Rev. B* **67**, 024104
24. NRC (1988) "Improved Embrittlement Correlations for Reactor Pressure Vessel Steels", *Office of Nuclear Regulatory Research, U.S. Nuclear Regulatory Commission*, Washington, DC.
25. Lifshitz, I., and Slyozov, V. (1961). *J. Phys. Chem. Solids* **19**, 35
26. Salje, G., and Feller-Kniepmeier, M. (1977) *J. Appl. Phys.* **48**, 1833
27. Server, W., Lott, R., and Rosinski, S. (2004) *Assessment of U.S. Embrittlement Trend Equations Considering the Latest Available Surveillance Data*, PVP-Vol. **490**, Storage Tank Integrity and Materials Evaluation, Jule 25-29, , San-Diego, California, USA, 19-24
28. Birkenheuer, U., Bergner, F. Ulbricht, A., and Gokhman, A. (2009) submitted to *J. of Nucl. Mater.*
29. Fisher, S., and Buswell, J. (1987) *Int. J. Pres. Ves. & Piping* **27**, 91

DYNAMICS OF SYSTEMS FOR MONITORING OF ENVIRONMENT

WALDEMAR NAWROCKI
Poznan University of Technology, Faculty of Electronics and Telecommunications, ul. Polanka 3, PL-60965 Poznan, Poland
e-mail: nawrocki@et.put.poznan.pl

Abstract: The paper describes system for monitoring important physical quantities of the environment, including meteorological information as well. Natural environment and technical infrastructure (artificial environment) can be consider like a metastable physical system. Monitoring system for environment consists of: sensors and optical cameras, communication interface and system controller with data acquisition, data processing, storage and presentation. Monitoring systems are always distributed systems. Communication channels (electrical and optical cables and wireless channels) play important role in dynamics and reliability of a monitoring system. For users it is necessary to know the dynamics of the whole monitoring system. Monitoring of infrastructure (road, power and communications networks, sewage systems) is as important as natural environment monitoring for population security.

Keywords: monitoring, environment, distributed measurement system, sensors

1. Introduction

Environmental monitoring is an important tool for keeping environmental security. Both the natural environment – comprising air, potable water reserves, soil, the natural electromagnetic field, etc. – and the manmade environment – referred to as infrastructure and including road, power and communications networks, sewage systems, electromagnetic fields generated by technical equipment, etc. – are important for humans. Therefore the environment must be monitored. Natural environment and infrastructure (artificial environment) can be consider like a metastable physical system. However time scale of processes in environment is much longer (hours, days or months) than in micro physical systems. A growing number of parameters of both these environments are monitored for environmental security. Natural environment parameters include physical quantities such as air temperature, river water level, metal content in water (including mercury, lead or other heavy metals), radiation, and many others. Important infrastructure parameters include, among others, road traffic intensity (especially the formation of bottlenecks), performance of power and water supply systems, and potable water quality parameters related to the water supply system. Note that in metropolitan areas an efficient infrastructure is as important as a clean natural environment. Big cities such as London, New York or many others could not function properly without power, water supply, sewage and communications systems. Therefore, infrastructure monitoring is as important as natural environment monitoring for population security.

An environmental monitoring system uses a number of sensors for physical or chemical quantity measurements, optical cameras, infrared cameras, a communications network, and a controller (Fig. 1).

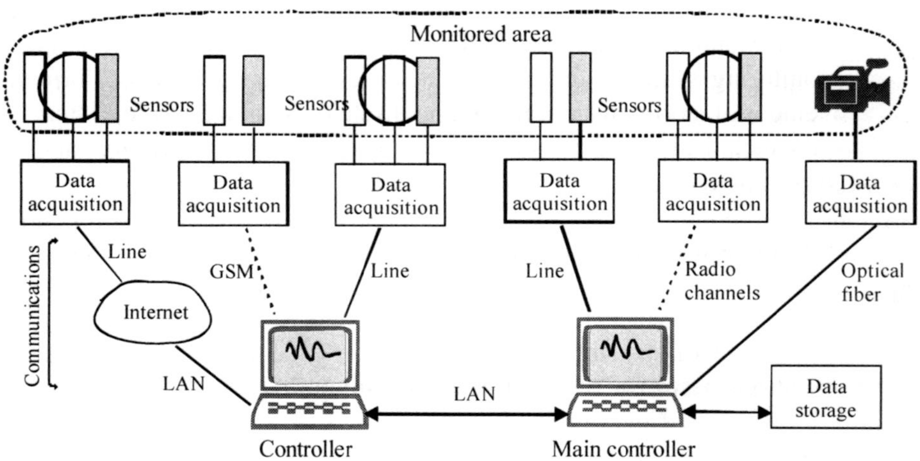

Figure 1. *Environment monitoring system.*

Sensors and cameras are used for data acquisition. Automated monitoring systems use sensors with electric output signals: examples include electric sensors for temperature, pressure, wind speed and wind direction; laser rangefinders; lightning detectors; vehicle detectors, and many others. Also, monitoring system sensors include image converters and infrared detectors built in visual and infrared cameras, respectively. Visual cameras are used for monitoring traffic intensity and detecting traffic jams or conditions of increased risk. They can also be useful in flood risk evaluation as well as in crime detection in public places. Infrared, or thermal, cameras are used for detecting high-temperature areas. Information acquired by means of such cameras is used in fire risk evaluation or for detecting a risk of machine failure in systems of social importance.

The acquired data are processed in the controller. The results are visualized, and the data stored for long-term analyses and statistics. Each segment of the system has an effect on the quality and usefulness of the information acquired.

Environmental monitoring systems are seldom fully automated. Some of the sensors used for environment parameter measurements are not electric and thus require visual reading, necessary for example in the case of non-electric rain or staff water gauges, or results of chemical analysis of water quality.

The dynamics of processes occurring in the natural environment tend to be low enough for available means of communications to provide on-line reading of all the data from a single measurement point. However, a monitoring system typically uses data acquired at a number of measurement points (e.g. collected from 1000 stations). The time necessary for communication with each station within a data collection cycle must be taken into account in the design of such multi-sensor monitoring systems. Enhanced communications are necessary in storm monitoring systems, designed to acquire information on the formation and movement of storms with atmospheric discharges, as well as in systems for monitoring technical infrastructure, especially power grid operation and city traffic in rush hours.

2. Communications in monitoring systems

Monitoring large surface areas, covering thousands of square kilometers, involves the use of distributed monitoring systems. Such systems can use communications networks based on various wire or wireless technologies. Data transfer in distributed measurement systems used for monitoring is performed on a bit-by-bit basis. The required rate of data transfer in a monitoring system depends on the dynamics of the monitored processes as well as on the amount of data acquired in and transferred from a single measurement point within a data collection cycle. The total amount of data depends on the number of measurement points and on the size (expressed in bytes) of a single data message.

The following means of communications (Nawrocki, 2005) are used in natural environment and infrastructure monitoring systems:

- public switched telephone network (PSTN);
- radiocommunication (based on radiomodems);
- computer network and Internet-based communication (currently tested, not yet in common use);
- GSM or UMTS mobile telecommunication network.

PSTN-based communication is suitable for systems monitoring areas covered by the telephone network. A measurement station and the controller can be connected by a leased line or by a switched one; in the latter case communication requires a dial-up similar to that preceding a phone call. Though using a switched line is cheaper, some systems use leased lines for security reasons. A leased line also has the advantage of providing enhanced reliability of operation of the measurement/monitoring system involved, by eliminating the risk of connection failure due to a busy signal, securing stable connection quality and saving the time otherwise necessary for switching.

The rate of digital data transfer in a telephone network depends on the network quality: analogue networks provide data rates up to 56 kilobits per second (kbps), while digital networks (ISDN) allow data transfer of hundreds of kilobits per second. The main standards for digital data transfer (through a modem) in telephone networks are RS-232C.

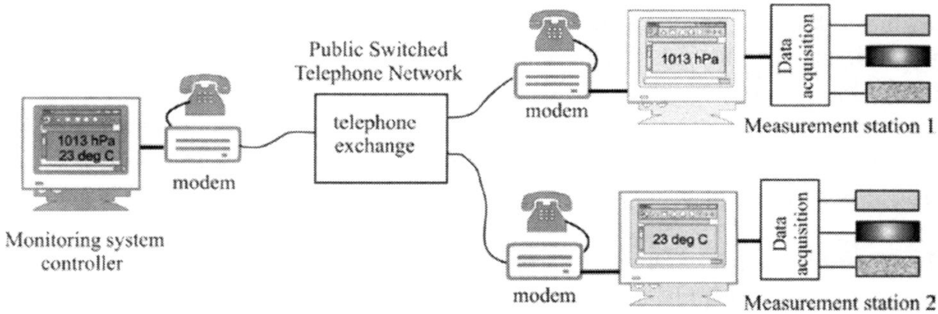

Figure 2. *Monitoring system using a PSTN for communication.*

Radiocommunications is a good solution when measurement stations are far from telephone lines or when access to them is difficult. A radio-communication-based monitoring system is shown in Fig. 3. The system consists of radiomodems, or transceivers equipped with modulators and de-modulators for processing output and input signals, respectively. Monitoring systems of national importance use special bands allocated to civil government institutions for radio-communication. Other systems use licensed bands allocated to industrial radiocommunications, with frequencies from the 430-860 MHz band being the most commonly used. Radiocommunication-based systems allow transfer rates up to 20 kbps, but actual rates tend to be lower. The communication range, typically up to 50 km, is determined by the radiotransmitter power output and the antenna height. However, both these parameters are limited by the license and the license fee. The problem of poor connection quality and limited system area can be solved by using some measurement stations as repeaters, as shown in Figure 3 (Nawrocki, 2005). A cellular telecommunication systems (GSM and UMTS) have many advantages when used as the interface in systems for monitoring of the environment see Fig. 4.

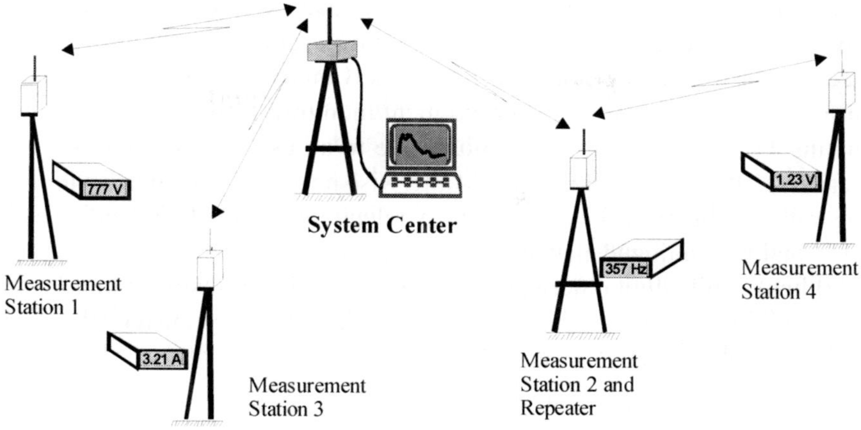

Figure 3. *Radiocommunication-based monitoring system.*

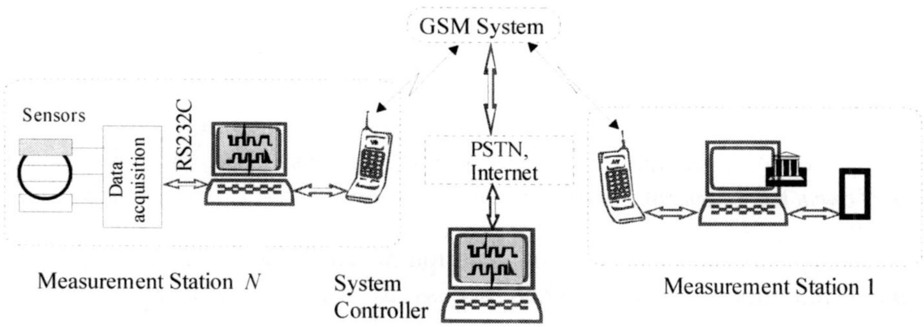

Figure 4. *GSM mobile phone network-based monitoring system.*

GSM and UMTS systems offer the following digital data transmission services:

- Short Message Service (SMS), which is the transmission of alphanumeric messages of up to 160 characters;
- Multimedia Messaging Service (MMS), which is the store-to-forward transmission of text, graphic and sound files;
- Circuit Switched Data (CSD) transmission, which is the switched transmission of digital data with speeds up to 9.6 kbps via traffic channel; an option of the CSD mode is a High Speed Circuit Switched Data (HSCSD) with speeds up to 57.6 kbps;

- General Packet Radio Service (GPRS), a packet data transmission mode, with the speed up to 42.8 kb/s (uplink) and 85.6 kb/s (downlink);
- Enhanced Data rates for GSM Evaluation (EDGE), a transmission with the speed up to 123 kb/s (uplink) and 247 kb/s (downlink);
- transmission of data in Universal Mobile Telecommunication System (UMTS), a transmission with the speed up to 384 kb/s (both uplink and downlink). In final form of UMTS its maximal transmission speed will be 1.9 Mbps (only inside a small cell called a picocell). UMTS has a terrestrial segment (operating since 2002) and a satellite segment (still under developing). UMTS with the satellite segment can be used for data transmission (and therefore for monitoring) whole territory of the Earth.

Where there is a very light channel load, SMS is the cheapest mode of data transmission. A delivery report can also be sent from recipient to sender, to verify that the message has been delivered. In the case of a lack of communication with the recipient at the moment of transmission, the message is stored in an SMS center, and forwarded to the recipient after communication is re-established. The typical message delivery time is a few seconds (6 to 9 s in our measurements – Nawrocki, 2005) from the moment of sending. However, delivery delays can be much longer. A message can be delivered after several hours or days, and in occasional cases can remain undelivered (which was the reason for creating the delivery report option). Our investigations shown that on New Year's Eve, several messages were not delivered (Nawrocki, 2005), which is for the reliability of the monitoring system.

MMS is a device-to-device transmission method, like SMS, of multimedia files via the GSM network. Files can be transferred between users (i.e., from one subscriber to another) or between devices. The MMS standard uses a Wireless Application Protocol as its transmission protocol. MMS can be used in monitoring systems such as road traffic monitoring or water level monitoring. All these applications would involve the transmission of image files. No common MMS standards, including a standard volume of transmitted MMS files, have yet been adopted by GSM network operators.

CDS and GPRS transmission technologies are used in monitoring systems when it is necessary to send not just a single result (or a single image) of a measurement but a stream of measurement data. The CDS transmission is online, without any delay, but its speed is rather slow – only 9.6 kbs. The GPRS is based on packet (file) switching instead. A GPRS session can be activated in the "always connected" mode, and data can be transferred during phone calls without interference. Each packet, or set of data transferred, is an integrated whole, and can be transmitted independently of the other packets. A great advantage of GPRS is the theoretical speed of data transmission – up to 115.2

kbps. As measurements show, GSM networks can currently transmit data at a maximum speed of 26.4 kbps (GPRS) and 28.8 kbps (HSCSD) for data transmission and at a speed of 53.6 kbps (GPRS) and 57.6 kbps (HSCSD) for data receiving (Mackowski, PhD Thesis, 2008).

Beside a maximal transmission speed in GSM a rate of transmitted samples (contained measurement results) is very important as well. In our experiment packets (each file containts one 2-byte sample) were transmitted with an effective speed of 2640 b/s using GPRD or HSCSD technology. However, if we transmitted packets consist 1000 samples each the effective speed was 10 times higher, up to 26 kb/s This data mean that single samples from measurement were transmitted in a live mode with a rate of 55 S/s (samples per second) but packets with 1000 samples were transmitted with a rate of 1600 S/s. (Mackowski, 2008).

3. Weather and water level monitoring in Poland

An example of a nationwide environment monitoring system is the system used for monitoring weather and water levels in Poland, operated by the state Institute of Meteorology and Water Management (IMWM, 2006), and covering the area of Poland, or 312,000 square kilometers.

The system comprises the following six subunits:

- weather monitoring system, with 211 weather stations;
- system for monitoring the water level in rivers, lakes and other water bodies, with 1000 gauging stations;
- radar system for monitoring clouds and waves of precipitation, with 8 radars;
- lightning monitoring subsystem for detecting and registering atmospheric discharges, with 9 measurement stations;
- system for monitoring air radioactivity, with 9 measurement stations;
- satellite system for monitoring based on images using waves in or beyond the visible light range (photographic or infrared imaging, respectively).

The network of 211 weather stations allows measurement of tens of basic weather parameters, the most important of which include air temperature, atmospheric pressure, air humidity, wind speed and wind direction. Among the other parameters measured are cloud base height (laser-measured), amount of precipitation (rain or snow), visibility, and air temperature near the ground. Measurements are performed in enclosed areas within the weather stations.

The basic equipment of a weather station is comprised of a facility referred to as Meteorological Automatic Weather Station (MAWS). A MAWS is equipped with electric sensors: a Pt100 resistance sensor for temperature, a capacitance sensor for barometric pressure, a semiconductor sensor for air humidity, sensors for solar radiation and for visibility, a pulse sensor for wind speed, and a wind direction indicator with a Gray code disk.

Beside a MAWS, a weather station uses a number of other devices: a laser ceilometer for determining the height of the cloud base; a rain gauge (electric or pulse), a radiation gauge, etc. Recognized manufacturers of meteorological instruments include Vaisala (Finland), Impulse Physik (Germany) and Young (USA). The weather stations in Poland use MAWS instruments manufactured by Vaisala.

Hydrological monitoring is based on data collected from 1000 gauging stations in rivers, lakes and water reservoirs. Information on the water levels is supplied in 10 minute cycles. Water level gauges used by hydrological stations are either electric sensors for hydrostatic pressure or optical sensors. The hydrological monitoring system is of key importance for flood risk evaluation and for prevention of flood effects.

A network of 8 radar stations provides information on the position and movement of clouds, as well as on the wind speed. These data are crucial for precipitation forecasts. The radars use the Doppler effect to measure clouds speed and by this way wind speed. Their observation range (the radius of the circle covered) is 200 kilometers. The resolution of cloud distribution measurements ranges from 1 kilometer to 4 kilometers, and that of wind speed is 0.5 kilometer per hour. The data are updated every 10 minutes.

The thunderstorm monitoring system is very important, and is used for determination of the location and intensity of atmospheric discharges, or lightning strikes. Its 9 stations distributed over the territory of Poland detect and register intracloud and cloud-to-ground discharges. The detection of intracloud discharges is of use for forecasting cloud-to-ground discharges. The site of the discharge is determined with resolution ranging from 1 kilometer to 4 kilometers, like that of radar system measurements.

Data on the intensity of lightning as well as on storm movement are of use for aviation, the power industry (for assessing the need of emergency power cut-off) as well as for high-altitude work management.

4. Data flow management in the national monitoring system

The general scheme of the Information System of the National Environment Monitoring (ISNEM) is presented in Fig. 5. National Environment Monitoring (NEM) collects, processes and provides access to such data as measurements of characteristics of particular environment components, measurements of some emissions and measurements of natural characteristics, e.g. measure indicators of air quality such as SO_2, NO_2, O_3, CO read from the Meteorological Automatic Weather Station; meteorological indicators e.g. noise or water surface indicators including water level in different river legs, which can be helpful in dealing with crisis situations. This information is collected by Provincial Environment Protection Inspectorates (PEPI), their sections and also research institutes and universities, which the Environment Protection Inspection (EPI) cooperates with. An Information System for National Environment Monitoring (ISNEM) for data and information storage should be created in Poland to provide wide, easy and inexpensive access to data, as well as efficient updating (Kraszewski, 2001).

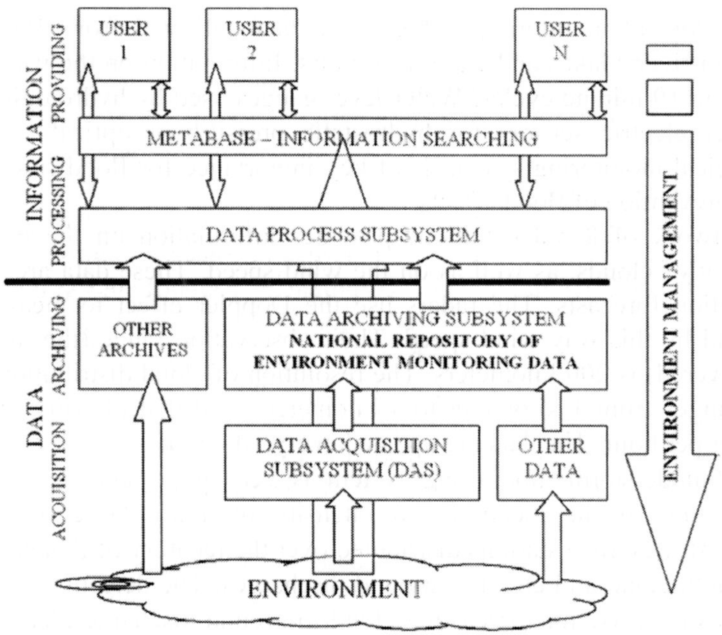

Figure 5. *General functional scheme of ISNEM.*

Below we present an ISNEM consisting of four interlinked subsystems, the correlation of which is schematically depicted in Fig. 5. The four subsystems include:

- Data Acquisition Subsystem (DAS);
- Archiving Subsystem;
- Process Data Subsystem;
- Data Provision Subsystem.

The Data Acquisition Subsystem, comprising mechanisms of data acquisition from the environment by means of measurement devices, observation or sampling and laboratory analyzes. A procedure of data input should be developed for every environmental component. All data should be stored in the system memory as acquired in the earliest phase of the measurement procedure. In many cases, this necessitates the creation of source databases. Measurement results can be verified within the subsystem and if accepted, forwarded to archives. These are designed for storing other data, not representing results of measurements performed within National Environment Monitoring programs, and providing information on parameters such as radioactive emission or river flow as well as meteorological data.

The other subsystems comprising the ISNEM include: the Archiving Subsystem with electronic databases for storing all the accumulated data; the Process Data Subsystem, comprising warehouse data and applications designed for data processing into a format required by end users; and the Data Provision Subsystem, to be used by public administration, public figures and the Central Statistical Office (Nawalaniec, 2003). Data will be provided in the form of reports published on the Internet.

The concept of ISNEM also involves two methods of data acquisition and archiving: one using a distributed database, the other a centralized one (see Figure 7). The former involves development of a distributed system for data acquisition and transfer to the Information System for National Environment Monitoring. The acquired information is to be verified by a PEPI administrator before being added to the Provincial Database and then to the Central Database. If environmental measurements and observations are only conducted on the national basis, a single centralized database is to be used.

Regardless of the organization of the Data Acquisition and Archiving Subsystems, part of the information will come from automatic measurements of air and water pollution concentrations at a number of measurement stations, and from automatic water level gauges in individual river segments, which are of use for flood management. Because of considerable distances between measurement stations and data acquisition centers in this distributed system, the best solution for communication is a nationwide 2G cellular phone network, covering over 96% of the country's area.

5. City traffic monitoring

Three kinds of monitoring subsystem have been integrated into a single SCADA system in the Ostrow Wielkopolski district of Poland. The system consists consists of the following units (Nawrocki, 2007):

- a subsystem for supervising junction traffic lights in the city (see Figure 7);
- a subsystem for monitoring environment data;
- a video monitoring subsystem.

The district SCADA System is mainly designed for traffic management. The traffic light subsystem allows on-line monitoring of the operation and status of all the traffic lights in the monitored area, analysis of coordinated signals and traffic intensity, as well as remote access to traffic controllers (see Fig. 6). It also gives a possibility of remotely blocking some junctions with all-red lights, if necessary. The environment data monitoring subsystem allows online monitoring of the status of environmental parameters such as air temperature, dew point, wind chill, air pressure, wind speed and direction, air humidity, or rainfall in the last hour or in the last 24 hours. Parameter history diagrams can be generated at any time as well. The video monitoring subsystem allows monitoring of the situation in a street or on a road and the adjacent area. This system can be of use for public security services.

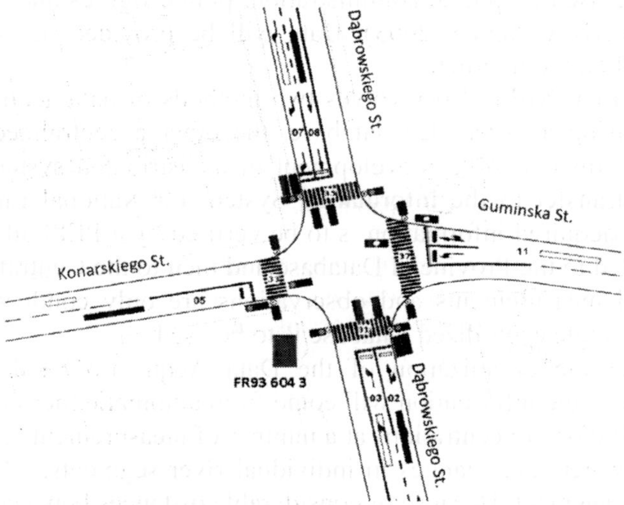

Figure 6. *Scheme of a road crossing with traffic control and monitoring.*

6. Conclusions

Environmental security is determined by parameters of the natural environment (such as the quantity and quality of water, the quality of air and soil, soil, or the weather conditions) as well as by those of the built environment, or technical infrastructure. Therefore, parameters of both types of environment should be involved in environmental monitoring aimed at maintaining environmental security and sustainability. Parameters such as road traffic intensity should be monitored along with air pollution, flood or hurricane indicators, etc.

Environment monitoring is based on the use of sensors and communication systems, and is vital for environmental security. Though the dynamics of environmental processes is assumed to be relatively slow in comparison with the average performance of high-speed communication systems, more thorough studies show that the operation of monitoring systems using a number of field measurement stations involves time limits that should be taken into account for the system to operate efficiently. Beside the number of field measurement stations, the dynamics of a monitoring system is also affected by the number of sensors used by each station, and, obviously, by the rate of data transfer in the communication system employed.

Natural environment and infrastructure monitoring is very important for environmental security. Note that only the infrastructure – the built component of the environment – besides being monitored, can also be controlled and modified by humans. The natural environment can only be monitored, and its status – for example, the weather conditions – taken into consideration in action planning. Natural environment and infrastructure can be consider like a metastable physical system. The decline of the stability of the natural environment (according to his parameters) can be a long process and expensive to the reconditioning.

References

1. Institute of Meteorology and Water Management (IMWM) (2006) Warsaw, www.imgw.pl
2. Kraszewski, A., and Lobocki, L. (2001) The Idea of Information System of the National Environment Monitoring. *Final Report v. 2.0 with supplement, Warsaw, Technical University of Warsaw 2001*, unpublished, (in Polish)
3. Mackowski, M. (2008) Application of the Measuring Systems with Data Transmission in GSM Cellular Network and Internet (in Polish), *PhD Thesis, Poznan University of Technology*
4. Nawalaniec, T., and Urbaniak A. (2003) Distributed System for Data Acquisition and Transfer to Information System of the Environment Monitoring, *Proceedings of ICC Conference, Tatranská Lomnica*, pp. 210-214.
5. Nawrocki, W., and Nawalaniec, T. (2007) Sensors and communications in environments monitoring systems, Chapter in "*Strategies to Enhance Environmental Security in Transition Countries*", Hall R.N. et al. (eds), Springer, Dordrecht. pp. 153-165.
6. Nawrocki, W. (2005) *Measurement Systems and Sensors*, Artech House, Boston – London.

Lightning Source UK Ltd.
Milton Keynes UK
16 March 2010

151358UK00006BB/8/P

9 789048 134069